WAVE PROPAGATION IN ELASTIC SOLIDS

WAVE PROPAGATION IN ELASTIC SOLIDS

BY

J. D. ACHENBACH

The Technological Institute,
Northwestern University, Evanston, Illinois

NORTH-HOLLAND PUBLISHING COMPANY
AMSTERDAM · NEW YORK · OXFORD

Library of Congress Catalog Card Number: 72-79720
North-Holland ISBN for this volume: 0 7204 0325 1

First edition : *First printing 1973*
First paperback edition: *First printing 1975*
Second printing 1976
Third printing 1980
Fourth printing 1984

Publishers:
ELSEVIER SCIENCE PUBLISHERS B.V.
P.O. Box 1991
1000 BZ Amsterdam
The Netherlands

Sole distributors for the U.S.A. and Canada:
ELSEVIER SCIENCE PUBLISHING COMPANY, INC.
52 Vanderbilt Avenue
New York, N.Y. 10017
U.S.A.

This book is the paperback edition of volume 16 in the Series:
Applied Mathematics and Mechanics
edited by: H. A. Lauwerier and W. T. Koiter.

Printed in The Netherlands

To Marcia

PREFACE

The propagation of mechanical disturbances in solids is of interest in many branches of the physical sciences and engineering. This book aims to present an account of the theory of wave propagation in *elastic* solids. The material is arranged to present an exposition of the basic concepts of mechanical wave propagation within a one-dimensional setting and a discussion of formal aspects of elastodynamic theory in three dimensions, followed by chapters expounding on typical wave propagation phenomena, such as radiation, reflection, refraction, propagation in waveguides, and diffraction. The treatment necessarily involves considerable mathematical analysis. The pertinent mathematical techniques are, however, discussed at some length.

I hope that the book will serve a dual purpose. In addition to being a reference book for engineers and scientists in the broad sense, it is also intended to be a textbook for graduate courses in elastic wave propagation. As a text the book should be suitable for students who have completed first-year graduate courses in mechanics and mathematics. To add to its utility as a textbook each chapter is supplemented by a set of problems, which provide a useful test of the reader's understanding, as well as further illustrations of the basic ideas.

The book was developed from notes for a course offered to graduate students at Northwestern University. In the spring of 1969 a substantial part of the text was prepared in the form of typewritten notes for a series of lectures, while I was a visiting member of the faculty at the University of California in La Jolla. I am pleased to record my thanks for that opportunity. I also wish to express my gratitude to the Rector Magnificus of the Technological University of Delft and the Trustees of the Ir. Cornelis Gelderman Fund for inviting me to act as visiting professor in the department of mechanical engineering at the Technological University, in 1970–1971. While I was in Delft the larger part of the manuscript was completed. A sabbatical leave from Northwestern University during that period is gratefully acknowledged.

For the material of Chapters 5 and 6 I should like to acknowledge my indebtedness to the lectures and publications of Professor R. D. Mindlin. Substantial parts of Chapter 3 are based on the dissertation of Professor A. T. de Hoop, and on the work of Professor E. Sternberg. I am also indebted to many colleagues who read chapters of the book, and who provided me with their constructive criticism. Needless to say, I alone am responsible for errors of fact and logic.

A special word of thanks goes to Mrs. Ruth H. Meier who for many years provided excellent secretarial assistance, and who typed and retyped most of the manuscript as the material was arranged and rearranged.

Let me close with the wish that this book may convey some of the fascinating aspects of wave propagation as a phenomenon, and that it may have done justice to the elegance of the mathematical methods that have been employed.

J. D. A.

TABLE OF CONTENTS

INTRODUCTION

The propagation of mechanical disturbances

The local excitation of a medium is not instantaneously detected at positions that are at a distance from the region of excitation. It takes time for a disturbance to propagate from its source to other positions. This phenomenon of propagation of disturbances is well known from physical experience, and some illustrative examples immediately come to mind. Thus an earthquake or an underground nuclear explosion is recorded in another continent well after it has occurred. The report of a distant gun is heard after the projectile has arrived, because the velocity of disturbances in air, i.e., the speed of sound, is generally smaller than the velocity of the projectile. More familiar manifestations of the propagation of disturbances are waves in a rope or propagating ripples on the surface of water. These examples illustrate mechanical wave motions or mechanical wave propagation.

Mechanical waves originate in the forced motion of a portion of a deformable medium. As elements of the medium are deformed the disturbance is transmitted from one point to the next and the disturbance, or wave, progresses through the medium. In this process the resistance offered to deformation by the consistency of the medium, as well as the resistance to motion offered by inertia, must be overcome. As the disturbance propagates through the medium it carries along amounts of energy in the forms of kinetic and potential energies. Energy can be transmitted over considerable distances by wave motion. The transmission of energy is effected because motion is passed on from one particle to the next and not by any sustained bulk motion of the entire medium. Mechanical waves are characterized by the transport of energy through motions of particles about an equilibrium position. Thus, bulk motions of a medium such as occur, for example, in turbulence in a fluid are not wave motions.

Deformability and inertia are essential properties of a medium for the transmission of mechanical wave motions. If the medium were not deformable any part of the medium would immediately experience a disturbance in the form of an internal force or an acceleration upon application of a localized excitation. Similarly, if a hypothetical medium were without inertia there would be no delay in the displacement of particles and the transmission of the disturbance from particle to particle would be effected instantaneously to the most distant particle. Indeed, in later chapters it will be shown analytically that the velocity of propagation of a mechanical disturbance always assumes the form of the square root of the ratio of a parameter defining the resistance to deformation and a parameter defining the inertia of the medium. All real materials are of course deformable and possess mass and thus all real materials transmit mechanical waves.

The inertia of a medium first offers resistance to motion, but once the medium is in motion inertia in conjunction with the resilience of the medium tends to sustain the motion. If, after a certain interval the externally applied excitation becomes stationary, the motion of the medium will eventually subside due to frictional lossess and a state of static deformation will be reached. The importance of dynamic effects depends on the relative magnitudes of two characteristic times: the time characterizing the external application of the disturbance and the characteristic time of transmission of disturbances across the body.

Suppose we consider a solid body subjected to an external disturbance $F(t)$ applied at a point P. The purpose of an analysis is to compute the deformation and the distribution of stresses as functions of the spatial coordinates and time. If the greatest velocity of propagation of disturbances is c, and if the external disturbance is applied at time $t = 0$, the disturbed regions at times $t = t_1$ and $t = t_2$ are bounded by spheres centered at the point P, with radii ct_1 and ct_2, respectively. Thus the whole of the body is disturbed at time $t = r/c$, where r is the largest distance within the body measured from the point P. Now suppose that the significant changes in $F(t)$ take place over a time t_a. It can then be stated that dynamic effects are of importance if t_a and r/c are of the same order of magnitude. If $t_a \gg r/c$, the problem is quasistatic rather than dynamic in nature and inertia effects can be neglected. Thus for bodies of small dimensions a wave propagation analysis is called for if t_a is small. If the excitation source is removed the body returns to rest after a certain time. For excitation sources that are applied and removed, the effects of wave motion are important if the time interval of application is of the same order of magnitude as a characteristic time of transmission of a disturbance across the body. For bodies of finite dimensions this is the case for loads of explosive origins or for impact loads. For sustained external disturbances the effects of wave motions need be considered if the externally applied disturbances are rapidly changing with time, i.e., if the frequency is high.

In mathematical terms a traveling wave in one dimension is defined by an expression of the type $f = f(x-ct)$, where f as a function of the spatial coordinate x and the time t represents a disturbance in the values of some physical quantity. For mechanical waves f generally denotes a displacement, a particle velocity or a stress component. The function $f(x-ct)$ is called a simple wave function, and the argument $x-ct$ is the phase of the wave function. If t is increased by any value, say Δt, and simultaneously x is increased by $c\Delta t$, the value of $f(x-ct)$ is clearly not altered. The function $f(x-ct)$ thus represents a disturbance advancing in the positive x-direction with a velocity c. The velocity c is termed the phase velocity. The propagating disturbance represented by $f(x-ct)$ is a special wave in that the shape of the disturbance is unaltered as it propagates through the medium.

Continuum mechanics

Problems of the motion and deformation of substances are rendered amenable to mathematical analysis by introducing the concept of a continuum or continuous medium. In this idealization it is assumed that properties averaged over a very small element, for example, the mean mass density, the mean displacement, the mean interaction force, etc., vary continuously with position in the medium, so that we may speak about *the* mass density, *the* displacement and *the* stress, as functions of position and time. Although it might seem that the microscopic structure of real materials is not consistent with the concept of a continuum, the idealization produces very useful results, simply because the lengths characterizing the microscopic structure of most materials are generally much smaller than any lengths arising in the deformation of the medium. Even if in certain special cases the microstructure gives rise to significant phenomena, these can be taken into account within the framework of the continuum theory by appropriate generalizations.

The analysis of disturbances in a medium within the context of the continuum concept belongs to the time-tested discipline of continuum mechanics. In achieving the traditional objective of determining the motion and deformation generated by external excitations the analysis passes through two major stages. In the first stage the body is idealized as a continuous medium and the physical phenomena are described in mathematical terms by introducing appropriate mathematical abstractions. Completion of this stage yields a system of partial differential equations with boundary and initial conditions. In the second stage the techniques of applied mathematics are employed to find solutions to the system of governing partial differential equations and to obtain the physical information which is desired. Usually the goal is to obtain analytical expressions for some of the field variables in terms of the position and time as well as in terms of the geometrical and material parameters.

Continuum mechanics is a classical subject which has been discussed in great generality in several treatises. The theory of continuous media is built upon the basic concepts of stress, motion and deformation, upon the laws of conservation of mass, linear momentum, moment of momentum, and energy, and on the constitutive relations. The constitutive relations characterize the mechanical and thermal response of a material while the basic conservation laws abstract the common features of all mechanical phenomena irrespective of the constitutive relations. The general system of equations governing the three-dimensional motions of elastic bodies is strongly nonlinear. As a consequence very few significant wave propagation problems can be solved analytically on the basis of this general system of equations. Fortunately it is a matter of wide experience that many wave propagation effects in solids can adequately be described by a linearized theory.

Outline of contents

A detailed discussion of the general three-dimensional theory of elasticity and the process of linearization which results in the equations governing the classical theory of linearized elasticity falls outside the scope of this book. It is, however, instructive to review briefly the nonlinear elastic theory within a one-dimensional geometry. This review is carried out in chapter 1, where the conditions justifying linearization of the one-dimensional theory for the express purpose of describing problems of wave propagation are also examined.

The three-dimensional equations governing isothermal linearized elasticity of a homogeneous isotropic medium are summarized in chapter 2. Except for the last chapter the remainder of the book is strictly concerned with linearized theory and perfectly elastic media. When, in the last chapter, the treatment goes beyond ideal elasticity and beyond isothermal conditions, the pertinent governing equations are introduced when needed.

Wave motions can be classified according to the trajectory of a particle of the medium as the disturbance passes by. Thus we distinguish a pulse or a single wave from a train of waves. A special case of the latter is a periodic train of waves in which each particle experiences a periodic motion. The simplest special case of periodic wave motion is a simple harmonic wave wherein each particle is displaced sinusoidally with time. Simple harmonic waves are treated in considerable detail in chapters 1, 5 and 6, not only because of their intrinsic interest but also because in the linear theory a general periodic disturbance or even a single pulse can be represented by a superposition of simple harmonic waves.

In considering wave propagation in three dimensions we can, at a certain instant of time, draw a surface through all points undergoing an identical disturbance. As time goes on, such a surface, which is called a wavefront, moves along showing how the disturbance propagates. The direction of propagation is always at right angles to the wavefront. The normals defining the direction of wave propagation are called the rays. For a homogeneous and isotropic medium the rays are straight lines. If the wave propagation is limited to a single direction the disturbance at a given instant will be the same at all points in a plane perpendicular to the direction of wave propagation and we speak of a plane wave. Other simple cases are spherical waves and cylindrical waves, where the wavefronts are spherical and cylindrical surfaces, respectively. It is shown in chapter 4 that there are two types of plane waves: transverse and longitudinal waves. In transverse waves the motion is normal to the direction of wave propagation. If the direction of motion coincides with the direction of wave propagation we speak of longitudinal waves.

Among the most important aspects of wave motion are the reflection and transmission of waves. When a wave encounters a boundary separating two media with different properties, part of the disturbance is reflected and part is transmitted into the second medium, as discussed in chapters 1 and 5. If a body has a finite cross-sectional dimension waves bounce back and forth between the bounding surfaces. Although it is then very difficult to trace the actual reflections it can be noted that the general direction of energy transmission is in a direction parallel to the bounding surfaces, and we say that the waves are propagating in a waveguide. The analysis of harmonic waves in waveguides leads to some new notions such as modes of wave propagation, the frequency spectrum, dispersion, and group velocity. Harmonic waves in waveguides are discussed in detail in chapter 6. Important wave propagation effects are surface waves propagating along a bounding surface. Rayleigh waves at a free boundary and Stoneley waves at an interface are discussed in chapter 5.

Pulses generated in elastic bodies by distributions of body forces or surface disturbances are analyzed in chapters 7 and 8 by means of integral transform techniques. Exact methods of inverting transforms such as the Cagniard-de Hoop method, as well as approximate means of evaluating integrals in terms of asymptotic expansions, are also discussed in these chapters.

In chapter 3 several formal aspects of the theory of dynamic elasticity are discussed. Among the theorems that are proven, the uniqueness theorem for the properly formulated boundary-initial value problem is of primary importance. The chapter also contains a discussion of the decomposition of the displacement vector in terms of derivatives of displacement potentials. The displacement potentials satisfy classical wave equations.

In the mathematical literature hyperbolic partial differential equations which govern wave propagation phenomena have been studied in great detail. It is well known that the general solution of the wave equation can be expressed in terms of the external disturbances by means of integrals over the bounding surface and the interior of a body. In chapter 3 the relevance of integral representations to problems of elastic wave propagation is examined.

A complete solution of a wave propagation problem involves a considerable amount of mathematical analysis. For transient waves information on discontinuities in the field variables at the moving surfaces separating the undisturbed from the disturbed regions of the body can be obtained in a fairly straightforward and simple manner by methods which are analogous to the ray tracing techniques of geometrical optics. In chapter 4 propagating discontinuities are analyzed within the context of the linear theory of elasticity.

There are several mathematical methods which are suitable for certain problems but not for others. For example, if a problem displays dynamic similarity, convenient and simple mathematical methods can be employed to obtain a solution, as shown in chapter 4. Also under certain conditions numerical schemes based on the method of characteristics offer an efficient means of obtaining information on the field variables. Some applications of the method of characteristics are discussed in chapter 4.

When a pulse propagating through an elastic medium encounters an irregularity such as a void or an inclusion, the pulse is diffracted. In chapter 9 the diffraction of waves is analyzed for the case that the diffracting surface is a semi-infinite slit. The analysis of diffraction problems requires the introduction of mathematical methods that are suitable for mixed boundary value problems. Particular attention is devoted to diffraction by a slit because a slit may be considered as representing a crack-type flaw in the medium. As the wave strikes the crack a stress singularity is generated at the crack tip, which may give rise to propagation of the crack and thus to fracture of the body.

In chapter 10 we relax some of the restrictions on material behavior inherent in the treatment of earlier chapters. In this chapter it is no longer required that the material behavior be identical in all directions. The field equations for an anisotropic medium are stated, and time-harmonic waves in such a medium are briefly discussed. All real materials exhibit some kind of damping of mechanical disturbances. This effect is included by modeling the constitutive behavior of the medium as linearly viscoelastic. The propagation of waves in a linearly viscoelastic medium is discussed in chapter 10. External disturbances are not necessarily of a mechanical nature. As an example we may think of the sudden deposition of heat in a medium. Since the heat deposition will give rise to thermal expansion, a mechanical wave can be generated. The interaction between thermal and mechanical effects is also examined in chapter 10. In the last section of the book we analyze a one-dimensional nonlinear problem and we explore some typically nonlinear effects such as the formation of shocks.

Historical sketch

The study of wave propagation in elastic solids has a long and distinguished history. The early work on elastic waves received its impetus from the view which was prevalent until the middle of the 19th century that light could be regarded as the propagation of a disturbance in an elastic aether. This view was espoused by such great mathematicians as Cauchy and Poisson and to a large extent motivated them to develop what is now generally known as the theory of elasticity. The early investigations on the propagation of waves in elastic solids carried out by Poisson, Ostrogradsky, Cauchy, Green, Lamé, Stokes, Clebsch and Christoffel are discussed in the Historical Introduction to Love's treatise of the mathematical theory of elasticity [1][1].

In the latter part of the 19th century interest in the study of waves in elastic solids gained momentum again because of applications in the field of geophysics. Several contributions of lasting significance, particularly as related to the discovery of specific wave propagation effects, stem from the

[1] Numbers in brackets refer to the bibliography at the end of this Introduction.

years between 1880 and 1910, and are due to Rayleigh, Lamb and Love. Since that time wave propagation in solids has remained a very active area of investigation in seismology because of the need for more accurate information on earthquake phenomena, prospecting techniques and the detection of nuclear explosions. Aspects of wave propagation that are of interest in seismology are discussed in the books by Bullen [2], Ewing et al. [3] and Cagniard [4].

As far as engineering applications are concerned a substantial interest in wave propagation effects manifested itself in the early forties, when the specific technological needs of the time required information on the performance of structures under high rates of loading. Since then interest in elastic waves has increased. This interest has been stimulated by technological developments related to high-speed machinery, ultrasonics, and piezoelectric phenomena, as well as to methods in materials science for measuring the properties of materials, and to, for example, such civil engineering practices as pile driving. By now the study of wave propagation effects has become well established in the field of applied mechanics.

With regard to other works specifically dealing with the propagation of waves in elastic solids we mention in the first place the book by Kolsky [5]. A thorough but rather brief discussion of elastodynamic theory was also presented by Schoch [6]. A review article which lists most of the contributions to the field until 1964 was published by Miklowitz [7].

Parallel to the study of waves in elastic solids the propagation of waves was investigated extensively within the context of applied mathematics, electromagnetic theory and acoustics. Especially the work in acoustics, beginning with the classical treatise of Lord Rayleigh [8], is closely related. In this regard we mention the books by Brekhovskikh [9], Lindsay [10] and Morse and Ingard [11], which actually contain chapters on waves in elastic solids. To conclude this brief bibliography we mention the monograph on ocean acoustics which was written by Tolstoy and Clay [12].

Bibliography

[1] A. E. H. Love, *The mathematical theory of elasticity*, 4th ed. New York, Dover Publications, Inc. (1944).

[2] K. E. Bullen, *An introduction to the theory of seismology*. Cambridge, University Press (1963).

[3] W. M. Ewing, W. S. Jardetzky and F. Press, *Elastic waves in layered media*. New York, McGraw-Hill Book Company (1957).

[4] L. Cagniard, *Reflection and refraction of progressive waves*, translated and revised by E. A. Flinn and C. H. Dix. New York, McGraw-Hill Book Company (1962).

[5] H. Kolsky, *Stress waves in solids*. New York, Dover Publications, Inc. (1963).
[6] A. Schoch, *Schallreflexion, Schallbrechung und Schallbeugung*, Ergebnisse der Exakten Naturwissenschaften, Vol. 23. Berlin, Springer-Verlag (1950).
[7] J. Miklowitz, *Elastic wave propagation*, Applied Mechanics Surveys. Spartan Books (1966).
[8] Lord Rayleigh, *The theory of sound*, Vols. I and II. New York, Dover Publications, Inc. (1945).
[9] L. M. Brekhovskikh, *Waves in layered media*. New York, Academic Press (1960).
[10] R. B. Lindsay, *Mechanical radiation*. New York, McGraw-Hill Book Company (1960).
[11] P. M. Morse and K. U. Ingard, *Theoretical acoustics*. New York, McGraw-Hill Book Company (1968).
[12] I. Tolstoy and C. S. Clay, *Ocean acoustics*. New York, McGraw-Hill Book Company (1966).

CHAPTER 1

ONE-DIMENSIONAL MOTION OF AN ELASTIC CONTINUUM

1.1. Introduction

Some of the characteristic features of wave motion in a continuum can be brought out by an analysis in a one-dimensional geometry. With one spatial variable the concepts and principles of continuum mechanics can be deployed without the encumbrance of geometrical complications to display the essential aspects of motion of a continuum and to derive the governing system of nonlinear equations. In one dimension it is subsequently straightforward to examine rigorously the conditions for linearization of the equations, particularly for elastic solids.

One-dimensional linearized elastic theory is governed by a simple partial differential equation whose general solution can be determined by the use of elementary mathematics. In section 1.3 it is shown analytically that an external disturbance applied to an initially quiescent medium generates a pulse which propagates with a distinct velocity. Any particle in the medium remains undisturbed until sufficient time has passed for the pulse to reach the particle. Reflection takes place when a pulse reaches an external boundary of the body. Strictly speaking, a pulse is completely reflected only at a boundary of an elastic body with a vacuum. At all other boundaries there is not only reflection but also transmission of the pulse across the interface into the neighboring medium. Transmission across interfaces between two media of comparable elastic moduli and mass densities can give rise to some interesting and typical wave propagation effects, which are discussed in section 1.4.

A substantial part of the wave propagation literature is devoted to the study of sinusoidal wave trains. This interest is meaningful because in a linearized theory a propagating pulse of arbitrary shape can be represented by a superposition of sinusoidal wave trains. The superposition is achieved by means of Fourier series or Fourier integrals. In sections 1.6–1.9 the salient aspects of Fourier analysis are displayed by means of one-dimensional examples.

1.2. Nonlinear continuum mechanics in one dimension

1.2.1. Motion

In a purely one-dimensional longitudinal motion all material particles move along parallel lines, and the motion is uniform in planes normal to the direction of motion. Clearly one length coordinate and time are sufficient to describe the one-dimensional longitudinal motion of a continuum. Suppose the position of a material point P at a certain time, say $t = 0$, is defined by the coordinate X. At a later time t the position of the particle can then be specified by

$$x = P(X, t). \tag{1.1}$$

The mapping $x = P(X, t)$ is called the material description of the motion. This description, where X and t are independent variables, is often also called the Lagrangian description. In eq. (1.1), a value of the independent variable X identifies the particle for all time and its value equals the reference position of that particle. A value of the dependent variable x specifies the present position of the particle whose reference position was X.

Alternatively, the motion may be described in the spatial description. In this description, which is often also called the Eulerian description, the independent variables are t and the *position* x. Values of x and t are related by

$$X = p(x, t). \tag{1.2}$$

In eq. (1.2), a value of the independent variable x specifies a place. The dependent variable X gives the reference position of the particle presently situated at position x. The two descriptions of the motion must of course be consistent with each other, i.e., eq. (1.2) can be obtained by solving (1.1) for X, and vice versa.

To identify field quantities relative to the independent variables we use upper case letters for quantities which are expressed in terms of the material variables X and t. Lower case letters are employed for the spatial description. For example, the displacement is denoted by $U(X, t)$ in the material description and it is denoted by $u(x, t)$ in the spatial description. We have

$$U(X, t) = P(X, t) - X, \tag{1.3}$$

and

$$u(x, t) = x - p(x, t). \tag{1.4}$$

1.2.2. Deformation

The purely one-dimensional motion of an element is depicted in figure 1.1.

As a consequence of the nonuniformity in the direction of the motion, the element undergoes a deformation. For the one-dimensional case, the simplest measure of the deformation is simply the extension divided by the original length of an element comprised of the same material particles. This

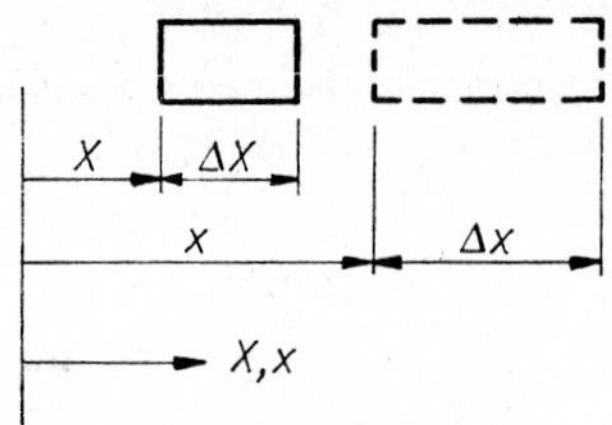

Fig. 1.1. Motion and deformation of an element.

deformation measure may thus be expressed by $(\Delta x - \Delta X)/\Delta X$. In the limit we have

$$\lim_{\Delta X \to 0} \frac{\Delta x - \Delta X}{\Delta X} = \frac{\partial U}{\partial X}, \tag{1.5}$$

and we obtain the displacement gradient $\partial U/\partial X$ as measure of the deformation in material coordinates. In three-dimensional deformations it is more convenient to take the difference between squares of length elements as measure of deformation. This leads to the Lagrangian strain tensor. For a one-dimensional geometry, the Lagrangian strain is

$$E = \tfrac{1}{2} \lim_{\Delta X \to 0} \frac{(\Delta x)^2 - (\Delta X)^2}{(\Delta X)^2} = \frac{\partial U}{\partial X} + \frac{1}{2}\left(\frac{\partial U}{\partial X}\right)^2. \tag{1.6}$$

Eqs. (1.5) and (1.6) refer the deformation to the undeformed configuration. The deformation can, of course, also be described in the system of spatial coordinates. For an elastic solid, the deformation measures (1.5) and (1.6) are more natural since there always is an undeformed reference state to which the material returns when the external loads are removed.

1.2.3. Time-rates of change

The velocity of a material particle is the time-rate of change of $x = P(X, t)$ for constant reference position X,

$$V(X, t) = \frac{\partial P(X, t)}{\partial t}. \tag{1.7}$$

This equation gives the material description of the particle velocity, i.e., it gives the velocity of a certain particle as a function of time. In the spatial description the particle velocity is obtained by substituting X from (1.2) into (1.7). We obtain

$$v(x, t) = V[p(x, t), t]. \tag{1.8}$$

For a fixed spatial position x, (1.8) defines the velocity of the particular particle which passes through position x at time t.

Often the instantaneous motion of the continuum is described by the displacement $u(x, t)$ of a particle instantaneously located at position x. Since the instantaneous velocity is defined for a fixed particle, we must use the chain rule of differentiation to obtain

$$v = \frac{\partial u}{\partial t} + v \frac{\partial u}{\partial x}. \tag{1.9}$$

Eq. (1.9) expresses the total or *material derivative* of $u(x, t)$. The operation expressed by (1.9) is usually denoted by $\mathrm{D}/\mathrm{D}t$, i.e.,

$$\frac{\mathrm{D}}{\mathrm{D}t} = \frac{\partial}{\partial t} + v \frac{\partial}{\partial x}. \tag{1.10}$$

In eqs. (1.7) and (1.9) the partial derivative $\partial/\partial t$ is the derivative with respect to time for a fixed value of the other variable. Generally it is evident from the context which one the other variable is, since upper case symbols are used for field quantities which are expressed in terms of the material variables X and t, while lower case symbols are employed for field quantities in the spatial description where x and t are the variables.

In the material description the acceleration is defined as

$$A = \frac{\partial V(X, t)}{\partial t}. \tag{1.11}$$

If the particle velocity is known as a function of position and time, we have in the spatial description

$$a = \frac{\mathrm{D}v}{\mathrm{D}t} = \frac{\partial v}{\partial t} + v \frac{\partial v}{\partial x}. \tag{1.12}$$

The second term in the right-hand side of (1.12) is called the convected part of the acceleration.

For the purpose of examining conservation of mass and balance of linear momentum it is necessary to evaluate time-rates of change of line integrals. In the spatial description we consider the fixed region $x_1 \leqq x \leqq x_2$, which instantaneously contains a moving mass system. The total of a global quantity $\mathrm{f}(x, t)$ instantaneously carried by the mass system is given by

$$\int_{x_1}^{x_2} \mathrm{f}(x, t)\mathrm{d}x.$$

The time-rate of change of this integral consists of two terms, namely, the rate of increase of $\mathrm{f}(x, t)$ instantaneously located inside the region and the net rate of outward flux of $\mathrm{f}(x, t)$. This is expressed in the form

$$\frac{\mathrm{d}}{\mathrm{d}t}\int_{x_1}^{x_2} \mathrm{f}(x, t)\mathrm{d}x = \int_{x_1}^{x_2} \frac{\partial \mathrm{f}}{\partial t}\mathrm{d}x + \mathrm{f}(x, t)v(x, t)\Big|_{x=x_1}^{x=x_2}. \tag{1.13}$$

The physical interpretation of the terms on the right-hand side of (1.13) thus is

$$\int_{x_1}^{x_2} \frac{\partial \mathrm{f}}{\partial t}\mathrm{d}x = \text{rate of increase inside the region,}$$

$$\mathrm{f}(x, t)v(x, t)\Big|_{x=x_1}^{x=x_2} = \text{net rate of outward flux.}$$

The left-hand side of (1.13) represents the time-rate of increase of the total of the quantity $\mathrm{f}(x, t)$ instantaneously located in the spatial region $x_1 \leqq x \leqq x_2$. Eq. (1.13) is the one-dimensional version of *Reynolds' transport theorem.* The theorem may, of course, also be expressed in the form

$$\frac{\mathrm{d}}{\mathrm{d}t}\int_{x_1}^{x_2} \mathrm{f}(x, t)\mathrm{d}x = \int_{x_1}^{x_2} \left[\frac{\partial \mathrm{f}}{\partial t} + \frac{\partial}{\partial x}(\mathrm{f}v)\right]\mathrm{d}x. \tag{1.14}$$

1.2.4. Conservation of mass

One of the fundamental principles of classical mechanics is that matter can be neither created nor destroyed. Let us consider a mass system which instantaneously occupies the fixed spatial region $x_1 \leqq x \leqq x_2$. Let $\rho(x, t)$ be the mass density at location x and time t in the spatial description, and let $\rho_0(X)$ be the mass density as a function of the reference configuration. Conservation of mass then implies

$$\int_{x_1}^{x_2} \rho(x, t)\mathrm{d}x = \int_{p(x_1, t)}^{p(x_2, t)} \rho_0(X)\mathrm{d}X.$$

where $p(x, t)$ is defined by eq. (1.2). In the left-hand side of this equality we now introduce a change of variables by the use of eq. (1.1). It then follows that $\rho_0(X)$ and $\rho(x, t)$ are related by

$$\rho_0(X) = \rho(x, t)\frac{\partial P(X, t)}{\partial X}. \tag{1.15}$$

The principle of conservation of mass also implies that the time-rate of change of the mass of a system of particles vanishes. Thus, if the particles instantaneously occupy the spatial region $x_1 \leqq x \leqq x_2$, we have

$$\frac{\mathrm{d}}{\mathrm{d}t}\int_{x_1}^{x_2} \rho(x, t)\mathrm{d}x = 0.$$

The use of (1.14) then yields

$$\frac{\partial \rho}{\partial t} + \frac{\partial}{\partial x}(\rho v) = 0. \tag{1.16}$$

This equation expresses conservation of mass in the spatial description.

Eq. (1.16) can conveniently be used to simplify the material derivatives of integrals over a product of the mass density and another function. According to the transport theorem (1.14), we can write

$$\frac{\mathrm{d}}{\mathrm{d}t}\int_{x_1}^{x_2} \rho \mathrm{f}\, \mathrm{d}x = \int_{x_1}^{x_2} \left[\frac{\partial}{\partial t}(\rho \mathrm{f}) + \frac{\partial}{\partial x}(\rho \mathrm{f} v)\right] \mathrm{d}x.$$

The right-hand side can be simplified by employing (1.16), and we obtain

$$\frac{\mathrm{d}}{\mathrm{d}t}\int_{x_1}^{x_2} \rho \mathrm{f}\, \mathrm{d}x = \int_{x_1}^{x_2} \rho \frac{\mathrm{Df}}{\mathrm{D}t}\mathrm{d}x, \tag{1.17}$$

where $\mathrm{D}/\mathrm{D}t$ is defined by eq. (1.10). This result will prove useful in the discussion of the balances of momentum and energy.

1.2.5. Balance of momentum

The principle of balance of linear momentum states that the instantaneous rate of change of the linear momentum of a system equals the resultant external force acting on the system at the particular instant of time. Considering a mass system of unit cross-sectional area instantaneously contained in the region $x_1 \leqq x \leqq x_2$, the principle implies that

$$\tau(x, t)|_{x=x_1}^{x=x_2} = \frac{\mathrm{d}}{\mathrm{d}t}\int_{x_1}^{x_2} \rho(x, t)v(x, t)\mathrm{d}x. \tag{1.18}$$

Here $\tau(x, t)$ defines the stress in the spatial description at position x, and body forces are not taken into account. Upon writing the left-hand side as an integral and simplifying the right-hand side by means of (1.17), the equation may be rewritten as

$$\int_{x_1}^{x_2} \frac{\partial \tau}{\partial x}\,\mathrm{d}x = \int_{x_1}^{x_2} \rho(x, t)\frac{\mathrm{D}v}{\mathrm{D}t}\,\mathrm{d}x. \tag{1.19}$$

Clearly, (1.19) is equivalent to

$$\frac{\partial \tau}{\partial x} = \rho \frac{\mathrm{D}v}{\mathrm{D}t}. \tag{1.20}$$

This is the equation of motion in spatial coordinates. To determine the equation of motion in material coordinates we return to (1.18), and we introduce a change of coordinates by means of the mapping (1.1). By employing the relation between $\rho_0(X)$ and $\rho(x, t)$, we then obtain

$$\frac{\partial T}{\partial X} = \rho_0 \frac{\partial V}{\partial t}. \tag{1.21}$$

where $T(X, t) = \tau[P(X, t), t]$ is the stress in the material description.

A stress-equation of motion must be supplemented by a relation between stress and deformation. For an ideally elastic solid, the stress is a function of the appropriate measure of deformation. In one dimension it is convenient to consider the stress $T(X, t)$ in the material description as a function of the displacement gradient $\partial U/\partial X$,

$$T(X, t) = \mathscr{S}(\partial U/\partial X). \tag{1.22}$$

The equation of motion in the material description may then be written as

$$C^2 \frac{\partial^2 U}{\partial X^2} = \frac{\partial^2 U}{\partial t^2}, \tag{1.23}$$

where

$$C^2 = \frac{1}{\rho_0} \frac{\mathrm{d}\mathscr{S}}{\mathrm{d}(\partial U/\partial X)}. \tag{1.24}$$

1.2.6. Balance of energy

The principle of conservation of energy states that the time-rate of change of the sum of the kinetic and internal energies of a mass system is equal to the rate of work of the external forces plus all other energies that enter or leave the system per unit time.

Denoting the instantaneous energy per unit mass by $\mathrm{e}(x, t)$, the internal energy instantaneously contained in the region $x_1 \leqq x \leqq x_2$ is

$$\int_{x_1}^{x_2} \rho e(x, t)\mathrm{d}x.$$

The instantaneous kinetic energy is

$$\frac{1}{2}\int_{x_1}^{x_2} \rho v^2 \mathrm{d}x.$$

The stresses transmit energy to the system in the form of mechanical work. The rate of work is called the power input P. We have

$$\begin{aligned} P &= -\tau(x_1, t)v(x_1, t)+\tau(x_2, t)v(x_2, t) \\ &= \int_{x_1}^{x_2} \frac{\partial}{\partial x}(\tau v)\mathrm{d}x. \end{aligned} \tag{1.25}$$

In this equation, body forces are again not taken into account.

Energy can also be transmitted in the form of heat, by conduction or by heat sources that are distributed inside the continuum. At this stage thermal effects will, however, be left out of consideration.

For a purely mechanical system the principle of conservation of energy can now be stated as

$$P = \frac{\mathrm{d}}{\mathrm{d}t}\int_{x_1}^{x_2} \rho e\,\mathrm{d}x + \frac{\mathrm{d}}{\mathrm{d}t}\int_{x_1}^{x_2} \tfrac{1}{2}\rho v^2 \mathrm{d}x. \tag{1.26}$$

The integrals on the right-hand side can be evaluated by employing (1.17). Using also the equation of motion (1.20), we find

$$\rho \frac{\mathrm{De}}{\mathrm{D}t} = \tau \frac{\partial v}{\partial x}, \tag{1.27}$$

which expresses local conservation of energy.

1.2.7. *Linearized theory*

Although it is possible to determine solutions for certain one-dimensional problems governed by the nonlinear theory, there are often rather substantial complications.[1] These complications disappear altogether when the theory is appropriately linearized.

A linearized theory is amenable to treatment by standard mathematical methods because the principle of superposition applies. Since a phenomenon described by linear equations is also intuitively easier to understand, an

[1] A nonlinear problem is discussed in section 10.7.

examination of a problem on the basis of linearized equations often leads to considerable insight into the actual physical situation. On the other hand one always has to make sure that the assumptions on which the linearization is based are satisfied, because small nonlinearities sometimes give rise to quite significant modifications of results obtained from a linearized theory. It is therefore worthwhile to examine the conditions justifying a linearization of the theory.

Let us first consider the constitutive relation (1.22) for the special case that $T(X, t)$ is proportional to $\partial U/\partial X$:

$$T = S_1 \frac{\partial U}{\partial X}. \tag{1.28}$$

For some materials this relation may be an approximation applicable only when $|\partial U/\partial X| \ll 1$. For other materials and within the context of a one-dimensional theory it may be exact in the sense that it may apply for quite large values of $|\partial U/\partial X|$. If (1.28) holds, eq. (1.23) reduces to the linear wave equation

$$\frac{\partial^2 U}{\partial X^2} = \frac{1}{C^2}\frac{\partial^2 U}{\partial t^2}, \tag{1.29}$$

where

$$C^2 = S_1/\rho_0. \tag{1.30}$$

A general solution to the linear wave equation can be obtained by introducing the substitutions

$$\alpha = t - \frac{X}{C}, \qquad \beta = t + \frac{X}{C}, \tag{1.31a, b}$$

whereby (1.29) reduces to

$$\frac{\partial^2 U}{\partial \alpha\, \partial \beta} = 0. \tag{1.32}$$

It follows that $\partial U/\partial \alpha$ is a function of α only. Integrating again, we find that $U(X, t)$ must be of the form

$$U = F(\alpha) + G(\beta) = F\left(t - \frac{X}{C}\right) + G\left(t + \frac{X}{C}\right). \tag{1.33}$$

Any arbitrary form of the functions $F(\ )$, $G(\ )$ employed in this equation will give a solution of the linear wave equation (1.29). If time t is increased by any value, say Δt, and simultaneously X is increased by $C\Delta t$, the argument

$t-X/C$ is clearly not altered. The term $F(t-X/C)$ thus represents a displacement disturbance propagating in the positive X-direction. Similarly, $G(t+X/C)$ represents a disturbance propagating in the negative X-direction. Since the shape of the propagating disturbances does not alter, the propagation is called distortionless and lossless.

It is rather remarkable that for the one-dimensional case a linear relation between the stress and the displacement gradient in the material description is all that is required for a linear wave equation in material coordinates. It should be emphasized, however, that the kinematics still are nonlinear in the spatial description of the motion. If an observer positioned at a fixed location x observes the displacement $u(x, t)$, the convective terms still must be included in computing the velocity and the acceleration of a particle. This becomes evident if we employ (1.9) and (1.12) to compute the following exact relations:

$$v = \frac{\partial u/\partial t}{1-\partial u/\partial x} \tag{1.34}$$

$$a = \frac{(1-\partial u/\partial x)^2\partial^2 u/\partial t^2 + 2(1-\partial u/\partial x)(\partial u/\partial t)(\partial^2 u/\partial x\,\partial t) + (\partial u/\partial t)^2\partial^2 u/\partial x^2}{(1-\partial u/\partial x)^3}. \tag{1.35}$$

From (1.3) and (1.4) it follows that

$$\frac{\partial u}{\partial x} = \frac{\partial U/\partial X}{1+\partial U/\partial X}. \tag{1.36}$$

If the linearization of the relation between T and $\partial U/\partial X$ is now justifiable only for $|\partial U/\partial X| \ll 1$, then the relations (1.36) and (1.34) can also be linearized. We find

$$\frac{\partial u}{\partial x} \simeq \frac{\partial U}{\partial X} \quad \text{and} \quad v \simeq \frac{\partial u}{\partial t}. \tag{1.37a, b}$$

It is, however, not yet possible to simplify (1.35), since the orders of magnitudes of $\partial u/\partial t$ and the second-order derivatives are not immediately apparent.

To determine the conditions for complete linearization, we write the equation of motion in spatial coordinates. By employing the results (1.15), (1.28), (1.35) and (1.36) in eq. (1.20), the following non-linear equation is obtained:

$$\left[C^2 - \left(\frac{\partial u}{\partial t}\right)^2\right]\frac{\partial^2 u}{\partial x^2} = \left(1-\frac{\partial u}{\partial x}\right)^2\frac{\partial^2 u}{\partial t^2} + 2\left(1-\frac{\partial u}{\partial x}\right)\frac{\partial u}{\partial t}\frac{\partial^2 u}{\partial x\,\partial t}, \tag{1.38}$$

where C is defined by (1.30). It is perhaps surprising that this equation also admits a solution of the form

$$u(x, t) = \mathrm{f}\left(t - \frac{x}{C}\right). \tag{1.39}$$

The validity of (1.39) can be checked by direct substitution, but it can also be concluded by constructing $u(x, t)$ from $F(t - X/C)$ via eq. (1.4). See Problem 1.2.

From (1.39) it follows that

$$\frac{\partial u}{\partial t} = \mathrm{f}', \qquad \frac{\partial u}{\partial x} = -\mathrm{f}'/C, \tag{1.40a, b}$$

where a prime denotes the derivative with respect to the argument. By inspecting (1.35) and (1.38) it can now be concluded on the basis of (1.39) and (1.40a, b) that

$$a = \frac{\partial^2 u}{\partial t^2} \tag{1.41}$$

$$C^2 \frac{\partial^2 u}{\partial x^2} = \frac{\partial^2 u}{\partial t^2} \tag{1.42}$$

provided that

$$|\mathrm{f}'| \ll C. \tag{1.43}$$

It should be noted that (1.43) implies $|\partial u/\partial x| \ll 1$ or equivalently $|\partial U/\partial X| \ll 1$. Thus, in the one-dimensional problem the rates of change of external disturbances must satisfy the restriction (1.43) in order that the motion can be described by a completely linearized theory.

Several additional observations on the linearization of the theory can be found in an article by Thurston.[2]

1.2.8. Notation for the linearized theory

When the problem is completely linearized the distinction between the material and spatial descriptions of the motion vanishes altogether. Thus, either the notation of the material description with upper case letters or the notation of the spatial description with lower case letters could be used. Over the years it has, however, become customary to use lower case symbols

[2] R. N. Thurston, *The Journal of the Acoustical Society of America* **45** (1969) 1329–1341.

for the linearized theory. Generally, the Greek symbols ε_x and τ_x are used for the displacement gradient $\partial u/\partial x$ and the stress component, respectively. According to (1.28), the stress component τ_x is proportional to ε_x. In the next chapter it will be shown that the proportionality constant can be expressed in terms of the Lamé elastic constants λ and μ as $S_1 = \lambda+2\mu$. Thus

$$\tau_x = (\lambda+2\mu)\frac{\partial u}{\partial x}. \tag{1.44}$$

The stress equation of motion may be written as

$$\frac{\partial \tau_x}{\partial x} = \rho \frac{\partial^2 u}{\partial t^2}, \tag{1.45}$$

where ρ is the *constant* mass density. Substitution of (1.44) into (1.45) yields the wave equation

$$\frac{\partial^2 u}{\partial x^2} = \frac{1}{c_L^2}\frac{\partial^2 u}{\partial t^2}, \tag{1.46}$$

where

$$c_L^2 = \frac{\lambda+2\mu}{\rho}. \tag{1.47}$$

1.3. Half-space subjected to uniform surface tractions

A simple example which displays many of the features of transient wave propagation according to the linearized theory is provided by the wave motion generated in an initially undisturbed, homogeneous, isotropic elastic half-space by the application of a spatially uniform surface pressure $p(t)$. Suppose that the half-space is defined by $x \geqq 0$ (see figure 1.2). Denoting

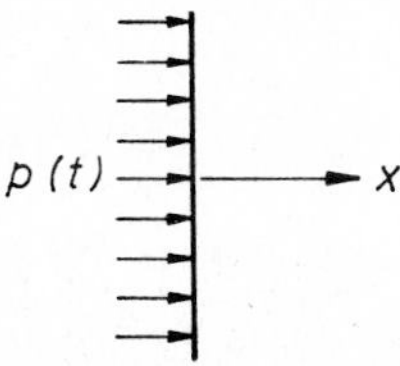

Fig. 1.2. Half-space subjected to surface tractions.

the normal stress in the x-direction by $\tau_x(x, t)$, we have at the boundary $x = 0$

$$\tau_x = -p(t), \text{ where } p(t) = 0 \text{ for } t < 0. \tag{1.48}$$

The other stress components vanish identically at $x = 0$.

Any plane parallel to the x-axis is clearly a plane of symmetry. As a consequence, transverse displacements are not possible and the motion of the half-space is described by the displacement in the x-direction, denoted by $u(x, t)$, which is a function of x and t only. The half-plane is evidently in a state of one-dimensional deformation, and the equations defined in section 1.2 apply. Thus the deformation of the half-space is completely described by the single strain component

$$\varepsilon_x = \frac{\partial u}{\partial x}. \tag{1.49}$$

We say that the half-plane is in a state of one-dimensional longitudinal strain. According to (1.44), the stress and the strain are related by

$$\tau_x = (\lambda+2\mu)\frac{\partial u}{\partial x}, \tag{1.50}$$

while the displacement equation of motion is

$$\frac{\partial^2 u}{\partial x^2} = \frac{1}{c_L^2}\frac{\partial^2 u}{\partial t^2}, \tag{1.51}$$

where c_L is defined by (1.47). Assuming that the half-space is at rest prior to time $t = 0$, eqs. (1.48) and (1.51) are supplemented by the initial conditions

$$u = \dot{u} \equiv 0 \text{ for } t = 0,\ x > 0. \tag{1.52}$$

The general solution of (1.51) is

$$u(x, t) = \mathrm{f}\left(t-\frac{x}{c_L}\right) + g\left(t+\frac{x}{c_L}\right). \tag{1.53}$$

Although it is intuitively rather obvious that the surface pressure $p(t)$ generates a wave propagating in the positive x-direction only, we will not *a priori* discard the function $g(t+x/c_L)$, but rather follow a strictly mathematical approach. Thus, employing the full form of (1.53), the initial conditions (1.52) require that for $x > 0$

$$f(-x/c_L)+g(x/c_L) = 0 \tag{1.54}$$

$$f'(-x/c_L)+g'(x/c_L) = 0, \tag{1.55}$$

where primes indicate differentiations with respect to the argument. The solutions to these equations are

$$f(-x/c_L) = -g(x/c_L) = A \qquad \text{for } x > 0, \tag{1.56}$$

where A is a constant. Since $t+x/c_L$ is always positive for $t \geqq 0$ and $x > 0$, eq. (1.53) reduces to

$$u(x, t) = \begin{cases} f\left(t - \dfrac{x}{c_L}\right) - A & \text{for } t > \dfrac{x}{c_L} \\ 0 & \text{for } t < \dfrac{x}{c_L} \end{cases} \tag{1.57}$$

which satisfies the initial conditions. This solution shows that a wavefront separating the disturbed from the undisturbed medium propagates through the material with the velocity c_L. A particle located at $x = \bar{x}$ remains at rest until time $t = \bar{t} = \bar{x}/c_L$.

The boundary condition at $x = 0$ yields

$$-\frac{\lambda+2\mu}{c_L} f'(t) = -p(t).$$

Upon integration of this equation, $f(t-x/c_L)$ follows as

$$f\left(t - \frac{x}{c_L}\right) = \frac{c_L}{\lambda+2\mu} \int_0^{t-x/c_L} p(s)\mathrm{d}s + B, \tag{1.58}$$

where B is a constant. According to (1.48), the function $p(s)$ vanishes identically for $s < 0$ and the integral over $p(s)$ thus disappears when $t < x/c_L$. In order that eq. (1.58) is consistent with (1.56) we should then have $B = A$. The expression for the displacement eq. (1.58) then becomes

$$u(x, t) = \frac{c_L}{\lambda+2\mu} \int_0^{t-x/c_L} p(s)\mathrm{d}s. \tag{1.59}$$

The corresponding normal stress $\tau_x(x, t)$ follows from (1.50) as

$$\tau_x = -p\left(t - \frac{x}{c_L}\right).$$

The normal stresses in the transverse directions, denoted by τ_y and τ_z, can be computed as

$$\tau_y = \tau_z = -\frac{\lambda}{\lambda+2\mu}\, p\left(t - \frac{x}{c_L}\right).$$

In the expressions for $u(x, t)$ and $\tau_x(x, t)$ it should be taken into account that $p(t) \equiv 0$ for $t < 0$, as stated by eq. (1.48).

The solutions for the displacement and the stresses show that the surface pressure generates a disturbance which propagates into the half-space with velocity c_L. A particle located at $x = \bar{x}$ remains at rest until the time $\bar{t} = \bar{x}/c_L$, when the wavefront arrives which separates the disturbed part from the undisturbed part of the half-space. The normal stress at $x = \bar{x}$ is compressive and has the value of the external pressure for the value of the argument $t - \bar{x}/c_L$. The displacement is proportional to the area under the curve representing the external pressure between arguments 0 and $t - \bar{x}/c_L$. The displacements and stresses describe a transient wave motion. It should be noted that for this simple problem the shape of the stress pulse does not alter as it propagates through the medium.

By employing (1.59) the particle velocity $\dot{u}(x, t) = \partial u/\partial t$ is computed as

$$\dot{u} = \frac{c_L}{\lambda + 2\mu} p\left(t - \frac{x}{c_L}\right). \tag{1.60}$$

Clearly then, for a wave propagating in the positive x-direction, the stress $\tau_x(x, t)$ and the particle velocity $\dot{u}(x, t)$ are related by

$$\tau_x = -\rho c_L \dot{u}. \tag{1.61}$$

The ratio of the stress and the particle velocity is called the *mechanical impedance*. In this case the mechanical impedance ρc_L is a material constant. Since it measures the stress that is required to generate motion it is often called the *wave resistance* of the material.

Considering a unit area element normal to the x-axis at a position x, the instantaneous rate of work of the traction acting on the element is the vector product of $\tau_x(x, t)$ and the particle velocity $\dot{u}(x, t)$. This instantaneous rate of work is called the *power* per unit area and it is denoted by $\mathscr{P}$. We have

$$\mathscr{P}(x, t) = -\tau_x \dot{u}. \tag{1.62}$$

The minus sign appears because a stress vector (which is positive in tension) and a velocity vector acting in the same direction yield a positive value for the power. By the use of (1.61) we find

$$\mathscr{P}(x, t) = \rho c_L \dot{u}^2. \tag{1.63}$$

The power defines the rate at which energy is communicated per unit time across a unit area. Clearly $\mathscr{P}$ represents the energy flux across the area element and it must, therefore, be related to the total energy density $\mathscr{H}$. The total energy per unit volume equals the sum of the kinetic energy density $\mathscr{K}$ and the strain energy density $\mathscr{U}$, thus

$$\mathcal{E} = \mathscr{H} = \mathscr{K} + \mathscr{U} = \tfrac{1}{2}\rho\dot{u}^2 + \tfrac{1}{2}(\lambda+2\mu)\left(\frac{\partial u}{\partial x}\right)^2. \tag{1.64}$$

If more energy flows across the point $x+\mathrm{d}x$ than flows across x, then the energy contained in length $\mathrm{d}x$ of the medium diminishes, i.e.,

$$\mathscr{P}(x+\mathrm{d}x) - \mathscr{P}(x) = -\mathrm{d}x\,\frac{\partial \mathscr{H}}{\partial t},$$

or

$$\frac{\partial \mathscr{P}}{\partial x} + \frac{\partial \mathscr{H}}{\partial t} = 0,$$

which is the equation of continuity for energy. Substituting (1.62) and (1.64) we find

$$-\left(\frac{\partial \tau_x}{\partial x} - \rho\ddot{u}\right)\dot{u} - \left[\tau_x - (\lambda+2\mu)\frac{\partial u}{\partial x}\right]\frac{\partial \dot{u}}{\partial x} = 0,$$

which is, of course, identically satisfied in view of eqs. (1.50) and (1.51).

Since energy is conserved it is required that at a certain time t the work done by the surface pressure $p(t)$ equals the sum of the kinetic and strain energies in the disturbed part of the half-space, i.e.,

$$\int_0^t \mathscr{P}(0, t)\mathrm{d}t = \int_0^{c_L t} \mathscr{H}(x, t)\mathrm{d}x. \tag{1.65}$$

The first integral assumes the form

$$\int_0^t \mathscr{P}(0, t)\mathrm{d}t = \frac{c_L}{\lambda+2\mu}\int_0^t [p(s)]^2 \mathrm{d}s.$$

By the use of (1.64) and (1.59) we find for the right-hand side of (1.65)

$$\int_0^{c_L t} \mathscr{H}(x, t)\mathrm{d}x = \frac{1}{\lambda+2\mu}\int_0^{c_L t}\left[p\left(t - \frac{x}{c_L}\right)\right]^2 \mathrm{d}x.$$

By introducing a change of variables in this equation, the equality (1.65) can easily be verified.

By neglecting the inertia term the equation governing the corresponding static problem is immediately obtained from (1.51). Upon integrating once, we find

$$\tau_x = (\lambda+2\mu)\frac{\mathrm{d}u}{\mathrm{d}x} = -p(t),$$

where we have used the boundary condition (1.48). Suppose we consider a monotonically increasing function $p(t)$, and we expand

$$p\left(t-\frac{x}{c_L}\right) \sim p(t)-\frac{x}{c_L}\,\dot{p}(t).$$

It is then clear that for given x and t, $t > x/c_L$, the difference between the static and dynamic stresses is small if

$$\frac{x}{c_L}\,\frac{\dot{p}(t)}{p(t)} \ll 1.$$

This result shows that dynamic effects are of most interest if either x is large or $\dot{p}(t)$ is appreciable.

It is finally noted that it follows from eqs. (1.39), (1.43) and (1.59) that for the present problem the linearized theory is valid if

$$\frac{p(t)}{\lambda+2\mu} \ll 1.$$

1.4. Reflection and transmission

When a propagating disturbance strikes the interface between two media of different material properties, part of the disturbance is reflected and part is transmitted across the interface. In the special case that the second medium cannot carry mechanical waves, i.e., if it is a vacuum, the incident wave is completely reflected. We will consider the latter case first.

Suppose the incident stress wave is denoted by

$$(\tau_x)_i = \mathrm{f}\left(t-\frac{x}{c_L}\right), \tag{1.66}$$

where $\mathrm{f}(s) \equiv 0$ for $s < 0$. Since the reflected stress wave propagates in the negative x-direction, it can be represented by

$$(\tau_x)_r = g\left(t+\frac{x}{c_L}\right). \tag{1.67}$$

At the free boundary $x = a$ the stress τ_x vanishes, i.e.

$$\tau_x = (\tau_x)_i+(\tau_x)_r = 0 \quad \text{at} \quad x = a.$$

For $t < a/c_L$ this equation implies $g(t+x/c_L) \equiv 0$. For $t \geqq a/c_L$ we find

$$g\left(t+\frac{a}{c_L}\right) = -\mathrm{f}\left(t-\frac{a}{c_L}\right).$$

Now setting $s = t + a/c_L$, we conclude

$$g(s) = -\mathrm{f}\left(s - 2\frac{a}{c_L}\right),$$

and the reflected stress wave may thus be represented by

$$(\tau_x)_r = -\mathrm{f}\left(t - \frac{a}{c_L} + \frac{x-a}{c_L}\right). \tag{1.68}$$

Eq. (1.68) shows that the reflected stress pulse has the same shape as the incident stress pulse, but the sign of the stress changes upon reflection. Thus, at a free surface a tensile pulse is reflected as a pressure pulse, and vice versa. The direction of the displacement remains unchanged, however, upon reflection.

The conversion of a pressure pulse into a tensile pulse can have some interesting implications for materials that are of relatively low tensile strength. As a compressive pulse of short enough length is reflected at a free surface, the resulting tensile stresses may cause fracture. This type of tensile fracture under rapid compressive loading is a typical wave propagation effect which is known as *spalling* or *scabbing*. The effect was first demonstrated experimentally by B. Hopkinson[3], who detonated an explosive charge in contact with a metal plate. The reflection of the pulse from the free surface produced tensile fractures which caused a disk of metal roughly in the shape of a spherical cap to break away from the surface directly opposite the explosive charge.

Let us now consider the reflection and transmission at an interface between two media of different material properties. The incident wave is of the form (1.66). The reflected and transmitted waves may be represented by

$$(\tau_x)_r = g\left(t - \frac{a}{c_L} + \frac{x-a}{c_L}\right)$$

and

$$(\tau_x)_t = h\left(t - \frac{a}{c_L} - \frac{x-a}{c_L^A}\right),$$

respectively. (See figure 1.3). By enforcing continuity of the stress and the particle velocity at the interface $x = a$, we find

[3] B. Hopkinson, *Collected scientific papers*. Cambridge, University Press (1921), p. 423.

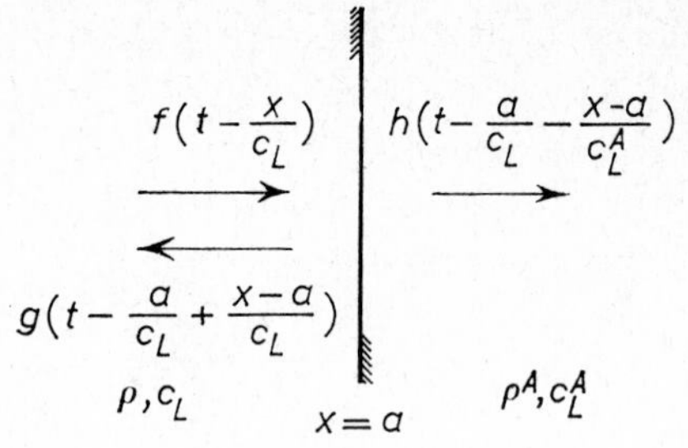

Fig. 1.3. Incident, reflected and transmitted waves.

$$(\tau_x)_r = C_r \mathrm{f}\left(t - \frac{a}{c_L} + \frac{x-a}{c_L}\right)$$

$$(\tau_x)_t = C_t \mathrm{f}\left(t - \frac{a}{c_L} - \frac{x-a}{c_L^A}\right).$$

The reflection coefficient C_r and the transmission coefficient C_t are

$$C_r = \frac{\rho^A c_L^A/\rho c_L - 1}{\rho^A c_L^A/\rho c_L + 1}$$

$$C_t = \frac{2\rho^A c_L^A/\rho c_L}{\rho^A c_L^A/\rho c_L + 1}.$$

These expressions show that the ratio of the mechanical impedances completely determines the nature of the reflection and the transmission at

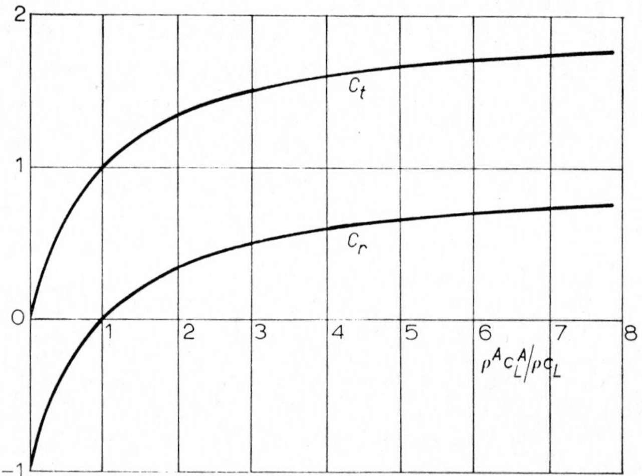

Fig. 1.4. Reflection and transmission coefficients.

the interface. The reflection and transmission coefficients are plotted versus the ratio of the mechanical impedances in figure 1.4. For $\rho^A c_L^A/\rho c_L = 0$, which corresponds to a free surface, $C_r = -1$ and $C_t = 0$, which agrees with the result (1.68). For $\rho^A c_L^A/\rho c_L = 1$, the pulse is completely transmitted. If $\rho^A c_L^A/\rho c_L > 1$, the reflected pulse is of the same sign as the incident wave. The transmitted wave always is of the same type as the incident wave.

If a number of layers follow each other, the pattern of transient waves can still be unraveled in a one-dimensional geometry. The succession of reflections and transmissions may give rise to somewhat unexpected effects, such as high tensile stresses under compressive loads. Thus it was shown by Achenbach et al.[4], by analysis and experiment, that tensile failure may occur in a segmented rod at the first interface adjacent to the end where a compressive load of short duration is applied. Failure occurs if segments with very different mechanical impedances are altered, particularly if the mechanical impedance of the first segment is relatively small.

1.5. Waves in one-dimensional longitudinal stress

Waves in one-dimensional longitudinal strain are not the only type of wave motion for a purely longitudinal disturbance. The second type is provided by wave motion in one-dimensional stress, whereby the longitudinal normal stress, say τ_x, which is a function of x and t only, is the one nonvanishing stress component. All other stress components vanish.

The deformation of an element in one-dimensional stress is sketched in figure 1.5. The difference with the deformation sketched in figure 1.1 is

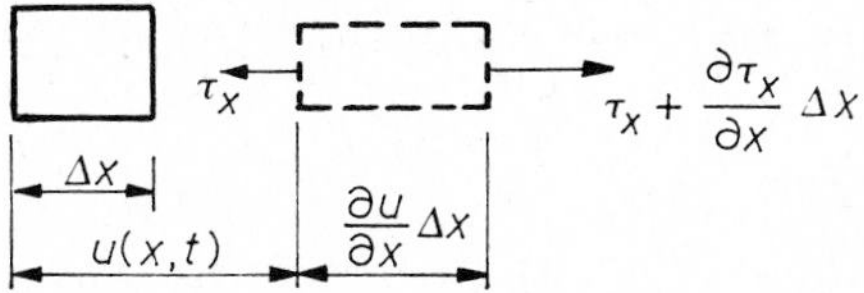

Fig. 1.5. Deformation in one-dimensional stress.

that in one-dimensional stress an element is not prevented from deforming in the transverse direction. In fact, as shown in figure 1.5, the cross section decreases if the element is in tension. For the case of one-dimensional stress, τ_x and ε_x are related by

$$\tau_x = E\varepsilon_x, \tag{1.69}$$

[4] J. D. Achenbach, J. H. Hemann and F. Ziegler, *AIAA Journal* **6** (1968) 2040–2043.

where E is Young's modulus. By writing the equation of motion for an element, we find

$$\frac{\partial \tau_x}{\partial x} = \rho \frac{\partial^2 u}{\partial t^2}. \tag{1.70}$$

Substitution of (1.69) into (1.70) yields

$$\frac{\partial^2 u}{\partial x^2} = \frac{1}{c_b^2} \frac{\partial^2 u}{\partial t^2}, \tag{1.71}$$

where

$$c_b^2 = \frac{E}{\rho}. \tag{1.72}$$

Waves in a one-dimensional state of stress approximate wave motion in a thin rod. The velocity c_b is usually called the bar velocity. If a semi-infinite thin rod ($x \geqq 0$) is subjected at $x = 0$ to a pressure $p(t)$, the resulting stress wave is

$$\tau_x = -p\left(t - \frac{x}{c_b}\right). \tag{1.73}$$

It should be emphasized that (1.73) is an approximate solution which is valid only for a very thin rod. If the rod is very thick, the deformation is expected to be closer to a state of one-dimensional strain. Often rods are neither very thin nor very thick, so that both approximations are unsatisfactory, and a more exact treatment allowing two- or three-dimensional variation of the field variable is required. A more exact treatment of wave propagation in a rod is discussed in chapter 8. It turns out, however, that the approximation of one-dimensional stress is generally very satisfactory.

1.6. Harmonic waves

1.6.1. Traveling waves

Let us consider an expression for the longitudinal displacement of the form

$$u(x, t) = A \cos [k(x - ct)], \tag{1.74}$$

where the amplitude A is independent of x and t. Eq. (1.74) is of the general form $f(x - ct)$ and thus clearly represents a traveling wave. The argument $k(x - ct)$ is called the phase of the wave; points of constant phase are propagated with the phase velocity c. At any instant t, $u(x, t)$ is a periodic function of x with wavelength Λ, where $\Lambda = 2\pi/k$. The quantity $k = 2\pi/\Lambda$, which counts the number of wavelengths over 2π, is termed the wavenumber. At any position the displacement $u(x, t)$ is time-harmonic with time period T, where $T = 2\pi/\omega$. The circular frequency ω follows from (1.74) as

$$\omega = kc. \tag{1.75}$$

It follows that an alternative representation of $u(x, t)$ is

$$u(x, t) = A \cos\left[\omega\left(\frac{x}{c} - t\right)\right]. \tag{1.76}$$

Eqs. (1.74) and (1.76) are called traveling harmonic waves. The expressions represent trains of sinusoidal waves, which disturb at any instant of time the complete (unbounded) extent of the medium. Harmonic waves are steady-state waves, as opposed to the transient waves (pulses) which were discussed in the preceding sections of this chapter.

By substituting (1.74) into the wave equation of sections 1.2.8 and 1.5, we obtain

$$c = c_L, \quad \text{and} \quad c = c_b, \tag{1.77a, b}$$

from eqs. (1.46) and (1.71), respectively. Eqs. (1.77a, b) show that the phase velocities of traveling harmonic waves in one-dimensional longitudinal strain and one-dimensional longitudinal stress are independent of the wavelength Λ. This implies that very short waves propagate with the same phase velocity as long waves. If the phase velocity does not depend on the wavelength we say that the system is nondispersive. If the material is not purely elastic but displays dissipative behavior, it is found that the phase velocity of harmonic waves depends on the wavelength, and the system is said to be dispersive. Dispersion is an important phenomenon because it governs the change of shape of a pulse as it propagates through a dispersive medium. Dispersion occurs not only in inelastic bodies but also in elastic waveguides, as will be discussed in chapter 6.

The phase velocity c should be clearly distinguished from the particle velocity $\dot{u}(x, t)$, which is obtained as

$$\dot{u}(x, t) = Akc \sin [k(x-ct)].$$

For one-dimensional longitudinal strain the maximum value of the ratio of the particle velocity and the phase velocity is thus obtained as

$$(\dot{u}/c_L)_{\max} = Ak = 2\pi A/\Lambda.$$

Within the range of validity of the linear theory the ratio A/Λ should be much less than unity.

1.6.2. *Complex notation*

For mathematical convenience we generally use, instead of eq. (1.74), the expression

$$u(x, t) = A \exp [ik(x-ct)], \tag{1.78}$$

where $i = \surd(-1)$. Without stating it explicitly, henceforth it is understood that the real or imaginary part of (1.78) is to be taken for the physical interpretation of the solution. For the case of one-dimensional longitudinal stress the corresponding stress is then represented by

$$\tau_x(x, t) = iEAk \exp[ik(x-c_b t)], \tag{1.79}$$

and the particle velocity is written as

$$\dot{u}(x, t) = -iAkc_b \exp [ik(x-c_b t)]. \tag{1.80}$$

1.6.3. *Standing waves*

Let us consider two displacement waves of the same frequency and wavelength, but traveling in opposite directions. Since the wave equation is linear the resultant displacement is

$$u(x, t) = A_+ e^{i(kx-\omega t+\gamma_+)} + A_- e^{i(kx+\omega t+\gamma_-)}, \tag{1.81}$$

where A_+ and A_- are real-valued amplitudes, and γ_+ and γ_- are phase angles. If the amplitudes of the two simple harmonic waves are equal, $A_+ = A_- = A$, we can write

$$\begin{aligned} u(x, t) &= Ae^{i(kx+\frac{1}{2}\gamma_+ +\frac{1}{2}\gamma_-)}[e^{-i(\omega t-\frac{1}{2}\gamma_+ +\frac{1}{2}\gamma_-)} + e^{i(\omega t-\frac{1}{2}\gamma_+ +\frac{1}{2}\gamma_-)}] \\ &= 2A \exp [i(kx+\tfrac{1}{2}\gamma_+ +\tfrac{1}{2}\gamma_-)] \cos (\omega t-\tfrac{1}{2}\gamma_+ +\tfrac{1}{2}\gamma_-). \end{aligned}$$

The real part of this expression is

$$u(x, t) = 2A \cos (kx+\tfrac{1}{2}\gamma_+ +\tfrac{1}{2}\gamma_-) \cos (\omega t-\tfrac{1}{2}\gamma_+ +\tfrac{1}{2}\gamma_-). \tag{1.82}$$

Eqs. (1.82) represents a *standing wave*, since the shape of the wave does not travel. At points where $\cos (kx+\frac{1}{2}\gamma_+ +\frac{1}{2}\gamma_-) = 0$, the two traveling waves always cancel each other and the medium is at rest. These points are called the nodal points. Halfway between each pair of nodal points are the antinodes, where the motion has the largest amplitude.

1.6.4. *Modes of free vibration*

Standing waves form certain modes of free vibration of an elastic body.

As an example we consider the vibrations of a rod. If we consider a semi-infinite rod and if we require that the displacement vanishes at $x = 0$, the possible harmonic motions are subject to restrictions. Eq. (1.78) can now not be used and we have to employ the standing wave form (1.82) with the angles γ_+ and γ_- chosen so that a nodal point coincides with the boundary $x = 0$, i.e.,

$$\gamma_+ + \gamma_- = \pi$$

and

$$u(x, t) = 2A \sin(kx) \sin(\omega t - \gamma_+). \tag{1.83}$$

When, as a second boundary condition we add $u = 0$ at $x = l$, the harmonic motion is still further limited, for now of all the harmonic motions represented by (1.83) only those which have a nodal point at $x = l$ can be used. Thus we require

$$\sin(kl) = 0,$$

which implies

$$kl = \frac{2\pi l}{\Lambda} = n\pi \qquad n = 1, 2, 3, 4, \ldots$$

The distance between nodal points is half the wavelength, and thus this distance must be l, $l/2$, $l/3$, etc. The corresponding circular frequencies are

$$\omega = kc_b = \frac{n\pi c_b}{l}.$$

The circular frequency of the lowest or *fundamental* mode, which is called the *fundamental frequency*, is $\omega_0 = \pi c_b/l$ radians per second; in cycles per second the fundamental frequency is $f_0 = c_b/2l$. The frequencies of the higher modes are in cycles per second, $f_2 = 2c_b/2l$, $f_3 = 3c_b/2l$, etc. The higher frequencies are called *overtones*. For the example of the rod with rigidly supported ends the overtones are integral multiples of the fundamental frequency. Overtones with this simple relation to the fundamental are called harmonics. Only for the simplest vibrating systems governed by the wave equation are the modes of vibration as simple as discussed in this section.

1.7. Flux of energy in time-harmonic waves

The rate at which energy is communicated per unit area is equal to the power per unit area, which can be computed by employing eq. (1.16). Since the product of the real parts is not the same as the real part of the product of two complex numbers, $\mathscr{R}(\dot{u})$ must be used in eq. (1.16). For the case of one-dimensional longitudinal stress, $\mathscr{P}(x, t)$ is obtained as

$$\begin{aligned}\mathscr{P} &= \rho c_b^3 k^2 A^2 \sin^2 [k(x-c_b t)] \\ &= \frac{EA^2\omega^2}{c_b} \sin^2 [k(x-c_b t)].\end{aligned} \tag{1.84}$$

It is noted that for harmonic waves the power per unit area is an infinite sequence of pulses traveling with the phase velocity c_b.

1.7.1. *Time-average power per unit area*

A useful representation of the intensity of the wave is expressed by an average of $\mathscr{P}$ over time, at an arbitrary position and at an arbitrary time. This time-average power per unit area is

$$\frac{EA^2\omega^2}{c_b} \frac{1}{t_a}\int_t^{t+t_a} \sin^2 (kx-\omega s)\mathrm{d}s, \tag{1.85}$$

which can easily be evaluated if we choose as range of integration the period $T = 2\pi/\omega$. By employing

$$\frac{1}{T}\int_t^{t+T} \sin^2 (kx-\omega s)\mathrm{d}s = \tfrac{1}{2},$$

the time average of $\mathscr{P}$ over a period, which is denoted by $\langle\mathscr{P}\rangle$, is obtained as

$$\langle\mathscr{P}\rangle = \frac{1}{2}\frac{EA^2\omega^2}{c_b}. \tag{1.86}$$

The same expression is found if the limit of (1.85) is taken for a t_a which increases beyond bounds.

In the cgs system, $\langle\mathscr{P}\rangle$ is measured in ergs/sec-cm^2, and in the mks system, $\langle\mathscr{P}\rangle$ is expressed in watts/sec-m^2. It is noted that $\langle\mathscr{P}\rangle$ is proportional to the squares of the frequency and the amplitude.

The time average of a product of the real parts of two complex functions F and f must often be evaluated in energy computations. If F and f are of the forms

$$F = F_0 e^{i(\omega t-\gamma_1)}, \qquad \mathrm{f} = \mathrm{f}_0 e^{i(\omega t-\gamma_2)},$$

where F_0 and f_0 are real-valued, the following relation holds

$$\langle\mathscr{R}(F)\times\mathscr{R}(\mathrm{f})\rangle = \tfrac{1}{2}\mathscr{R}(F\bar{\mathrm{f}}), \tag{1.87}$$

where $\bar{\mathrm{f}}$ is the complex conjugate of f. In the book by Brillouin[5], eq. (1.87) is proven by substituting f and F into the left-hand side of (1.87). Using eq. (1.87), the expression (1.86) for $\langle \mathscr{P} \rangle$ can be written immediately.

1.7.2. Velocity of energy flux

The average power $\langle \mathscr{P} \rangle$ represents the average energy transmission per unit time and per unit area. We can thus speak of a "flow" of energy and we can introduce a velocity of energy flux c_e. The energy velocity is defined as the time-average energy transmission divided by the time-average of the total energy density

$$\langle \mathscr{P} \rangle = \langle \mathscr{H} \rangle c_e. \tag{1.88}$$

The total energy per unit volume consists of kinetic energy and strain energy. By employing (1.87), the time average of the kinetic energy per unit volume is computed as

$$\langle \mathscr{K} \rangle = \frac{1}{2}\frac{1}{T}\int_t^{t+T} \rho(\dot{u})^2 \mathrm{d}t = \tfrac{1}{4}\rho A^2\omega^2. \tag{1.89}$$

Similarly, the time-average strain energy density is obtained as

$$\langle \mathscr{U} \rangle = \frac{1}{2}\frac{1}{T}\int_t^{t+T} E\left(\frac{\partial u}{\partial x}\right)^2 \mathrm{d}t = \tfrac{1}{4}EA^2k^2. \tag{1.90}$$

Since $k = \omega/c_b$ and $c_b^2 = E/\rho$, we conclude $\langle \mathscr{K} \rangle = \langle \mathscr{U} \rangle$. Thus, for plane time-harmonic waves the time-average energy density is equally divided between the time-averages of the kinetic and strain energy densities. Taking the sum of (1.89) and (1.90), we obtain

$$\langle \mathscr{H} \rangle = \tfrac{1}{2}\rho A^2\omega^2. \tag{1.91}$$

By substituting (1.86) and (1.91) into (1.88), the velocity c_e is found as

$$c_e = c_b.$$

From the foregoing it follows that there are essentially two ways of calculating the average rate of flow of energy in a plane time-harmonic progressive wave. The first consists of forming the vector product of the traction and the particle velocity and taking the time-average of this product. The second proceeds by calculating the time-average of either the kinetic or

[5] L. Brillouin, *Wave propagation in periodic structures*, New York, Dover Publications, Inc. (1953), p. 70.

the strain energy per unit volume; twice either quantity multiplied by the velocity of energy transmission yields the time-average energy transmission.

For the example discussed in this section the velocity of energy transmission equals the phase velocity. This is generally true for non-dispersive wave propagation. For dispersive wave propagation c_e differs, however, from the phase velocity, as will be shown in chapter 6.

1.7.3. Energy transmission for standing waves

Let us consider the power per unit area for the case that the displacement consists of the superposition of two waves propagating in opposite directions, as described by eq. (1.81). The spatial and the time derivatives of $u(x, t)$ are

$$\frac{\partial u}{\partial x} = A_+ k\theta_+ + A_- k\theta_-$$

$$\frac{\partial u}{\partial t} = -A_+ \omega\theta_+ + A_- \omega\theta_-,$$

where

$$\theta_+ = \exp\left[i\left(kx-\omega t+\gamma_+ + \frac{\pi}{2}\right)\right]$$

$$\theta_- = \exp\left[i\left(kx+\omega t+\gamma_- + \frac{\pi}{2}\right)\right].$$

By employing the relation $\mathscr{P} = -\tau_x \dot{u}$, the power per unit area is obtained as

$$\mathscr{P} = Ek\omega[(A_+)^2\mathscr{R}(\theta_+)\mathscr{R}(\theta_+)-(A_-)^2\mathscr{R}(\theta_-)\mathscr{R}(\theta_-)].$$

It is noted that for waves traveling in opposite directions $\mathscr{P}$ has no cross terms, so that even the instantaneous values of the energy transmission are simply the differences between the individual fluxes. By employing (1.87), the time average is immediately obtained as

$$\langle\mathscr{P}\rangle = \tfrac{1}{2}Ek\omega[(A_+)^2-(A_-)^2].$$

If the amplitudes A_+ and A_- are equal there is no net flow of energy. As discussed in section 1.6, this case corresponds to a standing wave. Indeed, in a standing wave energy cannot be transmitted past nodal points, and the energy can thus flow only back and forth between the nodes, whereby at any position the net flow over a period T vanishes.

1.8. Fourier series and Fourier integrals

It is well known that any physically reasonable function, though not all mathematically conceivable functions, can be split up into a collection of components. In the most common case the components are sinusoidal functions, or alternatively, exponential functions with imaginary exponents. If the function repeats periodically with period T and circular frequency $\omega = 2\pi/T$, it can be expressed as a Fourier series of cosine and sine terms having frequencies ω, 2ω, . . ., or as a series of exponentials with the same frequencies. If the disturbance is not periodic it can be expressed as a Fourier integral over sinusoidal or exponential terms. In this section we briefly summarize the salient aspects of the representations by Fourier series and Fourier integrals.

1.8.1. Fourier series

A function is periodic in time if its values are repeated through all time at an interval T. This implies $f(t+T) = f(t)$, where T is the period. A periodic function is depicted in figure 1.6. Similarly a function may be periodic in space with periodic length (wavelength) Λ, i.e., $f(x+\Lambda) = f(x)$.

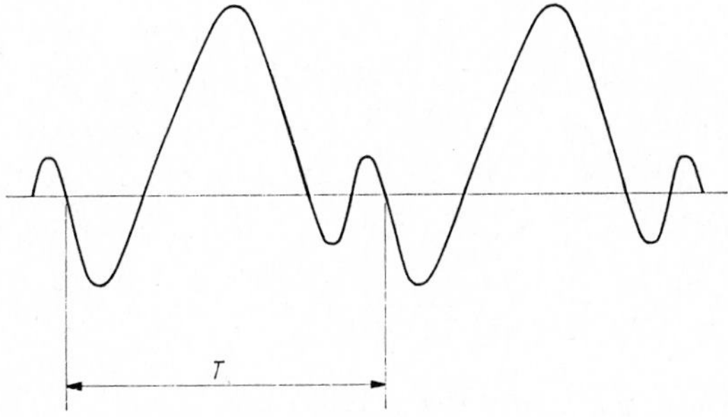

Fig. 1.6. A periodic function.

Subject to rather unrestrictive conditions a periodic function of period T can be represented by a Fourier series

$$f(t) = \tfrac{1}{2}a_0 + \sum_{n=1}^{\infty} \left[a_n \cos\left(\frac{2\pi nt}{T}\right) + b_n \sin\left(\frac{2\pi nt}{T}\right) \right]. \tag{1.92}$$

It is not at all a trivial matter to determine the conditions that this series is convergent and converges to $f(t)$. The conditions are, however, satisfied if $f(t)$ and its first derivative are continuous except for a finite number of discontinuities within each period. The smoother the function the more rapidly the series converges.

By employing the orthogonality relations of the sines and the cosines,

$$\int_{-\frac{1}{2}T}^{\frac{1}{2}T} \cos\left(\frac{2\pi nt}{T}\right)\cos\left(\frac{2\pi mt}{T}\right)\mathrm{d}t = \tfrac{1}{2}T\delta_{nm}$$

$$\int_{-\frac{1}{2}T}^{\frac{1}{2}T} \sin\left(\frac{2\pi nt}{T}\right)\sin\left(\frac{2\pi mt}{T}\right)\mathrm{d}t = \tfrac{1}{2}T\delta_{nm}$$

$$\int_{-\frac{1}{2}T}^{\frac{1}{2}T} \cos\left(\frac{2\pi nt}{T}\right)\sin\left(\frac{2\pi mt}{T}\right)\mathrm{d}t = 0 \qquad n \geqq 0,\ m > 0,$$

where the Kronecker delta symbol δ_{nm} is zero if $n \neq m$ and unity if $n = m$, the coefficients of the Fourier series are obtained as

$$a_n = \frac{2}{T}\int_{-\frac{1}{2}T}^{\frac{1}{2}T} f(t)\cos\left(\frac{2\pi nt}{T}\right)\mathrm{d}t \tag{1.93}$$

$$b_n = \frac{2}{T}\int_{-\frac{1}{2}T}^{\frac{1}{2}T} f(t)\sin\left(\frac{2\pi nt}{T}\right)\mathrm{d}t \tag{1.94}$$

$$a_0 = \frac{2}{T}\int_{-\frac{1}{2}T}^{\frac{1}{2}T} f(t)\mathrm{d}t. \tag{1.95}$$

By means of eq. (1.92) the recurrent function is represented by a non-periodic component of magnitude $\frac{1}{2}a_0$ and by a harmonic function of frequency $1/T$, together with an infinite series of higher harmonics.

A useful alternative of the sine and cosine series is obtained by writing

$$\cos(n\omega t) = \tfrac{1}{2}(e^{in\omega t} + e^{-in\omega t})$$

$$\sin(n\omega t) = \frac{1}{2i}(e^{in\omega t} - e^{-in\omega t}).$$

This gives in place of (1.92)

$$f(t) = \sum_{n=-\infty}^{\infty} c_n \exp(-in\omega t), \tag{1.96}$$

where we have used $\omega = 2\pi/T$, and

$$c_0 = \tfrac{1}{2}a_0, \qquad c_{-n} = \tfrac{1}{2}(a_n - ib_n), \qquad c_n = \tfrac{1}{2}(a_n + ib_n).$$

The exponential functions form an orthogonal set and the coefficients c_n can, therefore, also be calculated directly as

$$c_n = \frac{1}{T}\int_{-\frac{1}{2}T}^{\frac{1}{2}T} f(t)\exp\left(\frac{2\pi int}{T}\right)\mathrm{d}t.$$

Fourier series can also be used for non-periodic functions, if we are concerned only with a limited range of the variable, say $0 \leqq t \leqq T$. The function may be represented by a collection of harmonics, taking T as the longest period. The collection combines, of course, to form a function which is periodic over all values of the variable. This does, however, not matter since we are considering only the range $0 \leqq t \leqq T$. The longer the range, the lower is the fundamental frequency of the Fourier series.

The coefficients of a Fourier series can be plotted versus the frequency as a pair of line spectra (see figure 1.7). The frequencies of successive terms in the representation of a non-periodic function in the range $0 \leqq t \leqq T$ are $\omega_1 = 2\pi/T$, $\omega_2 = 2\omega_1, \ldots, \omega_n = n\omega_1$.

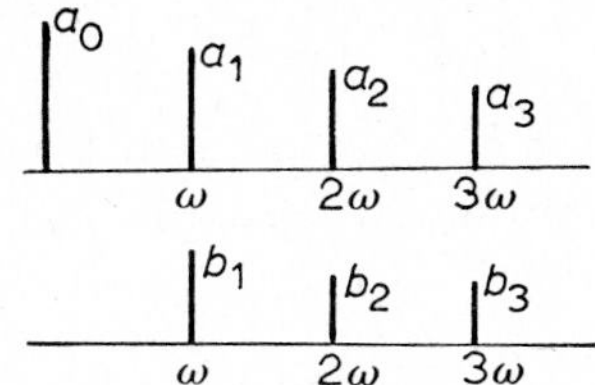

Fig. 1.7. Line spectra of Fourier coefficients.

Thus the spacing in the line spectrum becomes more closely packed when the range of representation is widened. The accuracy with which details are represented depends on the highest frequency components that are included. This follows from the consideration that a local variation which is of "duration" T_1 cannot be represented by terms of characteristic period much greater than T_1, and thus possessing frequencies less than $2\pi/T_1$. The number of Fourier terms required thus increases as the detail required increases.

1.8.2. Fourier integrals

If a function is not recurrent and must be represented over the *whole* range of the variable, i.e., if it consists of an isolated pulse, it can be represented by a Fouıier *integral*. The integral representation of a function $f(t)$ is of the form

$$f(t) = \frac{1}{2\pi}\int_{-\infty}^{\infty} e^{-i\omega t} f^*(\omega)\mathrm{d}\omega, \tag{1.97}$$

where

$$f^*(\omega) = \int_{-\infty}^{\infty} e^{i\omega t} f(t) \mathrm{d}t. \tag{1.98}$$

The function $f^*(\omega)$ is usually called the Fourier transform of $f(t)$. In that terminology (1.97) defines the inverse transform.

An appealing heuristic way of deriving the integral relations (1.97) and (1.98) is to consider the limiting case of a Fourier series whose interval of definition grows without limit. Indeed, the structure of the frequency spectrum of a Fourier series suggests that the range of representation of a function can be increased indefinitely by packing the terms closely in frequency space. In the limit as the range increases beyond bounds, the discrete spectrum should, in fact, convert into a continuous spectrum represented by $f^*(\omega)$. On the basis of this argument $f^*(\omega)\mathrm{d}\omega$ is the contribution to the amplitude at ω from the indefinitely small frequency band $\mathrm{d}\omega$.

Let us consider a function $f(t)$ defined in the interval $-\frac{1}{2}T < t < \frac{1}{2}T$. The function can be represented by an exponential Fourier series of the form (1.96)

$$f(t) = \sum_{-\infty}^{\infty} \left[\frac{1}{T} \int_{-\frac{1}{2}T}^{\frac{1}{2}T} f(s) \exp\left(\frac{2\pi ins}{T}\right) \mathrm{d}s \right] \exp\left(-\frac{2\pi int}{T}\right) \tag{1.99}$$

If we set $\omega_n = 2\pi n/T$, and observe that $\omega_{n+1} - \omega_n = 2\pi/T$, it can be stated that for large values of T a small frequency range $\Delta\omega$ embraces $\Delta\omega/(2\pi/T)$ terms. The contribution made by these to the sum (1.99) is

$$\frac{T\Delta\omega}{2\pi} \left[\frac{1}{T} \int_{-\frac{1}{2}T}^{\frac{1}{2}T} f(s) e^{i\omega_n s} \mathrm{d}s \right] e^{-i\omega_n t}.$$

Formally proceeding to the limit $T \to \infty$, the summation in eq. (1.99) becomes an integration, and we obtain

$$f(t) = \frac{1}{2\pi} \int_{-\infty}^{\infty} e^{-i\omega t} \mathrm{d}\omega \int_{-\infty}^{\infty} f(s) e^{i\omega s} \mathrm{d}s. \tag{1.100}$$

Eq. (1.100) is the well-known Fourier integral theorem.

The form of (1.100) suggests the Fourier transform $f^*(\omega)$ as given by (1.98), and the inverse transform as given by (1.97).

1.9. The use of Fourier integrals

As a consequence of the linearity of the wave propagation problems that are under discussion, it is allowable to express the total response to a number of separate excitations as the superposition of the individual responses. Linear superposition, in conjunction with integral representations of forcing functions provide us with the means of determining solutions to problems of elastic wave propagation.

Suppose we wish to determine the stress wave generated in a half-space by a surface traction at $x = 0$ of the form

$$\tau_x = -p_0 e^{-\eta t} H(t), \tag{1.101}$$

where $H(t)$ is the Heaviside step function. By means of (1.97) and (1.98) the surface traction can be represented by the following Fourier integral

$$\tau_x(0, t) = \frac{p_0}{2\pi i} \int_{-\infty}^{\infty} \frac{e^{-i\omega t}}{\omega + i\eta} \mathrm{d}\omega. \tag{1.102}$$

Now consider a time-harmonic stress wave of the form

$$\tau_x(x, t) = \frac{1}{\omega + i\eta} e^{-i\omega(t - x/c_L)}. \tag{1.103}$$

Clearly this wave is generated in a half-space by a surface traction of the form $(\omega + i\eta)^{-1} \exp(-i\omega t)$. Since linear superposition is allowable, and since the response to the individual components of the integral representation (1.102) is given by (1.103), the stress due to a surface traction of the form (1.101) may be expressed as

$$\tau_x = \frac{p_0}{2\pi i} \int_{-\infty}^{\infty} e^{-i\omega(t - x/c_L)} \frac{\mathrm{d}\omega}{\omega + i\eta}. \tag{1.104}$$

Eq. (1.104) provides us with a formal representation of the stress due to a surface traction of the form (1.101). The integral in eq. (1.104) can be evaluated by means of the technique of contour integration in the complex plane. Integrals appearing in solutions of the form (1.104) are exemplified by

$$I = \int_{-\infty}^{\infty} e^{ia\zeta} \mathrm{f}(\zeta) \mathrm{d}\zeta, \tag{1.105}$$

where $\mathrm{f}(\zeta)$ is single-valued and a is real. These integrals are evaluated by the *residue theorem*, which states that for counterclockwise integration

$$\frac{1}{2\pi i} \int_{\Gamma} e^{ia\zeta} \mathrm{f}(\zeta) \mathrm{d}\zeta = \text{sum of the residues inside } \Gamma. \tag{1.106}$$

To evaluate integrals of the form (1.105) we select a contour Γ consisting of the real axis and a semicircle of radius R about the origin. For $a > 0$, the semicircle is taken in the upper half-plane. In the limit $R \to \infty$, the integral along the semicircle vanishes if the following condition applies: $F(R) \to 0$ as $R \to \infty$, where $|f(Re^{i\theta})| \leqq F(R)$. This result is known as *Jordan's lemma.* The residue theorem and Jordan's lemma are discussed in books on functions of a complex variable.[6]

For the particular case of (1.104) we close the contour in the upper half-plane for $t - x/c_L < 0$, and in the lower half-plane for $t - x/c_L > 0$, in order that the integration over the semicircle vanishes. Since there are no poles in the upper half-plane, we find

$$\tau_x(x, t) \equiv 0 \quad \text{for} \quad c_L t < x. \tag{1.107}$$

In the lower half-plane there is a pole at $\omega = -i\eta$, whose contribution yields

$$\tau_x = -p_0 \exp\left[-\eta\left(t - \frac{x}{c_L}\right)\right] \qquad \text{for} \quad c_L t \geqq x. \tag{1.108}$$

This is, of course, just the solution obtained in section 1.3.

Although the foregoing example displays the features of Fourier integral analysis, the example is rather simple. Another example will be discussed in chapter 4, while a more complete treatment of Fourier transform techniques is given in chapter 7.

1.10. Problems

1.1. Derive the one-dimensional equations of motion in the material and spatial descriptions for the case when the medium is subjected to a distribution of body forces which depends on position and time.

1.2. Suppose that in the material description a propagating displacement pulse is defined by

$$U(X, t) = F\left(t - \frac{X}{C}\right).$$

For a specific time t the displacement pulse is shown in the figure. Since $x = X + U$, the corresponding pulse in the spatial description can be obtained by a shifting of abscissa. What is the form of $u(x, t)$? Carry out the geometrical construction of $u(x, t)$ and observe that there is a limitation

[6] See, e.g., G. F. Carrier, M. Krook and C. E. Pearson, *Functions of a complex variable.* New York, McGraw-Hill, Inc. (1966), pp. 57, 81.

on $F'(\ )$, where a prime denotes a derivative with respect to the argument. What is the physical significance of this limitation?

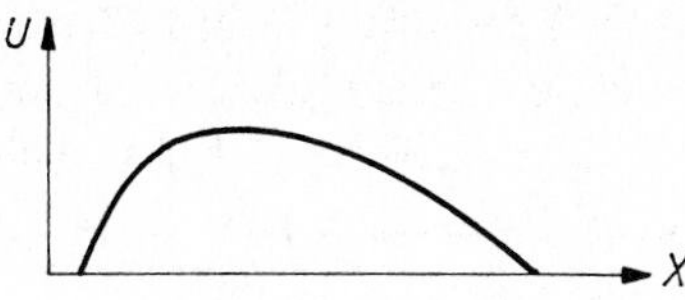

1.3. It can be verified by direct substitution that $f(t-x/c)$ and $g(t+x/c)$ satisfy eq. (1.38). The sum of these two expressions is, however, generally not a solution of eq. (1.38). Show that if $f+g$ is to satisfy eq. (1.38), the following condition must be met:

$$f''g'(1+g'/c)+g''f'(1+f'/c) = 0.$$

For what type of displacement distribution $F(t-X/C)$ is this relation automatically satisfied?

1.4. Consider an elastic rod (Young's modulus E, mass density ρ) of length l, which is rigidly clamped at $x = l$, as shown in the figure. The rod is initially at rest. At time $t = 0$ the end $x = 0$ is subjected to a pressure $p(t)$.

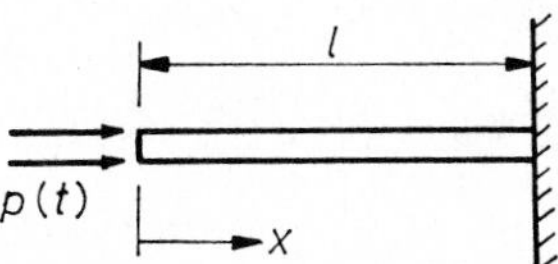

Assuming that the linearized one-dimensional stress approximation is valid,

(a) Determine the reflection at $x = l$ of the primary stress wave.

(b) Find an expression in the form of a series for the stress at a position x at an arbitrary time t.

(c) Suppose $p(t)$ is a square pulse of length a/c_L and magnitude p_0, where $a < l$. If the material is brittle, and much weaker in tension than in compression, and if p_0 exceeds the tensile strength of the rod, at what location would you expect the rod to break?

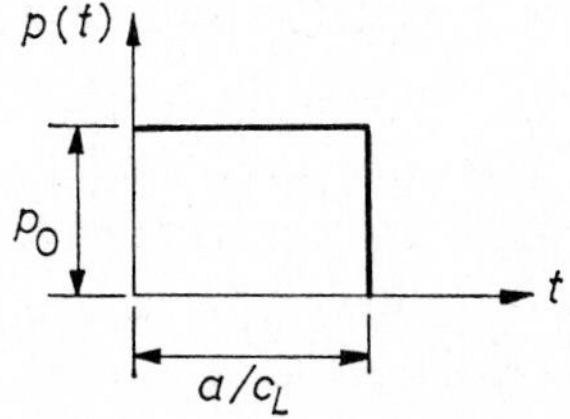

1.5. An initially undisturbed rod of cross-sectional area A is subjected to a concentrated load $F_0H(t)$, where $H(t)$ is the Heaviside step function. At $x = l$ a point mass m is fixed to the rod. Determine the reflected and transmitted waves for the linearized theory.

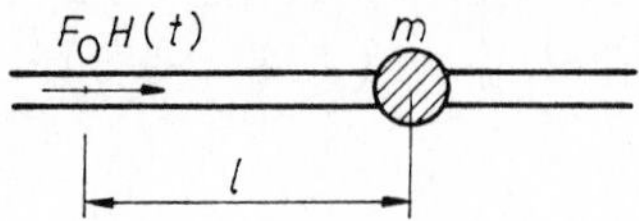

1.6. A rod of length l and square cross section (cross-sectional area h^2) is resting on one of its lateral sides on a smooth plane surface. At time $t = 0$ the rod is subjected at one end to an axial force $F_0H(t)$. There is no friction between the plane surface and the rod. The total mass of the rod is $h^2l\rho$, where ρ is the mass density. If the rod were infinitely rigid the motion of the center of gravity would follow from an application of Newton's law. Consider a rod of Young's modulus E, and compare in the time interval $0 \leqq t \leqq 5l/c_b$ the motions of the centers of gravity of the rigid and the deformable rods.

1.7. A split-Hopkinson bar is a device to measure mechanical properties of a material. A very thin slice of the material is placed between a main rod and an extension rod as shown in the figure. The inertia of the specimen

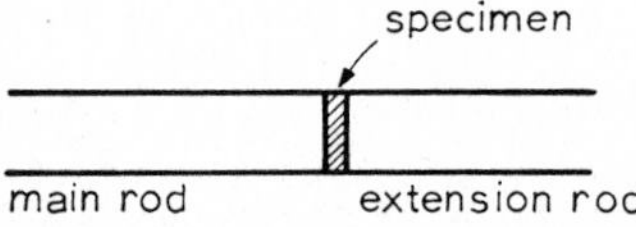

is neglected. Suppose that an incident and a reflected pulse are measured at a position in the main rod, and a transmitted pulse is measured in the extension rod. How can this information be used to determine the relation between τ_x and ε_x in the specimen?

1.8. An initially undisturbed rod of cross-sectional area A is subjected to a concentrated load $F_0H(t)$. In the interval $l \leqq x \leqq l+a$, an inclusion of

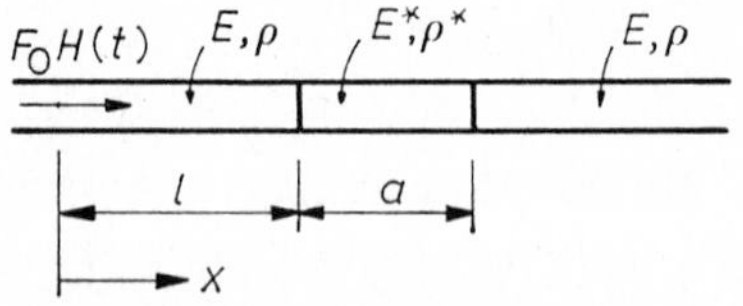

a different material is placed, as shown in the figure. Determine the transmitted waves into the region $x > l+a$. Show that the result of problem 1.5 is a limitcase of the result of the present problem.

1.9. A very long rod of cross-sectional area A with an attached (rigid) mass M is traveling to the right with constant velocity V_0, without stress. A remote section S is stopped in a time which is very short, but not zero. S is then held.

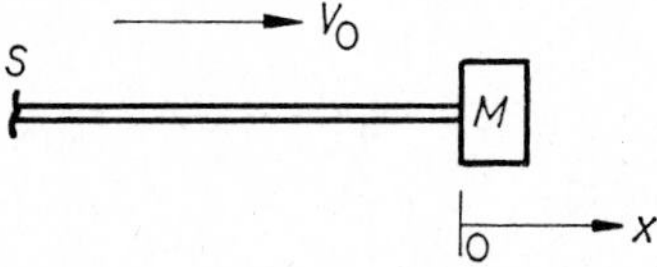

(a) Sketch the form of the particle velocity wave $f'(t-x/c_b)$ which runs along the rod from S, for a time t before arrival at M. Here $c_b = (E/\rho)^{\frac{1}{2}}$.

(b) The time of arrival at M is chosen as $t = 0$. Show that for $t > 0$ the (absolute) velocity of M is given by

$$v_M(t) = V_0 + f'(t) + e^{-\mu t}\int_0^t e^{\mu\tau}[\mu f'(\tau) - f''(\tau)]d\tau,$$

where $\mu = A\rho c_b/M$.

1.10. In problem 1.9 the time of stopping at S is now taken extremely short, approaching zero as a limit.

(a) Examine the limiting forms of $f'(t)$ and of the integrals in 1.9b, and show that in the limit

$$v_M(t) = V_0(2e^{-\mu t} - 1).$$

(b) Show that the displacement of M to the right after the arrival of the wave is

$$\Delta = \frac{V_0}{\mu}(1 - \ln 2).$$

(c) Consider the extremes $M \to \infty$ and $M \to 0$. State separately (i) what values of $v_M(t)$ you would expect, and why; and (ii) what values are obtained from 1.10a.

CHAPTER 2

THE LINEARIZED THEORY OF ELASTICITY

2.1. Introduction

An elastic body has a natural undeformed state to which it returns when all external loads are removed. There are, therefore, conceptual advantages in describing the deformation of an elastic body within the context of the material description, where the independent variables are time and the reference position of a particle in the undeformed state. A detailed derivation would show that the general system of equations governing the three-dimensional motion of an elastic body in the material description is strongly nonlinear. It is, however, a matter of wide experience that many wave propagation effects in elastic solids can adequately be described by a linearized theory.

For a one-dimensional geometry the general theory, as well as the linearized theory and the conditions for linearization, were discussed in chapter 1. To treat in some detail the nonlinear theory and the corresponding linearization in a three-dimensional setting falls outside the scope of this book. We will thus restrict ourselves to the remark that for the material description the linearization is justifiable if the spatial gradients of the displacement components are much smaller than unity and if all the components of the Cauchy stress tensor are of the same order of magnitude. If, moreover, the time derivatives of the displacement components are small enough, the convective terms in the spatial description of the velocity and the acceleration may be neglected. The differences between the material and the spatial descriptions of the motion then disappear and it suffices to employ one system of dependent and independent field variables.

The linearized theory of elasticity has been the subject of several treatises. For a detailed treatment we refer to the book by Sokolnikoff.[1] The basic equations are, however, briefly summarized in this chapter for the purpose of reference. Some topics which are particularly relevant to dynamic

[1] I. S. Sokolnikoff, *Mathematical theory of elasticity*. New York, Mc-Graw-Hill Book Co. (1956).

problems are discussed in more detail. We also include a summary of the linearized equations in rectangular, spherical and cylindrical coordinates. The chapter is concluded with a brief discussion of the governing equations for an ideal fluid.

2.2. Notation and mathematical preliminaries

Physical quantities are mathematically represented by tensors of various orders.[2] The equations describing physical laws are tensor equations. Quantities that are not associated with any special direction and are measured by a single number are represented by scalars, or tensors of order zero. Tensors of order one are vectors, which represent quantities that are characterized by a direction as well as a magnitude. More complicated physical quantities are represented by tensors of order greater than one. Throughout this book light-faced Roman or Greek letters stand for scalars, Roman letters in boldface denote vectors, while lower case Greek letters in boldface denote second-order tensors.

2.2.1. Indicial notation

A system of fixed rectangular Cartesian coordinates is sufficient for the presentation of the theory. In indicial notation, the coordinate axes may be denoted by x_j and the base vectors by $\boldsymbol{i}_j$, where $j = 1, 2, 3$. In the sequel, subscripts assume the values 1, 2, 3 unless explicitly otherwise specified. If the components of a vector $\boldsymbol{u}$ are denoted by u_j, we have

$$\boldsymbol{u} = u_1 \boldsymbol{i}_1 + u_2 \boldsymbol{i}_2 + u_3 \boldsymbol{i}_3. \tag{2.1}$$

Since summations of the type (2.1) frequently occur in the mathematical description of the mechanics of a continuous medium, we introduce the *summation convention*, whereby a repeated subscript implies a summation. Eq. (2.1) may then be rewritten as

$$\boldsymbol{u} = u_j \boldsymbol{i}_j. \tag{2.2}$$

As another example of the use of the summation convention, the scalar product of two vectors is expressed as

$$\boldsymbol{u} \cdot \boldsymbol{v} = u_j v_j = u_1 v_1 + u_2 v_2 + u_3 v_3. \tag{2.3}$$

[2] As a reference for the elements of vector and tensor analysis used in this section we refer to H. Jeffreys, *Cartesian tensors*. Cambridge, University Press (1931).

As opposed to the free index in u_j, which may assume any one of the values 1, 2, 3, the index j in (2.2) and (2.3) is a bound index or a dummy index, which must assume all three values 1, 2 and 3.

Quantities with two free indices as subscripts, such as τ_{ij}, denote components of a tensor of second rank $\boldsymbol{\tau}$, and similarly three free indices define a tensor of rank three. A well-known special tensor of rank two is the *Kronecker delta*, whose components are defined as

$$\delta_{ij} = \begin{matrix} 1 & \text{if } i = j \\ 0 & \text{if } i \neq j. \end{matrix} \tag{2.4}$$

A frequently-used special tensor of rank three is the *alternating tensor*, whose components are defined as follows:

$$e_{ijk} = \begin{matrix} +1 \text{ if } ijk \text{ represents an even permutation of } 123 \\ 0 \text{ if any two of the } ijk \text{ indices are equal} \\ -1 \text{ if } ijk \text{ represents an odd permutation of } 123. \end{matrix} \tag{2.5}$$

By the use of the alternating tensor and the summation convention, the components of the cross product $\boldsymbol{h} = \boldsymbol{u} \wedge \boldsymbol{v}$ may be expressed as

$$h_i = e_{ijk} u_j v_k. \tag{2.6}$$

In extended notation the components of $\boldsymbol{h}$ are

$$\begin{aligned} h_1 &= u_2 v_3 - u_3 v_2 \\ h_2 &= u_3 v_1 - u_1 v_3 \\ h_3 &= u_1 v_2 - u_2 v_1. \end{aligned}$$

2.2.2. Vector operators

Particularly significant in vector calculus is the *vector operator del* (or nabla) denoted by ∇,

$$= \boldsymbol{i}_1 \frac{\partial}{\partial x_1} + \boldsymbol{i}_2 \frac{\partial}{\partial x_2} + \boldsymbol{i}_3 \frac{\partial}{\partial x_3}. \tag{2.7}$$

When applied to the scalar field $f(x_1, x_2, x_3)$, the vector operator ∇ yields a vector field which is known as the *gradient* of the scalar field,

$$\text{grad f} = \nabla \text{f} = \boldsymbol{i}_1 \frac{\partial \text{f}}{\partial x_1} + \boldsymbol{i}_2 \frac{\partial \text{f}}{\partial x_2} + \boldsymbol{i}_3 \frac{\partial \text{f}}{\partial x_3}.$$

In indicial notation, partial differentiation is commonly denoted by a comma, and thus

$$\text{grad f} = \nabla \text{f} = \boldsymbol{i}_p \text{f}_{,p}. \tag{2.8}$$

The appearance of the single subscript in $\text{f}_{,p}$ indicates that $\text{f}_{,p}$ are the components of a tensor of rank one, i.e., a vector.

In a vector field, denoted by $\boldsymbol{u}(\boldsymbol{x})$, the components of the vector are functions of the spatial coordinates. The components are denoted by $u_i(x_1, x_2, x_3)$. Assuming that functions $u_i(x_1, x_2, x_3)$ are differentiable, the nine partial derivatives $\partial u_i(x_1, x_2, x_3)/\partial x_j$ can be written in indicial notation as $u_{i,j}$. It can be shown that $u_{i,j}$ are the components of a second-rank tensor.

When the vector operator ∇ operates on a vector in a manner analogous to scalar multiplication, the result is a scalar field, termed the *divergence* of the vector field $\boldsymbol{u}(\boldsymbol{x})$

$$\text{div}\, \boldsymbol{u} = \nabla \cdot \boldsymbol{u} = u_{i,i}. \tag{2.9}$$

By taking the cross product of ∇ and $\boldsymbol{u}$, we obtain a vector termed the *curl* of $\boldsymbol{u}$, denoted by curl $\boldsymbol{u}$ or $\nabla \wedge \boldsymbol{u}$. If $\boldsymbol{q} = \nabla \wedge \boldsymbol{u}$, the components of $\boldsymbol{q}$ are

$$q_i = e_{ijk} u_{k,j}. \tag{2.10}$$

The *Laplace operator* ∇^2 is obtained by taking the divergence of a gradient. The Laplacian of a twice differentiable scalar field is another scalar field,

$$\text{div grad f} = \nabla \cdot \nabla \text{f} = \text{f}_{,ii}. \tag{2.11}$$

The Laplacian of a vector field is another vector field denoted by $\nabla^2 \boldsymbol{u}$

$$\nabla^2 \boldsymbol{u} = \nabla \cdot \nabla \boldsymbol{u} = u_{p,jj} \boldsymbol{i}_p. \tag{2.12}$$

2.2.3. *Gauss' theorem*

We close this section with the statement of the most important integral theorem of tensor analysis. This theorem, which is known as *Gauss' theorem*, relates a volume integral to a surface integral over the bounding surface of the volume. Consider a convex region B of volume V, bounded by a surface S which possesses a piecewise continuously turning tangent plane. Such a region is said to be regular. Now let us consider a tensor field $\tau_{jkl\ldots p}$, and let every component of $\tau_{jkl\ldots p}$ be continuously differentiable in B. Then Gauss' theorem states

$$\int_V \tau_{jkl\ldots p,i} \mathrm{d}V = \int_S n_i \tau_{jkl\ldots p} \mathrm{d}A, \tag{2.13}$$

where n_i are the components of the unit vector along the outer normal to the surface S. If eq. (2.13) is written with the three components of a vector $\boldsymbol{u}$ successively substituted for $\tau_{jkl\ldots p}$, and if the three resulting equations are added, the result is

$$\int_V u_{i,i}\,\mathrm{d}V = \int_S n_i u_i\,\mathrm{d}A. \tag{2.14}$$

Eq. (2.14) is the well-known divergence theorem of vector calculus which states that the integral of the outer normal component of a vector over a closed surface is equal to the integral of the divergence of the vector over the volume bounded by the closed surface.

2.2.4. Notation

The equations governing the linearized theory of elasticity are presented in the following commonly used notation:

position vector:	$\boldsymbol{x}$ (coordinates x_i)	(2.15)
displacement vector:	$\boldsymbol{u}$ (components u_i)	(2.16)
small strain tensor:	$\boldsymbol{\varepsilon}$ (components ε_{ij})	(2.17)
stress tensor:	$\boldsymbol{\tau}$ (components τ_{ij})	(2.18)

2.3. Kinematics and dynamics

2.3.1. Deformation

Let the field defining the displacements of particles be denoted by $\boldsymbol{u}(\boldsymbol{x}, t)$. As a direct implication of the notion of a continuum, the deformation of the medium can be expressed in terms of the gradients of the displacement vector. Within the restrictions of the linearized theory the deformation is described in a very simple manner by the small-strain tensor $\boldsymbol{\varepsilon}$, with components

$$\varepsilon_{ij} = \tfrac{1}{2}(u_{i,j}+u_{j,i}). \tag{2.19}$$

It is evident that $\varepsilon_{ij} = \varepsilon_{ji}$, i.e., $\boldsymbol{\varepsilon}$ is a symmetric tensor of rank two. It is also useful to introduce the rotation tensor $\boldsymbol{\omega}$ whose components are defined as

$$\omega_{ij} = \tfrac{1}{2}(u_{i,j}-u_{j,i}). \tag{2.20}$$

We note that $\boldsymbol{\omega}$ is an antisymmetric tensor, $\omega_{ij} = -\omega_{ji}$.

2.3.2. *Linear momentum and the stress tensor*

A basic postulate in the theory of continuous media is that the mechanical action of the material points which are situated on one side of an arbitrary material surface within a body upon those on the other side can be completely accounted for by prescribing a suitable surface traction on this surface. Thus if a surface element has a unit outward normal $\boldsymbol{n}$ we introduce the surface tractions $\boldsymbol{t}$, defining a force per unit area. The surface tractions generally depend on the orientation of $\boldsymbol{n}$ as well as on the location $\boldsymbol{x}$ of the surface element.

Suppose we remove from a body a closed region $V+S$, where S is the boundary. The surface S is subjected to a distribution of surface tractions $\boldsymbol{t}(\boldsymbol{x}, t)$. Each mass element of the body may be subjected to a body force per unit mass, $\boldsymbol{f}(\boldsymbol{x}, t)$. According to the principle of balance of linear momentum, the instantaneous rate of change of the linear momentum of a body is equal to the resultant external force acting on the body at the particular instant of time. In the linearized theory this leads to the equation

$$\int_S \boldsymbol{t}\,\mathrm{d}A + \int_V \rho \boldsymbol{f}\,\mathrm{d}V = \int_V \rho \ddot{\boldsymbol{u}}\,\mathrm{d}V. \tag{2.21}$$

By means of the "tetrahedron argument," eq. (2.21) subsequently leads to the stress tensor $\boldsymbol{\tau}$ with components τ_{kl}, where

$$t_l = \tau_{kl} n_k. \tag{2.22}$$

Eq. (2.22) is the Cauchy stress formula. Physically τ_{kl} is the component in the x_l-direction of the traction on the surface with the unit normal $\boldsymbol{i}_k$.

By substitution of $t_l = \tau_{kl} n_k$, eq. (2.21) is rewritten in indicial notation as

$$\int_S \tau_{kl} n_k\,\mathrm{d}A + \int_V \rho f_l\,\mathrm{d}V = \int_V \rho \ddot{u}_l\,\mathrm{d}V. \tag{2.23}$$

The surface integral can be transformed into a volume integral by Gauss' theorem, eq. (2.13), and we obtain

$$\int_V (\tau_{kl,k} + \rho f_l - \rho \ddot{u}_l)\,\mathrm{d}V = 0. \tag{2.24}$$

Since V may be an arbitrary part of the body it follows that wherever the integrand is continuous, we have

$$\tau_{kl,k} + \rho f_l = \rho \ddot{u}_l. \tag{2.25}$$

This is Cauchy's first law of motion.

2.3.3. *Balance of moment of momentum*

For the linearized theory the principle of moment of momentum states

$$\int_S (\boldsymbol{x} \wedge \boldsymbol{t})\mathrm{d}A + \int_V (\boldsymbol{x} \wedge \boldsymbol{f})\rho\,\mathrm{d}V = \int_V \rho \frac{\partial}{\partial t}(\boldsymbol{x} \wedge \dot{\boldsymbol{u}})\mathrm{d}V.$$

Simplifying the right-hand side and introducing indicial notation, this equation can be written as

$$\int_S e_{klm} x_l t_m \mathrm{d}A + \int_V \rho e_{klm} x_l f_m \mathrm{d}V = \int_V \rho e_{klm} x_l \ddot{u}_m \mathrm{d}V. \qquad (2.26)$$

Elimination of t_m from the surface integral and the use of Gauss' theorem result in

$$\int_S e_{klm} x_l \tau_{nm} n_n \mathrm{d}A = \int_V e_{klm}(\delta_{ln}\tau_{nm} + x_l \tau_{nm,n})\mathrm{d}V.$$

By virtue of the first law of motion, eq. (2.26) then reduces to

$$\int_V e_{klm}\delta_{ln}\tau_{nm}\mathrm{d}V = 0$$

or

$$e_{klm}\tau_{lm} = 0.$$

This result implies that

$$\tau_{lm} = \tau_{ml}, \qquad (2.27)$$

i.e., the stress tensor is symmetric.

2.4. The homogeneous, isotropic, linearly elastic solid

2.4.1. *Stress-strain relations*

In general terms, the linear relation between the components of the stress tensor and the components of the strain tensor is

$$\tau_{ij} = C_{ijkl}\varepsilon_{kl},$$

where

$$C_{ijkl} = C_{jikl} = C_{klij} = C_{ijlk}.$$

Thus, 21 of the 81 components of the tensor C_{ijkl} are independent. The medium is elastically homogeneous if the coefficients C_{ijkl} are constants. The material is *elastically isotropic* when there are no preferred directions in the material, and the elastic constants must be the same whatever the orientation of the Cartesian coordinate system in which the components of τ_{ij} and ε_{ij} are evaluated. It can be shown that elastic isotropy implies that the constants C_{ijkl} may be expressed as

$$C_{ijkl} = \lambda\delta_{ij}\delta_{kl}+\mu(\delta_{ik}\delta_{jl}+\delta_{il}\delta_{jk}).$$

Hooke's law then assumes the well-known form

$$\tau_{ij} = \lambda\varepsilon_{kk}\delta_{ij}+2\mu\varepsilon_{ij}. \tag{2.28}$$

Eq. (2.28) contains two elastic constants λ and μ, which are known as Lamé's elastic constants.

Putting $j = i$ in eq. (2.28), thus implying a summation, we obtain

$$\tau_{ii} = (3\lambda+2\mu)\varepsilon_{ii}, \tag{2.29}$$

where we have used $\delta_{ii} = \delta_{11}+\delta_{22}+\delta_{33} = 3$. By substituting $\varepsilon_{ii} = \tau_{ii}/(3\lambda+2\mu)$ into (2.28) and solving for ε_{ij}, we obtain the strains in terms of the stresses as

$$\varepsilon_{ij} = -\frac{\lambda\delta_{ij}}{2\mu(3\lambda+2\mu)}\tau_{kk}+\frac{1}{2\mu}\tau_{ij}. \tag{2.30}$$

It is clear that ε_{ij} can be uniquely determined by τ_{ij} only if

$$\mu \neq 0 \text{ and } 3\lambda+2\mu \neq 0.$$

In order not to have zero strain for a finite stress we should also have

$$|\mu| < \infty \text{ and } |3\lambda+2\mu| < \infty.$$

By considering the special state of stress defined by $\tau_{12} \neq 0$, all other $\tau_{ij} \equiv 0$, which defines a *state of simple shear*, we identify μ as the shear modulus which relates τ_{12} and ε_{12} by $\tau_{12} = 2\,\mu\varepsilon_{12}$. Since experimental observations show that for small deformations τ_{12} and ε_{12} have the same direction, we may state

$$\mu > 0.$$

Another special state of stress, known as *hydrostatic pressure*, is defined by $\tau_{ij} = -p\delta_{ij}$. By employing eq. (2.29) we find $p = -B\varepsilon_{kk}$, where $B = \lambda+\frac{2}{3}\mu$ is known as the *modulus of compression* or the *bulk modulus*. For infinitesimal deformation ε_{kk} denotes the volume change of an element. Since a hydrostatic pressure should reduce the volume of a body, we should have $B > 0$, or

$$3\lambda+2\mu > 0.$$

The foregoing observations on the elastic constants μ and $3\lambda+2\mu$ may now be summarized as

$$0 < 3\lambda+2\mu < \infty,\ 0 < \mu < \infty. \tag{2.31}$$

2.4.2. *Stress and strain deviators*

The stress tensor can be written as the sum of two tensors, one representing a spherical or hydrostatic stress in which each normal stress component is $\frac{1}{3}\tau_{kk}$ and all shear stresses vanish. The complementary tensor is called the *stress deviator*, denoted by s_{ij}. Thus, the components of the stress deviator are defined by

$$s_{ij} = \tau_{ij} - \tfrac{1}{3}\tau_{kk}\delta_{ij}. \tag{2.32}$$

In the same manner we can define the *strain deviator* e_{ij} by

$$e_{ij} = \varepsilon_{ij} - \tfrac{1}{3}\varepsilon_{kk}\delta_{ij}. \tag{2.33}$$

From eq. (2.28) it can now quite easily be shown that the following simple relation exists between s_{ij} and e_{ij}:

$$s_{ij} = 2\mu e_{ij}, \tag{2.34}$$

where μ is the shear modulus. In addition we also have, according to (2.29),

$$\tfrac{1}{3}\tau_{kk} = B\varepsilon_{kk}, \tag{2.35}$$

where

$$B = \lambda + \tfrac{2}{3}\mu \tag{2.36}$$

is the bulk modulus, which came up earlier in the discussion of the state of hydrostatic pressure. Eqs. (2.34) and (2.35) are completely equivalent to Hooke's law (2.28), and these equations may thus also be considered as the constitutive equations for a homogeneous, isotropic, linearly elastic solid.

Other elastic constants that often appear in linear elasticity are Young's modulus E and the Poisson's ratio ν. A number of useful relationships among the isotropic elastic constants are summarized in table 2.1.

TABLE 2.1
Relationships among isotropic elastic constants

	E, ν	E, μ	λ, μ
λ	$\dfrac{E\nu}{(1+\nu)(1-2\nu)}$	$\dfrac{\mu(E-2\mu)}{3\mu-E}$	λ
μ	$\dfrac{E}{2(1+\nu)}$	μ	μ
E	E	E	$\dfrac{\mu(3\lambda+2\mu)}{\lambda+\mu}$
B	$\dfrac{E}{3(1-2\nu)}$	$\dfrac{\mu E}{3(3\mu-E)}$	$\lambda+\dfrac{2}{3}\mu$
ν	ν	$\dfrac{E-2\mu}{2\mu}$	$\dfrac{\lambda}{2(\lambda+\mu)}$

2.4.3. *Strain energy*

By the definition of the strain energy density $\mathscr{U}$, we have

$$\mathrm{d}\mathscr{U} = \tau_{ij}\mathrm{d}\varepsilon_{ij}.$$

In terms of the stress and strain deviators $\mathrm{d}\mathscr{U}$ assumes the form

$$\mathrm{d}\mathscr{U} = (s_{ij}+\tfrac{1}{3}\tau_{kk}\delta_{ij})\mathrm{d}(e_{ij}+\tfrac{1}{3}\varepsilon_{ll}\delta_{ij}),$$

which reduces to

$$\mathrm{d}\mathscr{U} = \tfrac{1}{3}\tau_{kk}\mathrm{d}\varepsilon_{ll}+s_{ij}\mathrm{d}e_{ij}.$$

Use of eqs. (2.34) and (2.35) then leads to the integrated form

$$\mathscr{U} = \tfrac{1}{2}B(\varepsilon_{kk})^2+\mu e_{ij}e_{ij},$$

where it is assumed that $\mathscr{U}$ vanishes in the undeformed reference state. It is now clear that the conditions on μ and $3\lambda+2\mu$ as stated by eq. (2.31) imply that $\mathscr{U}$ is positive semi-definite

$$\mathscr{U} \geqq 0. \tag{2.37}$$

One can also take a different point of view and state (2.31) as a necessary and sufficient condition for the *required* positive definiteness of the strain energy function.

By consulting table 2.1 it can be checked that the following conditions are equivalent to (2.31):

$$E > 0 \text{ and } -1 < \nu < \tfrac{1}{2}.$$

The isotropic strain energy density function can be written in the alternative form

$$\mathscr{U} = \tfrac{1}{2}\lambda(\varepsilon_{kk})^2+\mu\varepsilon_{ij}\varepsilon_{ij}. \tag{2.38}$$

2.5. Problem statement in dynamic elasticity

We consider a body B occupying a regular region V in space, which may be bounded or unbounded, with interior V, closure $\bar{V}$ and boundary S. The system of equations governing the motion of a homogeneous, isotropic, linearly elastic body consists of the stress equations of motion, Hooke's law and the strain-displacement relations:

$$\tau_{ij,j}+\rho f_i = \rho\ddot{u}_i \tag{2.39}$$

$$\tau_{ij} = \lambda\varepsilon_{kk}\delta_{ij}+2\mu\varepsilon_{ij} \tag{2.40}$$

and

$$\varepsilon_{ij} = \tfrac{1}{2}(u_{i,j}+u_{j,i}), \tag{2.41}$$

respectively. If the strain-displacement relations are substituted into Hooke's law and the expressions for the stresses are subsequently substituted in the stress-equations of motion, we obtain the displacement equations of motion

$$\mu u_{i,jj}+(\lambda+\mu)u_{j,ji}+\rho f_i = \rho \ddot{u}_i. \tag{2.42}$$

Eqs. (2.39)–(2.42) must be satisfied at every interior point of the undeformed body B, i.e., in the domain V. In general, we require

$$u_i(\boldsymbol{x}, t) \in \mathscr{C}^2(V\times T) \cap \mathscr{C}^1(\bar{V}\times T) \tag{2.43}$$

$$f_i(\boldsymbol{x}, t) \in \mathscr{C}(\bar{V}\times T), \tag{2.44}$$

where T is an arbitrary interval of time. The class of functions defined by $\mathscr{C}(R)$ consists of all tensor-valued functions of any order that are defined and continuous on a subset defined by R. For a positive integer, $\mathscr{C}^n(R)$ consists of all functions in $\mathscr{C}(R)$ whose partial derivatives of order up to and including n exist on the interior of R, and there coincide with functions belonging to $\mathscr{C}(R)$. If the displacements do not satisfy the smoothness requirements (2.43), separate relations must be satisfied by the discontinuities.

On the surface S of the undeformed body, boundary conditions must be prescribed. The following boundary conditions are most common:

(i) *Displacement boundary conditions*: the three components u_i are prescribed on the boundary.

(ii) *Traction boundary conditions*: the three traction components t_i are prescribed on the boundary with unit normal $\boldsymbol{n}$. Through Cauchy's formula

$$t_i = \tau_{ji} n_j, \tag{2.45}$$

this case actually corresponds to conditions on three components of the stress tensor.

(iii) *Displacement* boundary conditions on *part* S_1 of the boundary and *traction* boundary conditions on the *remaining part* $S-S_1$.

Other conditions are possible on the boundary of the body. In the discussion of the uniqueness theorem in chapter 3 the boundary conditions will be stated in detail.

To complete the problem statement we define initial conditions; in V we have at time $t = 0$

$$u_i(\boldsymbol{x}, 0) = \mathring{u}_i(\boldsymbol{x})$$

$$\dot{u}_i(\boldsymbol{x}, 0^+) = \mathring{v}_i(\boldsymbol{x}).$$

2.6. One-dimensional problems

If the body forces and the components of the stress tensor depend on one spatial variable, say x_1, the stress-equations of motion reduce to

$$\tau_{i1,1}+\rho f_i = \rho \ddot{u}_i. \tag{2.46}$$

Three separate cases can be considered.

Longitudinal strain. Of all displacement components only the longitudinal displacement $u_1(x_1, t)$ does not vanish. The one strain component is $\varepsilon_{11} = \partial u_1/\partial x_1$. By employing (2.28) the components of the stress tensor are obtained as

$$\tau_{11} = (\lambda+2\mu)u_{1,1}, \tau_{22} = \tau_{33} = \lambda u_{1,1}, \tag{2.47}$$

and the equation of motion is

$$(\lambda+2\mu)u_{1,11}+\rho f_1 = \rho \ddot{u}_1. \tag{2.48}$$

Longitudinal stress. The longitudinal normal stress τ_{11}, which is a function of x_1 and t only, is the one nonvanishing stress component. Equating the transverse normal stresses τ_{22} and τ_{33} to zero, we obtain the following relations

$$\varepsilon_{22} = \varepsilon_{33} = -\frac{\lambda}{2(\lambda+\mu)}\varepsilon_{11} = -\nu\varepsilon_{11}, \tag{2.49}$$

where ν is Poisson's ratio. Subsequent substitution of these results in the expression for τ_{11} yields

$$\tau_{11} = E\varepsilon_{11}, \tag{2.50}$$

where

$$E = \frac{\mu(3\lambda+2\mu)}{\lambda+\mu}. \tag{2.51}$$

The constant E is known as Young's modulus. The equation of motion follows by substitution of (2.50) into (2.46). Wave propagation in one-dimensional stress was considered in section 1.5.

Shear. In this case the displacement is in a plane normal to the x_1-axis,

$$\boldsymbol{u} = u_2(x_1, t)\boldsymbol{i}_2+u_3(x_1, t)\boldsymbol{i}_3.$$

The stresses are

$$\tau_{21} = \mu u_{2,1}, \; \tau_{31} = \mu u_{3,1}.$$

Clearly, the equations of motion reduce to uncoupled wave equations for u_2 and u_3, respectively.

2.7. Two-dimensional problems

In two-dimensional problems the body forces and the components of the stress tensor are independent of one of the coordinates, say x_3. The stress equations of motion can be derived from (2.39) by setting $\partial/\partial x_3 \equiv 0$. We find that the system of equations splits up into two uncoupled systems. These are

$$\tau_{3\beta,\beta} + \rho f_3 = \rho \ddot{u}_3 \tag{2.52}$$

and

$$\tau_{\alpha\beta,\beta} + \rho f_\alpha = \rho \ddot{u}_\alpha. \tag{2.53}$$

In eqs. (2.52) and (2.53), and throughout this section, Greek indices can assume the values 1 and 2 only.

2.7.1. Antiplane shear

A deformation described by a displacement distribution $u_3(x_1, x_2, t)$ is called an antiplane shear deformation. The corresponding stress components follow from Hooke's law as

$$\tau_{3\beta} = \mu u_{3,\beta}. \tag{2.54}$$

Eliminating $\tau_{3\beta}$ from eqs. (2.52) and (2.54), we find that $u_3(x_1, x_2, t)$ is governed by the scalar wave equation

$$\mu u_{3,\beta\beta} + \rho f_3 = \rho \ddot{u}_3. \tag{2.55}$$

Pure shear motions governed by (2.55) are usually called horizontally polarized shear motions.

2.7.2. In-plane motions

It follows from eq. (2.53) that the in-plane displacements u_α depend on x_1, x_2 and t only. With regard to the dependence of u_3 on the spatial coordinates and time, two separate cases are described by eq. (2.53).

Plane strain. In a deformation in plane strain *all* field variables are independent of x_3 and the displacement in the x_3-direction vanishes identically. Hooke's law then yields the following relations:

$$\tau_{\alpha\beta} = \lambda u_{\gamma,\gamma}\delta_{\alpha\beta}+\mu(u_{\alpha,\beta}+u_{\beta,\alpha}) \qquad (2.56)$$

$$\tau_{33} = \lambda u_{\gamma,\gamma}, \qquad (2.57)$$

where Greek indices can assume the values 1 and 2 only.

Elimination of $\tau_{\alpha\beta}$ from (2.53) and (2.56) leads to

$$\mu u_{\alpha,\beta\beta}+(\lambda+\mu)u_{\beta,\beta\alpha}+\rho f_\alpha = \rho\ddot{u}_\alpha. \qquad (2.58)$$

Eq. (2.58) can of course also be derived directly from (2.42) by setting $u_3 \equiv 0$ and $\partial/\partial x_3 \equiv 0$.

Plane stress. A two-dimensional stress field is called plane stress if τ_{33}, τ_{23} and τ_{13} are identically zero. From Hooke's law it follows that ε_{33} is related to $\varepsilon_{11}+\varepsilon_{22}$ by

$$\varepsilon_{33} = -\frac{\lambda}{\lambda+2\mu}u_{\gamma,\gamma}. \qquad (2.59)$$

Substitution of (2.59) into the expressions for $\tau_{\alpha\beta}$ yields

$$\tau_{\alpha\beta} = \frac{2\mu\lambda}{\lambda+2\mu}u_{\gamma,\gamma}\delta_{\alpha\beta}+\mu(u_{\alpha,\beta}+u_{\beta,\alpha}). \qquad (2.60)$$

Substituting (2.60) into (2.53), we obtain the displacement equations of motion. As far as the governing equations are concerned, the difference between plane strain and plane stress is merely a matter of different constant coefficients. It should be noted that (2.59) implies a linear dependence of u_3 on the coordinate x_3. The case of plane stress is often used for an approximate description of in-plane motions of a thin sheet.

The results of this section show that wave motions in two dimensions are the superposition of horizontally polarized motions and inplane motions. These motions are governed by uncoupled equations.

2.8. The energy identity

Surface tractions and body forces transmit mechanical energy to a body. The rate of work of external forces is called the power input. For a body

B occupying a regular region V with boundary S, the power input may be expressed in the form

$$P = \int_S t_i \dot{u}_i \mathrm{d}A + \int_V \rho f_i \dot{u}_i \mathrm{d}V, \tag{2.61}$$

where the summation convention must be invoked. By employing Cauchy's stress formula $t_i = \tau_{ij} n_j$, and by a subsequent application of Gauss' theorem, this expression may be rewritten as

$$P = \int_V [(\tau_{ij} \dot{u}_i)_{,j} + \rho f_i \dot{u}_i] \mathrm{d}V. \tag{2.62}$$

To further evaluate (2.62) we decompose $\dot{u}_{i,j}$ into

$$\dot{u}_{i,j} = \dot{\varepsilon}_{ij} + \dot{\omega}_{ij}, \tag{2.63}$$

where ε_{ij} and ω_{ij} are defined by eqs. (2.19) and (2.20), respectively. Since the contraction of a symmetric and an antisymmetric tensor vanishes, $\tau_{ij}\dot{u}_{i,j}$ may be written as

$$\tau_{ij}\dot{u}_{i,j} = \tau_{ij}\dot{\varepsilon}_{ij}. \tag{2.64}$$

By virtue of (2.64) and the equation of motion (2.39), the expression for the power input then reduces to

$$P = \int_V \rho \ddot{u}_i \dot{u}_i \mathrm{d}V + \int_V \tau_{ij} \dot{\varepsilon}_{ij} \mathrm{d}V. \tag{2.65}$$

The kinetic energy of the body is defined as

$$K = \tfrac{1}{2} \int_V \rho \dot{u}_i \dot{u}_i \mathrm{d}V, \tag{2.66}$$

and it thus follows that the power input may be rewritten as

$$P = \frac{\mathrm{d}K}{\mathrm{d}t} + \int_V \tau_{ij} \dot{\varepsilon}_{ij} \mathrm{d}V.$$

If the material is linearly elastic the total strain energy of the body is

$$U = \tfrac{1}{2} \int_V \tau_{ij} \varepsilon_{ij} \mathrm{d}V. \tag{2.67}$$

The time derivative of U can be evaluated as

$$\frac{\mathrm{d}U}{\mathrm{d}t} = \tfrac{1}{2} \int_V (\dot{\tau}_{ij} \varepsilon_{ij} + \tau_{ij} \dot{\varepsilon}_{ij}) \mathrm{d}V = \int_V \tau_{ij} \dot{\varepsilon}_{ij} \mathrm{d}V,$$

where we have used Hooke's law. For a linearly elastic body we thus find

$$P = \frac{\mathrm{d}K}{\mathrm{d}t} + \frac{\mathrm{d}U}{\mathrm{d}t}. \tag{2.68}$$

Eq. (2.68) shows that for an elastic body the energy flowing into the body through the activity of the surface tractions and the body forces is converted into kinetic energy and strain energy.

For a body with a quiescent past, (2.68) can be integrated over time between the limits 0 and t, to yield the energy identity

$$\int_0^t \int_S \boldsymbol{t}(\boldsymbol{x}, s) \cdot \dot{\boldsymbol{u}}(\boldsymbol{x}, s) \mathrm{d}A \, \mathrm{d}s + \int_0^t \int_V \rho \boldsymbol{f}(\boldsymbol{x}, s) \cdot \dot{\boldsymbol{u}}(\boldsymbol{x}, s) \mathrm{d}V \mathrm{d}s = K(t) + U(t), \tag{2.69}$$

where $K(t)$ and $U(t)$ are the kinetic and the strain energies of the body, respectively.

2.9. Hamilton's principle

2.9.1. *Statement of the principle*

The dynamic behavior of homogeneous systems in space, including continua, can be specified by a single function, a Lagrangian density $\mathscr{L}$, which is a function of, say, n local dependent variables $q_1, q_2, \ldots, q_n$, and their first derivatives

$$\dot{q}_i = \frac{\partial q_i}{\partial t}, \quad \text{and} \quad q_{i,j} = \frac{\partial q_i}{\partial x_j}. \tag{2.70}$$

Thus,

$$\mathscr{L} = \mathscr{L}(q_i, \dot{q}_i, q_{i,j}). \tag{2.71}$$

Generally there is no direct dependence of $\mathscr{L}$ on the independent variables x_j and t; there is only an indirect dependence since q_i, $\dot{q}_i$ and $q_{i,j}$ are functions of x_j and t.

Hamilton's principle states that of all possible paths of motion between two instants t_1 and t_2, the actual path taken by the system is such that the integral over time and space of the Lagrangian density $\mathscr{L}$ is stationary. An analogous but more usual statement of the principle is that the variation of the integral vanishes for any changes δq_i which vanish at $t = t_1$ and $t = t_2$, and on the boundary of the arbitrary volume V,

$$\delta \int_{t_1}^{t_2} \int_V \mathscr{L} \mathrm{d}x_1 \mathrm{d}x_2 \mathrm{d}x_3 \mathrm{d}t = 0. \tag{2.72}$$

It is well established in the calculus of variations[3] that the condition that the integral

$$\int_{t_1}^{t_2}\int_V \mathscr{L}\,\mathrm{d}x_1\,\mathrm{d}x_2\,\mathrm{d}x_3\,\mathrm{d}t \tag{2.73}$$

shall have a stationary value with respect to all possible values for which the variables t, x_1, x_2 and x_3 have an unchanged range of integration is the existence of the set of Euler equations

$$\frac{\partial}{\partial t}\left[\frac{\partial \mathscr{L}}{\partial \dot{q}_i}\right]+\sum_{j=1}^{3}\frac{\partial}{\partial x_j}\left[\frac{\partial \mathscr{L}}{\partial (q_{i,j})}\right]-\frac{\partial \mathscr{L}}{\partial q_i}=0. \tag{2.74}$$

There is one equation for each value of i ($i = 1, 2, 3$). This set of equations comprises the equations of motion of the system.

If matter is continuously distributed and if the system is conservative, the Lagrangian density $\mathscr{L}$ equals the kinetic energy density minus the potential energy density,

$$\mathscr{L} = \mathscr{K} - \mathscr{U}. \tag{2.75}$$

Thus, for the linearized theory of elasticity we have

$$\mathscr{L} = \tfrac{1}{2}\rho \dot{u}_i \dot{u}_i - [\tfrac{1}{2}\lambda(\varepsilon_{kk})^2 + \mu\varepsilon_{ij}\varepsilon_{ij}], \tag{2.76}$$

and $\mathscr{L}$ depends on $\dot{u}_i$ and $u_{i,j}$ only.

For an elastic body of finite dimensions subjected to body forces and surface tractions the statement of Hamilton's principle must be modified to

$$\delta\int_{t_1}^{t_2}\int_V (\mathscr{K}-\mathscr{U})\mathrm{d}V\mathrm{d}t+\int_{t_1}^{t_2}\delta W_e\,\mathrm{d}t = 0. \tag{2.77}$$

Here, δW_e denotes the work done by the body forces and surface tractions when the displacement is varied.

Hamilton's principle, as enunciated by eq. (2.72), is usually employed to obtain a system of equations of motion from given energy densities. Thus, by means of the Euler eqs. (2.74), the displacement equations of motion of a homogeneous, isotropic, linearly elastic solid can be derived by employing (2.76). An application of the principle is presented in section 6.11. It is, however, also instructive and of interest to work in the opposite direction and to construct Hamilton's principle by taking the stress equations of motion, eq. (2.39), as point of departure.

[3] P. M. Morse and H. Feshbach, *Methods of theoretical physics*. New York, McGraw-Hill Book Company, Inc. (1953), p. 275.

2.9.2. *Variational equation of motion*

We consider an elastodynamic problem for a body subjected to specified body forces and specified surface tractions. The boundary surface S consists of two parts, S_t and S_u, with the following boundary conditions:

over S_t: the surface traction $\boldsymbol{t}$ is prescribed,

over S_u: the displacement $\boldsymbol{u}$ is prescribed.

Let us now consider a class of displacements $\delta\boldsymbol{u}$ that are consistent with the external constraints on the body but that are otherwise arbitrary. Thus $\delta\boldsymbol{u}$ must vanish on S_u, but $\delta\boldsymbol{u}$ is arbitrary over S_t. Furthermore, $\delta\boldsymbol{u}$ as a function of x_i and t is assumed to be thrice differentiable. The arbitrary displacements $\delta\boldsymbol{u}$ are called *virtual* displacements. The terminology "virtual" implies that the virtual displacements are *not actual* displacements.

The virtual displacements that are imposed on the body cause the external forces and the surface tractions to do virtual work. If $\rho\boldsymbol{f}$ are the body forces per unit volume, the virtual work is

$$\delta W = \int_V \rho f_i \delta u_i \mathrm{d}V + \int_{S_t} t_i \delta u_i \mathrm{d}A. \tag{2.78}$$

The surface integral in this expression can be transformed into a volume integral by employing Cauchy's stress formula $t_i = \tau_{ij} n_j$, and by a subsequent application of Gauss' theorem, see eq. (2.13). We obtain

$$\begin{aligned} \int_{S_t} t_i \delta u_i \mathrm{d}A &= \int_V (\tau_{ij}\delta u_i)_{,j} \mathrm{d}V \\ &= \int_V (\tau_{ij,j}\delta u_i + \tau_{ij}\delta u_{i,j}) \mathrm{d}V. \end{aligned} \tag{2.79}$$

By employing the decomposition

$$\delta u_{i,j} = \tfrac{1}{2}(\delta u_{i,j} + \delta u_{j,i}) + \tfrac{1}{2}(\delta u_{i,j} - \delta u_{j,i}) = \delta\varepsilon_{ij} + \delta\omega_{ij},$$

and by taking note that $\delta\varepsilon_{ij}$ and $\delta\omega_{ij}$ are symmetric and antisymmetric, respectively, we conclude in view of the symmetry of τ_{ij}

$$\tau_{ij}\delta u_{i,j} = \tau_{ij}\delta\varepsilon_{ij}.$$

By virtue of the stress equations of motion, the virtual work may then be written as

$$\delta W = \int_V \rho \ddot{u}_i \delta u_i \mathrm{d}V + \int_V \tau_{ij}\delta\varepsilon_{ij} \mathrm{d}V. \tag{2.80}$$

Equating the two expressions for the virtual work, (2.78) and (2.80), yields the variational equation of motion

$$\int_V \rho f_i \delta u_i \mathrm{d}V + \int_{S_t} t_i \delta u_i \mathrm{d}A = \int_V \rho \ddot{u}_i \delta u_i \mathrm{d}V + \int_V \tau_{ij} \delta \varepsilon_{ij} \mathrm{d}V. \tag{2.81}$$

2.9.3. Derivation of Hamilton's principle

Let us integrate the variational equation of motion with respect to time between two arbitrary instants t_0 and t_1,

$$\int_{t_0}^{t_1} \delta W \mathrm{d}t = \int_{t_0}^{t_1} \mathrm{d}t \int_V \tau_{ij} \delta \varepsilon_{ij} \mathrm{d}V + \int_{t_0}^{t_1} \mathrm{d}t \int_V \rho \ddot{u}_i \delta u_i \mathrm{d}V. \tag{2.82}$$

By inverting the order of integration, and integrating by parts over t, the second term can be written as

$$I = \int_V \rho \dot{u}_i \delta u_i \Big|_{t_0}^{t_1} \mathrm{d}V - \int_V \mathrm{d}V \int_{t_0}^{t_1} \frac{\partial}{\partial t}(\rho \delta u_i) \dot{u}_i \mathrm{d}t.$$

Now we impose the restriction that δu_i vanishes identically at all points of the body at times $t = t_0$ and $t = t_1$. Then

$$\begin{aligned} I &= -\int_V \mathrm{d}V \int_{t_0}^{t_1} \dot{u}_i \frac{\partial}{\partial t}(\rho \delta u_i) \mathrm{d}t = -\int_V \rho \mathrm{d}V \int_{t_0}^{t_1} \dot{u}_i \delta \dot{u}_i \mathrm{d}t \\ &= -\int_{t_0}^{t_1} \delta K \mathrm{d}t, \end{aligned}$$

where K is the kinetic energy

$$K = \int_V \mathscr{K} \mathrm{d}V = \tfrac{1}{2} \int_V \rho \dot{u}_i \dot{u}_i \mathrm{d}V.$$

If the body is perfectly elastic we have a strain energy density $\mathscr{U}(\varepsilon_{ij})$, such that

$$\tau_{ij} = \frac{\partial \mathscr{U}}{\partial \varepsilon_{ij}}. \tag{2.83}$$

Then

$$\int_V \tau_{ij} \delta \varepsilon_{ij} \mathrm{d}V = \delta \int_V \mathscr{U} \mathrm{d}V = \delta U,$$

and eq. (2.82) reduces to

$$\delta \int_{t_0}^{t_1} (U-K)\mathrm{d}t = \int_{t_0}^{t_1} \delta W_e \,\mathrm{d}t. \tag{2.84}$$

Eq. (2.84) is Hamilton's principle for a perfectly elastic body.

2.10. Displacement potentials

In the absence of body forces the displacement equations of motion follow from (2.42) as

$$\mu u_{i,jj} + (\lambda+\mu) u_{j,ji} = \rho \ddot{u}_i. \tag{2.85}$$

As usual, the summation convention is implied. This system of equations has a disadvantageous feature in that it couples the three displacement components. The system of equations can of course be uncoupled by eliminating two of the three displacement components through two of the three equations, but this results in partial differential equations of the sixth order. A far more convenient approach is to express the components of the displacement vector in terms of derivatives of potentials. These potentials satisfy uncoupled wave equations.

In vector notation the displacement-equation of motion (2.85) can be written as

$$\mu \nabla^2 \boldsymbol{u} + (\lambda+\mu)\nabla\nabla \cdot \boldsymbol{u} = \rho \ddot{\boldsymbol{u}}. \tag{2.86}$$

Let us consider a decomposition of the displacement vector of the form

$$\boldsymbol{u} = \nabla\varphi + \nabla \wedge \boldsymbol{\psi}. \tag{2.87}$$

Substitution of the displacement representation (2.87) into eq. (2.86) yields

$$\mu\nabla^2[\nabla\varphi + \nabla \wedge \boldsymbol{\psi}] + (\lambda+\mu)\nabla\nabla \cdot [\nabla\varphi + \nabla \wedge \boldsymbol{\psi}] = \rho \frac{\partial^2}{\partial t^2}[\nabla\varphi + \nabla \wedge \boldsymbol{\psi}].$$

Since $\nabla \cdot \nabla\varphi = \nabla^2\varphi$ and $\nabla \cdot \nabla \wedge \boldsymbol{\psi} = 0$, we obtain upon rearranging terms

$$\nabla[(\lambda+2\mu)\nabla^2\varphi - \rho\ddot{\varphi}] + \nabla \wedge [\mu\nabla^2\boldsymbol{\psi} - \rho\ddot{\boldsymbol{\psi}}] = 0. \tag{2.88}$$

Clearly, the displacement representation (2.87) satisfies the equation of motion if

$$\nabla^2\varphi = \frac{1}{c_L^2}\ddot{\varphi} \tag{2.89}$$

and

$$\nabla^2\boldsymbol{\psi} = \frac{1}{c_T^2}\ddot{\boldsymbol{\psi}}, \tag{2.90}$$

where

$$c_L^2 = \frac{\lambda+2\mu}{\rho} \quad \text{and} \quad c_T^2 = \frac{\mu}{\rho}. \tag{2.91a, b}$$

Eqs. (2.89) and (2.90) are uncoupled wave equations.

Although the scalar potential φ and the components of the vector potential $\boldsymbol{\psi}$ are generally coupled through the boundary conditions, which still causes substantial mathematical complications, the use of the displacement decomposition generally simplifies the analysis. To determine the solution of a boundary-initial value problem one may simply select appropriate particular solutions of eqs. (2.89) and (2.90) in terms of arbitrary functions or integrals over arbitrary functions. If these functions can subsequently be chosen so that the boundary conditions and the initial conditions are satisfied, then the solution to the problem has been found. The solution is unique by virtue of the uniqueness theorem, which will be discussed in chapter 3.

It should be noted that eq. (2.87) relates the three components of the displacement vector to four other functions: the scalar potential and the three components of the vector potential. This indicates that φ and the components of $\boldsymbol{\psi}$ should be subjected to an additional constraint condition. Generally the components of $\boldsymbol{\psi}$ are taken to be related in some manner. Usually, but not always, the relation

$$\nabla \cdot \boldsymbol{\psi} = 0$$

is taken as the additional constraint condition. This relation has the advantage that it is consistent with the Helmholtz decomposition of a vector, which is discussed in section 3.5. Moreover, it will be shown in section 3.4 that for an unbounded medium subjected to a distribution of body forces, and for arbitrary initial conditions, the condition $\nabla \cdot \boldsymbol{\psi} = 0$ is a sufficient condition for the elastodynamic displacement to be of the form

$$\boldsymbol{u} = \nabla\varphi + \nabla \wedge \boldsymbol{\psi}.$$

2.11. Summary of equations in rectangular coordinates

In indicial notation the equations governing the linearized theory of elasticity for a homogeneous, isotropic medium are given in section 2.5. Many contributions to the field of elastic wave propagation employ, however, x, y and z as coordinates rather than x_1, x_2 and x_3. In terms of an x, y, z system, where the displacements in the coordinate directions are denoted by u, v and w, respectively, the strain-displacement relations are

$$\varepsilon_x = \frac{\partial u}{\partial x}, \qquad \varepsilon_y = \frac{\partial v}{\partial y}, \qquad \varepsilon_z = \frac{\partial w}{\partial z}, \tag{2.92a, b, c}$$

$$2\varepsilon_{xy} = 2\varepsilon_{yx} = \frac{\partial u}{\partial y} + \frac{\partial v}{\partial x}, \tag{2.93}$$

$$2\varepsilon_{yz} = 2\varepsilon_{zy} = \frac{\partial v}{\partial z} + \frac{\partial w}{\partial y}, \tag{2.94}$$

$$2\varepsilon_{zx} = 2\varepsilon_{xz} = \frac{\partial w}{\partial x} + \frac{\partial u}{\partial z}. \tag{2.95}$$

The stress-strain relations, as represented by Hooke's law, result in the following expressions

$$\tau_x = \lambda\left(\frac{\partial u}{\partial x} + \frac{\partial v}{\partial y} + \frac{\partial w}{\partial z}\right) + 2\mu\,\frac{\partial u}{\partial x} \tag{2.96}$$

$$\tau_y = \lambda\left(\frac{\partial u}{\partial x} + \frac{\partial v}{\partial y} + \frac{\partial w}{\partial z}\right) + 2\mu\,\frac{\partial v}{\partial y} \tag{2.97}$$

$$\tau_z = \lambda\left(\frac{\partial u}{\partial x} + \frac{\partial v}{\partial y} + \frac{\partial w}{\partial z}\right) + 2\mu\,\frac{\partial w}{\partial z} \tag{2.98}$$

$$\tau_{xy} = \tau_{yx} = \mu\left(\frac{\partial u}{\partial y} + \frac{\partial v}{\partial x}\right) \tag{2.99}$$

$$\tau_{yz} = \tau_{zy} = \mu\left(\frac{\partial v}{\partial z} + \frac{\partial w}{\partial y}\right) \tag{2.100}$$

$$\tau_{zx} = \tau_{xz} = \mu\left(\frac{\partial w}{\partial x} + \frac{\partial u}{\partial z}\right). \tag{2.101}$$

The relations between the components of the displacement vector and the scalar and vector potentials are represented by (2.87). In the *xyz*-system we have

$$u = \frac{\partial \varphi}{\partial x} + \frac{\partial \psi_z}{\partial y} - \frac{\partial \psi_y}{\partial z}, \tag{2.102}$$

$$v = \frac{\partial \varphi}{\partial y} - \frac{\partial \psi_z}{\partial x} + \frac{\partial \psi_x}{\partial z}, \tag{2.103}$$

$$w = \frac{\partial \varphi}{\partial z} + \frac{\partial \psi_y}{\partial x} - \frac{\partial \psi_x}{\partial y}. \tag{2.104}$$

The scalar potential φ and the components ψ_x, ψ_y and ψ_z of the vector potential ψ satisfy the equations

$$\nabla^2\varphi = \frac{1}{c_L^2}\frac{\partial^2\varphi}{\partial t^2} \tag{2.105}$$

$$\nabla^2\psi_x = \frac{1}{c_T^2}\frac{\partial^2\psi_x}{\partial t^2}, \qquad \nabla^2\psi_y = \frac{1}{c_T^2}\frac{\partial^2\psi_y}{\partial t^2}, \qquad \nabla^2\psi_z = \frac{1}{c_T^2}\frac{\partial^2\psi_z}{\partial t^2} \tag{2.106a, b, c}$$

where

$$\nabla^2 = \frac{\partial^2}{\partial x^2}+\frac{\partial^2}{\partial y^2}+\frac{\partial^2}{\partial z^2}. \tag{2.107}$$

By substituting (2.102)–(2.104) into (2.96)–(2.101), the stresses may be written in terms of the displacement potentials as

$$\tau_x = \lambda\nabla^2\varphi+2\mu\left[\frac{\partial^2\varphi}{\partial x^2}+\frac{\partial}{\partial x}\left(\frac{\partial\psi_z}{\partial y}-\frac{\partial\psi_y}{\partial z}\right)\right] \tag{2.108}$$

$$\tau_y = \lambda\nabla^2\varphi+2\mu\left[\frac{\partial^2\varphi}{\partial y^2}-\frac{\partial}{\partial y}\left(\frac{\partial\psi_z}{\partial x}-\frac{\partial\psi_x}{\partial z}\right)\right] \tag{2.109}$$

$$\tau_z = \lambda\nabla^2\varphi+2\mu\left[\frac{\partial^2\varphi}{\partial z^2}+\frac{\partial}{\partial z}\left(\frac{\partial\psi_y}{\partial x}-\frac{\partial\psi_x}{\partial y}\right)\right] \tag{2.110}$$

$$\tau_{xy} = \tau_{yx} = \mu\left[2\frac{\partial^2\varphi}{\partial x\,\partial y}+\frac{\partial}{\partial y}\left(\frac{\partial\psi_z}{\partial y}-\frac{\partial\psi_y}{\partial z}\right)-\frac{\partial}{\partial x}\left(\frac{\partial\psi_z}{\partial x}-\frac{\partial\psi_x}{\partial z}\right)\right] \tag{2.111}$$

$$\tau_{yz} = \tau_{zy} = \mu\left[2\frac{\partial^2\varphi}{\partial y\,\partial z}-\frac{\partial}{\partial z}\left(\frac{\partial\psi_z}{\partial x}-\frac{\partial\psi_x}{\partial z}\right)+\frac{\partial}{\partial y}\left(\frac{\partial\psi_y}{\partial x}-\frac{\partial\psi_x}{\partial y}\right)\right] \tag{2.112}$$

$$\tau_{zx} = \tau_{xz} = \mu\left[2\frac{\partial^2\varphi}{\partial x\,\partial z}+\frac{\partial}{\partial z}\left(\frac{\partial\psi_z}{\partial y}-\frac{\partial\psi_y}{\partial z}\right)+\frac{\partial}{\partial x}\left(\frac{\partial\psi_y}{\partial x}-\frac{\partial\psi_x}{\partial y}\right)\right]. \tag{2.113}$$

2.12. Orthogonal curvilinear coordinates

For the analysis of specific problems of elastic wave propagation, orthogonal curvilinear coordinates often lead to simplifications of the mathematical treatment. The simplifications materialize if in a suitably chosen system of curvilinear coordinates one of the coordinates is constant on a bounding surface of the body. For example, for waves emanating from a spherical cavity of radius a in an unbounded medium, the forcing conditions on the surface of the cavity are in spherical coordinates simply prescribed at $r = a$.

By means of general tensor calculus, nonorthogonal curvilinear coordinate systems of any dimension can be treated. Here we restrict the discussion, however, to three-dimensional orthogonal curvilinear coordinates. Let us consider a set of three independent functions q_i of the Cartesian variables x_j

$$q_i = q_i(x_1, x_2, x_3), \tag{2.114}$$

and let us assume that these equations may be solved for x_j in terms of q_i, or

$$x_j = x_j(q_1, q_2, q_3). \tag{2.115}$$

The three equations $q_i = c_i$, where c_i are constants, represent three families of surfaces whose lines of intersection form three families of curved lines. These lines of intersection will be used as the coordinate lines in our curvilinear coordinate system. Thus, the position of a point in space can be defined by the values of three coordinates q_1, q_2 and q_3. The local coordinate directions at a point are tangent to the three coordinate lines intersecting at the point.

In an orthogonal coordinate system, which is considered in this section, the three coordinate directions are mutually perpendicular. We choose an orthonormal right-handed basis whose unit vectors $\boldsymbol{e}_1$, $\boldsymbol{e}_2$ and $\boldsymbol{e}_3$ are respectively directed in the sense of increase of the coordinates q_1, q_2 and q_3. The following well-known relations hold

$$\boldsymbol{e}_i \cdot \boldsymbol{e}_j = \delta_{ij} \tag{2.116}$$

$$\boldsymbol{e}_i \wedge \boldsymbol{e}_j = \boldsymbol{e}_k, \tag{2.117}$$

where in (2.117) the indices i, j and k are in cyclic order. A major difference between curvilinear coordinates and Cartesian coordinates is that the coordinates q_1, q_2 and q_3 are not necessarily measured in lengths. For example, in cylindrical coordinates $q_1 = r$, $q_2 = \theta$, and $q_3 = x_3$. This difference manifests itself in the appearance of scale factors in the relation between the infinitesimal displacement vector $\mathrm{d}\boldsymbol{r}$ and the infinitesimal variations $\mathrm{d}q_1$, $\mathrm{d}q_2$ and $\mathrm{d}q_3$, namely,

$$\mathrm{d}\boldsymbol{r} = \boldsymbol{e}_1 h_1 \mathrm{d}q_1 + \boldsymbol{e}_2 h_2 \mathrm{d}q_2 + \boldsymbol{e}_3 h_3 \mathrm{d}q_3. \tag{2.118}$$

The scale factors h_i are in general functions of the coordinates q_j.

The unit vectors $\boldsymbol{e}_i$ generally vary in direction from point to point in space, and a careful examination of the partial derivatives $\partial/\partial q_j$ of the unit vectors $\boldsymbol{e}_i$ is consequently required. From the expression for $\mathrm{d}\boldsymbol{r}$, eq. (2.118), we have

$$\frac{\partial \boldsymbol{r}}{\partial q_i} = \boldsymbol{e}_i h_i \quad \text{(no summation, } i = 1, 2 \text{ or } 3\text{).}$$

Since the order of differentiation may be changed in $\partial^2 \boldsymbol{r}/\partial q_1 \partial q_2$, we have

$$\frac{\partial}{\partial q_1}(\boldsymbol{e}_2 h_2) = \frac{\partial}{\partial q_2}(\boldsymbol{e}_1 h_1),$$

or

$$\boldsymbol{e}_2 \frac{\partial h_2}{\partial q_1} + h_2 \frac{\partial \boldsymbol{e}_2}{\partial q_1} = \boldsymbol{e}_1 \frac{\partial h_1}{\partial q_2} + h_1 \frac{\partial \boldsymbol{e}_1}{\partial q_2}. \tag{2.119}$$

By taking the scalar product with $\boldsymbol{e}_1$ we obtain

$$h_2 \boldsymbol{e}_1 \cdot \frac{\partial \boldsymbol{e}_2}{\partial q_1} = \frac{\partial h_1}{\partial q_2}, \tag{2.120}$$

where we have used (2.116), as well as the fact that in an orthogonal system $\boldsymbol{e}_1 \cdot (\partial \boldsymbol{e}_1/\partial q_2) = 0$. By multiplying (2.119) scalarly by $\boldsymbol{e}_3$ we find

$$h_2 \boldsymbol{e}_3 \cdot \frac{\partial \boldsymbol{e}_2}{\partial q_1} = h_1 \boldsymbol{e}_3 \cdot \frac{\partial \boldsymbol{e}_1}{\partial q_2}. \tag{2.121}$$

By permutation of the indices, two expressions analogous to (2.121) are found as

$$h_3 \boldsymbol{e}_1 \cdot \frac{\partial \boldsymbol{e}_3}{\partial q_2} = h_2 \boldsymbol{e}_1 \cdot \frac{\partial \boldsymbol{e}_2}{\partial q_3}, \qquad h_1 \boldsymbol{e}_2 \cdot \frac{\partial \boldsymbol{e}_1}{\partial q_3} = h_3 \boldsymbol{e}_2 \cdot \frac{\partial \boldsymbol{e}_3}{\partial q_1}. \tag{2.122a, b}$$

Now, by taking the derivatives of $\boldsymbol{e}_3 \cdot \boldsymbol{e}_2 = 0$, $\boldsymbol{e}_1 \cdot \boldsymbol{e}_2 = 0$ and $\boldsymbol{e}_1 \cdot \boldsymbol{e}_3 = 0$ with respect to q_1, q_3 and q_2, respectively, and using the resulting equations as well as eqs. (2.122a, b), the following manipulations are directly verifiable:

$$\begin{aligned} \boldsymbol{e}_3 \cdot \frac{\partial \boldsymbol{e}_2}{\partial q_1} &= -\boldsymbol{e}_2 \cdot \frac{\partial \boldsymbol{e}_3}{\partial q_1} = -\frac{h_1}{h_3} \boldsymbol{e}_2 \cdot \frac{\partial \boldsymbol{e}_1}{\partial q_3} = \frac{h_1}{h_3} \boldsymbol{e}_1 \cdot \frac{\partial \boldsymbol{e}_2}{\partial q_3} \\ &= \frac{h_1}{h_2} \boldsymbol{e}_1 \cdot \frac{\partial \boldsymbol{e}_3}{\partial q_2} = -\frac{h_1}{h_2} \boldsymbol{e}_3 \cdot \frac{\partial \boldsymbol{e}_1}{\partial q_2} = -\boldsymbol{e}_3 \cdot \frac{\partial \boldsymbol{e}_2}{\partial q_1}. \end{aligned}$$

The result shows that the inner product of $\boldsymbol{e}_3$ and $\partial \boldsymbol{e}_2/\partial q_1$ vanishes. Since it is also clear that the inner product of $\boldsymbol{e}_2$ and $\partial \boldsymbol{e}_2/\partial q_1$ vanishes, it is concluded that $\partial \boldsymbol{e}_2/\partial q_1$ does not have components in the directions of $\boldsymbol{e}_2$ and $\boldsymbol{e}_3$, and it follows from (2.120) that

$$\frac{\partial \boldsymbol{e}_2}{\partial q_1} = \frac{\boldsymbol{e}_1}{h_2} \frac{\partial h_1}{\partial q_2}.$$

Five other relations of this form can be derived by permutation of the indices, and the six results can be summarized as

$$i \neq j: \qquad \frac{\partial \boldsymbol{e}_i}{\partial q_j} = \frac{\boldsymbol{e}_j}{h_i} \frac{\partial h_j}{\partial q_i} \quad \text{(no summation).} \tag{2.123}$$

By employing (2.117) we can also write

$$\frac{\partial \boldsymbol{e}_1}{\partial q_1} = \frac{\partial}{\partial q_1}(\boldsymbol{e}_2 \wedge \boldsymbol{e}_3) = \boldsymbol{e}_2 \wedge \frac{\partial \boldsymbol{e}_3}{\partial q_1} - \boldsymbol{e}_3 \wedge \frac{\partial \boldsymbol{e}_2}{\partial q_1}$$
$$= -\frac{\boldsymbol{e}_3}{h_3}\frac{\partial h_1}{\partial q_3} - \frac{\boldsymbol{e}_2}{h_2}\frac{\partial h_1}{\partial q_2}.$$

Two similar expressions can be derived for $\partial \boldsymbol{e}_2/\partial q_2$ and $\partial \boldsymbol{e}_3/\partial q_3$. In indicial notation we have

$$\frac{\partial \boldsymbol{e}_i}{\partial q_i} = -\frac{\boldsymbol{e}_j}{h_j}\frac{\partial h_i}{\partial q_j} - \frac{\boldsymbol{e}_k}{h_k}\frac{\partial h_i}{\partial q_k} \quad \text{(no summation),} \tag{2.124}$$

where i, j and k must be taken in cyclic order.

All equations in Cartesian coordinates which do not involve space derivatives and which pertain to properties at a point carry over unchanged into curvilinear coordinates. If space derivatives are involved, however, equations do not directly carry over, since the differential operators such as the gradient, divergence, curl and the Laplacian assume different forms.

We consider first the *gradient operator* ∇. When applied to a scalar φ it gives a vector $\nabla\varphi$, with components which we call f_1, f_2 and f_3. Thus

$$\nabla\varphi = \mathrm{f}_1\boldsymbol{e}_1 + \mathrm{f}_2\boldsymbol{e}_2 + \mathrm{f}_3\boldsymbol{e}_3.$$

The increment of φ due to a change of position $\mathrm{d}\boldsymbol{r}$ is

$$\mathrm{d}\varphi = \nabla\varphi \cdot \mathrm{d}\boldsymbol{r} = h_1\mathrm{f}_1\mathrm{d}q_1 + h_2\mathrm{f}_2\mathrm{d}q_2 + h_3\mathrm{f}_3\mathrm{d}q_3,$$

where (2.118) has been used. The increment $\mathrm{d}\varphi$ can also be written as

$$\mathrm{d}\varphi = \frac{\partial\varphi}{\partial q_1}\mathrm{d}q_1 + \frac{\partial\varphi}{\partial q_2}\mathrm{d}q_2 + \frac{\partial\varphi}{\partial q_3}\mathrm{d}q_3,$$

whence it can be concluded that

$$\nabla = \frac{\boldsymbol{e}_1}{h_1}\frac{\partial}{\partial q_1} + \frac{\boldsymbol{e}_2}{h_2}\frac{\partial}{\partial q_2} + \frac{\boldsymbol{e}_3}{h_3}\frac{\partial}{\partial q_3}. \tag{2.125}$$

If the operation (2.125) is applied to a vector $\boldsymbol{u}$, we obtain

$$\nabla\boldsymbol{u} = \frac{\boldsymbol{e}_1}{h_1}\frac{\partial\boldsymbol{u}}{\partial q_1} + \frac{\boldsymbol{e}_2}{h_2}\frac{\partial\boldsymbol{u}}{\partial q_2} + \frac{\boldsymbol{e}_3}{h_3}\frac{\partial\boldsymbol{u}}{\partial q_3}, \tag{2.126}$$

which can be written out in more detail by using (2.123) and (2.124).

The *divergence* of a vector $\boldsymbol{u}$ is

$$\begin{aligned} \operatorname{div} \boldsymbol{u} &= \nabla \cdot \boldsymbol{u} = \nabla \cdot (u_1 \boldsymbol{e}_1 + u_2 \boldsymbol{e}_2 + u_3 \boldsymbol{e}_3) \\ &= \nabla \cdot (u_1 \boldsymbol{e}_1) + \nabla \cdot (u_2 \boldsymbol{e}_2) + \nabla \cdot (u_3 \boldsymbol{e}_3). \end{aligned} \tag{2.127}$$

Since q_i are independent coordinates we find, by using (2.125)

$$\nabla q_i = \frac{\boldsymbol{e}_i}{h_i} \quad \text{(no summation)}. \tag{2.128}$$

By writing the vector product of ∇q_1 and ∇q_2 we then find

$$\boldsymbol{e}_i = h_j h_k \nabla q_j \wedge \nabla q_k \qquad \text{(no summation)},$$

where i, j and k are in cyclic order. Substituting this result for $i = 1, j = 2, k = 3$ into the first of (2.127), we obtain

$$\begin{aligned} \nabla \cdot (u_1 \boldsymbol{e}_1) &= \nabla (u_1 h_2 h_3) \cdot (\nabla q_2 \wedge \nabla q_3) + u_1 h_2 h_3 \nabla \cdot (\nabla q_2 \wedge \nabla q_3) \\ &= \nabla (u_1 h_2 h_3) \cdot \left(\frac{\boldsymbol{e}_1}{h_2 h_3} \right) + 0 \\ &= \frac{1}{h_1 h_2 h_3} \frac{\partial}{\partial q_1} (u_1 h_2 h_3). \end{aligned}$$

The two other terms can be worked out in a similar manner, and thus

$$\nabla \cdot \boldsymbol{u} = \frac{1}{h_1 h_2 h_3} \left[\frac{\partial}{\partial q_1} (u_1 h_2 h_3) + \frac{\partial}{\partial q_2} (u_2 h_3 h_1) + \frac{\partial}{\partial q_3} (u_3 h_1 h_2) \right]. \tag{2.129}$$

The *curl operator* of a vector $\boldsymbol{u}$ is

$$\nabla \wedge \boldsymbol{u} = \nabla \wedge (u_1 h_1 \nabla q_1) + \nabla \wedge (u_2 h_2 \nabla q_2) + \nabla \wedge (u_3 h_3 \nabla q_3),$$

where (2.128) has been used. By employing the rule on the curl of a product of a scalar and a vector we obtain for the first term

$$\nabla \wedge (u_1 h_1 \nabla q_1) = \frac{\boldsymbol{e}_2}{h_3 h_1} \frac{\partial}{\partial q_3} (u_1 h_1) - \frac{\boldsymbol{e}_3}{h_2 h_1} \frac{\partial}{\partial q_2} (u_1 h_1).$$

The other two terms can be worked out in a similar manner, and thus

$$\nabla \wedge \boldsymbol{u} = \frac{1}{h_1 h_2 h_3} \begin{vmatrix} h_1 \boldsymbol{e}_1 & h_2 \boldsymbol{e}_2 & h_3 \boldsymbol{e}_3 \\ \dfrac{\partial}{\partial q_1} & \dfrac{\partial}{\partial q_2} & \dfrac{\partial}{\partial q_3} \\ h_1 u_1 & h_2 u_2 & h_3 u_3 \end{vmatrix} \tag{2.130}$$

The *Laplace operator* of a scalar can easily be derived by using (2.129) and (2.125). Thus

$$\nabla^2\varphi = \nabla\cdot\nabla\varphi = \frac{1}{h_1 h_2 h_3}\sum_{n=1}^{3}\frac{\partial}{\partial q_n}\left[\frac{h_1 h_2 h_3}{h_n^2}\frac{\partial\varphi}{\partial q_n}\right]. \tag{2.131}$$

The Laplacian of a vector is more complicated, and we will not write out the expression. It is noted, however, that by using

$$\nabla^2 \boldsymbol{u} = \nabla(\nabla\cdot\boldsymbol{u}) - \nabla\wedge\nabla\wedge\boldsymbol{u},$$

and by employing the expressions for the gradient, the divergence and the curl, $\nabla^2\boldsymbol{u}$ can be obtained.

2.13. Summary of equations in cylindrical coordinates

In cylindrical coordinates we choose

$$q_1 = r,\ q_2 = \theta,\ q_3 = z.$$

The corresponding scale factors and unit base vectors are

$$h_1 = 1, \qquad h_2 = r, \qquad h_3 = 1$$
$$\boldsymbol{e}_1 = \boldsymbol{e}_r, \qquad \boldsymbol{e}_2 = \boldsymbol{e}_\theta, \qquad \boldsymbol{e}_3 = \boldsymbol{k}.$$

The following equations then follow from (2.123) and (2.124):

$$\frac{\partial \boldsymbol{e}_1}{\partial q_1} = 0, \qquad \frac{\partial \boldsymbol{e}_1}{\partial q_2} = \boldsymbol{e}_\theta, \qquad \frac{\partial \boldsymbol{e}_1}{\partial q_3} = 0$$
$$\frac{\partial \boldsymbol{e}_2}{\partial q_1} = 0, \qquad \frac{\partial \boldsymbol{e}_2}{\partial q_2} = -\boldsymbol{e}_r, \qquad \frac{\partial \boldsymbol{e}_2}{\partial q_3} = 0$$
$$\frac{\partial \boldsymbol{e}_3}{\partial q_1} = 0, \qquad \frac{\partial \boldsymbol{e}_3}{\partial q_2} = 0, \qquad \frac{\partial \boldsymbol{e}_3}{\partial q_3} = 0.$$

These relations can now be used to write the expressions for the differential operators, which are given by (2.125)–(2.131).

Denoting the displacement components in the r, θ and z directions by u, v and w, respectively, the relations between the displacement components and the potentials follow from (2.87) as

$$u = \frac{\partial\varphi}{\partial r} + \frac{1}{r}\frac{\partial\psi_z}{\partial\theta} - \frac{\partial\psi_\theta}{\partial z} \tag{2.132}$$

$$v = \frac{1}{r}\frac{\partial\varphi}{\partial\theta} + \frac{\partial\psi_r}{\partial z} - \frac{\partial\psi_z}{\partial r} \tag{2.133}$$

$$w = \frac{\partial \varphi}{\partial z} + \frac{1}{r}\frac{\partial(\psi_\theta r)}{\partial r} - \frac{1}{r}\frac{\partial \psi_r}{\partial \theta}, \tag{2.134}$$

where

$$\nabla^2 \varphi = \frac{1}{c_L^2}\frac{\partial^2 \varphi}{\partial t^2}, \tag{2.135}$$

and the Laplacian is defined as

$$\nabla^2 = \frac{\partial^2}{\partial r^2} + \frac{1}{r}\frac{\partial}{\partial r} + \frac{1}{r^2}\frac{\partial^2}{\partial \theta^2} + \frac{\partial^2}{\partial z^2}. \tag{2.136}$$

The components of the vector potential ψ satisfy the equations

$$\nabla^2 \psi_r - \frac{\psi_r}{r^2} - \frac{2}{r^2}\frac{\partial \psi_\theta}{\partial \theta} = \frac{1}{c_T^2}\frac{\partial^2 \psi_r}{\partial t^2} \tag{2.137}$$

$$\nabla^2 \psi_\theta - \frac{\psi_\theta}{r^2} + \frac{2}{r^2}\frac{\partial \psi_r}{\partial \theta} = \frac{1}{c_T^2}\frac{\partial^2 \psi_\theta}{\partial t^2} \tag{2.138}$$

$$\nabla^2 \psi_z = \frac{1}{c_T^2}\frac{\partial^2 \psi_z}{\partial t^2}. \tag{2.139}$$

In cylindrical coordinates the strain-displacement relations are given by

$$\varepsilon_r = \frac{\partial u}{\partial r}, \qquad \varepsilon_\theta = \frac{u}{r} + \frac{1}{r}\frac{\partial v}{\partial \theta}, \qquad \varepsilon_z = \frac{\partial w}{\partial z} \tag{2.140a, b, c}$$

$$2\varepsilon_{r\theta} = 2\varepsilon_{\theta r} = \frac{\partial v}{\partial r} - \frac{v}{r} + \frac{1}{r}\frac{\partial u}{\partial \theta} \tag{2.141}$$

$$2\varepsilon_{\theta z} = 2\varepsilon_{z\theta} = \frac{1}{r}\frac{\partial w}{\partial \theta} + \frac{\partial v}{\partial z} \tag{2.142}$$

$$2\varepsilon_{zr} = 2\varepsilon_{rz} = \frac{\partial u}{\partial z} + \frac{\partial w}{\partial r}. \tag{2.143}$$

The stress-strain relations are of the forms

$$\tau_r = \lambda\left(\frac{\partial u}{\partial r} + \frac{u}{r} + \frac{1}{r}\frac{\partial v}{\partial \theta} + \frac{\partial w}{\partial z}\right) + 2\mu\frac{\partial u}{\partial r} \tag{2.144}$$

$$\tau_\theta = \lambda\left(\frac{\partial u}{\partial r} + \frac{u}{r} + \frac{1}{r}\frac{\partial v}{\partial \theta} + \frac{\partial w}{\partial z}\right) + 2\mu\left(\frac{u}{r} + \frac{1}{r}\frac{\partial v}{\partial \theta}\right) \tag{2.145}$$

$$\tau_z = \lambda \left(\frac{\partial u}{\partial r} + \frac{u}{r} + \frac{1}{r} \frac{\partial v}{\partial \theta} + \frac{\partial w}{\partial z} \right) + 2\mu \frac{\partial w}{\partial z} \tag{2.146}$$

$$\tau_{r\theta} = \mu \left(\frac{\partial v}{\partial r} - \frac{v}{r} + \frac{1}{r} \frac{\partial u}{\partial \theta} \right) \tag{2.147}$$

$$\tau_{\theta z} = \mu \left(\frac{1}{r} \frac{\partial w}{\partial \theta} + \frac{\partial v}{\partial z} \right) \tag{2.148}$$

$$\tau_{zr} = \mu \left(\frac{\partial u}{\partial z} + \frac{\partial w}{\partial r} \right) . \tag{2.149}$$

The stresses can be written in terms of the displacement potentials by substituting (2.132)–(2.134) into the relations (2.144)–(2.149).

2.14. Summary of equations in spherical coordinates

In spherical coordinates the orthogonal surfaces are the spheres r = const, the circular cones θ = const, and the planes χ = const (see figure 2.1). We

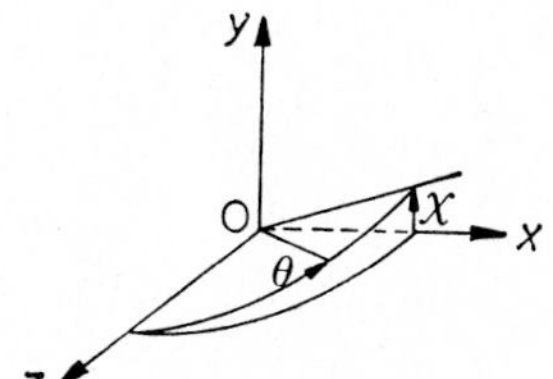

Fig. 2.1. Spherical coordinates.

choose the coordinates

$$q_1 = r,\ q_2 = \theta,\ q_3 = \chi,$$

with scale factors

$$h_1 = 1,\ h_2 = r,\ h_3 = r \sin \theta,$$

and unit base vectors

$$\boldsymbol{e}_1 = \boldsymbol{e}_r, \qquad \boldsymbol{e}_2 = \boldsymbol{e}_\theta, \qquad \boldsymbol{e}_3 = \boldsymbol{e}_\chi .$$

Eqs. (2.123) and (2.124) then yield

$$\frac{\partial \boldsymbol{e}_1}{\partial q_1} = 0, \qquad \frac{\partial \boldsymbol{e}_1}{\partial q_2} = \boldsymbol{e}_\theta, \qquad \frac{\partial \boldsymbol{e}_1}{\partial q_3} = \boldsymbol{e}_\chi \sin\theta$$

$$\frac{\partial \boldsymbol{e}_2}{\partial q_1} = 0, \qquad \frac{\partial \boldsymbol{e}_2}{\partial q_2} = -\boldsymbol{e}_r, \qquad \frac{\partial \boldsymbol{e}_2}{\partial q_3} = \boldsymbol{e}_\chi \cos\theta$$

$$\frac{\partial \boldsymbol{e}_3}{\partial q_1} = 0, \qquad \frac{\partial \boldsymbol{e}_3}{\partial q_2} = 0, \qquad \frac{\partial \boldsymbol{e}_3}{\partial q_3} = -\boldsymbol{e}_r \sin\theta - \boldsymbol{e}_\theta \cos\theta.$$

Substitution of these expressions into (2.125)–(2.131) yields the expressions for the differential operators in spherical coordinates.

Denoting the displacement components in the directions of increasing r, θ and χ by u, v and w, respectively, the relations between displacements and potentials become

$$u = \frac{\partial \varphi}{\partial r} + \frac{1}{r\sin\theta}\left[\frac{\partial}{\partial\theta}(\psi_\chi \sin\theta) - \frac{\partial\psi_\theta}{\partial\chi}\right] \tag{2.150}$$

$$v = \frac{1}{r}\frac{\partial\varphi}{\partial\theta} + \frac{1}{r}\left[\frac{1}{\sin\theta}\frac{\partial\psi_r}{\partial\chi} - \frac{\partial}{\partial r}(r\psi_\chi)\right] \tag{2.151}$$

$$w = \frac{1}{r\sin\theta}\frac{\partial\varphi}{\partial\chi} + \frac{1}{r}\left[\frac{\partial}{\partial r}(r\psi_\theta) - \frac{\partial\psi_r}{\partial\theta}\right], \tag{2.152}$$

where

$$\nabla^2\varphi = \frac{1}{c_L^2}\frac{\partial^2\varphi}{\partial t^2} \tag{2.153}$$

and the Laplacian ∇^2 is defined as

$$\nabla^2 = \frac{1}{r^2}\frac{\partial}{\partial r}\left(r^2\frac{\partial}{\partial r}\right) + \frac{1}{r^2\sin\theta}\frac{\partial}{\partial\theta}\left(\sin\theta\frac{\partial}{\partial\theta}\right) + \frac{1}{r^2\sin^2\theta}\frac{\partial^2}{\partial\chi^2}. \tag{2.154}$$

Also

$$\nabla^2\psi_r - \frac{2}{r^2}\psi_r - \frac{2}{r^2\sin\theta}\frac{\partial}{\partial\theta}(\psi_\theta\sin\theta) - \frac{2}{r^2\sin\theta}\frac{\partial\psi_\chi}{\partial\chi} = \frac{1}{c_T^2}\frac{\partial^2\psi_r}{\partial t^2} \tag{2.155}$$

$$\nabla^2\psi_\theta - \frac{\psi_\theta}{r^2\sin^2\theta} + \frac{2}{r^2}\frac{\partial\psi_r}{\partial\theta} - \frac{2\cos\theta}{r^2\sin^2\theta}\frac{\partial\psi_\chi}{\partial\chi} = \frac{1}{c_T^2}\frac{\partial^2\psi_\theta}{\partial t^2} \tag{2.156}$$

$$\nabla^2\psi_\chi - \frac{\psi_\chi}{r^2\sin^2\theta} + \frac{2}{r^2\sin\theta}\frac{\partial\psi_r}{\partial\chi} + \frac{2\cos\theta}{r^2\sin^2\theta}\frac{\partial\psi_\theta}{\partial\chi} = \frac{1}{c_T^2}\frac{\partial^2\psi_\chi}{\partial t^2}. \tag{2.157}$$

The strain-displacement relations are

$$\varepsilon_r = \frac{\partial u}{\partial r} \tag{2.158}$$

$$\varepsilon_\theta = \frac{1}{r}\frac{\partial v}{\partial \theta} + \frac{u}{r} \tag{2.159}$$

$$\varepsilon_\chi = \frac{1}{r \sin\theta}\left(\frac{\partial w}{\partial \chi} + u \sin\theta + v \cos\theta\right) \tag{2.160}$$

$$2\varepsilon_{r\theta} = 2\varepsilon_{\theta r} = \frac{\partial v}{\partial r} + \frac{1}{r}\left(\frac{\partial u}{\partial \theta} - v\right) \tag{2.161}$$

$$2\varepsilon_{\theta\chi} = 2\varepsilon_{\chi\theta} = \frac{1}{r}\frac{\partial w}{\partial \theta} + \frac{1}{r \sin\theta}\left(\frac{\partial v}{\partial \chi} - w\cos\theta\right) \tag{2.162}$$

$$2\varepsilon_{\chi r} = 2\varepsilon_{r\chi} = \frac{\partial w}{\partial r} + \frac{1}{r \sin\theta}\left(\frac{\partial u}{\partial \chi} - w\sin\theta\right) \tag{2.163}$$

From Hooke's law we find

$$\tau_r = \lambda\varepsilon + 2\mu\frac{\partial u}{\partial r} \tag{2.164}$$

$$\tau_\theta = \lambda\varepsilon + 2\mu\left(\frac{1}{r}\frac{\partial v}{\partial \theta} + \frac{u}{r}\right) \tag{2.165}$$

$$\tau_\chi = \lambda\varepsilon + \frac{2\mu}{r \sin\theta}\left(\frac{\partial w}{\partial \chi} + u \sin\theta + v\cos\theta\right) \tag{2.166}$$

$$\tau_{r\theta} = \tau_{\theta r} = \mu\left[\frac{\partial v}{\partial r} + \frac{1}{r}\left(\frac{\partial u}{\partial \theta} - v\right)\right] \tag{2.167}$$

$$\tau_{\theta\chi} = \tau_{\chi\theta} = \mu\left[\frac{1}{r}\frac{\partial w}{\partial \theta} + \frac{1}{r \sin\theta}\left(\frac{\partial v}{\partial \chi} - w\cos\theta\right)\right] \tag{2.168}$$

$$\tau_{\chi r} = \tau_{r\chi} = \mu\left[\frac{\partial w}{\partial r} + \frac{1}{r \sin\theta}\left(\frac{\partial u}{\partial \chi} - w\sin\theta\right)\right], \tag{2.169}$$

where

$$\varepsilon = \frac{\partial u}{\partial r} + \frac{1}{r}\frac{\partial v}{\partial \theta} + \frac{u}{r} + \frac{1}{r \sin\theta}\left(\frac{\partial w}{\partial \chi} + u\sin\theta + v\cos\theta\right). \tag{2.170}$$

The stresses are obtained in terms of the displacement potentials by substituting (2.150)–(2.152) into the relations (2.164)–(2.169).

2.15. The ideal fluid

It is a matter of experience that a fluid at rest or in uniform flow cannot sustain shear stresses. Hence the state of stress is purely hydrostatic. An ideal (nonviscous) fluid is a fluid which cannot sustain shear stresses, even when it is in motion. The state of stress in an ideal fluid is thus described by

$$\tau_{ij} = -p\delta_{ij}, \tag{2.171}$$

where p is the pressure, which satisfies the equation of state. When the equation of state is $p = p(\rho)$, where ρ is the mass density, the ideal fluid is called an elastic fluid. For small disturbances from equilibrium which we are considering here, the equations of state follow directly from Hooke's law, eq. (2.28), by setting $\mu \equiv 0$. We find

$$\tau_{11} = \tau_{22} = \tau_{33} = \lambda\varepsilon_{kk} = -p \tag{2.172}$$

$$\tau_{12} = \tau_{13} = \tau_{23} \equiv 0. \tag{2.173}$$

In liquids λ is very large, whereas it has only moderate values for gases. Ignoring body forces the equations of motion for an ideal elastic fluid are in the linear approximation

$$\lambda\nabla\nabla \cdot \boldsymbol{u} = \rho\ddot{\boldsymbol{u}}. \tag{2.174}$$

It is convenient to introduce a scalar velocity potential φ by the relation

$$\dot{\boldsymbol{u}} = \nabla\varphi. \tag{2.175}$$

Clearly eq. (2.174) is satisfied if

$$\nabla^2\varphi = \frac{1}{c_F^2}\ddot{\varphi}, \tag{2.176}$$

where

$$c_F^2 = \frac{\lambda}{\rho}. \tag{2.177}$$

This wave equation holds for small disturbances propagating in an ideal elastic fluid. It finally follows from (2.172) and (2.176) that

$$p = -\lambda\nabla \cdot \boldsymbol{u} = -\rho\dot{\varphi}. \tag{2.178}$$

CHAPTER 3

ELASTODYNAMIC THEORY

3.1. Introduction

This chapter is concerned with the discussion of several formal aspects of the theory of dynamic elasticity and with general methods of solution of elastodynamic problems. Among the theorems that are proven, the uniqueness theorem for the properly formulated boundary-initial value problem of elastodynamics is of primary importance. Another theorem states the dynamic reciprocal identity which relates two elastodynamic states of the same body. The dynamic reciprocal identity is of interest both as a vehicle for the development of further theoretical results and for the generation of solutions of problems. We also investigate in more detail the decomposition of the displacement vector which was introduced in section 2.10, and we prove a completeness theorem for the scalar and vector potentials.

The scalar and vector potentials for the displacement field are governed by classical wave equations which have been studied in great detail in the mathematical literature. In sections 3.6 and 3.7 general integral representations for the solution of the classical wave equation are examined with a view toward determining the elastic wave motion generated by body forces in an unbounded medium. The displacement and stress fields due to a time-dependent point load are determined in section 3.8. These fields which comprise the basic singular solution of elastodynamics are subsequently employed to derive general integral representations for the field variables in a bounded elastic body subjected to surface disturbances. The integral representations are, however, mainly of formal interest because the elastodynamic response to surface disturbances is generally determined in a more efficient manner by direct applications of methods of analysis to the system of governing equations. Several methods of analysis are introduced by means of examples in the next chapter. The particularly useful methods based on the application of integral transform techniques are discussed in considerable detail in chapter 7.

A general statement of the elastodynamic problem for a body B with interior V, closure $\bar{V}$ and boundary S was given in section 2.5. For economy of presentation of the material of this chapter it is convenient to employ the definition of elastodynamic states, which was introduced by Wheeler and Sternberg.[1]

Elastodynamic state: Consider a vector-valued function $\boldsymbol{u}(\boldsymbol{x}, t)$ and a tensor-valued function $\boldsymbol{\tau}(\boldsymbol{x}, t)$, both defined on $\bar{V} \times T$, where T is an arbitrary interval of time. We call the ordered pair $\mathscr{S} = [\boldsymbol{u}, \boldsymbol{\tau}]$ an elastodynamic state on $\bar{V} \times T$, with the displacement field $\boldsymbol{u}$ and the stress field $\boldsymbol{\tau}$, corresponding to the body-force density $\boldsymbol{f}$, the mass density ρ and the Lamé elastic constants λ and μ, if

(a) $\boldsymbol{u} \in \mathscr{C}^2(V \times T) \cap \mathscr{C}^1(\bar{V} \times T)$ (3.1)

$\boldsymbol{\tau} \in \mathscr{C}(\bar{V} \times T)$ (3.2)

$\boldsymbol{f} \in \mathscr{C}(\bar{V} \times T)$ (3.3)

(b) $\rho > 0,\ 0 < 3\lambda + 2\mu < \infty,\ 0 < \mu < \infty$ (3.4)

(c) on $V \times T$, $\boldsymbol{u}$, $\boldsymbol{\tau}$, $\boldsymbol{f}$, ρ, λ and μ satisfy the equations

$$\tau_{ij,j} + \rho f_i = \rho \ddot{u}_i \tag{3.5}$$

$$\tau_{ij} = \lambda u_{k,k}\delta_{ij} + \mu(u_{i,j} + u_{j,i}). \tag{3.6}$$

We write

$$\mathscr{S} = [\boldsymbol{u}, \boldsymbol{\tau}] \in \mathscr{E}(\boldsymbol{f}, \rho, \lambda, \mu; \bar{V} \times T).$$

If in particular

$$\boldsymbol{u} = 0 \text{ on } V \times (-\infty, 0],$$

where $(-\infty, 0]$ is a subinterval of T, we say that $\mathscr{S}$ is an elastodynamic state with a quiescent past, and we write

$$\mathscr{S} = [\boldsymbol{u}, \boldsymbol{\tau}] = \mathscr{E}_0(\boldsymbol{f}, \rho, \lambda, \mu; \bar{V}).$$

3.2. Uniqueness of solution

It is not difficult to show that the solution of the elastodynamic problem formulated in section 2.5 satisfies the criterion of uniqueness for an appropriate set of boundary conditions. The proof of uniqueness given in this section is essentially due to Neumann; it can also be found in this form in the book by Sokolnikoff.[2] The proof is based on energy considerations.

We consider a *bounded* body B occupying the regular region V with

[1] L. T. Wheeler and E. Sternberg, *Archive for Rational Mechanics and Analysis* **31** (1968) 51.

[2] I. S. Sokolnikoff, *Mathematical theory of elasticity*. New York, McGraw-Hill Book Co., Inc. (1956), p. 87.

boundary S. The time interval is the half-open interval $T^+ = [0, \infty)$. We state the following uniqueness theorem.

Theorem 3.1: Let $\mathscr{S}'$ and $\mathscr{S}''$ be two elastodynamic states with the following properties:

$$\text{(a)}\ \mathscr{S}' = [\boldsymbol{u}', \boldsymbol{\tau}'] \in \mathscr{E}(\boldsymbol{f}, \rho, \lambda, \mu; \bar{V} \times T^+)$$
$$\mathscr{S}'' = [\boldsymbol{u}'', \boldsymbol{\tau}''] \in \mathscr{E}(\boldsymbol{f}, \rho, \lambda, \mu; \bar{V} \times T^+),$$

i.e., both sets of displacements and stresses satisfy eqs. (3.5) and (3.6) and the requirements (3.1)–(3.4).

(b) Both elastodynamic states have the same initial conditions, i.e.,

$$\mathring{\boldsymbol{u}}'(\boldsymbol{x}) = \mathring{\boldsymbol{u}}''(\boldsymbol{x}),\ \mathring{\boldsymbol{v}}'(\boldsymbol{x}) = \mathring{\boldsymbol{v}}''(\boldsymbol{x}) \quad \text{for every } \boldsymbol{x} \in V.$$

(c) The boundary conditions of the two elastodynamic states satisfy the condition

$$\int_S (\boldsymbol{t}' - \boldsymbol{t}'') \cdot (\dot{\boldsymbol{u}}' - \dot{\boldsymbol{u}}'') \mathrm{d}A = 0 \quad \text{on} \quad S \times T^+$$

Then

$$\mathscr{S}' = \mathscr{S}'' \text{ on } \bar{V} \times T^+$$

or

$$\boldsymbol{u}'(\boldsymbol{x}, t) = \boldsymbol{u}''(\boldsymbol{x}, t) \text{ and } \boldsymbol{\tau}'(\boldsymbol{x}, t) = \boldsymbol{\tau}''(\boldsymbol{x}, t) \text{ on } \bar{V} \times T^+.$$

Proof. By virtue of the linearity of the problem it is clear that the set of solutions defined by

$$\boldsymbol{u} = \boldsymbol{u}' - \boldsymbol{u}'', \quad \boldsymbol{\tau} = \boldsymbol{\tau}' - \boldsymbol{\tau}'' \tag{3.7}$$

will satisfy eqs. (3.5) and (3.6) with $\boldsymbol{f} \equiv 0$. In view of supposition (b), the initial conditions on $\boldsymbol{u}$ and $\dot{\boldsymbol{u}}$ are

$$\boldsymbol{u} = \dot{\boldsymbol{u}} \equiv 0 \qquad \text{for } t \leqq 0. \tag{3.8}$$

We now turn to the energy identity for a body with a quiescent past, which was derived as eq. (2.69) in section 2.8, and which states

$$\int_0^t \int_S \boldsymbol{t} \cdot \dot{\boldsymbol{u}} \,\mathrm{d}A\,\mathrm{d}s + \int_0^t \int_V \boldsymbol{f} \cdot \dot{\boldsymbol{u}} \,\mathrm{d}V\mathrm{d}s = K(t) + U(t),$$

where $K(t)$ and $U(t)$ are the kinetic and the strain energies, respectively. From supposition (c) it now follows that

$$\int_0^t \int_S \boldsymbol{t} \cdot \dot{\boldsymbol{u}} \,\mathrm{d}A\,\mathrm{d}s = 0.$$

Since we also have $\mathbf{f} \equiv 0$, it is concluded that

$$K(t)+U(t) \equiv 0.$$

Both the kinetic and the strain energies are, however, positive semidefinite, and thus

$$K = U \equiv 0.$$

The first of these implies

$$\dot{\boldsymbol{u}} \equiv 0.$$

Since $\boldsymbol{u}(\boldsymbol{x}, 0) = 0$, we conclude

$$\boldsymbol{u}(\boldsymbol{x}, t) \equiv 0.$$

Hence the two solutions are identical and the proof of the theorem is completed.

For a discussion of the uniqueness theorem for an unbounded domain we refer to the previously cited paper by Wheeler and Sternberg.

The boundary conditions that will lead to a unique solution are implied in supposition (c). Obviously, if either the tractions or the displacements are prescribed, (c) is satisfied and the solution is unique. These boundary conditions are stated under (i), (ii) and (iii) in section 2.5. It is noted, however, that mixed boundary conditions whereby over the whole or part of S mutually orthogonal displacements and surface tractions are prescribed will also lead to a unique solution.

3.3. The dynamic reciprocal identity

The dynamic reciprocal identity, sometimes also referred to as the dynamic Betti-Rayleigh theorem presents a relation between two elastodynamic states of the same body. The identity relates two sets of displacements and stresses both satisfying (3.5) and (3.6), but with possibly different distributions of body forces, different initial conditions and different boundary conditions. The present discussion of the dynamic reciprocal identity is based on the previously cited work of Wheeler and Sternberg.

The identity can be stated and proved in a convenient manner by employing some results of the theory of Riemann convolutions. Consider two continuous scalar functions $g(\boldsymbol{x}, t)$ and $h(\boldsymbol{x}, t)$

$$g(\boldsymbol{x}, t) \in \mathscr{C}(V \times T^+),\ h(\boldsymbol{x}, t) \in \mathscr{C}(V \times T^+),$$

where V is a region in space, and T^+ is the half-open interval $[0, \infty)$. The half-open interval $(-\infty, 0]$ is denoted by T^-. The Riemann convolution of

$g(\boldsymbol{x}, t)$ and $h(\boldsymbol{x}, t)$, which is denoted by $[g * h](\boldsymbol{x}, t)$, is then defined as

$$\begin{aligned}[g * h](\boldsymbol{x}, t) &= 0 && \text{for all} \quad (\boldsymbol{x}, t) \in V \times T^- \\ &= \int_0^t g(\boldsymbol{x}, t-s) h(\boldsymbol{x}, s) \mathrm{d}s && \text{for all} \quad (\boldsymbol{x}, t) \in V \times T^+ .\end{aligned} \tag{3.9}$$

For brevity we usually write

$$g * h \text{ instead of } [g * h](\boldsymbol{x}, t).$$

Eq. (3.9) represents the convolution of two scalar quantities.

The Riemann convolution of two vectors $\boldsymbol{u}(\boldsymbol{x}, t)$ and $\boldsymbol{u}'(\boldsymbol{x}, t)$ is defined as

$$\boldsymbol{u} * \boldsymbol{u}' = u_i * u_i' = u_1 * u_1' + u_2 * u_2' + u_3 * u_3' . \tag{3.10}$$

Similarly the convolution of a tensor $\boldsymbol{\tau}$ and a vector $\boldsymbol{u}$ is defined as

$$\boldsymbol{\tau} * \boldsymbol{u} = \tau_{ij} * u_j , \tag{3.11}$$

where the summation convention must, of course, be invoked.

A frequently useful property of Riemann convolutions is the property of commutativity,

$$g * h = h * g . \tag{3.12}$$

If $g(\boldsymbol{x}, t)$ and $h(\boldsymbol{x}, t)$ are at least once continuously differentiable with respect to time, then $g * h$ is also at least once continuously differentiable with respect to time. The analogous statement also holds with regard to differentiation with respect to a spatial coordinate. The derivatives may be expressed as

$$\frac{\partial}{\partial t}(g * h) = \dot{g} * h + g(\boldsymbol{x}, 0) h(\boldsymbol{x}, t) \tag{3.13}$$

$$\frac{\partial}{\partial x_i}(g * h) = g_{,i} * h + g * h_{,i} . \tag{3.14}$$

We now turn to the reciprocal identity for a bounded body B occupying the regular region V in space, with boundary S.

Theorem 3.2: Suppose

$$\begin{aligned}\mathscr{S} &= [\boldsymbol{u}, \ \boldsymbol{\tau}\ ] \in \mathscr{E}(\boldsymbol{f}, \ \rho, \lambda, \mu; \ \bar{V} \times T^+) \\ \mathscr{S}' &= [\boldsymbol{u}', \boldsymbol{\tau}'] \in \mathscr{E}(\boldsymbol{f}', \rho, \lambda, \mu; \ \bar{V} \times T^+),\end{aligned}$$

with surface tractions

$$t_i = \tau_{ji} n_j \quad \text{and} \quad t'_i = \tau'_{ji} n_j, \tag{3.15a, b}$$

respectively, and further for every $\boldsymbol{x} \in V$

$$\begin{aligned} \boldsymbol{u}(\boldsymbol{x}, 0) &= \mathring{\boldsymbol{u}}(\boldsymbol{x}), \qquad & \dot{\boldsymbol{u}}(\boldsymbol{x}, 0^+) &= \mathring{\boldsymbol{v}}(x) \\ \boldsymbol{u}'(\boldsymbol{x}, 0) &= \mathring{\boldsymbol{u}}'(\boldsymbol{x}), \qquad & \dot{\boldsymbol{u}}'(\boldsymbol{x}, 0^+) &= \mathring{\boldsymbol{v}}'(x). \end{aligned}$$

Then, for every $t > 0$

$$\begin{aligned} &\int_S \boldsymbol{t} * \boldsymbol{u}' \mathrm{d}A + \rho \int_V \{\boldsymbol{f} * \boldsymbol{u}' + \dot{\boldsymbol{u}}'(\boldsymbol{x}, t) \cdot \mathring{\boldsymbol{u}}(\boldsymbol{x}) + \boldsymbol{u}'(\boldsymbol{x}, t) \cdot \mathring{\boldsymbol{v}}(\boldsymbol{x})\} \mathrm{d}V \\ &\quad = \int_S \boldsymbol{t}' * \boldsymbol{u} \mathrm{d}A + \rho \int_V \{\boldsymbol{f}' * \boldsymbol{u} + \dot{\boldsymbol{u}}(\boldsymbol{x}, t) \cdot \mathring{\boldsymbol{u}}'(\boldsymbol{x}) + \boldsymbol{u}(\boldsymbol{x}, t) \cdot \mathring{\boldsymbol{v}}'(\boldsymbol{x})\} \mathrm{d}V. \end{aligned} \tag{3.16}$$

Proof: We consider a time t in the interval $(0, \infty)$ and we define the vector-valued function $\boldsymbol{p}(\boldsymbol{x}, t)$ through

$$p_i(\boldsymbol{x}, t) = \tau_{ji} * u'_j - \tau'_{ji} * u_j. \tag{3.17}$$

Since $p_i(\boldsymbol{x}, t)$ is continuously differentiable the divergence of $p_i(\boldsymbol{x}, t)$ may be written as

$$p_{i,i} = \tau_{ji,i} * u'_j + \tau_{ji} * u'_{j,i} - \tau'_{ji,i} * u_j - \tau'_{ji} * u_{j,i},$$

where (3.14) has been used. By employing the stress-equation of motion, and by splitting the terms $u'_{j,i}$ and $u_{j,i}$ in symmetric and antisymmetric parts, we find

$$\begin{aligned} p_{i,i} = {} & \rho(\ddot{u}_j * u'_j) - \rho(f_j * u'_j) + \tau_{ji} * \varepsilon'_{ji} \\ & - \rho(\ddot{u}'_j * u_j) + \rho(f'_j * u_j) - \tau'_{ji} * \varepsilon_{ji}, \end{aligned} \tag{3.18}$$

where ε_{ji} is the strain tensor. By employing Hooke's law and the distributivity and the commutativity of convolutions we easily find

$$\tau_{ji} * \varepsilon'_{ji} = \tau'_{ji} * \varepsilon_{ji}. \tag{3.19}$$

Application of the rule on differentiation with respect to time, see (3.13), twice in succession to $u_i * u'_i$ yields the result

$$\ddot{\boldsymbol{u}} * \boldsymbol{u}' - \ddot{\boldsymbol{u}}' * \boldsymbol{u} = \mathring{\boldsymbol{v}}' \cdot \boldsymbol{u} + \mathring{\boldsymbol{u}}' \cdot \dot{\boldsymbol{u}} - \mathring{\boldsymbol{v}} \cdot \boldsymbol{u}' - \mathring{\boldsymbol{u}} \cdot \dot{\boldsymbol{u}}' \quad \text{on} \quad V \times (0, \infty).$$

Substitution of this result and (3.19) into (3.18) yields in vector notation

$$\boldsymbol{\nabla} \cdot \boldsymbol{p} = \rho\{\boldsymbol{f}' * \boldsymbol{u} + \dot{\boldsymbol{u}} \cdot \mathring{\boldsymbol{u}}' + \boldsymbol{u} \cdot \mathring{\boldsymbol{v}}'\} - \rho\{\boldsymbol{f} * \boldsymbol{u}' + \dot{\boldsymbol{u}}' \cdot \mathring{\boldsymbol{u}} + \boldsymbol{u}' \cdot \mathring{\boldsymbol{v}}\}. \tag{3.20}$$

In view of the divergence theorem we have

$$\begin{aligned}\int_V \nabla \cdot \boldsymbol{p}\,\mathrm{d}V &= \int_S p_i n_i \mathrm{d}A \\ &= \int_S (t_j * u_j' - t_j' * u_j)\mathrm{d}A,\end{aligned} \tag{3.21}$$

where (3.17) and (3.15a, b) have been used. Substitution of (3.20) into the left-hand side of (3.21) completes the proof.

An extension of this theorem to unbounded bodies was presented by Wheeler and Sternberg.[3]

3.4. Scalar and vector potentials for the displacement field

3.4.1. *Displacement representation*

In section 2.10 it was shown that a vector field of the form

$$\boldsymbol{u}(\boldsymbol{x}, t) = \nabla\varphi + \nabla \wedge \boldsymbol{\psi} \tag{3.22}$$

satisfies the displacement equations of motion provided that $\varphi(\boldsymbol{x}, t)$ and $\boldsymbol{\psi}(\boldsymbol{x}, t)$ are solutions of wave equations with characteristic velocities c_L and c_T, respectively.

The question of the completeness of the representation (3.22) was raised by Clebsch, who asserted that every solution of the displacement equation of motion admits the representation (3.22). The work of Clebsch and others, particularly Duhem, was discussed by Sternberg.[4]

3.4.2. *Completeness theorem*

In the formulation and the proof of the completeness theorem we will include body forces. The completeness theorem may then be stated as

Theorem 3.3: Let $\boldsymbol{u}(\boldsymbol{x}, t)$ and $\boldsymbol{f}(\boldsymbol{x}, t)$ satisfy the conditions

$$\boldsymbol{u} \in \mathscr{C}^2(V \times T)$$
$$\boldsymbol{f} \in \mathscr{C}(V \times T),$$

and meet the equation

$$\mu\nabla^2\boldsymbol{u} + (\lambda+\mu)\nabla\nabla \cdot \boldsymbol{u} + \rho\boldsymbol{f} = \rho\ddot{\boldsymbol{u}} \tag{3.23}$$

[3] L. T. Wheeler and E. Sternberg, *ibid.*, p. 80.
[4] E. Sternberg, *Archive Rational Mechanics and Analysis* **6** (1960), 34.

in a region of space V and in a closed time interval T. Also let[5]

$$\boldsymbol{f} = c_L^2 \nabla F + c_T^2 \nabla \wedge \boldsymbol{G}. \tag{3.24}$$

Then there exists a scalar function $\varphi(\boldsymbol{x}, t)$ and a vector-valued function $\boldsymbol{\psi}(\boldsymbol{x}, t)$ such that $\boldsymbol{u}(\boldsymbol{x}, t)$ is represented by

$$\boldsymbol{u} = \nabla\varphi + \nabla \wedge \boldsymbol{\psi}, \tag{3.25}$$

where

$$\nabla \cdot \boldsymbol{\psi} = 0, \tag{3.26}$$

and where $\varphi(\boldsymbol{x}, t)$ and $\boldsymbol{\psi}(\boldsymbol{x}, t)$ satisfy the inhomogeneous wave equations

$$\nabla^2\varphi + F = \frac{1}{c_L^2}\ddot{\varphi} \quad \left(c_L^2 = \frac{\lambda+2\mu}{\rho}\right) \tag{3.27}$$

$$\nabla^2\boldsymbol{\psi} + \boldsymbol{G} = \frac{1}{c_T^2}\ddot{\boldsymbol{\psi}} \quad \left(c_T^2 = \frac{\mu}{\rho}\right). \tag{3.28}$$

Proof: We start by eliminating $\nabla^2\boldsymbol{u}$ from eq. (3.23) by employing the well-known vector identity

$$\nabla^2\boldsymbol{u} = \nabla\nabla \cdot \boldsymbol{u} - \nabla \wedge \nabla \wedge \boldsymbol{u}.$$

The displacement equation of motion then may be written in the form

$$\ddot{\boldsymbol{u}} = c_L^2 \nabla\nabla \cdot \boldsymbol{u} - c_T^2 \nabla \wedge \nabla \wedge \boldsymbol{u} + \boldsymbol{f}.$$

Integrating this equation twice with respect to t, we obtain

$$\boldsymbol{u} = c_L^2 \nabla \int_0^t \int_0^\tau (\nabla \cdot \boldsymbol{u}) \mathrm{d}s\, \mathrm{d}\tau - c_T^2 \nabla \wedge \int_0^t \int_0^\tau (\nabla \wedge \boldsymbol{u}) \mathrm{d}s\, \mathrm{d}\tau$$
$$+ \int_0^t \int_0^\tau \boldsymbol{f} \mathrm{d}s\, \mathrm{d}\tau + \mathring{\boldsymbol{v}}(\boldsymbol{x})t + \mathring{\boldsymbol{u}}(\boldsymbol{x}),$$

where $\mathring{\boldsymbol{u}}(\boldsymbol{x})$ and $\mathring{\boldsymbol{v}}(\boldsymbol{x})$ are the initial conditions on $\boldsymbol{u}(\boldsymbol{x}, t)$ and $\dot{\boldsymbol{u}}(\boldsymbol{x}, t)$, respectively. We now proceed to define

$$\varphi = c_L^2 \int_0^t \int_0^\tau (\nabla \cdot \boldsymbol{u}) \mathrm{d}s\, \mathrm{d}\tau + c_L^2 \int_0^t \int_0^\tau F \,\mathrm{d}s\, \mathrm{d}\tau + \dot{\varphi}_0(\boldsymbol{x})t + \varphi_0(\boldsymbol{x}) \tag{3.29}$$

$$\boldsymbol{\psi} = -c_T^2 \int_0^t \int_0^\tau (\nabla \wedge \boldsymbol{u}) \mathrm{d}s\, \mathrm{d}\tau + c_T^2 \int_0^t \int_0^\tau \boldsymbol{G} \,\mathrm{d}s\, \mathrm{d}\tau + \dot{\boldsymbol{\psi}}_0(\boldsymbol{x})t + \boldsymbol{\psi}_0(\boldsymbol{x}), \tag{3.30}$$

[5] It will be shown in the next section that any reasonably well-behaved vector $\boldsymbol{f}(\mathbf{x}, t)$ can be expressed in the form (3.24).

where eq. (3.24) and the following representations for $\mathring{\boldsymbol{v}}$ and $\mathring{\boldsymbol{u}}$ have been used:

$$\mathring{\boldsymbol{v}} = \nabla\dot{\varphi}_0+\nabla\wedge\dot{\boldsymbol{\psi}}_0 \tag{3.31}$$

$$\mathring{\boldsymbol{u}} = \nabla\varphi_0+\nabla\wedge\boldsymbol{\psi}_0. \tag{3.32}$$

It remains to be proven that φ and $\boldsymbol{\psi}$ satisfy eqs. (3.26)–(3.28). Differentiation of (3.29) and (3.30) twice with respect to time yields

$$\ddot{\varphi} = c_L^2\nabla\cdot\boldsymbol{u}+c_L^2 F \tag{3.33}$$

$$\ddot{\boldsymbol{\psi}} = -c_T^2\nabla\wedge\boldsymbol{u}+c_T^2\boldsymbol{G}. \tag{3.34}$$

By applying the $\nabla\cdot$ operation to (3.25) we obtain

$$\nabla\cdot\boldsymbol{u} = \nabla^2\varphi. \tag{3.35}$$

Substitution of (3.35) into (3.33) shows that φ satisfies (3.27). Next we apply the operation $\nabla\wedge$ to (3.25) to obtain

$$\begin{aligned}\nabla\wedge\boldsymbol{u} &= \nabla\wedge\nabla\wedge\boldsymbol{\psi}\\ &= -\nabla^2\boldsymbol{\psi}+\nabla\nabla\cdot\boldsymbol{\psi}.\end{aligned} \tag{3.36}$$

From (3.36) and (3.34) it is now evident that $\boldsymbol{\psi}$ will satisfy the wave equation (3.28) provided that

$$\nabla\cdot\boldsymbol{\psi} = 0. \tag{3.37}$$

This completes the proof of theorem 3.3. The proof as given here is essentially due to Somigliana.

The initial conditions on φ and $\boldsymbol{\psi}$ follow from eqs. (3.29) and (3.30). We have

$$\varphi(\boldsymbol{x},0) = \varphi_0(\boldsymbol{x}), \qquad \dot{\varphi}(\boldsymbol{x},0) = \dot{\varphi}_0(\boldsymbol{x}) \tag{3.38a, b}$$

$$\boldsymbol{\psi}(\boldsymbol{x},0) = \boldsymbol{\psi}_0(\boldsymbol{x}), \qquad \dot{\boldsymbol{\psi}}(\boldsymbol{x},0) = \dot{\boldsymbol{\psi}}_0(\boldsymbol{x}). \tag{3.39a, b}$$

It should also be pointed out that well-defined vectors such as $\boldsymbol{f}$, $\mathring{\boldsymbol{v}}$ and $\mathring{\boldsymbol{u}}$, which form the given data of a problem can always be resolved into the forms shown by eqs. (3.24), (3.31) and (3.32). The representation of a vector $\boldsymbol{p}$ by

$$\boldsymbol{p} = \nabla P+\nabla\wedge\boldsymbol{Q}, \tag{3.40}$$

where

$$\nabla\cdot\boldsymbol{Q} = 0, \tag{3.41}$$

is called the Helmholtz decomposition of the vector $\boldsymbol{p}$. The construction of the scalar P and the vector $\boldsymbol{Q}$ in terms of the vector $\boldsymbol{p}$ is discussed in the next section.

As a last comment on the completeness theorem it is noted from (3.30) and (3.41) that all that is required for the condition $\nabla \cdot \boldsymbol{\psi} = 0$ to be satisfied is that the vector potentials $\boldsymbol{G}$, $\boldsymbol{\psi}_0$ and $\dot{\boldsymbol{\psi}}_0$ are constructed as Helmholtz potentials according to the procedure which is discussed in the next section.

3.5. The Helmholtz decomposition of a vector

Let the vector $\boldsymbol{p}(\boldsymbol{x})$ be piecewise differentiable in a finite open region V of space. With each point of space we now associate the vector

$$\boldsymbol{W}(\boldsymbol{x}) = -\frac{1}{4\pi}\iiint_V \frac{\boldsymbol{p}(\xi)}{|\boldsymbol{x}-\xi|}\,\mathrm{d}V_\xi, \tag{3.42}$$

where

$$\mathrm{d}V_\xi = \mathrm{d}\xi_1\,\mathrm{d}\xi_2\,\mathrm{d}\xi_3,$$

and

$$|\boldsymbol{x}-\xi| = [(x_1-\xi_1)^2+(x_2-\xi_2)^2+(x_3-\xi_3)^2]^{\frac{1}{2}}.$$

It is well known that $\boldsymbol{W}(\boldsymbol{x})$ then satisfies the vector equation

$$\nabla^2 \boldsymbol{W} = \boldsymbol{p}(\boldsymbol{x}) \tag{3.43}$$

at interior points where $\boldsymbol{p}$ is continuous, and

$$\nabla^2 \boldsymbol{W} = 0$$

at points outside the region V.

Now we employ the vector identity

$$\nabla^2 \boldsymbol{W} = \nabla\nabla\cdot \boldsymbol{W} - \nabla\wedge\nabla\wedge \boldsymbol{W}. \tag{3.44}$$

Let $\boldsymbol{W}(\boldsymbol{x}, t)$ be defined by eq. (3.42). By using (3.43), eq. (3.44) can then be rewritten as

$$\boldsymbol{p} = \nabla[\nabla\cdot \boldsymbol{W}]+\nabla\wedge[-\nabla\wedge \boldsymbol{W}]. \tag{3.45}$$

Eq. (3.45) is of the form

$$\boldsymbol{p} = \nabla P+\nabla\wedge \boldsymbol{Q} \tag{3.46}$$

if we set

$$P = \nabla\cdot \boldsymbol{W} \tag{3.47}$$

$$\boldsymbol{Q} = -\nabla\wedge \boldsymbol{W}. \tag{3.48}$$

It can be shown that P and $\boldsymbol{Q}$ are everywhere definite and continuous, and are differentiable at interior points where $\boldsymbol{p}$ is continuous.

To prove the Helmholtz decomposition we have thus provided a recipe for the construction of P and $\boldsymbol{Q}$. Given the vector field $\boldsymbol{p}(\boldsymbol{x})$, the vector $\boldsymbol{W}(\boldsymbol{x})$ can be constructed from (3.42), whereupon $P(\boldsymbol{x})$ and $\boldsymbol{Q}(\boldsymbol{x})$ are obtained from (3.47) and (3.48), respectively. Since the divergence of a curl vanishes we observe that

$$\nabla \cdot \boldsymbol{Q} = 0. \tag{3.49}$$

The Helmholtz decomposition is also valid for an infinite domain provided that $\boldsymbol{p} = O(r^{-2})$, i.e., provided that $|\boldsymbol{p}|$ decreases to zero at large distances r from the origin at least as rapidly as a constant times r^{-2}.[6]

3.6. Wave motion generated by body forces

3.6.1. *Radiation*

A class of interesting problems is concerned with an initially undisturbed body which in its interior, and at a specified time, say $t = 0$, is subjected to external disturbances. The external disturbances give rise to wave motions propagating away from the disturbed region. Problems defined in this manner may be called radiation problems. A typical elastodynamic radiation problem concerns the motion generated in an unbounded medium by body forces distributed over a finite region V of the medium.

Wave motions in homogeneous, isotropic, linearly elastic media can be analyzed in a convenient manner by employing scalar and vector potentials for the displacement field, as discussed in sections 2.10 and 3.4. For the problem at hand it follows from eqs. (3.27) and (3.28) that the governing equations for the displacement potentials are classical wave equations with inhomogeneous terms and with homogeneous initial conditions. To determine the motion generated in an initially quiescent unbounded medium by a distribution of body forces it is thus only required to find solutions in unbounded domains for (at most) four inhomogeneous wave equations.

Let us consider the scalar potential $\varphi(\boldsymbol{x}, t)$ in some detail. We need to determine the solution of the inhomogeneous wave equation

$$\nabla^2\varphi - \frac{1}{c_L^2}\frac{\partial^2\varphi}{\partial t^2} = -F(\boldsymbol{x}, t), \tag{3.50}$$

with the homogeneous initial conditions

[6] Cf. H. B. Phillips, *Vector analysis*. New York, John Wiley & Sons, Inc. (1933), p. 187.

$$\varphi(\boldsymbol{x}, 0) = 0, \qquad \dot{\varphi}(\boldsymbol{x}, 0) = 0. \tag{3.51a, b}$$

In the sequel it will be convenient to employ a generalized function, which is known as the Dirac delta function. In one dimension the delta function is defined by the property

$$\left.\begin{aligned}\int_a^b h(s)\delta(t-s)\mathrm{d}s &= h(t) \qquad \text{for} \quad t \in (a, b)\\ &= 0 \qquad\quad \text{for} \quad t \notin (a, b)\end{aligned}\right\}. \tag{3.52}$$

The delta function has any number of derivatives $\delta'(t-s)$, $\delta''(t-s)$, . . ., the prime indicating a derivative with respect to the argument. The derivatives are defined by the following property:

$$\left.\begin{aligned}\int_a^b h(s)\delta^{(n)}(t-s)\mathrm{d}s &= (-1)^n h^{(n)}(t) \qquad \text{for} \quad t \in (a, b)\\ &= 0 \qquad\qquad\qquad\quad \text{for} \quad t \notin (a, b)\end{aligned}\right\}. \tag{3.53}$$

In three-dimensional space the delta function $\delta|\boldsymbol{x}-\boldsymbol{\xi}|$ is defined in the following manner:

$$\left.\begin{aligned}\int_V h(\boldsymbol{\xi})\delta|\boldsymbol{x}-\boldsymbol{\xi}|\mathrm{d}V_\xi &= h(\boldsymbol{x}) \qquad \text{for} \quad \boldsymbol{x} \in V\\ &= 0 \qquad\quad \text{for} \quad \boldsymbol{x} \notin V\end{aligned}\right\}. \tag{3.54}$$

where V is a regular region in space, and $\mathrm{d}V_\xi = \mathrm{d}\xi_1\mathrm{d}\xi_2\mathrm{d}\xi_3$. In rectangular coordinates we may write

$$\delta|\boldsymbol{x}-\boldsymbol{\xi}| = \delta(x_1-\xi_1)\delta(x_2-\xi_2)\delta(x_3-\xi_3).$$

A delta function centered at the origin of the coordinate system may in spherical coordinates be represented by

$$\frac{\delta(r)}{4\pi r^2}, \qquad \text{where} \quad r = (x_1^2+x_2^2+x_3^2)^{\frac{1}{2}}. \tag{3.55}$$

The fundamental solution $\hat{\varphi}(r, t)$ describing radiation from a point, which we take as the origin of the coordinate system, is singular at $r = 0$ in such a way that

$$\lim_{\varepsilon\to 0}\iint \frac{\partial\hat{\varphi}}{\partial r}\,\mathrm{d}A = \mathrm{f}(t). \tag{3.56}$$

In eq. (3.56) the integration is carried out at time t, and over the surface of a sphere with radius ε, where $\mathrm{d}A$ is the surface element. The function $\mathrm{f}(t)$ is the intensity of radiation as a function of time. The fundamental solution satisfies homogeneous initial conditions and, except at the origin, satisfies the homogeneous wave equation in spherical coordinates with polar symmetry

$$\frac{1}{r^2}\frac{\partial}{\partial r}\left(r^2\frac{\partial\hat{\varphi}}{\partial r}\right)-\frac{1}{c_L^2}\frac{\partial^2\hat{\varphi}}{\partial t^2}=0, \qquad r>0. \tag{3.57}$$

A formulation which is equivalent to (3.57) together with the condition (3.56) is

$$\frac{1}{r^2}\frac{\partial}{\partial r}\left(r^2\frac{\partial\hat{\varphi}}{\partial r}\right)-\frac{1}{c_L^2}\frac{\partial^2\hat{\varphi}}{\partial t^2}=\mathrm{f}(t)\frac{\delta(r)}{4\pi r^2}, \tag{3.58}$$

or in rectangular coordinates

$$\nabla^2\hat{\varphi}-\frac{1}{c_L^2}\frac{\partial^2\hat{\varphi}}{\partial t^2}=\mathrm{f}(t)\delta|\boldsymbol{x}|. \tag{3.59}$$

The general solution of (3.57) can be obtained by introducing the substitution

$$\hat{\varphi}(r,t)=\frac{1}{r}\,\Phi(r,t).$$

We find that $\Phi(r, t)$ is governed by the one-dimensional wave equation

$$\frac{\partial^2\Phi}{\partial r^2}-\frac{1}{c_L^2}\frac{\partial^2\Phi}{\partial t^2}=0.$$

This equation represents one of the rare cases when one can immediately write out the solution in general form in terms of arbitrary functions. The general solution, which was given by d'Alembert, was derived in section 1.2 as

$$\Phi(r,t)=\mathrm{f}\left(t-\frac{r}{c_L}\right)+\mathrm{g}\left(t+\frac{r}{c_L}\right).$$

A general solution of (3.57) is thus obtained as

$$\hat{\varphi}(r,t)=\frac{1}{r}\,\mathrm{f}\left(t-\frac{r}{c_L}\right)+\frac{1}{r}\,\mathrm{g}\left(t+\frac{r}{c_L}\right).$$

The two terms represent an outgoing and an incoming wave, respectively, whose amplitudes steadily change.

For the radiation problem defined by (3.56) and (3.57) the wave obviously is an outgoing wave, and we consider

$$\hat{\varphi}(r, t) = -\frac{1}{4\pi r} \mathrm{f}\left(t - \frac{r}{c_L}\right), \tag{3.60}$$

where $\mathrm{f}(\tau) \equiv 0$ for $\tau < 0$. It is easily verified that (3.60) meets the condition (3.56). Consequently (3.60) is also the solution of the equivalent eqs. (3.58) and (3.59). Eq. (3.60) shows that a point at a distance r from the origin is at rest until time $t = r/c_L$, when the particle experiences a single impulse of the same duration as the external disturbance which was applied at the origin at time $t = 0$.

If the source point is located at $x_i = \xi_i$, we have instead of eq. (3.59)

$$\nabla^2 \hat{\varphi} - \frac{1}{c_L^2} \frac{\partial^2 \hat{\varphi}}{\partial t^2} = \mathrm{f}(t)\delta|\boldsymbol{x} - \boldsymbol{\xi}|. \tag{3.61}$$

It is now obvious that the solution of (3.61) is of the form

$$\hat{\varphi}(\boldsymbol{x}, t; \boldsymbol{\xi}) = -\frac{1}{4\pi|\boldsymbol{x} - \boldsymbol{\xi}|} \mathrm{f}\left(t - \frac{|\boldsymbol{x} - \boldsymbol{\xi}|}{c_L}\right), \tag{3.62}$$

where

$$|\boldsymbol{x} - \boldsymbol{\xi}| = [(x_1 - \xi_1)^2 + (x_2 - \xi_2)^2 + (x_3 - \xi_3)^2]^{\frac{1}{2}}. \tag{3.63}$$

If $\mathrm{f}(t)$ is a delta function applied at $t = s$, i.e., $\mathrm{f}(t) = \delta(t - s)$, the corresponding solution evidently is

$$g(\boldsymbol{x}, t; \boldsymbol{\xi}, s) = -\frac{1}{4\pi|\boldsymbol{x} - \boldsymbol{\xi}|} \delta\left(t - s - \frac{|\boldsymbol{x} - \boldsymbol{\xi}|}{c_L}\right). \tag{3.64}$$

In physical terms this expression is the field at time t at the point of observation $\boldsymbol{x}$ due to an impulsive unit point source applied at time s at the source point $\boldsymbol{\xi}$. Eq. (3.64) is called the Green's function for the unbounded domain. The field due to a distribution of sources $-F(\boldsymbol{x}, t)$ can be obtained by adding the effects due to each elementary portion of source. By this argument $\varphi(\boldsymbol{x}, t)$ takes the form of an integral over s and ξ_i of the product

$$-F(\boldsymbol{\xi}, s) g(\boldsymbol{x}, t; \boldsymbol{\xi}, s),$$

i.e.,

$$\varphi(\boldsymbol{x}, t) = \frac{1}{4\pi} \int_0^t \mathrm{d}s \int_V \frac{F(\boldsymbol{\xi}, s)}{|\boldsymbol{x} - \boldsymbol{\xi}|} \delta\left(t - s - \frac{|\boldsymbol{x} - \boldsymbol{\xi}|}{c_L}\right) \mathrm{d}V_\xi, \tag{3.65}$$

where V is the domain over which the distribution of body forces is defined. By employing (3.52), eq. (3.65) can be simplified to

$$\varphi(\boldsymbol{x}, t) = \frac{1}{4\pi} \int_{B_L} \frac{F(\boldsymbol{\xi}, t - |\boldsymbol{x} - \boldsymbol{\xi}|/c_L)}{|\boldsymbol{x} - \boldsymbol{\xi}|} \mathrm{d}V_\xi, \tag{3.66}$$

where $dV_\xi = d\xi_1 d\xi_2 d\xi_3$, and the integration is carried out over the sphere B_L with center at $\boldsymbol{x}$ and radius $c_L t$. A convenient shorthand notation for this integral expression is

$$\varphi(\boldsymbol{x}, t) = \frac{1}{4\pi}\int_{B_L} \frac{\{F\}_L}{|\boldsymbol{x}-\boldsymbol{\xi}|}\, dV_\xi . \tag{3.67}$$

The accolades indicate that we must take the value of the inhomogeneous term at time $t-|\boldsymbol{x}-\boldsymbol{\xi}|/c_L$. In physical terms this indicates that the effect of a source at $\boldsymbol{\xi}$ needs a finite time $|\boldsymbol{x}-\boldsymbol{\xi}|/c_L$ to reach the position $\boldsymbol{x}$. The expression (3.66) is called a *retarded potential.* It may be verified by direct substitution that (3.65) indeed satisfies eq. (3.50).

3.6.2. *Elastodynamic solution*

It is evident that in an analogous manner the solution of (3.28) can be expressed as

$$\psi(\boldsymbol{x}, t) = \frac{1}{4\pi}\int_{B_T} \frac{\{\boldsymbol{G}\}_T}{|\boldsymbol{x}-\boldsymbol{\xi}|}\, dV_\xi,$$

where the accolades indicate that we must take the value of $\boldsymbol{G}$ at time $t-|\boldsymbol{x}-\boldsymbol{\xi}|/c_T$, and B_T is the sphere with center at $\boldsymbol{x}$ and radius $c_T t$. In view of eq. (3.25), the displacement may then be written

$$4\pi\boldsymbol{u}(\boldsymbol{x}, t) = \int_{B_L} \nabla_x \frac{\{F\}_L}{|\boldsymbol{x}-\boldsymbol{\xi}|}\, dV_\xi + \int_{B_T} \nabla_x \wedge \frac{\{\boldsymbol{G}\}_T}{|\boldsymbol{x}-\boldsymbol{\xi}|}\, dV_\xi,$$

where the ∇-operator is with respect to the x_i-coordinates.

3.7. Radiation in two dimensions

The scheme of constructing the solution to the inhomogeneous problem as a superposition of impulses can also be used to treat the radiation problem in two dimensions. We will again focus our attention on the scalar potential which is governed by

$$\frac{\partial^2\varphi}{\partial x_1^2} + \frac{\partial^2\varphi}{\partial x_2^2} - \frac{1}{c_L^2}\frac{\partial^2\varphi}{\partial t^2} = -F(x_1, x_2, t) \tag{3.68}$$

and

$$\varphi(x_1, x_2, 0) = 0, \qquad \dot{\varphi}(x_1, x_2, 0) = 0. \tag{3.69}$$

Analogously to eq. (3.61) the fundamental solution describing radiation from a point source must meet the equation

$$\frac{\partial^2 \hat{\varphi}}{\partial x_1^2} + \frac{\partial^2 \hat{\varphi}}{\partial x_2^2} - \frac{1}{c_L^2}\frac{\partial^2 \hat{\varphi}}{\partial t^2} = \mathrm{f}(t)\delta(x_1-\xi_1)\delta(x_2-\xi_2). \tag{3.70}$$

In a convenient manner the solution of (3.70) can be constructed by linear superposition of the three-dimensional solution, which is defined by eq. (3.62). Thus we express the solution of (3.70) as

$$\hat{\varphi}(\boldsymbol{x}, t; \xi) = -\frac{1}{4\pi}\int_{-\infty}^{\infty} \mathrm{f}\left(t - \frac{|\boldsymbol{x}-\xi|}{c_L}\right)\frac{\mathrm{d}\xi_3}{|\boldsymbol{x}-\xi|}, \tag{3.71}$$

where $|\boldsymbol{x}-\xi|$ is defined by eq. (3.63). Since $\mathrm{f}(\tau) \equiv 0$ for $\tau < 0$, the actual limits of integration are

$$\xi_3 = x_3 \pm [c_L^2 t^2 - (x_1-\xi_1)^2 - (x_2-\xi_2)^2]^{\frac{1}{2}}.$$

In view of the symmetry with respect to x_3, eq. (3.71) may then also be written as twice the integral between the limits $x_3 - [c_L^2 t^2 - (x_1-\xi_1)^2 - (x_2-\xi_2)^2]^{\frac{1}{2}}$ and x_3. Upon introducing the new variable

$$\zeta = t - \frac{[R^2 + (x_3-\xi_3)^2]^{\frac{1}{2}}}{c_L},$$

where

$$R = [(x_1-\xi_1)^2 + (x_2-\xi_2)^2]^{\frac{1}{2}},$$

eq. (3.71) subsequently reduces to

$$\hat{\varphi}(x_i, t; \xi_i) = -\frac{c_L}{2\pi}\int_0^{t-R/c_L} \frac{\mathrm{f}(\zeta)\mathrm{d}\zeta}{[c_L^2(t-\zeta)^2 - R^2]^{\frac{1}{2}}} \tag{3.72}$$

for $t > R/c_L$. Note that $\hat{\varphi}$ is independent of x_3.

Comparing the fundamental solutions (3.62) and (3.72), we note a basic difference in the manner in which localized external disturbances generate the fields in three and two dimensions, respectively. In the three-dimensional field a particular position experiences a single impulse, of the same time-dependence as the external disturbance which was applied at the origin at time $t = 0$. The disturbance arrives at time

$$t = t_1 = [(x_1-\xi_1)^2 + (x_2-\xi_2)^2 + (x_3-\xi_3)^2]^{\frac{1}{2}}/c_L,$$

and the material returns to the quiescent state at time t_1 after the external disturbance at the origin has been removed. The two-dimensional solution does not depend on an external impulse applied at a single point, but rather on an infinite number of points which form the line defined by $x_1 = \xi_1$, $x_2 = \xi_2$, $-\infty < \xi_3 < \infty$. As a consequence, the disturbance experienced at a position takes the form of a superposition integral as shown in eq. (3.72). The disturbance arrives at time $t = t_1 = [x_1-\xi_1)^2+(x_2-\xi_2)^2]^{\frac{1}{2}}/c_L$, but does not cease at time t_1 after the external disturbance has been removed. For a long time there will be a tail effect decaying with time.

Graphically, an explanation of the different effects of a point-disturbance and a line-disturbance, i.e., the difference in radiation in three and two dimensions, is shown in figure 3.1. The observation that in three dimensions an external disturbance of finite duration is observed later at a different point as an effect which timewise is equally delimited, was made by Huyghens.

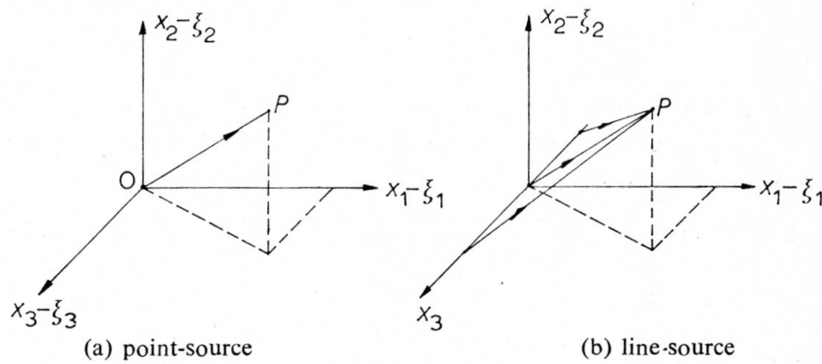

(a) point-source (b) line-source

Fig. 3.1. Radiation from a point-source and a line-source.

It is known as Huyghens' principle. In two dimensions, an external disturbance of finite duration produces a persisting signal, although of decaying amplitude, implying that Huyghens' principle is not valid in two dimensions.

If $\mathrm{f}(t)$ is a delta function applied at $t = s$, i.e., $\mathrm{f}(t) = \delta(t-s)$, eq. (3.72) reduces to the two-dimensional Green's function

$$g(x_i, t; \xi_i, s) = -\frac{c_L}{2\pi}\frac{H(t-s-R/c_L)}{[c_L^2(t-s)^2-R^2]^{\frac{1}{2}}}, \tag{3.73}$$

where $H(\ )$ is the Heaviside step function. This expression represents the field at time t at position x_1, x_2 due to an impulsive unit line source applied at time s at $\xi_1, \xi_2, -\infty < \xi_3 < \infty$. By linear superposition the solution to eq. (3.68) may then be written as

$$\varphi(x_1, x_2, t) = \frac{c_L}{2\pi}\iiint_K \frac{F(\xi_1, \xi_2, s)\mathrm{d}\xi_1\,\mathrm{d}\xi_2\,\mathrm{d}s}{[c_L^2(t-s)^2-(x_1-\xi_1)^2-(x_2-\xi_2)^2]^{\frac{1}{2}}}, \tag{3.74}$$

where K is a cone in (ξ_1, ξ_2, s)-space defined by

$$0 \leqq s \leqq t, \qquad (x_1-\xi_1)^2+(x_2-\xi_2)^2 \leqq c_L^2(t-s)^2. \tag{3.75}$$

It hardly needs to be stated that the components of the vector potential can be computed in the same manner.

3.8. The basic singular solution of elastodynamics

3.8.1. Point load

In this section we will examine the displacements and the stresses in an unbounded medium due to the sudden application of a rather special distribution of body forces, namely, a time-dependent point load. These displacements and stresses comprise the fundamental singular solution of the field equations of elastodynamics. The displacements generated by a time-dependent point load were first presented by Stokes.[7]

Stokes' solution can be deduced by means of a limit process from a family of time-dependent body-force fields that tends to a point load. Here we present a somewhat more formal approach in that we immediately view the point load as a special distribution of body forces per unit mass, namely, as a distribution in the form of a Dirac delta function. In three-dimensional space the delta function $\delta|\boldsymbol{x}-\boldsymbol{\xi}|$ was defined by eq. (3.54).

In the presence of a system of body forces per unit mass the displacement equations of motion become

$$\mu\nabla^2\boldsymbol{u}+(\lambda+\mu)\nabla\nabla\cdot\boldsymbol{u}+\rho\boldsymbol{f} = \rho\ddot{\boldsymbol{u}}. \tag{3.76}$$

We consider a concentrated load of magnitude $g(t)$, where $g(t)$ is twice continuously differentiable. The load is directed along the constant unit vector $\boldsymbol{a}$. Without loss of generality we may place the origin of the coordinate system at the point of application of the concentrated load. In that case we have

$$\boldsymbol{f}(\boldsymbol{x}, t) = \boldsymbol{a}g(t)\delta|\boldsymbol{x}|. \tag{3.77}$$

In accordance with eq. (3.24) we wish to decompose the body-force vector as

$$\boldsymbol{f} = c_L^2\nabla F+c_T^2\nabla\wedge\boldsymbol{G}.$$

The decomposition can be achieved by means of the Helmholtz resolution which was discussed in section 3.5. Thus, employing eqs. (3.47), (3.48) and (3.42), we find

$$c_L^2 F(\boldsymbol{x}, t) = -\nabla\cdot\left(\frac{\boldsymbol{a}}{4\pi r}\right) g(t)$$

and

[7] G. G. Stokes, *Transactions of the Cambridge Philosophical Society* **9** (1849) 1.

$$c_T^2 \boldsymbol{G}(\boldsymbol{x}, t) = \nabla \wedge \left(\frac{\boldsymbol{a}}{4\pi r}\right) g(t),$$

where

$$r^2 = x_1^2 + x_2^2 + x_3^2. \tag{3.78}$$

Writing the displacement vector in the form (3.25), i.e.,

$$\boldsymbol{u}(\boldsymbol{x}, t) = \nabla\varphi + \nabla \wedge \boldsymbol{\psi}, \tag{3.79}$$

it follows from eqs. (3.27) and (3.28) that $\varphi(\boldsymbol{x}, t)$ and $\boldsymbol{\psi}(\boldsymbol{x}, t)$ must satisfy the inhomogeneous wave equations

$$\nabla^2\varphi - \frac{1}{c_L^2}\ddot{\varphi} = \frac{g(t)}{c_L^2}\nabla \cdot \left(\frac{\boldsymbol{a}}{4\pi r}\right) \tag{3.80}$$

$$\nabla^2\boldsymbol{\psi} - \frac{1}{c_T^2}\ddot{\boldsymbol{\psi}} = -\frac{g(t)}{c_T^2}\nabla \wedge \left(\frac{\boldsymbol{a}}{4\pi r}\right). \tag{3.81}$$

For convenience we define

$$\varphi = \nabla \cdot \boldsymbol{\Phi}, \qquad \boldsymbol{\psi} = -\nabla \wedge \boldsymbol{\Psi}. \tag{3.82a, b}$$

By setting

$$\boldsymbol{\Phi} = \Phi \boldsymbol{a}, \qquad \boldsymbol{\Psi} = \Psi \boldsymbol{a}, \tag{3.83a, b}$$

Eqs. (3.80) and (3.81) are satisfied if Φ and Ψ are solutions of the following inhomogeneous scalar wave equations

$$\nabla^2\Phi - \frac{1}{c_L^2}\ddot{\Phi} = \frac{g(t)}{c_L^2}\frac{1}{4\pi r} \tag{3.84}$$

$$\nabla^2\Psi - \frac{1}{c_T^2}\ddot{\Psi} = \frac{g(t)}{c_T^2}\frac{1}{4\pi r}, \tag{3.85}$$

where r is defined by (3.78).

Appropriate solutions of (3.84) and (3.85) are readily derived. Since the inhomogeneous term shows polar symmetry, the solution is most conveniently obtained by employing spherical coordinates. Eq. (3.84) may then be rewritten as

$$\frac{1}{r^2}\frac{\partial}{\partial r}\left(r^2\frac{\partial \Phi}{\partial r}\right) - \frac{1}{c_L^2}\frac{\partial^2\Phi}{\partial t^2} = \frac{g(t)}{4\pi c_L^2}\frac{1}{r}. \tag{3.86}$$

Now we introduce the substitution

$$\Phi(r, t) = \frac{\chi(r, t)}{r}, \tag{3.87}$$

whereupon for $r > 0$ eq. (3.86) reduces to the inhomogeneous one-dimensional wave equation

$$\frac{\partial^2 \chi}{\partial r^2} - \frac{1}{c_L^2}\frac{\partial^2 \chi}{\partial t^2} = \frac{g(t)}{4\pi c_L^2}. \tag{3.88}$$

A particular solution of this equation is

$$\chi_p(t) = -\frac{1}{4\pi}\int_0^t s g(t-s)\mathrm{d}s,$$

where we have used that $g(t) \equiv 0$ for $t < 0$. The general solutions of the homogeneous equation are arbitrary functions of the arguments $t - r/c_L$ and $t + r/c_L$, respectively. Clearly the solution of (3.88) satisfying the conditions that $\chi(r, t)$ vanishes at $r = 0$ may then be written

$$4\pi\chi(r, t) = \int_0^{t \pm r/c_L} s g\left(t \pm \frac{r}{c_L} - s\right)\mathrm{d}s - \int_0^t s g(t-s)\mathrm{d}s.$$

The plus signs correspond to a wave approaching $r = 0$; the minus signs correspond to a wave propagating away from the point $r = 0$. For the problem at hand we discard the waves converging on $r = 0$, and returning to (3.87), the pertinent solution of (3.86), and thus of (3.84), is obtained as

$$4\pi\boldsymbol{\Phi}(\boldsymbol{x}, t) = \frac{1}{r}\int_0^{t - r/c_L} s g\left(t - \frac{r}{c_L} - s\right)\mathrm{d}s - \frac{1}{r}\int_0^t s g(t-s)\mathrm{d}s, \tag{3.89}$$

where r is defined by eq. (3.78). The pertinent solution of (3.85) is analogously obtained as

$$4\pi\boldsymbol{\Psi}(\boldsymbol{x}, t) = \frac{1}{r}\int_0^{t - r/c_T} s g\left(t - \frac{r}{c_T} - s\right)\mathrm{d}s - \frac{1}{r}\int_0^t s g(t-s)\mathrm{d}s. \tag{3.90}$$

Since $g(t) \equiv 0$ for $t < 0$, the first integrals in (3.89) and (3.90) do not yield contributions for $t < r/c_L$ and $t < r/c_T$, respectively.

In view of (3.79) and (3.82a, b), the displacement vector is written as

$$\boldsymbol{u} = \nabla\nabla \cdot \boldsymbol{\Phi} - \nabla \wedge \nabla \wedge \boldsymbol{\Psi}.$$

By virtue of the vector identity

$$\nabla^2 \boldsymbol{\Psi} = \nabla\nabla \cdot \boldsymbol{\Psi} - \nabla \wedge \nabla \wedge \boldsymbol{\Psi}$$

the displacement vector can also be expressed as

$$\boldsymbol{u} = \nabla\nabla \cdot (\boldsymbol{\Phi} - \boldsymbol{\Psi}) + \nabla^2 \boldsymbol{\Psi}. \tag{3.91}$$

Substitution of (3.89) and (3.90) into (3.91) yields

$$4\pi \boldsymbol{u}(\boldsymbol{x}, t) = \nabla\nabla \cdot \left\{ \frac{\boldsymbol{a}}{r} \int_0^{t-r/c_L} sg\left(t - \frac{r}{c_L} - s\right) \mathrm{d}s - \frac{\boldsymbol{a}}{r} \int_0^{t-r/c_T} sg\left(t - \frac{r}{c_T} - s\right) \mathrm{d}s \right\} + \frac{\boldsymbol{a}}{c_T^2} \frac{1}{r} g\left(t - \frac{r}{c_T}\right). \tag{3.92}$$

Suppose now that the concentrated load is acting in the direction of the x_k-axis, i.e., $\boldsymbol{a} = \boldsymbol{i}_k$. The displacement in the direction of the x_i-axis is then of the form

$$4\pi u_i(\boldsymbol{x}, t) = D_i^k[\boldsymbol{0}; g(t)], \tag{3.93}$$

where $D_i^k[\;]$ is a linear operator, wherein $\boldsymbol{0}$ indicates that the load is applied at the origin, and $g(t)$ defines the function subjected to the operation. The linear operator $D_i^k[\;]$ follows from (3.92) as

$$D_i^k[\boldsymbol{0}; g(t)] = \frac{\partial}{\partial x_i} \frac{\partial}{\partial x_k} \left\{ \frac{1}{r} \int_0^{t-r/c_L} sg\left(t - \frac{r}{c_L} - s\right) \mathrm{d}s - \frac{1}{r} \int_0^{t-r/c_T} sg\left(t - \frac{r}{c_T} - s\right) \mathrm{d}s \right\} + \frac{\delta_{ik}}{c_T^2} \frac{1}{r} g\left(t - \frac{r}{c_T}\right), \tag{3.94}$$

where r is defined as $r = (x_1^2 + x_2^2 + x_3^2)^{\frac{1}{2}}$, see eq. (3.78). After a few manipulations, including changes of the integration variables, the operator may be rewritten as

$$D_i^k[\boldsymbol{0}; g(t)] = \frac{\partial^2 r^{-1}}{\partial x_i \partial x_k} \int_{r/c_L}^{r/c_T} sg(t-s)\mathrm{d}s + \frac{1}{2} \left(\frac{\partial r}{\partial x_i}\right)\left(\frac{\partial r}{\partial x_k}\right) \left[\frac{1}{c_L^2} g\left(t - \frac{r}{c_L}\right) - \frac{1}{c_T^2} g\left(t - \frac{r}{c_T}\right)\right] + \frac{\delta_{ik}}{c_T^2} \frac{1}{r} g\left(t - \frac{r}{c_T}\right). \tag{3.95}$$

For a force acting parallel to the x_1-axis this expression yields the same displacement as given by Love.[8]

The position dependence of the integration limits can be eliminated through a change of the integration variable. Also evaluating the derivatives (3.95) may then be rewritten as

[8] A. E. H. Love, *The mathematical theory of elasticity*. New York, Dover Publications (1944), p. 305.

$$D_i^k[\mathbf{0}; g(t)] = \left(\frac{3x_i x_k}{r^3} - \frac{\delta_{ik}}{r}\right)\int_{1/c_L}^{1/c_T} sg(t-rs)\mathrm{d}s$$

$$+\frac{x_i x_k}{r^3}\left[\frac{1}{c_L^2} g\left(t-\frac{r}{c_L}\right) - \frac{1}{c_T^2} g\left(t-\frac{r}{c_T}\right)\right] + \frac{\delta_{ik}}{c_T^2}\frac{1}{r} g\left(t-\frac{r}{c_T}\right). \quad (3.96)$$

The corresponding stresses can be computed by the use of Hooke's law. We obtain

$$4\pi\tau_{ij}(\mathbf{x}, t) = S_{ij}^k[\mathbf{0}; g(t)], \quad (3.97)$$

where the operator $S_{ij}^k[\ ]$ is given by

$$\rho^{-1}S_{ij}^k[\mathbf{0}; g(t)] = -6c_T^2\left[\frac{5x_i x_j x_k}{r^5} - \frac{\delta_{ij}x_k + \delta_{ik}x_j + \delta_{jk}x_i}{r^3}\right]\int_{1/c_L}^{1/c_T} sg(t-rs)\mathrm{d}s$$

$$+2\left[\frac{6x_i x_j x_k}{r^5} - \frac{\delta_{ij}x_k + \delta_{ik}x_j + \delta_{jk}x_i}{r^3}\right]\left[g\left(t-\frac{r}{c_T}\right) - \left(\frac{c_T}{c_L}\right)^2 g\left(t-\frac{r}{c_L}\right)\right]$$

$$+\frac{2x_i x_j x_k}{r^4 c_T}\left[\dot{g}\left(t-\frac{r}{c_T}\right) - \left(\frac{c_T}{c_L}\right)^3 \dot{g}\left(t-\frac{r}{c_L}\right)\right]$$

$$-\frac{x_k\delta_{ij}}{r^3}\left[1-2\left(\frac{c_T}{c_L}\right)^2\right]\left[g\left(t-\frac{r}{c_L}\right) + \frac{r}{c_L}\dot{g}\left(t-\frac{r}{c_L}\right)\right]$$

$$-\frac{\delta_{ik}x_j + \delta_{jk}x_i}{r^3}\left[g\left(t-\frac{r}{c_T}\right) + \frac{r}{c_T}\dot{g}\left(t-\frac{r}{c_T}\right)\right]. \quad (3.98)$$

The displacement and stress fields were listed in the forms (3.96) and (3.98) by Wheeler and Sternberg.[9]

If the point load is applied at $\mathbf{x} = \xi$ rather than at the origin we have instead of (3.77).

$$\mathbf{f}(\mathbf{x}, t) = \mathbf{a}g(t)\delta|\mathbf{x}-\xi|.$$

It is apparent that for a load in the x_k direction the displacement and stress fields can be obtained from (3.93) and (3.97) by writing $x_i-\xi_i$ instead of x_i in the operators $D_i^k[\ ]$ and $S_{ij}^k[\ ]$, and by defining r as

$$r = |\mathbf{x}-\xi| = [(x_1-\xi_1)^2+(x_2-\xi_2)^2+(x_3-\xi_3)^2]^{\frac{1}{2}}.$$

These solutions are denoted by

$$4\pi u_i(\mathbf{x}, t) = D_i^k[\xi; g(t)] \quad (3.99)$$

and

$$4\pi\tau_{ij}(\mathbf{x}, t) = S_{ij}^k[\xi; g(t)]. \quad (3.100)$$

[9] L. T. Wheeler and E. Sternberg, *loc. cit.*, p. 80.

3.8.2. *Center of compression*

Let a force of magnitude $h^{-1}\rho g(t)$ be applied at the origin in the direction of the x_1-axis, and let an equal and opposite force be applied at the point $x_1 = h$, $x_2 = x_3 = 0$.

If we pass to the limit by supposing that h is diminished indefinitely, the components of the displacements become the derivatives with respect to x_1 of eq. (3.93),

$$4\pi u_i(\boldsymbol{x}, t) = \frac{\partial}{\partial x_1} \{D_i^1[\boldsymbol{0}; g(t)]\}$$

These displacements may be referred to as the displacements due to a "double force without moment". Now we combine three double forces without moment, applied in the directions of x_1, x_2 and x_3, respectively. The components of the total displacement are

$$4\pi u_i(\boldsymbol{x}, t) = \frac{\partial}{\partial x_1} \{D_i^1[\boldsymbol{0}; g(t)]\} + \frac{\partial}{\partial x_2} \{D_i^2[\boldsymbol{0}; g(t)]\} + \frac{\partial}{\partial x_3} \{D_i^3[\boldsymbol{0}; g(t)]\}. \tag{3.101}$$

From (3.94) we can, however, easily conclude that

$$D_i^2[\boldsymbol{0}; g(t)] = D_2^i[\boldsymbol{0}; g(t)],$$

and (3.101) may be replaced by

$$4\pi u_i(\boldsymbol{x}, t) = \frac{\partial}{\partial x_1} \{D_1^i[\boldsymbol{0}; g(t)]\} + \frac{\partial}{\partial x_2} \{D_2^i[\boldsymbol{0}; g(t)]\} + \frac{\partial}{\partial x_3} \{D_3^i[\boldsymbol{0}; g(t)]\}. \tag{3.102}$$

The displacement component $u_i(\boldsymbol{x}, t)$ defined in this manner is recognized as the divergence of the displacement vector for a concentrated load acting in the direction of x_i. From (3.79) we conclude

$$\nabla \cdot \boldsymbol{u} = \nabla^2 \varphi.$$

By employing (3.82a) and (3.83a) we then find the divergence due to the concentrated load in the x_i-direction as

$$\nabla \cdot \boldsymbol{u} = \nabla^2 \left(\frac{\partial \Phi}{\partial x_i}\right) = \frac{\partial}{\partial x_i} \nabla^2 \Phi.$$

Substitution of (3.89) subsequently yields

$$\nabla \cdot \boldsymbol{u} = \frac{1}{4\pi c_L^2} \frac{\partial}{\partial x_i} \left[\frac{1}{r} g\left(t - \frac{r}{c_L}\right)\right],$$

and thus (3.102) becomes

$$u_i = \frac{1}{4\pi c_L^2} \frac{\partial}{\partial x_i} \left[\frac{1}{r} g \left(t - \frac{r}{c_L} \right) \right] . \tag{3.103}$$

Eq. (3.103) represents the displacement components due to a center of compression.

3.9. Three-dimensional integral representation

The object of this section is to obtain an integral representation for the elastodynamic displacement distribution analogous to Kirchhoff's formula for the wave equation.

3.9.1. Kirchhoff's formula

For a regular domain V bounded by a surface S, a general solution of the inhomogeneous wave equation

$$\nabla^2 \varphi - \frac{1}{c^2} \frac{\partial^2 \varphi}{\partial t^2} = -\mathrm{f}(\boldsymbol{x}, t)$$

was obtained by Kirchhoff in the form[10]

$$\varphi(\boldsymbol{x}, t) = \frac{1}{4\pi} \int_V \frac{1}{|\boldsymbol{x}-\boldsymbol{\xi}|} \{\mathrm{f}\} \mathrm{d}V_\xi$$
$$+ \frac{1}{4\pi} \int_S \left[\frac{1}{|\boldsymbol{x}-\boldsymbol{\xi}|} \left\{ \frac{\partial \varphi}{\partial n_\xi} \right\} - \{\varphi\} \frac{\partial}{\partial n_\xi} \frac{1}{|\boldsymbol{x}-\boldsymbol{\xi}|} + \frac{1}{c} \frac{1}{|\boldsymbol{x}-\boldsymbol{\xi}|} \left\{ \frac{\partial \varphi}{\partial t} \right\} \frac{\partial |\boldsymbol{x}-\boldsymbol{\xi}|}{\partial n_\xi} \right] \mathrm{d}A_\xi ,$$

where n_ξ is the outward unit normal to S, and $\boldsymbol{x}$ is, of course, inside the domain V. Also, the accolade notation was defined earlier with reference to eq. (3.67); thus $\{\partial\varphi/\partial n_\xi\}$ means calculate $(\partial/\partial n_\xi)\varphi(\xi, s)$ and then replace s by the retarded time $t - |\boldsymbol{x}-\boldsymbol{\xi}|/c$. The volume integral in the Kirchhoff formula is analogous to eq. (3.67) and represents the contribution to the scalar potential φ of the distribution of sources inside V. The surface integral provides the contribution of sources outside V that are necessary to yield the required conditions on S. If φ and its derivatives are known on S, φ is completely determined. It is, of course, not possible to assign both φ and $\partial\varphi/\partial n_\xi$ arbitrarily on the surface S.

[10] Proofs can be found in B. B. Baker and E. T. Copson, *The mathematical theory of Huygen's principle*. London, Oxford University Press (1953), p. 36, or D. S. Jones, *The theory of electromagnetism*. London, The Macmillan Company (1964), p. 40.

3.9.2. *Elastodynamic representation theorem*

Following the work of Wheeler and Sternberg[11] the theorem is derived by an application of the dynamic reciprocal identity to suitably chosen elastodynamic states.

Theorem 3.4 (integral identity for the displacement field): Let V be a regular region with boundary S. Suppose

$$\mathscr{S} = [\boldsymbol{u}, \boldsymbol{\tau}] \in \mathscr{E}_0(\boldsymbol{f}, \rho, \lambda, \mu; \bar{V}). \tag{3.104}$$

Then the displacement components can be expressed in terms of the tractions and the displacements on S and the body forces in V, by

$$4\pi\rho u_k(\xi, t) = \sum_{i=1}^{3} \int_S \{D_i^k[\xi; t_i(\boldsymbol{x}, t)] - n_j S_{ij}^k[\xi; u_i(\boldsymbol{x}, t)]\} \mathrm{d}A + \rho \sum_{i=1}^{3} \int_V D_i^k[\xi; f_i(\boldsymbol{x}, t)] \mathrm{d}V, \tag{3.105}$$

where the integrations are carried out over x_1, x_2 and x_3, and where the operators $D_i^k[\ ]$ and $S_{ij}^k[\ ]$ are defined by eqs. (3.96), (3.99) and (3.98), (3.100), respectively.

Proof: As point of departure we take the dynamic reciprocal identity which was stated by eq. (3.16). We consider two elastodynamic states, namely, the state defined by eq. (3.104) and the state defined by

$$u_i' = D_i^k[\xi; g(t)] \tag{3.106}$$

$$\tau_{ij}' = S_{ij}^k[\xi; g(t)]. \tag{3.107}$$

The latter distribution of displacements and stresses is due to a concentrated load of magnitude $g(t)$, acting in the x_k-direction, and applied at the position $\boldsymbol{x} = \xi$ at time $t = 0$. Since both states have a quiescent past the terms containing the initial values disappear from the dynamic reciprocal identity, and we can write

$$\int_S \tau_{ij} n_j * u_i' \mathrm{d}A + \rho \int_V f_i * u_i' \mathrm{d}V = \int_S \tau_{ij}' n_j * u_i \mathrm{d}A + \rho \int_V f_i' * u_i \mathrm{d}V. \tag{3.108}$$

The body-force distribution $\boldsymbol{f}'(\boldsymbol{x}, t)$ is, however, of the special form

$$\boldsymbol{f}'(\boldsymbol{x}, t) = 4\pi \boldsymbol{i}_k g(t) \delta|\boldsymbol{x} - \xi|.$$

By employing the sifting property of the Dirac delta function the last term

[11] L. T. Wheeler and E. Sternberg, *loc. cit.*, p. 80.

of (3.108) can then immediately be evaluated, yielding the result

$$4\pi\rho[g * u_k](\xi, t) = \rho \int_V f_i * u_i' \mathrm{d}V + \int_S [\tau_{ij} n_j * u_i' - \tau_{ij}' n_j * u_i] \mathrm{d}A.$$

We introduce the operator relations (3.106) and (3.107) to obtain

$$4\pi\rho[g * u_k](\xi, t) = \rho \int_V f_i * D_i^k[\xi; g(t)]\, \mathrm{d}V + \int_S \left\{ t_i * D_i^k[\xi; g(t)] - n_j u_i * S_{ij}^k[\xi; g(t)] \right\} \mathrm{d}A. \tag{3.109}$$

If the function $g(t)$ is twice continuously differentiable we have

$$f_i * D_i^k[\xi; g(t)] = \sum_{i=1}^{3} g * D_i^k[\xi; f_i(\boldsymbol{x}, t)]. \tag{3.110}$$

This equality can readily be verified by employing eq. (3.96) and by introducing appropriate changes of the integration variables in the convolution integrals. Similarly we have

$$t_i * D_i^k[\xi; g(t)] = \sum_{i=1}^{3} g * D_i^k[\xi; t_i(\boldsymbol{x}, t)] \tag{3.111}$$

$$u_i * S_{ij}^k[\xi; g(t)] = \sum_{i=1}^{3} g * S_{ij}^k[\xi; u_i(\boldsymbol{x}, t)]. \tag{3.112}$$

Eqs. (3.110)–(3.112) are substituted into (3.109). In the resulting integrals the order of the spatial integrations and the convolutions can be interchanged, and we obtain

$$[g * w_k](\xi, t) = 0, \tag{3.113}$$

where

$$w_k(\xi, t) = 4\pi\rho u_k(\xi, t) - \rho \sum_{i=1}^{3} \int_V D_i^k[\xi; f_i(\boldsymbol{x}, t)] \mathrm{d}V - \sum_{i=1}^{3} \int_S \{D_i^k[\xi; t_i(\boldsymbol{x}, t)] - n_j S_{ij}^k[\xi; u_i(\boldsymbol{x}, t)]\} \mathrm{d}A. \tag{3.114}$$

Since ξ and t were chosen arbitrarily in $V \times (0, \infty)$, eq. (3.114) holds for all $(\xi, t) \in V \times (0, \infty)$. Since the terms in the convolution (3.113) are continuous, it follows that either $g(t)$ or $w_k(\xi, t)$ must vanish. Since we have chosen $g(t) \neq 0$ we must have

$$w_k(\xi, t) \equiv 0,$$

and (3.105) immediately follows from (3.114). This completes the proof.

Eq. (3.105) is valid for any point of observation inside the region V. The first two terms involve the tractions and the displacements on the bounding surface S. The third term gives the displacement due to the distribution of body forces per unit mass of density $f_i(\mathbf{x}, t)$. The integral identity is the extension to elastodynamics of Kirchhoff's formula for the wave equation. The representation theorem (3.105) was apparently first derived in essentially the present form by De Hoop.[12]

The integral identity (3.105) is useful in the solution of certain elastodynamic diffraction problems in an unbounded medium. Since it involves *both* the surface displacements and the surface tractions on the boundary of the body, it is not suited to solve the first and the second fundamental boundary-initial value problems in classical elastodynamics, since the boundary data consist of prescribed surface displacements in the first problem and of prescribed surface tractions in the second problem. To obtain integral identities for the fundamental boundary-initial value problems it is thus necessary to eliminate from the integrals in (3.105) the surface tractions for the first problem and the surface displacements in connection with the second problem. This work was carried out in the previously cited paper by Wheeler and Sternberg. These authors proved the integral representations for the solutions of the first and the second boundary-initial value problems for states with a quiescent past.

3.10. Two-dimensional integral representations

If the field variables are independent of the x_3-coordinate, the displacement equations of motion reduce to the following system of uncoupled equations:

$$\mu u_{3,\beta\beta} + \rho f_3 = \rho \ddot{u}_3 \tag{3.115}$$

$$\mu u_{\alpha,\beta\beta} + (\lambda+\mu) u_{\beta,\beta\alpha} + \rho f_\alpha = \rho \ddot{u}_\alpha . \tag{3.116}$$

In eqs. (3.115) and (3.116), and throughout this section, Greek indices can assume the values 1 and 2 only. Eqs. (3.115) and (3.116), which describe antiplane shear motions and plane strain motions, respectively, were derived in section 2.7.

3.10.1. *Basic singular solutions*

In two dimensions the basic singular solutions are the displacements

[12] A. T. de Hoop, *Representation theorems for the displacement in an elastic solid and their application to elastodynamic diffraction theory*. Doctoral Dissertation, Delft (1958).

and the stresses due to the following distributions of body forces:

$$f_3(x_1, x_2, t) = g(t)\delta(x_1)\delta(x_2), \tag{3.117}$$

and

$$f_\alpha(x_1, x_2, t) = g(t)\delta(x_1)\delta(x_2), \tag{3.118}$$

where $\alpha = 1, 2$. We can, of course, directly determine the solutions of eqs. (3.115) and (3.116) for the body-force distributions (3.117) and (3.118), respectively. Because of the linearity of the problem it is, however, also possible to obtain the two-dimensional solution as a superposition integral over the solutions for the point load. Suppose a point load of magnitude $g(t)$ in the x_k-direction is applied at the position $x_1 = 0$, $x_2 = 0$, $x_3 = \xi_3$. According to eq. (3.99) the corresponding displacements may then be expressed as

$$4\pi u_i(\mathbf{x}, t) = D_i^k[\xi_3 \mathbf{i}_3; g(t)], \tag{3.119}$$

where we may employ any of the forms (3.94), (3.95) or (3.96) for the operator $D_i^k[\ ]$, provided x_3 is replaced by $x_3 - \xi_3$. It should also be realized that r is now defined by

$$r = [x_1^2 + x_2^2 + (x_3 - \xi_3)^2]^{\frac{1}{2}}. \tag{3.120}$$

The solution to a line load is obtained by integrating (3.119) between the limits $\xi_3 = -\infty$ and $\xi_3 = +\infty$. As we shall see soon, the dependence on x_3 vanishes in this process. The two-dimensional solution again is of the form of an operator, which we denote by $\Delta_i^k[\ ]$, i.e.

$$4\pi u_i(x_1, x_2, t) = \Delta_i^k[\mathbf{0}; g(t)], \tag{3.121}$$

where the symbol $\mathbf{0}$ indicates that the line load is applied at $x_1 = 0$, $x_2 = 0$, and $\rho g(t)$ indicates the magnitude of the load. We have

$$\Delta_i^k[\mathbf{0}; g(t)] = \int_{-\infty}^{+\infty} D_i^k[\xi_3 \mathbf{i}_3; g(t)] \mathrm{d}\xi_3. \tag{3.122}$$

Similary the corresponding stresses are denoted by

$$4\pi\tau_{ij}(x_1, x_2, t) = \Sigma_{ij}^k[\mathbf{0}; g(t)], \tag{3.123}$$

where

$$\Sigma_{ij}^k[\mathbf{0}; g(t)] = \int_{-\infty}^{+\infty} S_{ij}^k[\xi_3 \mathbf{i}_3; g(t)] \mathrm{d}\xi_3. \tag{3.124}$$

At a position defined by $\mathbf{x}$, signals propagating with velocities c_L and c_T do, however, not arrive until $c_L t = r$ and $c_T t = r$, respectively, which implies that the limits of integration of (3.122) and (3.124) will be appropriately modified. For example, by employing the form of $D_i^k[\ ]$ given by (3.94), eq. (3.122) assumes the form

$$\Delta_i^k[\mathbf{0}; g(t)] = \int_{x_3-[c_L{}^2t^2-x_1{}^2-x_2{}^2]^{\frac{1}{2}}}^{x_3+[c_L{}^2t^2-x_1{}^2-x_2{}^2]^{\frac{1}{2}}} \mathrm{d}\xi_3 \frac{\partial}{\partial x_i} \frac{\partial}{\partial x_k} \left\{\frac{1}{r}\int_0^{t-r/c_L} sg\left(t-\frac{r}{c_L}-s\right)\mathrm{d}s\right\}$$
$$-\int_{x_3-[c_T{}^2t^2-x_1{}^2-x_2{}^2]^{\frac{1}{2}}}^{x_3+[c_T{}^2t^2-x_1{}^2-x_2{}^2]^{\frac{1}{2}}} \mathrm{d}\xi_3 \frac{\partial}{\partial x_i} \frac{\partial}{\partial x_k} \left\{\frac{1}{r}\int_0^{t-r/c_T} sg\left(t-\frac{r}{c_T}-s\right)\mathrm{d}s\right\}$$
$$+\frac{\delta_{ik}}{c_T^2}\int_{x_3-[c_T{}^2t^2-x_1{}^2-x_2{}^2]^{\frac{1}{2}}}^{x_3+[c_T{}^2t^2-x_1{}^2-x_2{}^2]^{\frac{1}{2}}} \mathrm{d}\xi_3 \frac{1}{r} g\left(t-\frac{r}{c_T}\right), \tag{3.125}$$

where r is defined by (3.120).

3.10.2. Antiplane line load

Let us first consider the case $k = 3$. For $i = 1$ or 2, the first two integrals vanish because they are integrations of odd functions of $x_3-\xi_3$ over intervals that are symmetric with respect to x_3. The last integral vanishes on account of the Kronecker delta. Thus

$$\Delta_\alpha^3[\mathbf{0}; g(t)] = 0. \tag{3.126}$$

For $k = 3$ and $i = 3$ the evaluation of the integrals is slightly more complicated. One way of achieving a quick evaluation is to bring one of the derivatives $\partial/\partial x_3$ in front of the integrals in both the first and the second integral of (3.125). This is allowable provided that we compensate for the contributions from the limits of integration to the derivative of the integrals with respect to x_3. The latter contributions are, however, zero since the integrand vanishes at the integration limits. Thus we can simply bring one derivative with respect to x_3 outside of the integration over ξ_3. The thus obtained integrals vanish, however, since they are integrations of odd functions. The only contribution thus comes from the last term of (3.125), and we have

$$\Delta_3^3[\xi_3 \mathbf{i}_3; g(t)] = \frac{2}{c_T^2}\int_{x_3}^{x_3+(c_T{}^2t^2-x_1{}^2-x_2{}^2)^{\frac{1}{2}}} \frac{1}{r} g\left(t-\frac{r}{c_T}\right)\mathrm{d}\xi_3, \tag{3.127}$$

where r is defined by (3.120). To further evaluate eq. (3.127) we introduce the change of variables

$$\tau = t-\frac{r}{c_T},$$

whereupon (3.127) reduces to

$$\Delta_3^3[\xi_3 \boldsymbol{i}_3; g(t)] = \frac{2}{c_T^2}\int_0^{t-R/c_T} \frac{g(\tau)\mathrm{d}\tau}{[(t-\tau)^2-R^2/c_T^2]^{\frac{1}{2}}}, \tag{3.128}$$

where

$$R^2 = x_1^2+x_2^2. \tag{3.129}$$

It should be noted that (3.128) is independent of x_3. Substitution of (3.128) into (3.119) shows that u_3 agrees with the fundamental solution of the wave equation which was previously derived as eq. (3.72).

Although it is, of course, possible to evaluate (3.124) it is, in fact easier to derive the stresses corresponding to the displacement distribution (3.128) by direct substitution of the displacement into Hooke's law. Since (3.128) is independent of x_3 we find that this displacement gives rise to stresses $\tau_{31}(x_1, x_2, t)$ and $\tau_{32}(x_1, x_2, t)$ only

$$\tau_{3\alpha}(x_1, x_2, t) = \frac{\mu}{2\pi c_T^2}\frac{\partial}{\partial x_\alpha}\int_0^{t-R/c_T} \frac{g(\tau)\mathrm{d}\tau}{[(t-\tau)^2-R^2/c_T^2]^{\frac{1}{2}}}, \tag{3.130}$$

where R is defined by (3.129). We conclude that a line load of body forces pointing in the x_3-direction gives rise to shear waves with the displacements polarized in the x_3-direction.

3.10.3. *In-plane line load*

We proceed to the case $k = 1, 2$, i.e., we set $k = \gamma$, where $\gamma = 1, 2$. The first two integrals in eq. (3.125) vanish for $i = 3$, because they comprise integrations of odd functions of $x_3-\xi_3$, over intervals that are symmetric with respect to x_3. The last integral vanishes on account of the Kronecker delta. Thus

$$\Delta_3^\gamma[\mathbf{0}; g(t)] = 0. \tag{3.131}$$

For $i = 1, 2$ we first bring the derivatives with respect to x_γ and x_α ($\alpha = 1, 2$) in front of the integral signs. This can be done because the integrands vanish at the limit of integration. Then we introduce changes of variables for the integration over ξ_3 by introducing

$$\tau = t - \frac{r}{c_L}$$

in the first integral, and

$$\tau = t - \frac{r}{c_T}$$

in the second and third integrals, where r is defined by (3.120). The results of these operations are

$$\begin{aligned} \Delta_\alpha^\gamma[\mathbf{0}; g(t)] = {} & 2\frac{\partial}{\partial x_\alpha}\frac{\partial}{\partial x_\gamma}\int_0^{t-R/c_L}\frac{\mathrm{d}\tau}{[(t-\tau)^2-R^2/c_L^2]^{\frac{1}{2}}}\int_0^\tau sg(\tau-s)\mathrm{d}s \\ & -2\frac{\partial}{\partial x_\alpha}\frac{\partial}{\partial x_\gamma}\int_0^{t-R/c_T}\frac{\mathrm{d}\tau}{[(t-\tau)^2-R^2/c_T^2]^{\frac{1}{2}}}\int_0^\tau sg(\tau-s)\mathrm{d}s \\ & +\frac{2\delta_{\alpha\gamma}}{c_T^2}\int_0^{t-R/c_T}\frac{g(\tau)\mathrm{d}\tau}{[(t-\tau)^2-R^2/c_T^2]^{\frac{1}{2}}}, \end{aligned} \tag{3.132}$$

where, as before $R = (x_1^2+x_2^2)^{\frac{1}{2}}$.

The stresses corresponding to (3.132) are obtained by substitution of $4\pi\Delta_\alpha^\gamma[\mathbf{0}; g(t)]$ into the stress-displacement relations. It is observed that (3.131) does not contain x_3 and that concentrated line loads in the x_1- and x_2-directions produce in-plane motions only.

3.10.4. Integral representations

The results of section (3.9) can be employed to derive a two-dimensional integral representation from the three-dimensional integral identity for the displacement field given by (3.105). The reduction from three to two dimensions by appropriate integration of the three-dimensional integral identity was shown by de Hoop[13], whose derivation is by and large reproduced here.

Let C be a simple closed curve in the (x_1, x_2)-plane, and let D be its interior. In the three-dimensional integral identity (3.105) we take for S the closed surface consisting of the plane portions $x_3 = \xi_3-L$, $x_3 = \xi_3+L$, $(x_1, x_2) \in D$, together with the cylindrical part $-L \leqq x_3-\xi_3 \leqq L$, $(x_1, x_2) \in C$, where $L > 0$. As can be seen from the definitions of $D_i^k[\ ]$ and $S_{ij}^k[\ ]$, the contributions to the surface integrals of $x_3 = \xi_3-L$ and $x_3 = \xi_3+L$, $(x_1, x_2) \in D$ vanish in the limit $L \to \infty$. In this way we obtain

$$\begin{aligned} 4\pi\rho u_k(\xi_1, \xi_2, t) = {} & \sum_{i=1}^3\int_C \{\Delta_i^k[\xi; t_i(\mathbf{x}, t)]-n_j\Sigma_{ij}^k[\xi; u_i(\mathbf{x}, t)]\}\mathrm{d}c \\ & +\rho\sum_{i=1}^3\int_D \Delta_i^k[\xi; f_i(\mathbf{x}, t)]\mathrm{d}x_1\,\mathrm{d}x_2, \end{aligned} \tag{3.133}$$

where dc is the element of the contour C. The operators are defined similarly to (3.122) and (3.124) except that the point of application is now at $\mathbf{x} = \xi$.

[13] A. T. de Hoop, *loc. cit.*, p. 105.

The result (3.125) still holds except that we must replace x_1 by $x_1-\xi_1$ and x_2 by $x_2-\xi_2$. In particular, $\Delta^3_\alpha[\xi; g(t)]$, $\Delta^3_3[\xi; g(t)]$, $\Delta^\gamma_3[\xi; g(t)]$ and $\Delta^\gamma_\alpha[\xi; g(t)]$ are defined by (3.126), (3.128), (3.131) and (3.132), respectively, provided we define R as

$$R = [(x_1-\xi_1)^2+(x_2-\xi_2)^2]^{\frac{1}{2}}. \tag{3.134}$$

The corresponding operators $\Sigma^k_{ij}[\ ]$ are easiest obtained by substituting the displacements into Hooke's law.

For $k = 3$ the integral identity (3.133) simplifies to

$$4\pi\rho u_3(\xi_1, \xi_2, t) = \int_C \{\Delta^3_3[\xi; t_3(\mathbf{x}, t)]-n_\alpha \Sigma^3_{3\alpha}[\xi; u_3(\mathbf{x}, t)]\}\mathrm{d}c$$
$$+\rho\int_D \Delta^3_3[\xi; f_3(\mathbf{x}, t)]\mathrm{d}x_1\,\mathrm{d}x_2. \tag{3.135}$$

For $k = 1$ or 2, we find

$$4\pi\rho u_\gamma(\xi_1, \xi_2, t) = \sum_{\alpha=1}^{2}\int_C \{\Delta^\gamma_\alpha[\xi; t_\alpha(\mathbf{x}, t)]-n_\beta \Sigma^\gamma_{\alpha\beta}[\xi; u_\alpha(\mathbf{x}, t)]\}\mathrm{d}c$$
$$+\rho\sum_{\alpha=1}^{2}\int_D \Delta^\gamma_\alpha[\xi; f_\alpha(\mathbf{x}, t)]\mathrm{d}x_1\,\mathrm{d}x_2. \tag{3.136}$$

Eq. (3.135) agrees with Volterra's solution of the two-dimensional wave equation.[14]

3.11. Boundary-value problems

In fairly general terms the elastodynamic boundary-value problem was formulated in section 2.5, where the following conditions on the boundary S were considered:

$$u_i = U_i(x_j, t) \text{ on } S_1 \tag{3.137}$$

and

$$\tau_{ji}n_j = t_i(x_j, t) \text{ on } S-S_1. \tag{3.138}$$

If $S_1 = S$, the displacements are prescribed over the whole of the boundary S. For $S_1 = 0$, the boundary conditions are on the tractions only. In the form (3.137) and (3.138) the boundary conditions define a mixed boundary--value problem. Except for bodies of a very simple shape, such as half-spaces, layers or cylinders, it is rather difficult to obtain solutions of boundary-value problems of elastodynamics.

[14] V. Volterra, *Acta Mathematica* **18**, (1894), 161.

Let us first consider the dynamic response of an initially undisturbed body of finite dimensions to body forces. For definiteness we consider the motion generated by a distribution of body forces $f_i(x_j, t)$ in a body which is rigidly clamped along its outer surface S, i.e., $u_i \equiv 0$ on S. In accordance with the approach of sections 3.6 and 3.7 it is conceptually possible to write the displacement as a superposition integral over $f_i(x_j, t)$ and a Green's function. The Green's function should satisfy the homogeneous equations except at the source point where an appropriate singularity must exist. Moreover, the Green's function should vanish on S. In physical terms the Green's function defines the response of the rigidly supported body to an impulsive point load. Unfortunately it is generally rather complicated to determine an elastodynamic Green's function even for a simple domain such as a half-space. Only in special cases is it possible to construct the function in a simple manner from the Green's function of the unbounded domain.

Some simple examples for which Green's functions can easily be obtained are concerned with two-dimensional motions in antiplane shear. These motions are governed by a single wave equation, which was stated in section 2.7 as

$$\frac{\partial^2 u_3}{\partial x_1^2} + \frac{\partial^2 u_3}{\partial x_2^2} + f_3 = \frac{1}{c_T^2} \frac{\partial^2 u_3}{\partial t^2}. \tag{3.139}$$

If we consider a half-space with traction free boundary, $u_3(x_1, x_2, t)$ must satisfy

$$\frac{\partial u_3}{\partial x_2} = 0 \quad \text{at} \quad x_2 = 0. \tag{3.140}$$

The Green's function is the displacement field generated by an antiplane impulsive line load applied at time $t = s$ at position $x_1 = \xi_1, x_2 = \xi_2$ (see figure 3.2). Suppose we consider an unbounded medium and suppose we also apply an impulsive line load at time $t = s$ at position $x_1 = \xi_1$, $x_2 = -\xi_2$. In the unbounded medium the derivative with respect to x_2 of the sum of the two displacement fields vanishes at $x_2 = 0$ because of symmetry with respect to $x_2 = 0$. In the half-plane $x_2 > 0$, the sum of the two displacement fields satisfies (3.139) and (3.140) as well as the proper condition at $x_1 = \xi_1, x_2 = \xi_2$. This sum thus defines the Green's function for the half-space. At a point defined by $x_1 > 0, x_2 > 0$, the first signal arrives directly from the primary source. After an additional time interval a signal arrives from the image of the primary source with respect to the plane $x_2 = 0$. This second signal is to be interpreted as the reflection from the plane $x_2 = 0$. The wavefronts are shown in figure 3.2.

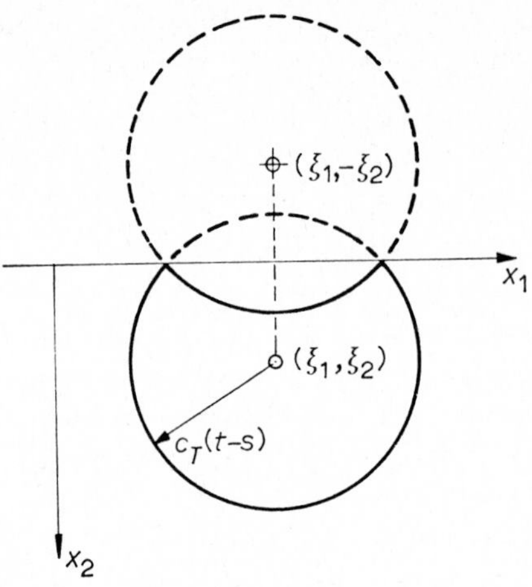

Fig. 3.2. Antiplane line load in a half-space.

The idea of imaging the primary source with respect to free surfaces can also be used to determine the Green's function for a region bounded by two planes. Suppose we have at $x_2 = 0$ and $x_2 = h$:

$$\frac{\partial u_3}{\partial x_2} = 0.$$

In this case we take the image of the primary source with respect to both $x_2 = 0$ and $x_2 = h$. The image with respect to $x_2 = h$ destroys, however, the symmetry of the primary source and its image with respect to $x_2 = 0$, and we thus must add another source in the region $x_2 < 0$ to restore symmetry. The system of sources must, however, also again be symmetric with respect to $x_2 = h$, which establishes the need for another source at $x_2 > h$, whereupon an additional source must be applied at $x_2 < 0$, and so forth. It is concluded that an infinite number of sources is needed to satisfy the conditions of $\partial u_3/\partial x_2 = 0$ at $x_2 = 0$ and $x_2 = h$. At a point $0 \leqq x_2 \leqq h$, an infinite sequence of signals is observed. The signals are interpreted as the primary signal and subsequent reflections from the two free surfaces. The sources and the pattern of waves are shown in figure 3.3. The arcs AB and $A'B'$ indicate the wavefronts which separate the disturbed from the undisturbed part of the layer.

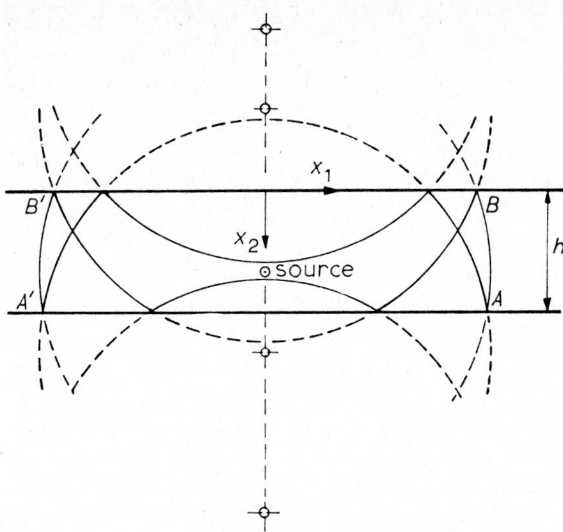

Fig. 3.3. Antiplane line load in a layer.

By employing displacement potentials the general elastodynamic problem is also governed by classical wave equations. The potentials are, however, coupled by the boundary conditions, at least for the physically significant boundary conditions covered by eqs. (3.137) and (3.138). As a consequence it is generally not possible to employ the scheme of imaging to construct

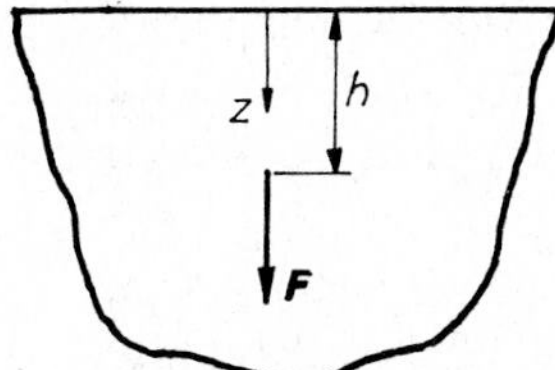

Fig. 3.4. Half-space subjected to an interior point load.

the Green's functions for simple domains from the Green's function for the unbounded domain. This can be exemplified by considering a half-space subjected to an internal point load, as shown in figure 3.4. In cylindrical coordinates the boundary conditions at $z = 0$ are

$$z = 0: \qquad \tau_z = 0, \qquad \tau_{zr} = 0, \qquad \tau_{z\theta} = 0.$$

Now let us compare these boundary conditions with the conditions that prevail in an unbounded medium with two symmetrically placed point loads as shown in figure 3.5. In view of axial symmetry we do have $\tau_{z\theta} = 0$ at $z = 0$. Symmetry with respect to $z = 0$ yields, however $\tau_{zr} = 0$ and $w = 0$ at $z = 0$. By the method of images it is apparently possible to satisfy the conditions of a smooth boundary at $z = 0$, but not the much more significant conditions of a traction free boundary.

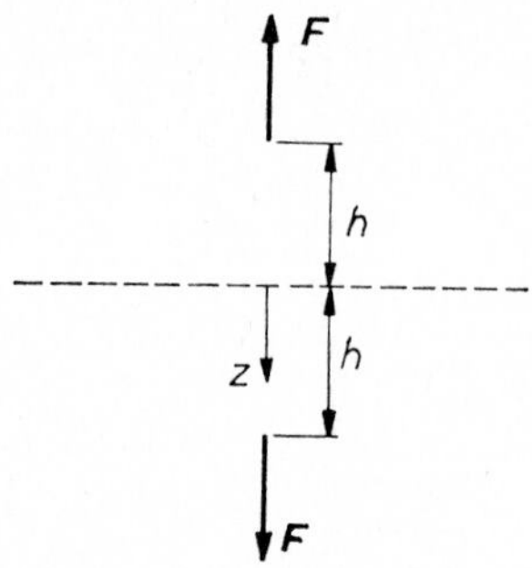

Fig. 3.5. Symmetrically placed point loads.

Motions that are generated by surface disturbances are equally difficult to express in a simple manner. For this case we may employ a Green's function for the boundary condition. Such a Green's function satisfies the governing equation inside the body and shows appropriate singular behavior at one point of the boundary. For example, if a body is subjected to a distribution of normal surface tractions, the appropriate Green's function is the displacement due to an impulsive normal point load on the surface. The displacement at any point can then be expressed in the form of a superposition integral over the distribution of surface tractions and the Green's function. For the general elastodynamic problem such Green's functions are again rather complicated. For the half-space and the layer some examples are worked out in chapters 7 and 8, respectively.

Green's function techniques are only one way of solving boundary-value problems. Solutions of elastodynamic boundary-value problems are usually obtained by direct application of methods of applied mathematics to the system of governing partial differential equations and the boundary and initial conditions. Several worked out examples are presented in chapters 4, 7, 8 and 9.

3.12. Steady-state time-harmonic response

3.12.1. *Time-harmonic source*

We will return to the radiation problems of sections 3.6 and 3.7 to consider the special case that the external disturbances vary in a simple harmonic manner with time. Suppose, for example, that the inhomogeneous term in the radiation problem for an unbounded medium, defined by (3.50), is time-harmonic with period of oscillation $2\pi/\omega$; i.e.,

$$\nabla^2\varphi - \frac{1}{c_L^2}\frac{\partial^2\varphi}{\partial t^2} = -F(\boldsymbol{x})\cos(\omega t+\alpha). \tag{3.141}$$

The general solution then follows from eq. (3.66) as

$$\varphi(\boldsymbol{x}, t) = \frac{1}{4\pi}\int_{B_L}\frac{F(\boldsymbol{\xi})}{|\boldsymbol{x}-\boldsymbol{\xi}|}\cos\left(\omega t - \frac{\omega}{c_L}|\boldsymbol{x}-\boldsymbol{\xi}|+\alpha\right)\mathrm{d}V_\xi, \tag{3.142}$$

where $\mathrm{d}V_\xi = \mathrm{d}\xi_1\,\mathrm{d}\xi_2\,\mathrm{d}\xi_3$, and the integration is carried out over the sphere B_L with center at $\boldsymbol{x}$ and with radius $c_L t$.

The inhomogeneous term in (3.141) may also be written as

$$\mathscr{R}F(\boldsymbol{x})e^{i(\omega t+\alpha)}, \tag{3.143}$$

where $i = \sqrt{-1}$ and $\mathscr{R}$ indicates that the real part should be taken. Eq. (3.142) then assumes the form

$$\varphi(\boldsymbol{x}, t) = \frac{1}{4\pi}\mathscr{R}e^{i(\omega t+\alpha)}\int_{B_L}\frac{F(\boldsymbol{\xi})}{|\boldsymbol{x}-\boldsymbol{\xi}|}e^{-i(\omega/c_L)|\boldsymbol{x}-\boldsymbol{\xi}|}\mathrm{d}V_\xi. \tag{3.144}$$

It must be emphasized that the time does not appear only in the exponential term of (3.144), but also in the limits of integration. The solution as it is represented by (3.144) is for an initial-value problem. Let us now consider this solution for large values of time. If t is large enough the integration over B will include the complete region over which the inhomogeneous term is defined. This region, which is independent of time, is denoted by V. Thus, for large t the solution (3.144) may be written as

$$\varphi(\boldsymbol{x}, t) = \mathscr{R}e^{i(\omega t+\alpha)}\Phi(\boldsymbol{x}), \tag{3.145}$$

where

$$\Phi(\boldsymbol{x}) = \frac{1}{4\pi}\int_V\frac{F(\boldsymbol{\xi})}{|\boldsymbol{x}-\boldsymbol{\xi}|}e^{-i(\omega/c_L)|\boldsymbol{x}-\boldsymbol{\xi}|}\mathrm{d}V_\xi. \tag{3.146}$$

Eq. (3.145) is called the steady-state solution for the problem at hand. For the special case that $F(\boldsymbol{x}) = \delta|\boldsymbol{x}-\boldsymbol{\zeta}|$ we find

$$\Phi(\boldsymbol{x};\zeta) = \frac{1}{4\pi}\,\frac{e^{-i(\omega/c_L)|\boldsymbol{x}-\zeta|}}{|\boldsymbol{x}-\zeta|}. \tag{3.147}$$

Clearly (3.147) in conjunction with (3.145) represents a wave propagating in the direction of increasing values of $|\boldsymbol{x}-\zeta|$. For a position $\boldsymbol{x} = \boldsymbol{x}_1$, this solution is valid for $t > |\boldsymbol{x}_1-\zeta|/c_L$. In the three-dimensional case the steady-state solution applies for a position defined by $\boldsymbol{x}_1$ if $c_L t$ is larger than the radius of the smallest sphere, centered at $\boldsymbol{x} = \boldsymbol{x}_1$, which includes the region of external disturbances V.

In a steady-state analysis we generally immediately assume a solution of the form (3.145). It is, however, convenient to omit the symbol $\mathscr{R}$ during the analysis, until the very last stage, when the real part must be taken to obtain the final result. Thus we consider steady-state solutions of the form

$$\varphi(\boldsymbol{x}, t) = \Phi(\boldsymbol{x})e^{i\omega t}, \tag{3.148}$$

with the understanding that we are interested in the real part. Here $\Phi(\boldsymbol{x})$ may be a complex function because it is now taken to include the phase angle α. This representation is permissible provided that only linear operations are carried out, i.e., operations in which it is immaterial whether taking the real part is done before or after the operation. Typical examples of such operations are addition, subtraction, integration and taking a derivative.

Time-harmonic motions described by (3.148) may be of two types, either progressive harmonic waves or standing harmonic waves. Standing waves are characterized by the appearance of stationary points of zero phase, called nodes. Progressive harmonic waves do not have stationary nodes, but moving nodes. Standing waves occur in bodies of finite extent. If the shape of the body is simple enough, standing waves can be analyzed by the method of separation of variables. Progressive harmonic waves are generally of interest for very large bodies.

3.12.2. Helmholtz's equation

If a steady-state solution of the form (3.148) is substituted into

$$\nabla^2\varphi - \frac{1}{c_L^2}\frac{\partial^2\varphi}{\partial t^2} = -F(\boldsymbol{x})e^{i\omega t},$$

we obtain

$$\nabla^2\Phi + k^2\Phi = -F(\boldsymbol{x}), \tag{3.149}$$

where we have introduced the wavenumber k as

$$k = \frac{\omega}{c}, \tag{3.150}$$

A particular solution of (3.149) is given by (3.146). The homogeneous form of (3.149) is called the *space form* of the wave equation, or *Helmholtz equation.*

3.12.3. Helmholtz's first (interior) formula

A formula analogous to Kirchhoff's formula, which was stated in section 3.9, can now be stated for the time-harmonic case. Considering a point defined by the position vector $\boldsymbol{x}$ inside a region V bounded by the surface S, we find

$$\begin{aligned}\Phi(\boldsymbol{x}) = &\frac{1}{4\pi}\int_V \frac{F(\xi)}{|\boldsymbol{x}-\xi|}\, e^{-ik|\boldsymbol{x}-\xi|} \mathrm{d}V_\xi \\ &+\frac{1}{4\pi}\int_S \left[\frac{e^{-ik|\boldsymbol{x}-\xi|}}{|\boldsymbol{x}-\xi|}\frac{\partial\Phi(\xi)}{\partial n_\xi} - \Phi(\xi)\frac{\partial}{\partial n_\xi}\left(\frac{e^{-ik|\boldsymbol{x}-\xi|}}{|\boldsymbol{x}-\xi|}\right)\right]\mathrm{d}A_\xi .\end{aligned} \tag{3.151}$$

As usual, n_ξ is the outward normal to S. The integral representation (3.151) is known as Helmholtz's interior formula.

3.12.4. Helmholtz's second (exterior) formula

Let us now consider a point defined by the position vector $\boldsymbol{x}$ outside the boundary S of the region V. The point is in the region T which is bounded internally by S and externally by another closed surface S_R. The surface in eq. (3.151) is now composed of $S+S_R$. It should be realized that $\boldsymbol{n}$ is the outward normal to the region of integration T. For the surface S_R we now take a spherical surface of (large) radius R centered at $\boldsymbol{x}$. Then as $R \to \infty$

$$\frac{\partial}{\partial n_\xi} \sim \frac{\partial}{\partial R} \quad \text{and} \quad |\boldsymbol{x}-\xi| = R.$$

Hence

$$\frac{e^{-ik|\boldsymbol{x}-\xi|}}{|\boldsymbol{x}-\xi|}\frac{\partial\Phi}{\partial n_\xi} - \Phi(\xi)\frac{\partial}{\partial n_\xi}\left[\frac{e^{-ik|\boldsymbol{x}-\xi|}}{|\boldsymbol{x}-\xi|}\right] \sim \left[\frac{\partial\Phi}{\partial R} + ik\Phi\right]\frac{e^{-ikR}}{R} + \Phi\frac{e^{-ikR}}{R^2}.$$

Now let it be required that

$$|R\Phi| < M \tag{3.152}$$

and

$$R\left(\frac{\partial \Phi}{\partial R}+ik\Phi\right) \to 0 \tag{3.153}$$

uniformly with respect to direction as $R \to \infty$, where R is the distance from position $\boldsymbol{x}$. These conditions are known as the *Sommerfeld radiation conditions*. Since the area of S is of order R^2, we have in view of (3.152)

$$\int_{S_R} \Phi \frac{e^{-ikR}}{R^2}\,\mathrm{d}A \to 0 \quad \text{as} \quad R \to \infty,$$

and in view of (3.153)

$$\int_{S_R} \left(\frac{\partial \Phi}{\partial R}+ik\Phi\right)\frac{e^{-ikR}}{R}\,\mathrm{d}A \to 0 \quad \text{as} \quad R \to \infty.$$

Hence when Φ satisfies the radiation conditions the integral representation for $\Phi(\boldsymbol{x})$ becomes

$$\Phi(\boldsymbol{x}) = \frac{1}{4\pi}\int_T \frac{F(\xi)}{|\boldsymbol{x}-\xi|}\,e^{-ik|\boldsymbol{x}-\xi|}\mathrm{d}V_\xi$$
$$+\frac{1}{4\pi}\int_S \left[\frac{e^{-ik|\boldsymbol{x}-\xi|}}{|\boldsymbol{x}-\xi|}\frac{\partial \Phi(\xi)}{\partial n'_\xi} - \Phi(\xi)\frac{\partial}{\partial n'_\xi}\left(\frac{e^{-ik|\boldsymbol{x}-\xi|}}{|\boldsymbol{x}-\xi|}\right)\right]\mathrm{d}A_\xi, \tag{3.154}$$

where T is the whole of space outside S, and $\boldsymbol{n}'$ is an inward normal to the closed surface S. Eq. (3.154) is Helmholtz's second formula, which is applied extensively in the investigation of scattering of acoustic waves by obstacles. For a discussion of applications of the formula to scattering and diffraction problems we refer to the monograph by Mow and Pao.[15]

It can be shown that Sommerfeld's radiation conditions are a consequence of the property of the solution of the initial value problem that waves propagate outward from their source.[16]

3.12.5. Steady-state solutions in two dimensions

In two dimensions, a solution of

$$\nabla^2\Phi+k^2\Phi = \delta|\boldsymbol{x}-\xi|$$

is given by

$$\Phi(\boldsymbol{x};\xi) = \frac{i}{4}H_0^{(2)}(kr), \tag{3.155}$$

[15] C. C. Mow and Y. H. Pao, *The diffraction of elastic waves and dynamic stress concentrations*. Report R-482-PR, The Rand Corporation (1971), p. 140.

[16] C. H. Wilcox, *Archive Rational Mechanics and Analysis* 3 (1959), 133.

where

$$r = |\boldsymbol{x}-\boldsymbol{\xi}| = [(x_1-\xi_1)^2+(x_2-\xi_2)^2]^{\frac{1}{2}}. \tag{3.156}$$

Here $H_0^{(2)}(kr)$ is the Hankel function of the second kind. In books on Bessel functions the following asymptotic representation can be found:

$$H_0^{(2)}(kr) \sim \left(\frac{2}{\pi kr}\right)^{\frac{1}{2}} e^{-i(kr-\pi/4)},$$

which shows that

$$\varphi(x_1, x_2, t) = \frac{i}{4} e^{i\omega t} H_0^{(2)}(kr)$$

does indeed represent a wave diverging from the axis $\boldsymbol{x} = \boldsymbol{\xi}$.

By employing (3.155), a particular integral of the two-dimensional Helmholtz's equation with inhomogeneous term $F(\boldsymbol{x})$ is obtained as

$$\Phi(r) = \frac{i}{4}\int_A F(\boldsymbol{\xi}) H_0^{(2)}(kr) \mathrm{d}A_\xi,$$

where r is defined by (3.156), and A is the area occupied by $F(\boldsymbol{x})$. The two-dimensional fundamental solution can be employed to derive an integral representation. For a point located inside the area A bounded by the curve C, which does not contain sources, we find

$$\Phi(r) = -\frac{i}{4}\int_C \left[\frac{\partial \Phi}{\partial n_\xi} H_0^{(2)}(kr) - \Phi \frac{\partial}{\partial n_\xi} H_0^{(2)}(kr)\right] \mathrm{d}C_\xi. \tag{3.157}$$

Here $\boldsymbol{n}$ is the outward normal to C. Eq. (3.157) is known as *Weber's interior formula*. An analogous formula can be derived for an exterior region.

3.13. Problems

3.1. Suppose a point load acting in the x_3-direction is applied at time $t = 0$ in an unbounded medium and is then maintained at a position which moves with a constant velocity v along the positive x_3-axis. Consider the case $v < c_L$. If the initial point of application is taken as the origin of the coordinate system the displacement equation of motion may be stated as ($t \geqq 0$)

$$c_T^2 \nabla^2 \boldsymbol{u} + (c_L^2 - c_T^2)\nabla(\nabla \cdot \boldsymbol{u}) - \ddot{\boldsymbol{u}} = Q\delta(x_1)\delta(x_2)\delta(x_3 - vt)\boldsymbol{i}_3.$$

Here Q is a constant measuring the strength of the moving force. The initial conditions are

$$\boldsymbol{u}(\boldsymbol{x}, 0) = \dot{\boldsymbol{u}}(\boldsymbol{x}, 0) \equiv 0.$$

The displacement distribution for this moving load problem can be determined in an elegant manner by employing the dynamic reciprocal identity.

(a) Show that the dynamic reciprocal identity reduces to

$$\int_V \boldsymbol{f} * \boldsymbol{u}' \mathrm{d}V_\xi = \int_V \boldsymbol{f}' * \boldsymbol{u} \,\mathrm{d}V_\xi,$$

where

$$\boldsymbol{f}(\xi, s) = Q\delta(\xi_1)\delta(\xi_2)\delta(\xi_3 - vs)\boldsymbol{i}_3.$$

(b) Determine the displacement in the x_1-direction by choosing appropriate expressions for $\boldsymbol{f}'$ and $\boldsymbol{u}'$.

3.2. Determine the displacement components u_1 and u_2 due to a two-dimensional center of compression.

3.3. Show that in cylindrical coordinates the radial displacement due to a point load of magnitude $\rho g(t)$ acting in the axial direction may be expressed in the form

$$u_r = \frac{1}{4\pi}\frac{3zr}{R^5}\int_{R/c_L}^{R/c_T} sg(t-s)\mathrm{d}s + \frac{1}{4\pi}\frac{zr}{R^3}\left[\frac{1}{c_L^2} g\left(t - \frac{R}{c_L}\right) - \frac{1}{c_T^2} g\left(t - \frac{R}{c_T}\right)\right],$$

where z and r are cylindrical coordinates and

$$R^2 = r^2 + z^2.$$

3.4. Use the expression for u_r of Problem 3.3 to write the displacement in the radial direction in the plane $z = 0$ for the case that two equal point loads are oriented as shown in figure 3.5.

3.5. An unbounded medium is subjected to a distribution of antiplane loads which are independent of the x_3-coordinate. The distribution is represented by

$$\boldsymbol{f}(x_1, x_2, t) = f_0 H(x_1)\delta(x_2)H(t)\boldsymbol{i}_3.$$

By employing eq. (3.130) determine the stress field $\tau_{31}(x_1, x_2, t)$ generated by the loads. What is the nature of the stress singularity in the vicinity of $x_1 = 0$, $x_2 = 0$?

3.6. An unbounded medium containing a semi-infinite crack is subjected to equal and opposite antiplane *line* loads on the faces of the crack, as shown in the figure. Observe that for $x_1 \geqq 0$ the displacement u_3 vanishes in the

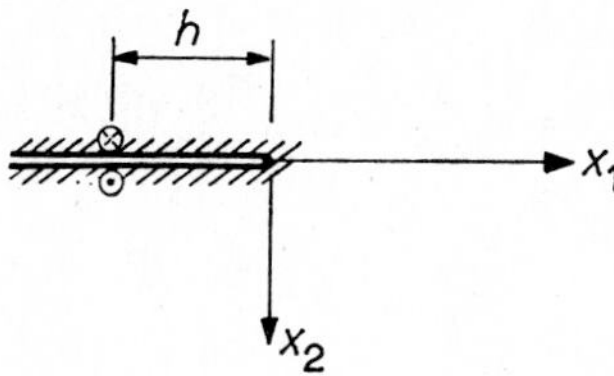

plane of the crack. The medium is not subjected to body forces. Consider the time interval $t \geqq h/c_T$, and employ the appropriate integral representation in conjunction with the fact that $u_3(x_1, 0, t) \equiv 0$ for $x_1 \geqq 0$, to obtain an integral equation for $u_3(x_1, 0, t)$ on the faces of the crack.

3.7. A quarterspace which is free of surface tractions is subjected to a concentrated line load, which is defined by

$$\boldsymbol{f} = f_0 \delta(x - \xi_1)\delta(x - \xi_2)\delta(t)\boldsymbol{i}_3.$$

Determine the displacement at $\bar{x}_1, \bar{x}_2$ as a function of time.

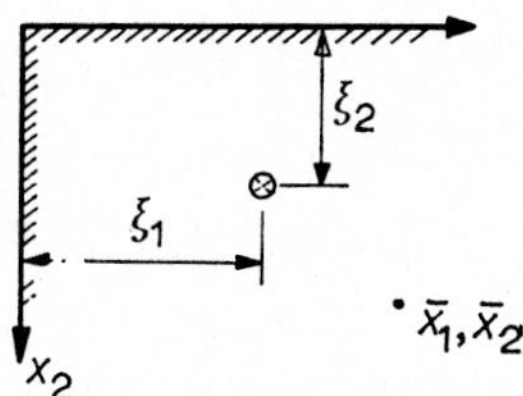

3.8. An unbounded medium is subjected to a time-harmonic point load. Use the results of section 3.12 to determine the steady-state displacement response of the medium.

3.9. For the two-dimensional case an equation analogous to eq. (3.154) can be derived. What are the pertinent radiation conditions?

3.10. Reexamine eqs. (3.155)–(3.157) for the case that the steady-state solution is assumed in the form

$$\varphi(\boldsymbol{x}, t) = \Phi(\boldsymbol{x})e^{-i\omega t}.$$

CHAPTER 4

ELASTIC WAVES IN AN UNBOUNDED MEDIUM

4.1. Plane waves

A plane displacement wave propagating with phase velocity c in a direction defined by the unit propagation vector $\boldsymbol{p}$ is represented by

$$\boldsymbol{u} = \mathrm{f}(\boldsymbol{x} \cdot \boldsymbol{p} - ct)\boldsymbol{d}. \tag{4.1}$$

In this equation $\boldsymbol{d}$ and $\boldsymbol{p}$ are unit vectors defining the directions of motion and propagation, respectively. The vector $\boldsymbol{x}$ denotes the position vector, and $\boldsymbol{x} \cdot \boldsymbol{p}$ = constant describes a plane normal to the unit vector $\boldsymbol{p}$. Eq. (4.1) thus represents a plane wave whose planes of constant phase are normal to $\boldsymbol{p}$ and propagate with velocity c.

By substituting the components of (4.1) into Hooke's law, see eq. (2.40), the components of the stress tensor are obtained as

$$\tau_{lm} = [\lambda \delta_{lm}(d_j p_j) + \mu(d_l p_m + d_m p_l)]\mathrm{f}'(x_n p_n - ct), \tag{4.2}$$

where the summation convention must be invoked, and a prime denotes a derivative of the function f( ) with respect to its argument.

In the absence of body forces the components of the displacement vector in a homogeneous, isotropic, linearly elastic medium are governed by the following system of partial differential equations:

$$\mu \nabla^2 \boldsymbol{u} + (\lambda + \mu)\nabla\nabla \cdot \boldsymbol{u} = \rho \ddot{\boldsymbol{u}}, \tag{4.3}$$

where λ and μ are Lamé's elastic constants, and ρ is the mass density. The vector operator ∇ is defined as

$$\nabla = \boldsymbol{i}_1 \frac{\partial}{\partial x_1} + \boldsymbol{i}_2 \frac{\partial}{\partial x_2} + \boldsymbol{i}_3 \frac{\partial}{\partial x_3}, \tag{4.4}$$

and ∇^2 is the Laplacian. We will substitute the expression for the plane wave, eq. (4.1), into the system of field equations (4.3). By employing the relations

$$\nabla \cdot \boldsymbol{u} = (\boldsymbol{p} \cdot \boldsymbol{d})\mathrm{f}'(\boldsymbol{x} \cdot \boldsymbol{p} - ct)$$

$$\nabla\nabla \cdot \boldsymbol{u} = (\boldsymbol{p} \cdot \boldsymbol{d})\mathrm{f}''(\boldsymbol{x} \cdot \boldsymbol{p} - ct)\boldsymbol{p}$$

$$\nabla^2 \boldsymbol{u} = \mathrm{f}''(\boldsymbol{x} \cdot \boldsymbol{p} - ct)\boldsymbol{d}$$

$$\ddot{\boldsymbol{u}} = c^2\mathrm{f}''(\boldsymbol{x} \cdot \boldsymbol{p} - ct)\boldsymbol{d},$$

we obtain

$$[\mu \boldsymbol{d} + (\lambda+\mu)(\boldsymbol{p} \cdot \boldsymbol{d})\boldsymbol{p} - \rho c^2 \boldsymbol{d}]\mathrm{f}''(\boldsymbol{x} \cdot \boldsymbol{p} - ct) = 0,$$

or

$$(\mu - \rho c^2)\boldsymbol{d} + (\lambda+\mu)(\boldsymbol{p} \cdot \boldsymbol{d})\boldsymbol{p} = 0. \tag{4.5}$$

Since $\boldsymbol{p}$ and $\boldsymbol{d}$ are two different unit vectors, eq. (4.5) can be satisfied in two ways only:

$$\text{either } \boldsymbol{d} = \pm \boldsymbol{p}, \quad \text{or} \quad \boldsymbol{p} \cdot \boldsymbol{d} = 0.$$

If $\boldsymbol{d} = \pm \boldsymbol{p}$, we have $\boldsymbol{d} \cdot \boldsymbol{p} = \pm 1$, and eq. (4.5) yields

$$c = c_L = \left(\frac{\lambda + 2\mu}{\rho}\right)^{\frac{1}{2}}. \tag{4.6}$$

In this case the motion is parallel to the direction of propagation and the wave is therefore called a *longitudinal* wave.

The components of the rotation $\nabla \wedge \boldsymbol{u}$ are

$$e_{klm}\partial_l u_m = e_{klm} p_l d_m \mathrm{f}',$$

and thus

$$\nabla \wedge \boldsymbol{u} = (\boldsymbol{p} \wedge \boldsymbol{d})\mathrm{f}'(\boldsymbol{x} \cdot \boldsymbol{p} - ct) = 0.$$

The rotation thus vanishes, which has motivated the alternative terminology *irrotational wave*. This type of wave is also often called a dilatational wave, a pressure wave, or a P-wave (primary, pressure).

If $\boldsymbol{p} \neq \pm \boldsymbol{d}$, both terms in (4.5) have to vanish independently, yielding

$$\boldsymbol{p} \cdot \boldsymbol{d} = 0 \quad \text{and} \quad c = c_T = \left(\frac{\mu}{\rho}\right)^{\frac{1}{2}}. \tag{4.7a, b}$$

Now the motion is normal to the direction of propagation, and the wave is called a *transverse* wave. It can easily be checked that in this case the divergence of the displacement vector vanishes, and we speak therefore also of an equivoluminal wave. This type of wave is also often called a *rotational wave*, a *shear wave*, or an *S-wave* (secondary, shear). The displacement can have any direction in a plane normal to the direction of propagation, but usually we choose the $(x_1 x_2)$-plane to contain the vector $\boldsymbol{p}$ and we consider motions which are in the $(x_1 x_2)$-plane or normal to the $(x_1 x_2)$-plane. These transverse motions are called "vertically" and "horizontally" polarized transverse waves, respectively.

From eqs. (4.6) and (4.7b) it follows that

$$\frac{c_L}{c_T} = \kappa = \left(\frac{\lambda+2\mu}{\mu}\right)^{\frac{1}{2}} = \left[\frac{2(1-\nu)}{1-2\nu}\right]^{\frac{1}{2}}, \tag{4.8}$$

where ν = Poisson's ratio, and where table 2.1 has been employed. Since $0 \leqq \nu \leqq 0.5$, it follows that $c_L > c_T$. For metals, the phase velocities of longitudinal and transverse waves are generally very large. Thus we find for structural steel, c_L = 590,000 cm/sec, and c_T = 320,000 cm/sec. For a few materials, representative values of ρ, c_L, c_T and κ are listed in table 4.1.

TABLE 4.1.

Approximate values of ρ, c_L, c_T and κ

Material	ρ (kg/m^3)	c_L (m/sec)	c_T (m/sec)	κ
air	1.2	340		
water	1000	1480		
steel	7800	5900	3200	1.845
copper	8900	4600	2300	2
aluminum	2700	6300	3100	2.03
glass	2500	5800	3400	1.707
rubber	930	1040	27	38.5

4.2. Time-harmonic plane waves

In chapter 1 it was already pointed out that the results of studies on traveling harmonic waves in a linearly elastic medium are of interest by virtue of the applicability of linear superposition. By the use of Fourier series, harmonic waves can be employed to describe the propagation of periodic disturbances. Propagating pulses can be described by superpositions of harmonic waves in Fourier integrals.

A plane harmonic displacement wave propagating with phase velocity c in a direction defined by the unit propagation vector $\boldsymbol{p}$ is represented by

$$\boldsymbol{u} = A\boldsymbol{d} \exp\left[ik(\boldsymbol{x} \cdot \boldsymbol{p} - ct)\right], \tag{4.9}$$

where $i = \sqrt{(-1)}$ and it is understood that the actual displacement components are the real or imaginary parts of the right-hand side. The amplitude A may be real-valued or complex, but it is independent of $\boldsymbol{x}$ and t. As defined in chapter 1, $\omega = kc$ is the circular frequency and k is the wavenumber. These quantities are related to the period T and the wavelength Λ by $\omega = 2\pi/T$ and $k = 2\pi/\Lambda$, respectively.

Eq. (4.9) clearly is a special case of (4.1). This implies that the results of the previous section are applicable. Thus, we have two types of plane harmonic waves, longitudinal and transverse waves, propagating with phase velocities c_L and c_T, respectively. Since the wavenumber k does not appear in the expressions for the phase velocities, plane harmonic waves in an unbounded homogeneous, isotropic, linearly elastic medium are not dispersive.

4.2.1. *Inhomogeneous plane waves*

The space and time coordinates in the expression for plane harmonic waves, (4.9), are intrinsically real. For simple harmonic time variation ω is also real, while p_j and k are usually taken as real-valued. As an interesting generalization of (4.9) we will now, however, consider the possibility of a complex-valued unit propagation vector $\boldsymbol{p}$,

$$\boldsymbol{p} = \boldsymbol{p}' + i\boldsymbol{p}''. \tag{4.10}$$

Since $\boldsymbol{p}$ is a unit vector we still require

$$\boldsymbol{p} \cdot \boldsymbol{p} = 1, \tag{4.11}$$

which implies

$$\boldsymbol{p}' \cdot \boldsymbol{p}' - \boldsymbol{p}'' \cdot \boldsymbol{p}'' = 1 \tag{4.12}$$

$$\boldsymbol{p}' \cdot \boldsymbol{p}'' = 0. \tag{4.13}$$

Substituting (4.10) into (4.9), we obtain

$$\boldsymbol{u} = A\boldsymbol{d} \exp\left[-k(\boldsymbol{x} \cdot \boldsymbol{p}'')\right] \exp\left[ik(\boldsymbol{x} \cdot \boldsymbol{p}' - ct)\right]. \tag{4.14}$$

This expression describes a wave with varying amplitude propagating in a direction defined by the vector $\boldsymbol{p}'$. Planes of constant amplitude are given by

$$\boldsymbol{x} \cdot \boldsymbol{p}'' = C_1, \tag{4.15}$$

and at any particular time t, planes of constant phase are given by

$$\boldsymbol{x} \cdot \boldsymbol{p}' = C_2, \tag{4.16}$$

where C_1 and C_2 are constants. It now follows from (4.13) that the planes expressed by (4.15) and (4.16) are orthogonal, i.e., the planes of constant phase are orthogonal to the planes of constant amplitude. This means that the amplitude remains constant along lines in the direction of propagation, but varies exponentially in planes perpendicular to the direction of propagation. When the components of $\boldsymbol{p}$ are real, then we have the familiar case of uniform amplitude throughout space. Plane waves with a unit propagation vector whose components are not real are called *inhomogeneous* plane waves.

Although the wave is harmonic and the vector $\boldsymbol{p}$ is complex-valued, the wave is still of the general form (4.1). The development leading to eq. (4.5) is not affected and it is concluded that we must again have

$$(\mu-\rho c^2)\boldsymbol{d}+(\lambda+\mu)(\boldsymbol{p}\cdot\boldsymbol{d})\boldsymbol{p} = 0. \tag{4.17}$$

It should, of course, be realized that the unit vector $\boldsymbol{d}$ is now also complex-valued. Just as for the case of real-valued vectors, (4.17) can be satisfied in two ways only: either $\boldsymbol{d} = \pm\boldsymbol{p}$, or $\boldsymbol{d}\cdot\boldsymbol{p} = 0$. These two cases yield $c = c_L$ and $c = c_T$, respectively.

For the subsequent discussion it is convenient to rewrite the expression for a plane wave as

$$\boldsymbol{u} = A\boldsymbol{d}\exp[i\omega(\boldsymbol{x}\cdot\boldsymbol{q}-t)], \tag{4.18}$$

where the *slowness* vector is defined as

$$\boldsymbol{q} = \boldsymbol{p}/c.$$

In view of (4.11) we have

$$q_1^2+q_2^2+q_3^2 = q^2,$$

where $q = 1/c_L$ for longitudinal waves, and $q = 1/c_T$ for transverse waves. In most cases we choose our coordinate system so that $p_3 \equiv 0$. Then if we assume that q_1 is real, q_2 must be either real or imaginary, depending on the magnitude of q_1 in comparison with q:

$$q_2 = (q^2-q_1^2)^{\frac{1}{2}} \tag{4.19}$$

or

$$q_2 = i(q_1^2-q^2)^{\frac{1}{2}} = i\beta. \tag{4.20}$$

The corresponding plane waves have the form

$$\boldsymbol{u} = A\boldsymbol{d}\exp[i\omega(x_1q_1+x_2q_2-t)] \tag{4.21}$$

and

$$\boldsymbol{u} = A\boldsymbol{d}\exp(-\omega x_2\beta)\exp[i\omega(x_1q_1-t)], \tag{4.22}$$

respectively. In the first of these the oscillation amplitude is uniform throughout the field, and planes of fixed phase advance with speed $c = 1/q$ in the direction of the slowness vector $\boldsymbol{q}$, which has real components. In eq. (4.22) the oscillation amplitude decays exponentially in the positive x_2-direction. For fixed x_2, however, there is a simple harmonic wave propagating in the x_1-direction with phase velocity $1/q_1$.

4.2.2. *Slowness diagrams*

As pointed out by Crandall[1], the relations (4.19) and (4.20) can be represented in a slowness diagram, figure 4.1, by a circle and an equilateral hyperbola, respectively. Given $q_1 < q$, the corresponding value of q_2 is on the circle of radius q. When $q < q_1$ the corresponding value of β is on the hyperbola.

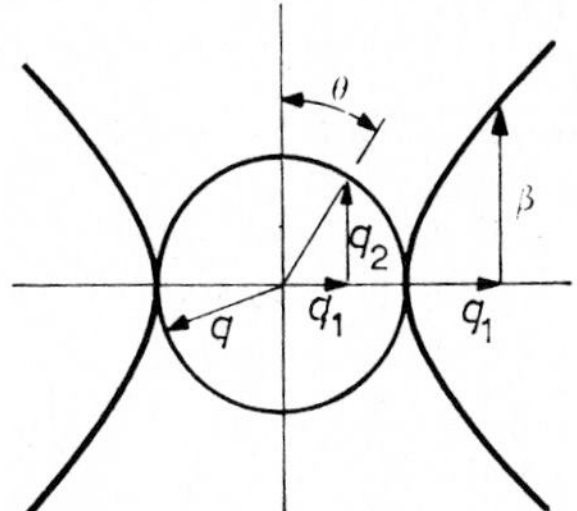

Fig. 4.1. Slowness diagram.

An alternative representation of eqs. (4.19) and (4.20) is obtained by introducing the angle θ by

$$p_1 = \sin\theta, \qquad p_2 = \cos\theta.$$

Then, as shown in figure 4.1,

$$q_1 = q\sin\theta, \qquad q_2 = q\cos\theta.$$

When $q_1 < q$, the angle θ is real. When $q < q_1$, it is convenient to set

$$\theta = \tfrac{1}{2}\pi - i\gamma,$$

and write

$$q_1 = q\cosh\gamma, \qquad \beta = q\sinh\gamma.$$

As we shall see in section 13 of chapter 5, slowness diagrams are useful to illustrate reflections and refractions of waves.

[1] S. H. Crandall, *Journal of the Acoustical Society of America* **47** (1970), 1338–1342.

4.3. Wave motions with polar symmetry

4.3.1. Governing equations

Within a system of spherical coordinates the field variables describing wave motions with polar symmetry depend on r and t only. The single displacement component, which is in the radial direction, is denoted by $u(r, t)$. The non-vanishing stresses are the radial stress

$$\tau_r = (\lambda+2\mu)\frac{\partial u}{\partial r} + 2\lambda\frac{u}{r}, \tag{4.23}$$

and the normal stress in any direction perpendicular to r

$$\tau_\theta = \lambda\frac{\partial u}{\partial r} + 2(\lambda+\mu)\frac{u}{r}. \tag{4.24}$$

The equation of motion takes the form

$$\frac{\partial \tau_r}{\partial r} + \frac{2(\tau_r - \tau_\theta)}{r} = \rho\frac{\partial^2 u}{\partial t^2}. \tag{4.25}$$

By substituting (4.23) and (4.24), the displacement equation of motion is obtained as

$$\frac{\partial^2 u}{\partial r^2} + \frac{2}{r}\frac{\partial u}{\partial r} - \frac{2u}{r^2} = \frac{1}{c_L^2}\frac{\partial^2 u}{\partial t^2}. \tag{4.26}$$

It is convenient to express the radial displacement in terms of a potential function $\varphi(r, t)$,

$$u = \frac{\partial \varphi}{\partial r}. \tag{4.27}$$

By substitution into (4.26) it is easily shown that $u(r, t)$ is a solution of (4.26) if the product $r\varphi$ satisfies the one-dimensional wave equation

$$\frac{\partial^2(r\varphi)}{\partial r^2} = \frac{1}{c_L^2}\frac{\partial^2(r\varphi)}{\partial t^2}. \tag{4.28}$$

Eq. (4.27) is consistent with the general representation given by eqs. (2.87) and (2.150) of chapter 2. The governing equation (4.28) could have been obtained directly from eq. (2.153) of chapter 2.

The general solution of (4.28) is

$$\varphi(r, t) = \frac{1}{r}\,\mathrm{f}\left(t - \frac{r}{c_L}\right) + \frac{1}{r}\,\mathrm{g}\left(t + \frac{r}{c_L}\right).$$

Clearly the two terms represent waves diverging from $r = 0$ and converging to $r = 0$, respectively.

4.3.2. *Pressurization of a spherical cavity*

Now let us consider the specific wave motions that are generated by the rapid pressurization of a spherical cavity in a homogeneous, isotropic, linearly elastic medium (see figure 4.2). These wave motions display polar symmetry with respect to the center of the cavity. The problem at hand is

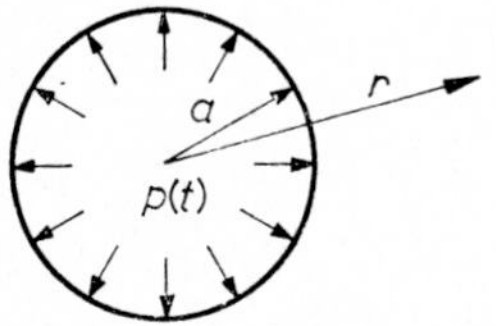

Fig. 4.2. Pressurized spherical cavity.

governed by (4.23) and (4.26), supplemented by the following boundary and initial conditions:

$$r = a,\ t \geqq 0: \qquad \tau_r = -p(t) \tag{4.29}$$

$$r \geqq a,\ t < 0: \qquad u(r, t) = \dot{u}(r, t) \equiv 0. \tag{4.30}$$

Eq. (4.30) indicates that the medium is assumed to be at rest prior to time $t = 0$.

The appropriate solution of (4.28), i.e., the one representing outgoing waves, may be expressed as

$$\varphi(r, t) = \frac{1}{r}\,\mathrm{f}(s), \tag{4.31}$$

where, instead of $t - r/c_L$, we employ the argument

$$s = t - \frac{r-a}{c_L}, \tag{4.32}$$

and where $\mathrm{f}(s) \equiv 0$ for $s < 0$. The functional form of $\mathrm{f}(s)$ depends on the boundary condition (4.29) and the initial conditions (4.30).

In terms of $\mathrm{f}(s)$, the displacements and stresses may be written as

$$u(r, t) = -\frac{\mathrm{f}'}{c_L r} - \frac{\mathrm{f}}{r^2} \tag{4.33}$$

$$\tau_r(r,t) = \frac{\rho c_L^2}{1-\nu}\left[\frac{(1-\nu)\mathrm{f}''}{c_L^2 r} + 2(1-2\nu)\left(\frac{\mathrm{f}'}{c_L r^2} + \frac{\mathrm{f}}{r^3}\right)\right] \tag{4.34}$$

$$\tau_\theta(r,t) = \frac{\rho c_L^2}{1-\nu}\left[\frac{\nu \mathrm{f}''}{c_L^2 r} - (1-2\nu)\left(\frac{\mathrm{f}'}{c_L r^2} + \frac{\mathrm{f}}{r^3}\right)\right]. \tag{4.35}$$

In eqs. (4.33)–(4.35), a prime denotes a differentiation of the function $\mathrm{f}(s)$ with respect to the argument s. We have eliminated λ and μ in favor of c_L^2 and the Poisson's ratio ν, whereby the relations of table 2.1 have been employed.

The actual functional form of $\mathrm{f}(s)$ remains to be determined. Since the displacement should be continuous at the wavefront, the initial conditions on $\mathrm{f}(s)$ are, in view of (4.30),

$$\mathrm{f}(0) = \mathrm{f}'(0) \equiv 0. \tag{4.36}$$

At $r = a$, the argument of $\mathrm{f}(s)$ reduces to $s = t$. The boundary condition (4.29) and the expression for the radial stress, (4.34), then combine to produce the following ordinary differential equation for $\mathrm{f}(s)$:

$$\frac{d^2\mathrm{f}}{ds^2} + 2\alpha\frac{d\mathrm{f}}{ds} + (\alpha^2+\beta^2)\mathrm{f} = -\frac{a}{\rho}\,p(s), \tag{4.37}$$

where

$$\alpha = \frac{1-2\nu}{1-\nu}\frac{c_L}{a}, \qquad \beta^2 = \frac{1-2\nu}{(1-\nu)^2}\frac{c_L^2}{a^2}.$$

By means of the substitution

$$\mathrm{f}(s) = g(s)\exp(-\alpha s), \tag{4.38}$$

Eq. (4.37) can be simplified to

$$\frac{d^2 g}{ds^2} + \beta^2 g = -\frac{a}{\rho}\,p(s)\exp(\alpha s). \tag{4.39}$$

The solution of (4.39) satisfying quiescent initial conditions is

$$g(s) = -\frac{a}{\rho}\frac{1}{\beta}\int_0^s p(\tau)\exp(\alpha\tau)\sin[\beta(s-\tau)]d\tau. \tag{4.40}$$

By substituting (4.40) into (4.38), by a change of the integration variable, and a subsequent substitution into (4.31), we find

$$\varphi(r,s) = -\frac{a}{\rho}\frac{1}{\beta}\frac{1}{r}\int_0^s p(s-\tau)e^{-\alpha\tau}\sin(\beta\tau)d\tau. \tag{4.41}$$

The radial displacement and the radial and the tangential stresses can be computed from (4.33), (4.34) and (4.35), respectively.

The corresponding quasistatic solution can be obtained by evaluating the limit of (4.41) for $\rho \to 0$.

$$\varphi_{st} = -\frac{a}{r}\lim_{\rho\to 0}\frac{1}{\rho\beta}\mathscr{I}\int_0^s p(s-\tau)e^{-(\alpha-i\beta)\tau}d\tau.$$

By integrating by parts we find

$$\varphi_{st} = -\frac{a^3}{r}\frac{p(t)}{\lambda+2\mu}\frac{1-\nu}{2(1-2\nu)},$$

which can also be written as

$$\varphi_{st} = -\frac{a^3}{r}\frac{p(t)}{4\mu}.$$

Eq. (4.41) can easily be evaluated for the special case

$$p(t) = p_0 H(t),$$

where $H(t)$ is the Heaviside step function. The result is

$$\varphi(r,t) = -\frac{1}{4\mu}\frac{a^3 p_0}{r}\left[1-(2-2\nu)^{\frac{1}{2}}e^{-\alpha s}\sin(\beta s+\gamma)\right]H(s), \tag{4.42}$$

where s is defined by (4.32), and

$$\gamma = \cot^{-1}(1-2\nu)^{\frac{1}{2}} \qquad (\tfrac{1}{4}\pi \leqq \gamma \leqq \tfrac{1}{2}\pi).$$

It is noted that $\varphi(r, t)$ approaches the quasistatic solution as t increases beyond bounds. The potential for a pulse of finite duration T may be found by superimposing on (4.42) the corresponding displacement potential for a pressure of the same magnitude but of opposite sign, and of indefinitely long duration starting at time T, so that the cavity surface is made free from applied pressure after time T.

For two positions the radial stresses and the circumferential normal stresses corresponding to (4.42) are plotted in figures 4.3 and 4.4, respectively. For a fixed position r, the wave arrives at time $t = (r-a)/c_L$. The static solutions are also shown in these figures and it is noted that both stresses show a dynamic overshoot of the static values. From eq. (4.42) it is seen that at a fixed position damped oscillations persist, which shows a distinction between spherical waves and plane waves. Another distinction

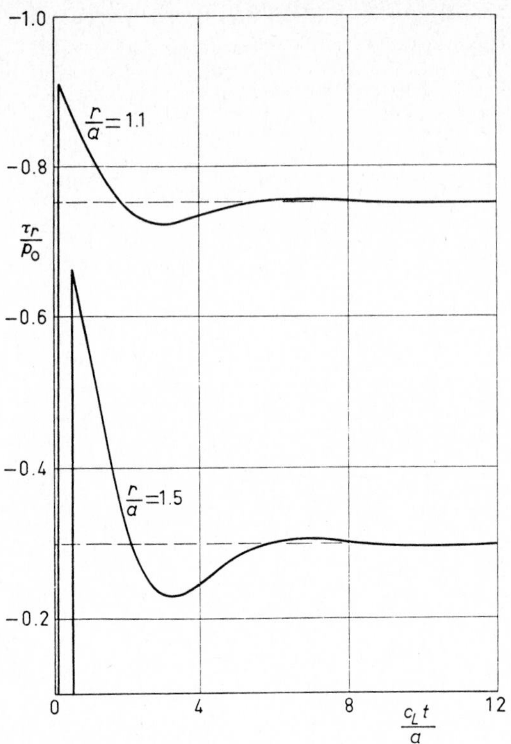

Fig. 4.3. Radial stress; – – – – denotes static solution.

is that in spherical geometry the field variables are subjected to geometrical attenuation which is at least of order r^{-1}. It is noteworthy that the circumferential normal stress changes sign as the wave passes by.

4.3.3. *Superposition of harmonic waves*

For a spherical geometry an expression representing a harmonic wave with polar symmetry follows immediately from eqs. (4.31) and (4.32) as

$$\varphi(r, t) = \frac{A}{r} \exp\left[i\omega\left(\frac{r-a}{c_L} - t\right)\right]. \tag{4.43}$$

By employing (4.34), the corresponding radial stress is found as

$$\tau_r = -\rho D(\omega, r) \frac{A}{r} \exp\left[i\omega\left(\frac{r-a}{c_L} - t\right)\right], \tag{4.44}$$

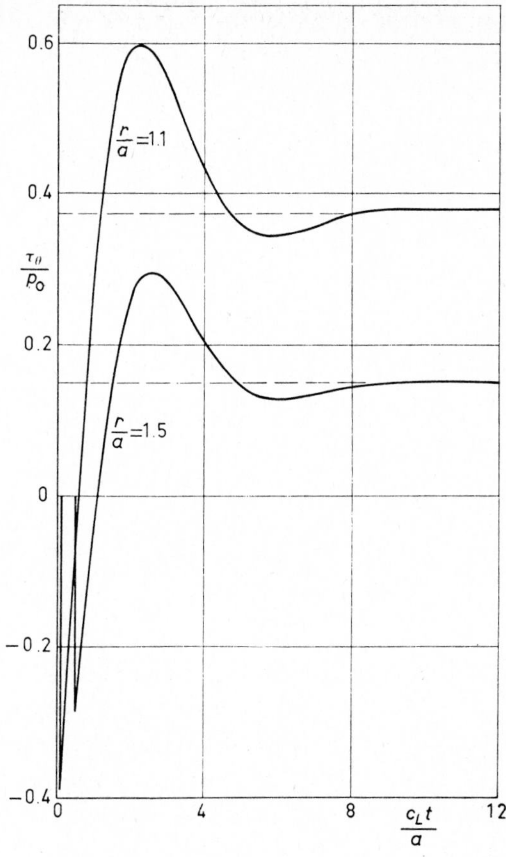

Fig. 4.4. Circumferential normal stress; — — — denotes static solution.

where

$$D(\omega, r) = \omega^2 + 2i\left(\frac{a}{r}\right)\alpha\omega - 2(1-\nu)\left(\frac{a}{r}\right)^2\beta^2, \tag{4.45}$$

and α and β are defined earlier in this section. Eqs. (4.43) and (4.44) describe outgoing waves.

Superposition of harmonic waves represented by (4.43) and (4.44) can be employed as an alternative means to obtain the wave motion due to pressurization of a spherical cavity. To this end we employ eq. (1.97) of section

1.8 and we express the boundary condition (4.29) at $r = a$ by

$$\tau_r(a, t) = -\frac{1}{2\pi}\int_{-\infty}^{\infty} p^*(\omega)e^{-i\omega t}\mathrm{d}\omega. \tag{4.46}$$

An outgoing time-harmonic spherical wave is given by eqs. (4.43) and (4.44). It follows that $\varphi(r, t)$ may be expressed as

$$\varphi(r, t) = \frac{1}{2\pi}\frac{a}{r}\frac{1}{\rho}\int_{-\infty}^{\infty}\frac{p^*(\omega)}{D(\omega)}e^{-i\omega[t-(r-a)/c_L]}\mathrm{d}\omega, \tag{4.47}$$

where $D(\omega)$ follows from (4.45) by setting $r = a$,

$$D(\omega) = \omega^2+2i\alpha\omega-2(1-\nu)\beta^2.$$

Eq. (4.47) can be evaluated by contour integration in the complex ω-plane. Let us consider the case that

$$p(t) = \frac{\varepsilon}{\pi(\varepsilon^2+t^2)}. \tag{4.48}$$

By employing eq. (1.98) of section 1.8, we find

$$p^*(\omega) = e^{-\varepsilon\omega}. \tag{4.49}$$

The pressure distribution given by (4.48) is a pulse. In the limit as $\varepsilon \to 0$, $p(t)$ is zero everywhere, except at $t = 0$, where $p(t)$ becomes unbounded. This limitcase is the well-known Dirac delta function, denoted by $\delta(t)$. From (4.49) we see that $p^*(\omega)$ reduces to unity as ε approaches zero. Although it may be considered as the limitcase of an impulse, the Dirac delta function has little physical significance as a forcing function. If it is possible to evaluate the response to a Dirac delta function, the response to other external disturbances can, however, be obtained by a superposition over time. Any function can be considered as an infinite sequence of delta functions since

$$\mathrm{f}(t) = \int_{-\infty}^{\infty}\mathrm{f}(s)\delta(t-s)\mathrm{d}s. \tag{4.50}$$

For $p(t) = \delta(t)$, eq. (4.47) reduces to

$$\varphi(r, t) = \frac{1}{2\pi}\frac{a}{r}\frac{1}{\rho}\int_{-\infty}^{\infty}\frac{1}{D(\omega)}e^{-i\omega[t-(r-a)/c_L]}\mathrm{d}\omega.$$

The roots of $D(\omega) = 0$ are computed as

$$\omega_{1,2} = \pm\frac{(1-2\nu)^{\frac{1}{2}}}{1-\nu}\frac{c_L}{a} - i\frac{1-2\nu}{1-\nu}\frac{c_L}{a}.$$

Thus the poles of the integrand are located in the lower half-plane at $\omega = \omega_1$ and $\omega = \omega_2$, respectively. For $t-(r-a)/c_L > 0$, the contour is closed in the lower half-plane and the computation of the residues yields

$$\varphi(r, t) = -\frac{a}{r}\frac{1}{\rho}\frac{1}{\beta}\exp(-\alpha s)\sin(\beta s). \tag{4.51}$$

For $t-(r-a)/c_L < 0$, the contour must be closed in the upper half-plane and we find $\varphi(r, t) \equiv 0$. Eq. (4.51) agrees with (4.41) if we substitute $p(t) = \delta(t)$ and employ the sifting property of the delta function expressed by (4.50).

4.4. Two-dimensional wave motions with axial symmetry

Transient wave motions with axial symmetry are much more difficult to analyze than wave motions with polar symmetry, even if the motion is also independent of the axial coordinate. The reason is the absence of a simple general solution for cylindrical waves.

4.4.1. Governing equations

In a system of cylindrical coordinates, motions that are independent of the axial coordinate z and the angular coordinate θ can be separated into three types of uncoupled wave motions:

Radial motions: These are governed by

$$u = \frac{\partial \varphi}{\partial r} \tag{4.52}$$

$$\tau_r = \lambda\left(\frac{\partial u}{\partial r} + \frac{u}{r}\right) + 2\mu\frac{\partial u}{\partial r} \tag{4.53}$$

$$\tau_\theta = \lambda\left(\frac{\partial u}{\partial r} + \frac{u}{r}\right) + 2\mu\frac{u}{r} \tag{4.54}$$

$$\tau_z = \lambda\left(\frac{\partial u}{\partial r} + \frac{u}{r}\right) \tag{4.55}$$

$$\frac{\partial^2 \varphi}{\partial r^2} + \frac{1}{r}\frac{\partial \varphi}{\partial r} = \frac{1}{c_L^2}\frac{\partial^2 \varphi}{\partial t^2}. \tag{4.56}$$

Rotary shear motions: These are governed by

$$v = \frac{\partial \psi_r}{\partial r} \tag{4.57}$$

$$\tau_{r\theta} = \mu \left(\frac{\partial v}{\partial r} - \frac{v}{r} \right) \tag{4.58}$$

$$\frac{\partial^2 \psi_r}{\partial r^2} + \frac{1}{r} \frac{\partial \psi_r}{\partial r} - \frac{\psi_r}{r^2} = \frac{1}{c_T^2} \frac{\partial^2 \psi_r}{\partial t^2}. \tag{4.59}$$

Axial shear motions: The governing equations are

$$w = \frac{1}{r} \frac{\partial(\psi_\theta r)}{\partial r} \tag{4.60}$$

$$\tau_{zr} = \mu \frac{\partial w}{\partial r} \tag{4.61}$$

$$\frac{\partial^2 \psi_\theta}{\partial r^2} + \frac{1}{r} \frac{\partial \psi_\theta}{\partial r} - \frac{\psi_\theta}{r^2} = \frac{1}{c_T^2} \frac{\partial^2 \psi_\theta}{\partial t^2}. \tag{4.62}$$

These sets of equations follow directly from the general system of equations stated in section 2.13.

4.4.2. *Harmonic waves*

Expressions representing time-harmonic wave motions of the radial, rotary shear or axial shear types can be obtained in a straightforward manner. To illustrate cylindrical waves of harmonic time dependence we consider radial motions.

Considering a solution of the general form

$$\varphi(r, t) = \Phi(r)e^{i\omega t}, \tag{4.63}$$

it follows from eq. (4.56) that $\Phi(r)$ must satisfy

$$\frac{d^2\Phi}{dr^2} + \frac{1}{r} \frac{d\Phi}{dr} + k_L^2 \Phi = 0, \tag{4.64}$$

where

$$k_L = \frac{\omega}{c_L}. \tag{4.65}$$

The general solution of (4.64) is

$$\Phi(r) = AH_0^{(1)}(k_L r) + BH_0^{(2)}(k_L r), \tag{4.66}$$

where $H_0^{(1)}(k_L r)$ and $H_0^{(2)}(k_L r)$ are Hankel functions of the second type. Thus, $\varphi(r, t)$ may be written as

$$\varphi(r, t) = Ae^{i\omega t}H_0^{(1)}(k_L r) + Be^{i\omega t}H_0^{(2)}(k_L r). \tag{4.67}$$

The nature of the wave motions represented by the two terms in (4.67) becomes evident by inspecting the asymptotic representations of the Hankel functions for large values of $k_L r$. In a text on Bessel functions the following asymptotic representations can be found:

$$H_0^{(1)}(k_L r) \sim \left(\frac{2}{\pi k_L r}\right)^{\frac{1}{2}} e^{i(k_L r - \pi/4)} \quad \text{as} \quad k_L r \to \infty \tag{4.68}$$

$$H_0^{(2)}(k_L r) \sim \left(\frac{2}{\pi k_L r}\right)^{\frac{1}{2}} e^{-i(k_L r - \pi/4)} \quad \text{as} \quad k_L r \to \infty. \tag{4.69}$$

It is now apparent that the first term in (4.67) represents a wave converging toward $r = 0$, while the second term represents a wave diverging from $r = 0$. Since most applications are concerned with diverging waves we will generally deal with

$$\varphi(r, t) = B e^{i\omega t} H_0^{(2)}(k_L r). \tag{4.70}$$

The Hankel function $H_0^{(2)}(k_L r)$ can be expressed in terms of Bessel functions of the first and second kinds as

$$H_0^{(2)}(k_L r) = J_0(k_L r) - i Y_0(k_L r).$$

Well-known integral representations of the Bessel functions are

$$J_0(x) = \frac{2}{\pi} \int_1^\infty \frac{\sin xs}{(s^2-1)^{\frac{1}{2}}} \, ds$$

$$Y_0(x) = -\frac{2}{\pi} \int_1^\infty \frac{\cos xs}{(s^2-1)^{\frac{1}{2}}} \, ds,$$

and thus $H_0^{(2)}(k_L r)$ may be represented by

$$H_0^{(2)}(k_L r) = \frac{2i}{\pi} \int_1^\infty \frac{e^{-ik_L rs}}{(s^2-1)^{\frac{1}{2}}} \, ds. \tag{4.71}$$

The general solution (4.70) can now of course be employed to solve a specific problem, for example, the wave motion generated by a time-harmonic pressure acting on the surface of a circular cylindrical cavity. By employing (4.53) and (4.52) the constant B can be obtained without difficulty. The details of this straightforward computation are left to the reader. Subsequently harmonic waves can be superposed to determine the wave motion generated by a pressure of arbitrary time dependence acting in a circular cylindrical cavity. Unfortunately it is rather difficult to evaluate the resulting superposition integrals.

Several authors have used Laplace transform techniques[2] to investigate transient cylindrical wave motions, see for example the paper by Kromm.[3] For an interesting problem of cylindrical waves in plane stress we refer to an article by Miklowitz[4], which is concerned with an application of the Laplace transform to examine plane-stress unloading waves emanating from a suddenly punched hole in a stretched elastic plate. The alternative to integral transform techniques are numerical methods based on the method of characteristics as discussed in section 4.8.

4.5. Propagation of wavefronts

If a medium is disturbed from a quiescent state by excitation at a boundary or within a restricted domain of the interior, neighboring domains are soon set in motion and put into states of deformation. The moving surface which separates the disturbed from the undisturbed part of the body is called the wavefront. Clearly the field quantities and/or their derivatives are discontinuous at the wavefront. In the previous sections some cases of transient wave motion were considered with plane and spherical wavefronts. These problems were amenable to an exact analysis. Generally it is not possible to determine exact solutions if the wavefront is not of a geometrically simple shape. Consider for example the wave motion generated by the sudden pressurization of a cavity of irregular shape in an unbounded medium. It is evident that it will be rather difficult to obtain expressions describing the details of the induced wave motion. For a problem of this type it is, however, possible to determine the field variables and/or their spatial and time derivatives at the wavefronts.

4.5.1. Propagating discontinuities

The computation of variables at wavefronts is based on some general results regarding propagating surfaces of discontinuity in continuous media. The basic techniques for the study of propagating surfaces of discontinuity in continuum mechanics were established toward the end of the last century. For elastic media a brief exposition can be found in the book by Love.[5] In recent years the theory was discussed in detail by T. Y. Thomas[6].

[2] Integral transform methods are discussed in chapter 7.

[3] A. Kromm, *Zeitschrift für angewandte Mathematik und Mechanik* **28** (1948), 297.

[4] J. Miklowitz, *Journal of Applied Mechanics* **27** (1960), 165.

[5] A. E. H. Love, *The mathematical theory of elasticity*, 4th edition. New York, Dover Publications (1944), p. 295.

[6] T. Y. Thomas, *Plastic flow and fracture in solids*. New York, Academic Press (1961), p. 37.

The procedures that are used are analogous to the ray-tracing and associated wavefront analysis of geometrical optics. For the scalar wave equation, wavefront analysis was discussed in great detail by Friedlander.[7] A theory for the propagation of stress discontinuities in inhomogeneous isotropic media was presented by H. Keller.[8]

The equations governing the motions of a homogeneous, isotropic, linearly elastic medium are stated in section 2.5. From the examples that have been treated in the preceding sections it can be concluded that we can impose initial and boundary conditions so that the stresses are continuous. On the other hand, we can also impose external conditions so that the stresses are discontinuous at the wavefront. It should be realized, however, that discontinuities of most kinds which occur in the mathematical analysis of physical situations are really idealizations of quantities which vary very rapidly in a small interval of space and time. Thus a discontinuous change is the mathematical description of a physical change which takes place in a thin layer.

A wavefront does not necessarily propagate into undisturbed material. In the remainder of this section it will be assumed that the material is already disturbed before the wavefront of an additional disturbance arrives.

We consider a nonstationary surface of discontinuity D, which divides a region V into two parts V_1 and V_2. The subscript 1 is used to denote the values of field variables on D when D is approached through V_1, and the subscript 2 is employed to denote the values when D is approached through V_2. If, say, the components of τ_{ij} are discontinuous across D, the jumps will be denoted by the standard bracket notation

$$[\tau_{ij}] = (\tau_{ij})_2 - (\tau_{ij})_1 . \tag{4.72}$$

We denote by $\boldsymbol{p}$ the unit normal to D pointing from V_1 to V_2 (see figure 4.5). It is assumed that the surface propagates with velocity c in the direction of the propagation vector $\boldsymbol{p}$.

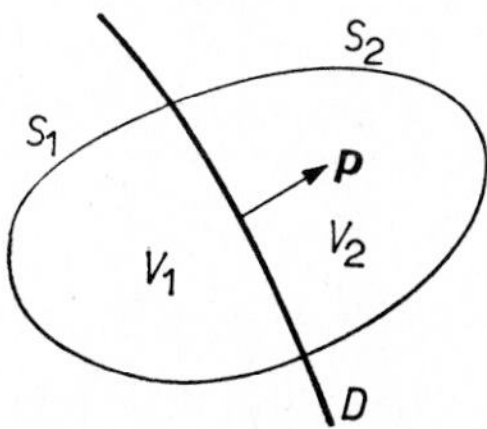

Fig. 4.5. Propagating surface of discontinuity.

[7] F. G. Friedlander, *Sound pulses*. Cambridge, University Press (1958).
[8] H. Keller, *SIAM Review* **6** (1964), 356.

At time t, the propagating surface of discontinuity is represented by an equation of the form $F(\boldsymbol{x}, t) = 0$. The position at a neighboring subsequent time $t+\Delta t$ can be derived from the position at a previous time by subjecting the points of D to a displacement, directed at each point along the normal $\boldsymbol{p}$. The equation $F(\boldsymbol{x}, t) = 0$ must then be satisfied to the first order in Δt when for x_i and t we substitute

$$x_i + cp_i \Delta t, \qquad t + \Delta t. \tag{4.73a, b}$$

It follows that

$$\frac{\partial F}{\partial t} + cF_{,i}\, p_i = 0. \tag{4.74}$$

If a surface of discontinuity is nonstationary we have $\partial F/\partial t \neq 0$, and $F(\boldsymbol{x}, t)$ may be solved for t. Thus, without loss of generality the surface may be represented by

$$D = F(\boldsymbol{x}, t) \equiv \chi(\boldsymbol{x}) - t = 0. \tag{4.75}$$

Consequently the surface in x_i-space

$$\chi(\boldsymbol{x}) = \text{constant} \tag{4.76}$$

is a surface across which certain field variables may be discontinuous at an appropriate instant of time; $\chi(\boldsymbol{x})$ is called the wave function.

By employing (4.75), eq. (4.74) becomes

$$c\chi_{,i}\, p_i = 1. \tag{4.77}$$

We also have, however, that

$$p_i = \frac{\chi_{,i}}{|\chi_{,i}|}, \tag{4.78}$$

where

$$|\chi_{,i}| = \left\{ \left(\frac{\partial \chi}{\partial x_1} \right)^2 + \left(\frac{\partial \chi}{\partial x_2} \right)^2 + \left(\frac{\partial \chi}{\partial x_3} \right)^2 \right\}^{\frac{1}{2}}. \tag{4.79}$$

It then follows from (4.77) and (4.78) that p_i may be written as

$$p_i = c\chi_{,i}. \tag{4.80}$$

4.5.2. Dynamical conditions at the wavefront

The dynamical conditions at the moving surface of discontinuity D are found by considering the impulse-momentum relation of a thin slice of the medium surrounding a small area D_0 of D. We consider the prismatic element which is bounded by D_0, by the normals to D at the edge of D_0 and by a surface parallel to D at a distance $c\Delta t$ behind the wavefront. Assuming that the medium is already in motion ahead of the wavefront, the element passes in the short time Δt from a motion defined by $(\dot{u}_i)_2$ to a motion defined by $(\dot{u}_i)_1$. The change is effected by the resultant force across D_0. The components of the resultant traction are

$$(\tau_{ij})_2 p_j - (\tau_{ij})_1 p_j, \tag{4.81}$$

where $\boldsymbol{p}$ is the normal to D. The resultant force is obtained by multiplying (4.81) by D_0, and the impulse is obtained by subsequently multiplying by Δt. The impulse-momentum relation is therefore

$$\rho D_0 c\Delta t\{(\dot{u}_i)_1 - (\dot{u}_i)_2\} = \{(\tau_{ij})_2 - (\tau_{ij})_1\} p_j D_0 \Delta t \tag{4.82}$$

or

$$[\tau_{ij}]p_j = -\rho c[\dot{u}_i], \tag{4.83}$$

where we have introduced the notation defined by (4.72).

4.5.3. *Kinematical conditions at the wavefront*

Let $\mathrm{f}_i(\boldsymbol{x}, t)$ denote a field quantity which may be discontinuous across the moving surface D. Certain relations between the time rate of change of the jump and the jumps of spatial and temporal derivatives of f_i follow directly from an examination of the motion of the moving surface.

Consider two successive positions of the moving surface D, one at time t, and the other at time $t+\Delta t$. Let the normal $\boldsymbol{p}$ at a generic point P of D intersect the surface D' at $t+\Delta t$ at a point P', where $\overline{PP'} = \Delta \boldsymbol{x}$. Denote by f'_i the one-sided limit of the function $\mathrm{f}_i(\boldsymbol{x}, t)$ at point P' at time $t+\Delta t$, and let f_i be the one-sided limit of the function $\mathrm{f}_i(\boldsymbol{x}, t)$ at point P at time t. For an observer moving with D at the normal velocity $\boldsymbol{c} = c\boldsymbol{p}$, the time rate of change of the field variable is

$$\begin{aligned}\left(\frac{\mathrm{df}_i}{\mathrm{d}t}\right)_D &= \lim_{\Delta t\to 0} \frac{\mathrm{f}'_i - \mathrm{f}_i}{\Delta t} \\ &= \frac{\partial \mathrm{f}_i}{\partial t} + \frac{\partial \mathrm{f}_i}{\partial x_j}\frac{\mathrm{d}x_j}{\mathrm{d}t} \\ &= \frac{\partial \mathrm{f}_i}{\partial t} + c\mathrm{f}_{i,j}p_j.\end{aligned} \tag{4.84}$$

Observing that

$$\mathrm{f}_{i,j}\,p_j = \frac{\partial \mathrm{f}_i}{\partial p} \tag{4.85}$$

is the derivative in the direction normal to the surface D, we find

$$\frac{\mathrm{d}_D \mathrm{f}_i}{\mathrm{d}t} = \frac{\partial \mathrm{f}_i}{\partial t} + c\,\frac{\partial \mathrm{f}_i}{\partial p}. \tag{4.86}$$

In (4.86) we have introduced the notation $\mathrm{d}_D/\mathrm{d}t$ for the time-rate of change of a quantity as observed by an observer who moves with the propagating surface D.

Assuming that f_i is discontinuous, let us replace f_i in (4.86) by $(\mathrm{f}_i)_2$ and $(\mathrm{f}_i)_1$, respectively. Subtracting the resulting equations, we obtain

$$\frac{\mathrm{d}_D[\mathrm{f}_i]}{\mathrm{d}t} = \left[\frac{\partial \mathrm{f}_i}{\partial t}\right] + c\left[\frac{\partial \mathrm{f}_i}{\partial p}\right]. \tag{4.87}$$

Eq. (4.87) is known as the kinematical condition of compatibility of the first order. If f_i is continuous across D, we have

$$\left[\frac{\partial \mathrm{f}_i}{\partial t}\right] = -c\left[\frac{\partial \mathrm{f}_i}{\partial p}\right]. \tag{4.88}$$

A first kinematical relation between spatial and temporal derivatives of the displacement components follows from the required continuity of the displacements. Since the material should maintain its integrity at the wavefront, we have

$$[u_i] = 0. \tag{4.89}$$

It then follows from (4.88) that

$$[\dot{u}_i] = -c\left[\frac{\partial u_i}{\partial p}\right]. \tag{4.90}$$

Since

$$[u_{i,j}] = \left[\frac{\partial u_i}{\partial p}\right] p_j,$$

we can also write

$$[\dot{u}_i]p_j = -c[u_{i,j}]. \tag{4.91}$$

4.5.4. Wavefronts and rays

Discontinuous derivatives of the displacement vector give rise to discontinuities of the components of the stress tensor across the surface D. From Hooke's law, eq. (2.40), we find

$$[\tau_{ij}] = \lambda\delta_{ij}[u_{k,k}]+\mu([u_{i,j}]+[u_{j,i}]).$$

Substitution of (4.91) yields

$$[\tau_{ij}] = -\frac{\lambda}{c}\delta_{ij}[\dot{u}_k]p_k - \frac{\mu}{c}([\dot{u}_i]p_j+[\dot{u}_j]p_i).$$

According to (4.83) we have

$$[\tau_{ij}]p_j = -\rho c[\dot{u}_i],$$

and thus

$$\rho c^2[\dot{u}_i] = \lambda\delta_{ij}[\dot{u}_k]p_kp_j+\mu([\dot{u}_i]p_jp_j+[\dot{u}_j]p_ip_j)$$

or

$$(\mu-\rho c^2)[\dot{u}_i]+(\lambda+\mu)[\dot{u}_k]p_kp_i = 0. \tag{4.92}$$

This equation is equivalent to eq. (4.5) for plane waves. Again we can distinguish two cases: either $[\dot{u}_i]$ is proportional to p_i, or $[\dot{u}_k]p_k = 0$. The first case defines a longitudinal wavefront. By taking the inner product of (4.92) with p_i we find

$$c^2 = c_L^2 = \frac{\lambda+2\mu}{\rho}.$$

The case $[\dot{u}_k]p_k = 0$ defines a transverse wavefront. We find

$$c^2 = c_T^2 = \frac{\mu}{\rho}.$$

The path of a point P which moves with the wavefront is always normal to the wavefront. Such an orthogonal trajectory of the family of wavefronts is called a ray. Let us consider the change of direction of the path by examining the time derivative of $\boldsymbol{p}$. By employing (4.80), we find

$$\frac{\mathrm{d}p_i}{\mathrm{d}t} = c\frac{\mathrm{d}}{\mathrm{d}t}(\chi_{,i})$$

or

$$\frac{\mathrm{d}p_i}{\mathrm{d}t} = c\frac{\partial}{\partial x_i}\left(\frac{\mathrm{d}\chi}{\mathrm{d}t}\right) = c\frac{\partial}{\partial x_i}(1) = 0,$$

where (4.75) has been used. Thus for a homogeneous material the rays are straight lines. With this information the wavefronts at times t subsequent to a time $t = t_0$ can be constructed by advancing from the wavefront at $t = t_0$ a distance $c(t-t_0)$ along a ray. This construction of wavefronts was proposed very early in the development of wave theory by Huyghens; it is known as Huyghens' principle.

It remains to determine the changes of the magnitudes as the discontinuities propagate through the medium. Here we state just the result.[9] If R and S are the principal radii of curvature at the wavefront we have for the magnitude f of a field variable at a wavefront

$$\mathrm{f}^2 RS = \text{constant} \tag{4.93}$$

along the ray. This shows that for spherical waves f decays as R^{-1}. If one of the radii, say S, is infinite, as in the case of two-dimensional wave motions, we have instead of (4.93)

$$\mathrm{f}^2 R = \text{constant}.$$

Thus for cylindrical waves f decays as $R^{-\frac{1}{2}}$.

4.6. Expansions behind the wavefront

It is possible to compute not only the magnitudes of propagating discontinuities of field variables, but also the discontinuities of their temporal derivatives. This then makes it feasible to construct Taylor expansions of field quantities at a fixed position for short times after the wavefront has passed. Denoting the position of the wavefront by $\chi(\boldsymbol{x}) = t$, the Taylor expansions are of the general form

$$\mathrm{f}_i(\boldsymbol{x}, t) = \sum_{n=0}^{\infty} \frac{1}{n!} \{t-\chi(\boldsymbol{x})\}^n \left[\frac{\partial^n \mathrm{f}_i}{\partial t^n}\right]_{t=\chi(\boldsymbol{x})}, \tag{4.94}$$

for $t \geqq \chi(\boldsymbol{x})$. The computation of the propagating discontinuities $[\partial^n \mathrm{f}_i/\partial t^n]$ will be illustrated by the example of transient rotatory shear motion.

We consider an elastic plate with a circular hole of radius a. Suppose at time $t = 0$ an in-plane shear traction is uniformly applied round the hole. The resulting motion is entirely rotary and involves shear stresses only. The one nonvanishing displacement component is $v(r, t)$, while the single component of the stress tensor is

$$\tau_{r\theta} = \mu \left(\frac{\partial v}{\partial r} - \frac{v}{r}\right). \tag{4.95}$$

[9] For a derivation we refer to H. Keller, *SIAM Review* 6 (1964) 356.

The corresponding equation of motion is of the form

$$\frac{\partial \tau_{r\theta}}{\partial r} + \frac{2\tau_{r\theta}}{r} = \rho \frac{\partial^2 v}{\partial t^2} . \tag{4.96}$$

The statement of the problem is completed by the initial condition

$$v(r, 0) = \dot{v}(r, \theta) \equiv 0, \qquad r > a, \tag{4.97}$$

and the boundary condition

$$r = a\colon \tau_{r\theta} = \tau_0 g(t). \tag{4.98}$$

The problem defined by eqs. (4.95)–(4.98) was treated by Goodier and Jahsman[10], who employed the Laplace transform technique.

For axially symmetric rotary shear waves the position of the wavefront is defined by an equation of the form

$$\chi(r) - t = 0. \tag{4.99}$$

The relation expressing conservation of linear momentum at the wavefront assumes the form

$$[\tau_{r\theta}] = -\rho c \left[\frac{\partial v}{\partial t}\right] . \tag{4.100}$$

In view of the continuity of the displacement $v(r, t)$ we have

$$\left[\frac{\partial v}{\partial r}\right] = -\frac{1}{c}\left[\frac{\partial v}{\partial t}\right], \tag{4.101}$$

where we have taken into account that the wavefront propagates in the radial direction. From the stress-strain relation (4.95) it is concluded that

$$[\tau_{r\theta}] = \mu \left[\frac{\partial v}{\partial r}\right] . \tag{4.102}$$

Combining (4.100), (4.101) and (4.102) it follows that

$$c = c_T = \left(\frac{\mu}{\rho}\right)^{\frac{1}{2}},$$

and $\chi(r)$ may thus be expressed as

$$\chi(r) = \frac{r-a}{c_T} . \tag{4.103}$$

[10] J. N. Goodier and W. E. Jahsman, *Journal of Applied Mechanics* **23** (1956) 284.

Eq. (4.99) then implies that the wavefront is located at

$$r = a + c_T t. \tag{4.104}$$

From the preceding section we also recall the kinematical condition of compatibility which for the present problem assumes the form

$$\frac{\mathrm{d}_D}{\mathrm{d}t}[\mathrm{f}] = \left[\frac{\partial \mathrm{f}}{\partial t}\right] + c_T \left[\frac{\partial \mathrm{f}}{\partial r}\right], \tag{4.105}$$

where $\mathrm{f}(r, t)$ is a discontinuous field variable.

The computational work simplifies somewhat if we introduce new variables τ and V by

$$\tau = r^2 \tau_{r\theta}, \qquad V = \frac{v}{r}. \tag{4.106a, b}$$

Eqs. (4.95) and (4.96) then reduce to

$$\tau = \mu r^3 \frac{\partial V}{\partial r} \tag{4.107}$$

$$\frac{\partial \tau}{\partial r} = \rho r^3 \frac{\partial^2 V}{\partial t^2}. \tag{4.108}$$

The conservation of momentum relation (4.100) yields

$$[\tau] = -\rho c_T r^3 \left[\frac{\partial V}{\partial t}\right].$$

In view of (4.104), this relation can also be written as

$$[\tau] = -\rho c_T (a + c_T t)^3 \left[\frac{\partial V}{\partial t}\right]. \tag{4.109}$$

Let us assume that the function $g(t)$ in eq. (4.98) can be expanded in a Maclaurin series. The condition on τ at $r = a$ can then be expressed as

$$\tau(a, t) = a^2 \tau_0 \sum_{n=0}^{\infty} g_n \frac{t^n}{n!}, \qquad t > 0. \tag{4.110}$$

Analogously to (4.94) we will seek solutions for $\tau(r, t)$ of the form

$$\tau(r, t) = \sum_{n=0}^{\infty} \frac{\{t - \chi(r)\}^n}{n!} \left[\frac{\partial^n \tau}{\partial t^n}\right]_{t=\chi(r)}. \tag{4.111}$$

It will evolve that the discontinuities $[\partial^n \tau / \partial t^n]$ satisfy simple ordinary differential equations.

Assuming that $\tau(r, t)$ is discontinuous at the wavefront, we have according to (4.105)

$$\frac{\mathrm{d}_D}{\mathrm{d}t}[\tau] = \left[\frac{\partial \tau}{\partial t}\right] + c_T \left[\frac{\partial \tau}{\partial r}\right]. \tag{4.112}$$

Employing (4.107) and (4.108), this expression can be rewritten as

$$\frac{\mathrm{d}_D}{\mathrm{d}t}[\tau] = \mu(a+c_T t)^3 \left[\frac{\partial^2 V}{\partial r\, \partial t}\right] + \rho c_T(a+c_T t)^3 \left[\frac{\partial^2 V}{\partial t^2}\right]. \tag{4.113}$$

The kinematical condition of compatibility yields for $\partial V/\partial t$

$$\frac{\mathrm{d}_D}{\mathrm{d}t}\left[\frac{\partial V}{\partial t}\right] = \left[\frac{\partial^2 V}{\partial t^2}\right] + c_T \left[\frac{\partial^2 V}{\partial r\, \partial t}\right]. \tag{4.114}$$

Combining (4.113) and (4.114) and employing (4.109), we find

$$\frac{\mathrm{d}_D}{\mathrm{d}t}[\tau] + \tfrac{1}{2}(a+c_T t)^3 \frac{\mathrm{d}_D}{\mathrm{d}t}\left\{\frac{1}{(a+c_T t)^3}[\tau]\right\} = 0. \tag{4.115}$$

The solution of (4.115) is

$$[\tau] = A_0(a+c_T t)^{\frac{3}{2}},$$

where A_0 follows from (4.110) as

$$A_0 = a^{\frac{1}{2}} \tau_0 g_0 .$$

The actual discontinuity in the shear stress follows from (4.106a) as

$$[\tau_{r\theta}] = \left(\frac{a}{a+c_T t}\right)^{\frac{1}{2}} \tau_0 g_0 . \tag{4.116}$$

As is usual for cylindrical waves the discontinuity decays with the square root of the radial distance from the cavity.

To compute the higher order discontinuities we apply the condition of compatibility (4.105) to $\partial^n \tau/\partial t^n$

$$\frac{\mathrm{d}_D}{\mathrm{d}t}\left[\frac{\partial^n \tau}{\partial t^n}\right] = \left[\frac{\partial^{n+1} \tau}{\partial t^{n+1}}\right] + c_T \left[\frac{\partial^{n+1} \tau}{\partial r\, \partial t^n}\right]. \tag{4.117}$$

We also apply (4.105) to $\partial^{n+1} V/\partial t^{n+1}$

$$\frac{\mathrm{d}_D}{\mathrm{d}t}\left[\frac{\partial^{n+1} V}{\partial t^{n+1}}\right] = \left[\frac{\partial^{n+2} V}{\partial t^{n+2}}\right] + c_T \left[\frac{\partial^{n+2} \tau}{\partial r\, \partial t^{n+1}}\right]. \tag{4.118}$$

In eq. (4.118), the derivatives can be eliminated in favor of derivatives of τ

by employing (4.107) and (4.108). On combining the result with (4.117) we find

$$\frac{\mathrm{d}_D}{\mathrm{d}t}\left[\frac{\partial^n \tau}{\partial t^n}\right] = c_T(a+c_T t)^3 \frac{\mathrm{d}_D}{\mathrm{d}t}\left\{\frac{1}{(a+c_T t)^3}\left[\frac{\partial^n \tau}{\partial r\, \partial t^{n-1}}\right]\right\}. \qquad (4.119)$$

Next we write the condition of compatibility (4.105) for $\partial^{n-1}\tau/\partial t^{n-1}$, and we eliminate $[\partial^n \tau/\partial r\, \partial t^{n-1}]$ from the result and eq. (4.119) to obtain

$$\frac{\mathrm{d}_D}{\mathrm{d}t}\left[\frac{\partial^n \tau}{\partial t^n}\right] - \frac{3}{2}\frac{c_T}{a+c_T t}\left[\frac{\partial^n \tau}{\partial t^n}\right] = F_n(t), \qquad (4.120)$$

where

$$F_n(t) = \tfrac{1}{2}(a+c_T t)^3 \frac{\mathrm{d}_D}{\mathrm{d}t}\left\{\frac{1}{(a+c_T t)^3}\frac{\mathrm{d}_D}{\mathrm{d}t}\left[\frac{\partial^{n-1}\tau}{\partial t^{n-1}}\right]\right\}.$$

The solution of eq. (4.119) is

$$\left[\frac{\partial^n \tau}{\partial t^n}\right] = (a+c_T t)^{\frac{3}{2}} \int_0^t F_n(s) \frac{\mathrm{d}s}{(a+c_T s)^{\frac{3}{2}}} + a^2 \tau_0 g_n .$$

This is a relatively simple relation expressing the discontinuity of order n in terms of the discontinuity of order $n-1$.

It was shown by Sun[11] that the method of this section can be extended to inhomogeneous and viscoelastic media.

4.7. Axial shear waves by the method of characteristics

A discussion of the characteristic surfaces of a system of partial differential equations comes best to its right if the equations are more complicated than the ones discussed in this chapter. The properties of characteristic surfaces indeed have many ramifications. These are discussed in considerable detail elsewhere.[12] In this and the next section we will very briefly touch on what is called the method of characteristics, with regard to the numerical solution of problems of elastic wave propagation in two dimensions and with axial symmetry.

We consider an unbounded elastic medium containing a circular cylindrical cavity of radius a. Suppose at time $t = 0$ an axial shear traction is

[11] C. T. Sun, *International Journal of Solids and Structures* 7 (1971), 25.

[12] R. Courant and D. Hilbert, *Methods of mathematical physics*, Vol. II. New York, Interscience Publishers (1962).

uniformly applied to the surface of the cavity. The resulting axial shear motion is governed by

$$\frac{\partial \tau_{rz}}{\partial r} + \frac{\tau_{rz}}{r} = \rho \frac{\partial^2 w}{\partial t^2} \tag{4.121}$$

and

$$\tau_{rz} = \mu \frac{\partial w}{\partial r}, \tag{4.122}$$

with the initial conditions

$$w(r, 0) = \dot{w}(r, 0) \equiv 0 \text{ for } r > a, \tag{4.123}$$

and the boundary conditions

$$r = a \qquad \tau_{rz} = \tau_0 \, g(t). \tag{4.124}$$

Eqs. (4.121) and (4.122) are analogous to eqs. (4.60)–(4.62).

For convenience of notation we introduce

$$\tau = \frac{\tau_{rz}}{\mu}, \qquad \dot{w} = \frac{\partial w}{\partial t}, \tag{4.125a, b}$$

Eq. (4.121) may then be rewritten as

$$\frac{\partial \tau}{\partial r} - \frac{1}{c_T^2} \frac{\partial \dot{w}}{\partial t} = -\frac{\tau}{r}, \tag{4.126}$$

while eq. (4.122) implies the relation

$$\frac{\partial \tau}{\partial t} - \frac{\partial \dot{w}}{\partial r} = 0. \tag{4.127}$$

Eqs. (4.126) and (4.127) form a system of coupled partial differential equations of the first order for $\tau(r, t)$ and $\dot{w}(r, t)$. Now suppose that τ and $\dot{w}$ are known on a curve defined by $C(r, t) = 0$, and suppose that we wish to find these quantities at another point, say a point close to the curve $C(r, t) = 0$. The desired values of τ and $\dot{w}$ can be computed by means of Taylor expansions if at any point of the curve $C(r, t) = 0$ we can determine the derivatives of τ and $\dot{w}$ with respect to r and t, by the use of (4.126) and (4.127). As we shall see, this is, however, not possible for certain curves, which are called the characteristic curves of (4.126) and (4.127).

We introduce new coordinates ξ and η, where $\xi = C(r, t)$. The expression for η is restricted only by the requirement that each curve η = constant shall intersect each curve ξ = constant once and once only in the domain under consideration. On the curve $\xi = 0$, the functions τ and $\dot{w}$ are given as functions of η, and thus $\partial\tau/\partial\eta$ and $\partial\dot{w}/\partial\eta$ are known. Also,

$$\frac{\partial\tau}{\partial\eta} = \frac{\partial\tau}{\partial r}\frac{\partial r}{\partial\eta} + \frac{\partial\tau}{\partial t}\frac{\partial t}{\partial\eta}$$

$$\frac{\partial\dot{w}}{\partial\eta} = \frac{\partial\dot{w}}{\partial r}\frac{\partial r}{\partial\eta} + \frac{\partial\dot{w}}{\partial t}\frac{\partial t}{\partial\eta}.$$

At any point of the curve $C(r, t) = 0$, these equations together with (4.126) and (4.127) form a system of four *algebraic* equations for the unknowns $\partial\tau/\partial r$, $\partial\tau/\partial t$, $\partial\dot{w}/\partial r$ and $\partial\dot{w}/\partial t$. In matrix notation this system may be written as

$$\begin{bmatrix} 1 & 0 & 0 & -1/c_T^2 \\ 0 & 1 & -1 & 0 \\ \partial r/\partial\eta & \partial t/\partial\eta & 0 & 0 \\ 0 & 0 & \partial r/\partial\eta & \partial t/\partial\eta \end{bmatrix} \begin{bmatrix} \partial\tau/\partial r \\ \partial\tau/\partial t \\ \partial\dot{w}/\partial r \\ \partial\dot{w}/\partial t \end{bmatrix} = \begin{bmatrix} -(1/r)\tau \\ 0 \\ \partial\tau/\partial\eta \\ \partial\dot{w}/\partial\eta \end{bmatrix} \tag{4.128}$$

With the inhomogeneous terms τ, $\partial\tau/\partial\eta$ and $\partial\dot{w}/\partial\eta$ being known functions, the system of algebraic equations will have unique solutions for the derivatives of τ and $\dot{w}$ with respect to r and t if the determinant of the coefficients does not vanish. In that case the derivatives will have the same values below and above the curve $C(r, t) = 0$. If the determinant vanishes, the derivatives cannot be determined uniquely from (4.126) and (4.127) and the values of τ and $\dot{w}$ on $C(r, t) = 0$. The curves along which the determinant vanishes are the *characteristic curves* of the system (4.126) and (4.127).

If the determinant of the coefficients is set equal to zero, the following equation is obtained:

$$\left(\frac{\partial t}{\partial\eta}\right)^2 - \frac{1}{c_T^2}\left(\frac{\partial r}{\partial\eta}\right)^2 = 0,$$

which implies

$$\frac{\partial r}{\partial t} = \pm c_T. \tag{4.129}$$

The lines in the $r-t$ plane defined by (4.129) are referred to as the c_T^+ and c_T^- characteristics, respectively. By multiplying (4.126) by dr and (4.127) by dt, and by adding the two equations, we find that (4.126) and (4.127) may be replaced by

Along c_T^+: $\mathrm{d}r/\mathrm{d}t = +c_T$

$$\mathrm{d}\tau - \frac{1}{c_T}\,\mathrm{d}\dot{w} = -\tau\frac{\mathrm{d}r}{r}, \tag{4.130}$$

Along c_T^-: $\mathrm{d}r/\mathrm{d}t = -c_T$

$$\mathrm{d}\tau + \frac{1}{c_T}\,\mathrm{d}\dot{w} = -\tau\frac{\mathrm{d}r}{r}. \tag{4.131}$$

It follows from (4.124) that at time $t = 0$ the initially undisturbed region is subjected to the following boundary condition at $r = a$:

$$\tau(a, t) = \frac{\tau_0}{\mu}\, g(t). \tag{4.132}$$

In the $r-t$ plane information is then prescribed along the lines $r = a$ and $t = 0$. Let us now consider the three points A, B and C (see figure 4.6). The lines $\overline{AC}$ and $\overline{BC}$ belong to the families of c_T^+ and c_T^- characteristics, respectively. We can thus employ (4.130) along $\overline{AC}$ and (4.131) along $\overline{BC}$.

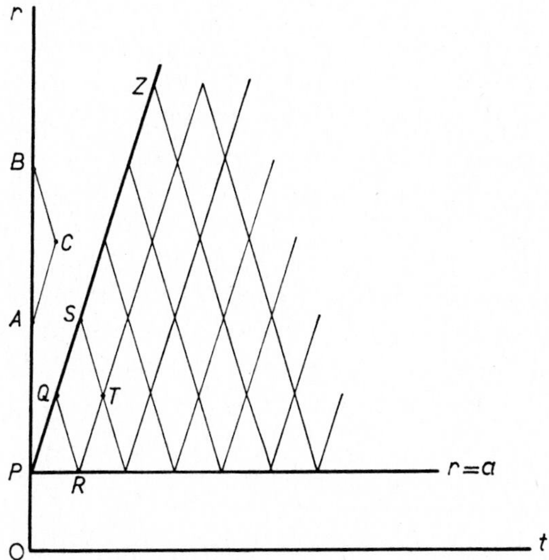

Fig. 4.6. Network of characteristic lines.

Since τ is zero at both A and B, the integration of these two equations yields, as result, that τ and $\dot{w}$ must vanish at the point C. In fact, it is now easily seen that τ and $\dot{w}$ are identically zero in the part of the $r-t$ plane defined by

$r > a + c_T t$. If the function $g(t)$ defined by (4.132) is discontinuous at $t = 0$, we must anticipate a discontinuity of the field quantities τ and $\dot{w}$ at $r = a + c_T t$.

The ordinary differential eqs. (4.130) and (4.131) can now be used to solve for $\tau(r, t)$ and $\dot{w}(r, t)$ in the domain $a \leqq r \leqq a + c_T t$. The solution is obtained by means of a numerical procedure of stepwise integration of (4.130) and (4.131) along the characteristic lines. We consider the grid system of characteristic lines in the $r-t$ plane, as shown in figure 4.6. The ordinate $r = a$ represents the boundary where the external disturbances are applied and where $\tau(a, t)$ is given by (4.132). If $\tau(a, t)$ is discontinuous in time, the discontinuity of τ along $r = a + c_T t$ and the corresponding discontinuity in $\dot{w}$ can be computed by the methods discussed in the previous sections. Alternatively, a jump in $g(t)$ can be replaced by a change with a (small) rise time, so that the stresses and the particle velocities are continuous at the wavefront. In any case we will know both τ and $\dot{w}$ along $\overline{PQS \ldots Z}$. We first compute $\dot{w}$ at point R by integration of (4.131) along $\overline{QR}$. Both τ and $\dot{w}$ are then known at R. Then we compute τ and $\dot{w}$ at point T by integrating (4.130) and (4.131) along $\overline{RT}$ and $\overline{ST}$, respectively, and solving the resulting system of two inhomogeneous algebraic equations. This process is continued. At every grid point in the region $a \leqq r \leqq a + c_T t$, τ and $\dot{w}$ can be computed in this manner. The computations require very little time on a digital computer. The smaller the characteristic grid, the more accurate the numerical results will be.

4.8. Radial motions

The method of characteristics can also conveniently be applied to investigate transient wave motions in plane strain and with axial symmetry generated by a pressure distribution in a circular cylindrical cavity. For this case the stress equations of motion are of the form

$$\frac{\partial \tau_r}{\partial r} + \frac{\tau_r - \tau_\theta}{r} = \rho \frac{\partial^2 u}{\partial t^2} \,. \tag{4.133}$$

The pertinent stress-strain relations are given by eqs. (4.53)–(4.55).

In terms of the particle velocity

$$\dot{u} = \frac{\partial u}{\partial t}, \tag{4.134}$$

the equation of motion (4.133) can be written as

$$\frac{\partial \tau_r}{\partial r} + \frac{\tau_r}{r} - \frac{\tau_\theta}{r} - \rho \frac{\partial \dot{u}}{\partial t} = 0. \tag{4.135}$$

After differentiation with respect to time, eq. (4.53) becomes

$$\frac{\partial \tau_r}{\partial t} - (\lambda + 2\mu) \frac{\partial \dot{u}}{\partial r} - \lambda \frac{\dot{u}}{r} = 0. \tag{4.136}$$

Also differentiating (4.54) and (4.55) with respect to time, and subsequently eliminating $\partial \dot{u}/\partial r$ with eq. (4.136), we find

$$\frac{\partial \tau_\theta}{\partial t} - \frac{\lambda}{\lambda+2\mu} \frac{\partial \tau_r}{\partial t} - 4 \frac{(\lambda+\mu)\mu}{\lambda+2\mu} \frac{\dot{u}}{r} = 0 \tag{4.137}$$

$$\frac{\partial \tau_z}{\partial t} - \frac{\lambda}{\lambda+2\mu} \frac{\partial \tau_r}{\partial t} - \frac{2\lambda\mu}{\lambda+2\mu} \frac{\dot{u}}{r} = 0. \tag{4.138}$$

The characteristic curves of eqs. (4.135)–(4.138) can be obtained in the manner shown in the previous section. We find

$$\frac{dr}{dt} = \pm c_L, \qquad \text{where} \quad c_L^2 = \frac{\lambda+2\mu}{\rho} \tag{4.139}$$

and

$$\frac{dr}{dt} = 0. \tag{4.140}$$

The lines in the $r-t$ plane defined by eq. (4.139) are referred to as the c_L^+ and c_L^- characteristics. Along these lines, (4.135) and (4.136) may be replaced by:

Along c_L^+: $dr/dt = c_L$

$$d\tau_r - \rho c_L d\dot{u} + \left(\tau_r - \tau_\theta - \lambda \frac{\dot{u}}{c_L}\right) \frac{dr}{r} = 0. \tag{4.141}$$

Along c_L^-: $dr/dt = -c_L$

$$d\tau_r + \rho c_L d\dot{u} - \left(\tau_r - \tau_\theta + \lambda \frac{\dot{u}}{c_L}\right) \frac{dr}{r} = 0. \tag{4.142}$$

Along the third characteristic line, which is defined by (4.140) and which is denoted the c^0 characteristic, we have, according to (4.137) and (4.138)

$$d\tau_\theta - \frac{\lambda}{\lambda+2\mu} d\tau_r - \frac{4(\lambda+\mu)\mu}{\lambda+2\mu} \frac{\dot{u}}{r} dt = 0 \tag{4.143}$$

$$d\tau_z - \frac{\lambda}{\lambda+2\mu} d\tau_r - \frac{2\lambda\mu}{\lambda+2\mu} \frac{\dot{u}}{r} dt = 0 \tag{4.144}$$

In terms of the more commonly used Young's modulus E and Poisson's ratio ν, the pertinent relations (4.141)–(4.143) may be rewritten as

$$\text{Along } c_L^+ : \mathrm{d}\tau_r - \rho c_L \mathrm{d}\dot{u} = -\left[\tau_r - \tau_\theta - \frac{\nu}{(1+\nu)(1-2\nu)} \frac{E}{c_L} \dot{u}\right] \frac{\mathrm{d}r}{r} \tag{4.145}$$

$$\text{Along } c_L^- : \mathrm{d}\tau_r + \rho c_L \mathrm{d}\dot{u} = -\left[\tau_r - \tau_\theta + \frac{\nu}{(1+\nu)(1-2\nu)} \frac{E}{c_L} \dot{u}\right] \frac{\mathrm{d}r}{r} \tag{4.146}$$

$$\text{Along } c^0 : \mathrm{d}\tau_r - \frac{1-\nu}{\nu} \mathrm{d}\tau_\theta = -\frac{E}{(1+\nu)\nu} \frac{\dot{u}}{r} \mathrm{d}t. \tag{4.147}$$

By writing eqs. (4.145)–(4.147) in finite difference form, solutions can be obtained for τ_r, τ_θ and $\dot{u}$ by numerical integration, analogously to the computations described in the previous section. Details of the numerical computations as well as several worked out examples can be found in the paper by Chou and Koenig.[13] We close this section with the remarks that a more general approach to the numerical solution of two-dimensional elastodynamic problems was presented by Clifton[14], and that the theory of characteristics was also applied to elastic wave propagation in two dimensions by Ziv.[15]

4.9. Homogeneous solutions of the wave equation

4.9.1. *Chaplygin's transformation*

A function $w(q_1, q_2, \ldots, q_n)$ is said to be homogeneous of degree m in a region R if the relation

$$w(\alpha q_1, \alpha q_2, \ldots, \alpha q_n) = \alpha^m w(q_1, q_2, \ldots, q_n) \tag{4.148}$$

holds identically for every point $(q_1, q_2, \ldots, q_n) \in R$. In this section we will investigate homogeneous functions which satisfy the two-dimensional wave equation. We will in particular consider homogeneous solutions of degree zero.

It follows from (4.148) that in Cartesian coordinates homogeneous solutions of degree zero are functions of the arguments x/t and y/t. For the present purpose it is, however, more convenient to consider the wave equation in polar coordinates

$$\frac{1}{r}\frac{\partial}{\partial r}\left(r\frac{\partial w}{\partial r}\right) + \frac{1}{r^2}\frac{\partial^2 w}{\partial \theta^2} = \frac{1}{c_T^2}\frac{\partial^2 w}{\partial t^2}. \tag{4.149}$$

[13] P. C. Chou and H. A. Koenig, *Journal of Applied Mechanics* **33** (1966), 159.
[14] R. J. Clifton, *Quarterly of Applied Mathematics* **XXV** (1967), 97.
[15] M. Ziv, *International Journal of Solids and Structures* **5** (1969), 1135.

For (4.149), homogeneous solutions of order zero are functions of r/t and θ. It is noted that eq. (4.149) governs antiplane shear motions, where $w(r/t, \theta)$ is the displacement normal to the (r, θ)-plane.

It is not immediately evident what kind of conditions on the geometry and the external excitation actually lead to solutions that are of the general form $w(r/t, \theta)$. In general terms it can be stated that the wave motion should emanate from one point, which can be a point of application of a concentrated load, a sharp corner which is struck by a wave, or another type of discontinuity which will give rise to cylindrical waves. The geometry should not include a characteristic length which would give rise to more than one center of cylindrical waves. This implies that in two dimensions the most general domain for which a homogeneous solution for all times can be found is an obtuse angled wedge. At the outset it is also not necessarily apparent what the time-dependence of the external excitation should be in order that a homogeneous solution is generated. It is therefore often necessary to proceed by a semi-inverse method, that is, a class of solutions is first constructed, and only then is the corresponding time-dependence of the external excitation determined. This is not a serious defect of the method because the response to arbitrary time-dependence can subsequently be written out in terms of a superposition integral.

The method of homogeneous solutions has been used extensively in supersonic aerodynamics, where it is known as the method of conical flows.[16]

Since we seek solutions of (4.149) depending on r/t and θ, it is expedient to introduce the new variable

$$s = \frac{r}{t}. \tag{4.150}$$

As a function of s and θ the displacement $w(s, \theta)$ must satisfy the equation

$$s^2\left(1 - \frac{s^2}{c_T^2}\right)\frac{\partial^2 w}{\partial s^2} + s\left(1 - \frac{2s^2}{c_T^2}\right)\frac{\partial w}{\partial s} + \frac{\partial^2 w}{\partial \theta^2} = 0. \tag{4.151}$$

For $s < c_T$, the following transformation

$$\beta = \cosh^{-1}\left(\frac{c_T}{s}\right), \tag{4.152}$$

which is known as *Chaplygin's transformation*, reduces eq. (4.151) to

[16] See G. N. Ward, *Linearized theory of steady high-speed flow*. Cambridge, University Press (1955), chapter 7.

Laplace's equation

$$\frac{\partial^2 w}{\partial \beta^2} + \frac{\partial^2 w}{\partial \theta^2} = 0. \tag{4.153}$$

There are many ways of obtaining solutions of Laplace's equation. A powerful method is to express $w(\beta, \theta)$ as the real part of an analytic function $\chi(\beta, \theta)$,

$$w(\beta, \theta) = \mathscr{R}\chi(\beta, \theta),$$

and to employ the theory of analytic functions to construct an appropriate analytic function of the complex variable

$$\zeta = \theta + i\beta.$$

The domain in the ζ-plane generally is a strip. From eq. (4.152) it is noted that $s = r/t = c_T$ corresponds to $\beta = 0$, while $r = 0$ corresponds to $\beta = \infty$. The strip is thus generally defined by $0 \leqq \theta \leqq \theta^*$, $0 \leqq \beta < \infty$.

For $s > c_T$, eq. (4.151) may be reduced to a simple wave equation by the transformation $s = c_T \sec \alpha$.

Now we will consider two examples.

4.9.2. *Line load*

Suppose at time $t = 0$ a concentrated antiplane line load is applied in an undisturbed medium. The load will give rise to an axially symmetric cylindrical wave with a wavefront defined by $r = c_T t$. Behind the wavefront we have $r/t < c_T$, which implies that eq. (4.153) applies. In view of the axial symmetry the dependence on θ vanishes, and the general solution of eq. (4.153) may be written as

$$w = A\beta + C. \tag{4.154}$$

Let us assume that the time-dependence of the load is such that the displacement is continuous at the wavefront. Since the material is undisturbed prior to arrival of the wavefront, the displacement is zero at $r = ct$, or equivalently

$$w = 0 \qquad \text{for} \qquad \beta = 0. \tag{4.155}$$

It follows from (4.154) and (4.155) that the constant C is zero. The displacement may thus be written as

$$w = A \cosh^{-1}\left(\frac{c_T t}{r}\right) = A \ln \frac{c_T t + (c_T^2 t^2 - r^2)^{\frac{1}{2}}}{r}. \tag{4.156}$$

To determine the constant A we compute the stress τ_{rz} as

$$\tau_{rz} = \mu \frac{\partial w}{\partial r} = -\frac{\mu A c_T t}{r(c_T^2 t^2 - r^2)^{\frac{1}{2}}}. \tag{4.157}$$

The magnitude of the concentrated force per unit length is defined as

$$P = -2\pi \lim_{r \to 0} r\tau_{rz}. \tag{4.158}$$

By inspection it now follows that eq. (4.156) is the solution for a *suddenly applied* antiplane line load of magnitude P per unit length. From (4.157) and (4.158) the constant A is computed as $A = P/2\pi\mu$. Thus, the displacement wave due to a suddenly applied antiplane line load is

$$w = \frac{P}{2\pi\mu} \ln \frac{c_T t + (c_T^2 t^2 - r^2)^{\frac{1}{2}}}{r} H(c_T t - r). \tag{4.159}$$

4.9.3. *Shear waves in an elastic wedge*

Next we consider a wedge of interior angle $\gamma\pi$ whose faces are defined by $\theta = 0$ and $\theta = \gamma\pi$, respectively. The geometry is shown in figure 4.7. On the face $\theta = 0$ the wedge is subjected to a uniform but time-dependent shear traction $\tau_{\theta z}$. No generality is lost by assuming $\gamma \geqq \frac{1}{2}$, since solutions for the case $\gamma < \frac{1}{2}$ can be obtained by symmetry considerations. The shear tractions generate horizontally polarized shear motion in the z-direction.

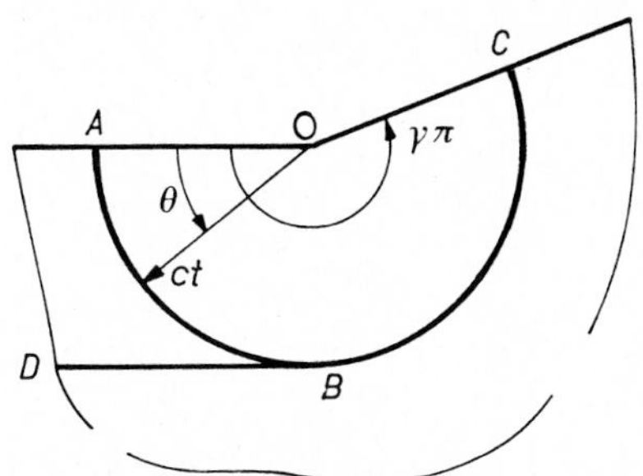

Fig. 4.7. Wavefronts at time t.

The displacement field generated by a uniform surface traction of arbitrary time-dependence can be obtained by linear superposition, once the displacement for a surface traction varying with time as the Dirac delta function has been found. As we shall see, the displacement is a homogeneous function of degree zero if the surface tractions are proportional to $\delta(t)$.

Thus we consider the following boundary conditions:

$$\theta = 0, \quad r \geqq 0: \qquad \tau_{\theta z} = \frac{\mu}{r}\frac{\partial w}{\partial \theta} = \tau_1\, \delta(t) \tag{4.160}$$

$$\theta = \gamma\pi, \quad r \geqq 0: \qquad \tau_{\theta z} = 0. \tag{4.161}$$

Some observations on the pattern of waves propagating into the wedge can be deduced from elementary principles of wave propagation. The surface traction (4.160) generates a plane wave with constant displacement

$$w_1 = -\frac{c_T \tau_1}{\mu}. \tag{4.162}$$

This wave is called the primary wave, and in figure 4.7 its wavefront at an arbitrary time t is indicated by BD. Since the wedge is at rest prior to time $t = 0$, the medium is undisturbed ahead of the wavefront BD, and as discussed above the displacement is constant behind it. In addition to the primary wave, the vertex of the wedge, as well as the nonuniformity of the surface traction across the vertex, generates a cylindrical wave with center at O. Since the displacement is continuous across the cylindrical wavefront it follows that $w \equiv 0$ along BC.

By means of eq. (4.152) the cylindrical domain $OABC$ of the wedge is mapped into a strip in the (β, θ)-plane defined by $0 \leqq \theta \leqq \gamma\pi$, $0 \leqq \beta < \infty$. Since the displacement remains bounded at $r = 0$, we have that w approaches zero as β increases. The conditions in the (β, θ)-plane are

$$\theta = 0, \quad \beta \geqq 0: \qquad \frac{\partial w}{\partial \theta} = 0 \tag{4.163}$$

$$\theta = \gamma\pi, \quad \beta \geqq 0: \qquad \frac{\partial w}{\partial \theta} = 0 \tag{4.164}$$

$$\beta = 0, \quad 0 \leqq \theta \leqq \frac{\pi}{2}: \qquad w = w_1 \tag{4.165}$$

$$\beta = 0, \quad \frac{\pi}{2} \leqq \theta \leqq \gamma\pi: \qquad w = 0. \tag{4.166}$$

In this case the appropriate solution of Laplace's equation can be obtained by elementary methods. A solution of (4.153) which satisfies the boundary conditions (4.163) and (4.164) may be written as

$$w(\beta, \theta) = \sum_{n=0}^{\infty} a_n e^{-n\beta/\gamma} \cos\left(\frac{n\theta}{\gamma}\right). \tag{4.167}$$

This solution also behaves properly, as β increases beyond bounds. From the orthogonality conditions of $\cos(n\theta/\gamma)$ over the interval $0 \leqq \theta \leqq \gamma\pi$ and the boundary conditions given by eqs. (4.165) and (4.166) the coefficients a_n follow as

$$a_n = \frac{2w_1}{n\pi} \sin\left(\frac{n\pi}{2\gamma}\right). \tag{4.168}$$

Hence the displacement is given by

$$w(\beta, \theta) = \frac{2w_1}{\pi} \sum_{n=0}^{\infty} \frac{1}{n} e^{-n\beta/\gamma} \cos\left(\frac{n\theta}{\gamma}\right) \sin\left(\frac{n\pi}{2\gamma}\right). \tag{4.169}$$

For $\gamma = \frac{1}{2}$, eq. (4.169) vanishes identically. This case corresponds to a quarter-space subjected to uniform anti-plane surface tractions, and we should indeed have just a plane primary wave. For $\gamma = 1$ and $\gamma = 2$, the series can be summed. For arbitrary values of γ it may be more convenient to obtain a closed form solution by employing a conformal mapping to map the region in the ζ-plane on an upper half-plane. This method was pursued by Achenbach.[17]

Let us consider the special case $\gamma = 1$, which corresponds to a half-space which is subjected to a uniform surface traction over half its surface. Eq. (4.169) then reduces to

$$w(\beta, \theta) = \frac{2w_1}{\pi} \sum_{n=1,3\ldots} \frac{1}{n} e^{-n\beta} \cos(n\theta) \sin(\tfrac{1}{2}n\pi).$$

This series can be rewritten in the form

$$w(\beta, \theta) = \frac{w_1}{\pi} \mathscr{I} \sum_{n=1,3\ldots} \frac{1}{n} [(p_1)^n + (p_2)^n], \tag{4.170}$$

where

$$p_1 = \exp\left[-\beta + i\left(\frac{\pi}{2} + \theta\right)\right]$$

$$p_2 = \exp\left[-\beta + i\left(\frac{\pi}{2} - \theta\right)\right].$$

Noting that each series in (4.170) is the expansion of $\tanh^{-1} p$ a closed form expression for $w(\beta, \theta)$ can be written as

$$w(\beta, \theta) = \frac{w_1}{\pi} \mathscr{I}[\tanh^{-1} p_1 + \tanh^{-1} p_2].$$

[17] J. D. Achenbach, *International Journal of Solids and Structures* **6** (1970), 379.

This expression can be simplified to

$$w(\beta, \theta) = \frac{w_1}{2\pi} \left\{ \tan^{-1} \left[\frac{\sin(\frac{1}{2}\pi + \theta)}{\sinh \beta} \right] + \tan^{-1} \left[\frac{\sin(\frac{1}{2}\pi - \theta)}{\sinh \beta} \right] \right\}. \quad (4.171)$$

From (4.152) it follows that

$$\sinh \beta = \left[\left(\frac{c_T t}{r} \right)^2 - 1 \right]^{\frac{1}{2}},$$

and eq. (4.171) thus further reduces to

$$w(r, \theta) = \frac{w_1}{\pi} \tan^{-1} \left\{ \frac{\cos \theta}{[(c_T t/r)^2 - 1]^{\frac{1}{2}}} \right\}. \quad (4.172)$$

The corresponding stress $\tau_{\theta z}$ is subsequently computed as

$$\tau_{\theta z} = \frac{[(c_T t)^2 - r^2]^{\frac{1}{2}} c_T \tau_1 \sin \theta}{(c_T t)^2 - r^2 \sin^2 \theta}, \quad (4.173)$$

where (4.162) has been used.

An equivalent method of finding homogeneous solutions of the wave equation was developed in the early 1930's by V. I. Smirnov and S. L. Sobolev. A general discussion of that method as well as a few examples can be found in the book by Smirnov.[18]

4.10. Problems

4.1. An elastic hollow sphere of inner radius a and outer radius b is filled with a rigid substance which prevents relative motion of the inner surface of the sphere. At time $t = 0$ the outer surface is subjected to a uniform pressure of magnitude p_0, which varies as a step function with time.

(a) Determine the transient wave motion in the time interval $0 \leqq t < (b-a)/c_L$.

(b) Determine the wave motion in the time interval $(b-a)/c_L \leqq t < 2(b-a)/c_L$.

(c) Discuss the limitcase of $a \to 0$.

4.2. A rigid sphere of radius a is embedded in an unbounded elastic medium, with perfect contact between the sphere and the surrounding medium. By

[18] V. I. Smirnov, *A course of higher mathematics*, English translation, Vol. III. New York, Pergamon Press and Addison-Wesley Publishing Co., Inc. (1964), p. 203.

an internal mechanism the sphere is forced to oscillate around an axis through its center. In terms of the system of spherical coordinates stated in section 2.14 the motion of a point on $r = a$ is defined by

$$u = 0, \qquad v = 0, \qquad w = w_0 \sin\theta \sin\omega t,$$

where ω is the circular frequency.

Observe that the motion induced in the surrounding medium is independent of the angular coordinate χ. Determine the steady-state displacement response of the medium.

Hint: Note that for $\partial/\partial\chi \equiv 0$ the displacement component $w(r, \theta, t)$ depends on ψ_r and ψ_θ only. Also note that the equations for ψ_r and ψ_θ do not contain φ and ψ_χ. Select solutions of the forms

$$\psi_r = \Psi_r(r) \cos\theta \sin\omega t$$

and

$$\psi_\theta = \Psi_\theta(r) \sin\theta \sin\omega t,$$

and obseıve that the ordinary differential equations for $\Psi_r(r)$ and $\Psi_\theta(r)$ can be satisfied by $\Psi_r(r) = -\Psi_\theta(r) = \Psi(r)$. Proceed to solve for $\Psi(r)$. An alternative approach proceeds directly from the displacement equations of motion in spherical coordinates.

4.3. A cylindrical cavity of radius a in an unbounded medium is lined with an elastic shell of thickness h, where $h/a \ll 1$. Contact between the shell and the wall of the cavity is perfect. The lined cavity is subjected to a spatially uniform but time-harmonic pressure distribution defined by $p(t) = p_0 (1+\sin\omega t)$. For axial symmetry the equation of motion of the thin shell may be taken as

$$\tau_r(a, t) + p(t) - \frac{h}{a^2} E' u(t) = \rho_s h \ddot{u},$$

where $\tau_r(a, t)$ is the stress in the medium at $r = a$, and $p(t)$ is the pressure inside the shell. The mass density of the shell is ρ_s and E' is defined by

$$E' = \frac{E_s}{1-\nu_s^2},$$

where E_s and ν_s are Young's modulus and Poisson's ratio for the material of the shell.

Considering the steady-state response,

(a) Determine the radial displacement at $r = a$ and at a point in the medium.

(b) Compare (a) with the corresponding result for a cavity without lining.

4.4. An elastic medium is bounded by a cylindrical parabolic surface which is defined by

$$x = ay^2,$$

as shown in the figure. At time $t = 0$ the surface is suddenly subjected to a uniform distribution of antiplane shear tractions of magnitude τ_0. The medium is at rest prior to time $t = 0$.

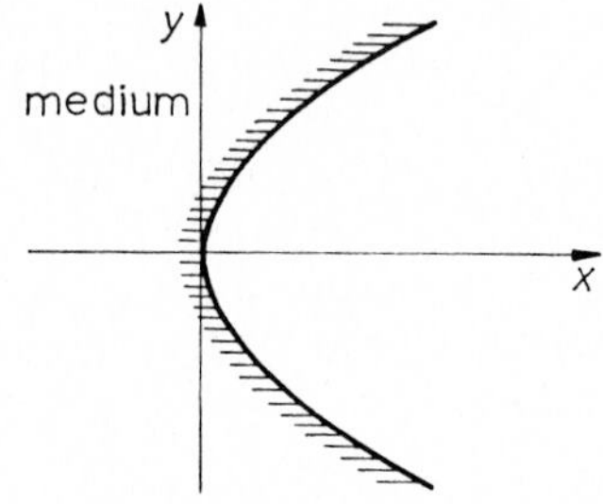

(a) Find an expression for the position of the wavefront at time t.

(b) Determine the magnitude of the discontinuity $[\tau_{nz}]$, where n is the normal to the wavefront.

4.5. The method of characteristics may be applied to investigate transient wave motions with polar symmetry generated by a pressure distribution in a spherical cavity. Derive the expressions defining the characteristics, and determine the differential equations along the characteristic curves.

4.6. Use eqs. (4.145)–(4.147) to examine the propagation of discontinuities in the stresses and the particle velocity.

4.7. A hollow sphere of inner radius a and outer radius b is embedded in an infinite medium. The inner surface of the sphere is subjected to a uniform pressure of the form $p_0 H(t)$. Contact between the sphere and the surrounding medium is perfect. The material properties of the sphere and the medium are λ, μ, ρ and λ^B, μ^B, ρ^B, respectively. Examine the transmission of waves into the surrounding medium.

4.8. A hollow elastic cylinder of inner radius a and outer radius b fits into another hollow cylinder of inner radius b and outer radius c. Contact at the surface $r = b$ is perfect. The inner surface $(r = a)$ is subjected to a spatially uniform pressure $p(t)$. The outer surface $(r = c)$ is free of tractions. The axially symmetric wave motion in the cylinders is to be analyzed numerically by the method of characteristics. Write out the equations defining the characteristics and state the differential equations along the characteristic curves. Sketch a network of characteristic lines analogous to the gridwork shown in figure 4.6. *Describe* the numerical procedure. Clearly state the computations that must be carried out for points on the boundaries $r = a$ and $r = c$, for points at the interface $r = b$, and for interior points.

4.9. Use eq. (4.169) to determine an expression for the shear stress $\tau_{\theta z}$, as a function of β and θ. Eqs. (4.150) and (4.152) imply

$$\beta = \ln \left\{ \frac{c_T t}{r} + \left[\left(\frac{c_T t}{r} \right)^2 - 1 \right]^{\frac{1}{2}} \right\}.$$

Use this result to show that the singularity of $\tau_{\theta z}$ as $r \to 0$ is of the form

$$\tau_{\theta z} \sim \frac{2 c_T \tau_1}{\pi \gamma} (2 c_T t)^{-1/\gamma} \sin\left(\frac{\pi}{2\gamma}\right) \sin\left(\frac{\theta}{\gamma}\right) r^{1/\gamma - 1}.$$

4.10. A rigid wedge of semi-vertex angle α moves in the z-direction with a constant velocity V. At time $t = 0$ the wedge strikes the surface of an ideal elastic fluid. The geometry at the instant of impact is shown in the figure.

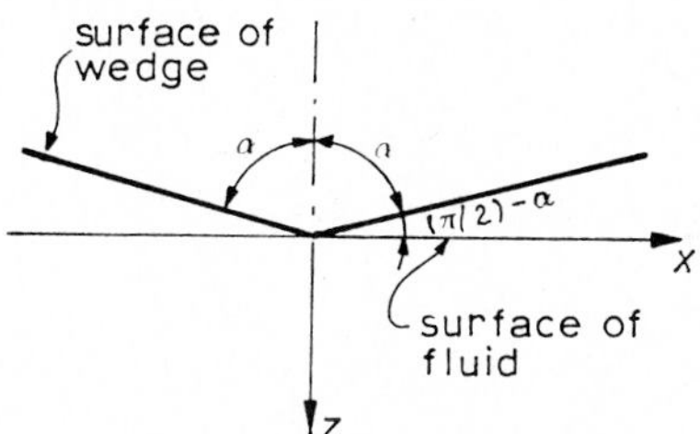

The motion of the fluid is governed by eqs. (2.175)–(2.178). The initial conditions are

$$t = 0 \qquad \varphi = \dot{\varphi} \equiv 0.$$

It is assumed that the penetration of the wedge into the fluid is small. The boundary conditions then are prescribed at $z = 0$ as

$$\frac{\partial \varphi}{\partial z}(x, 0, t) = V \qquad \text{for} \quad |x| < Vt \tan \alpha$$

$$\varphi(x, 0, t) = 0 \qquad \text{for} \quad |x| > Vt \tan \alpha.$$

Consider the case

$$M = \frac{V}{c_F} \tan \alpha > 1,$$

i.e., the region over which the velocity is prescribed moves out supersonically.

The motion of the fluid can be analyzed by employing the method of section 4.9. Show that on the wedge surface $z = 0$ the pressure is given by

$$\frac{p(x, 0, t)}{\rho V c_F} = \frac{M}{\sqrt{M^2 - 1}} \qquad \text{for} \quad c_F t < |x| < Vt \tan \alpha$$

$$\frac{p(x, 0, t)}{\rho V c_F} = \frac{2}{\pi} \frac{M}{\sqrt{M^2 - 1}} \tan^{-1} \sqrt{\frac{(M^2 - 1)(Vt \tan \alpha)^2}{(Vt \tan \alpha)^2 - (xM)^2}} \qquad \text{for} \quad |x| < c_F t.$$

CHAPTER 5

PLANE HARMONIC WAVES IN ELASTIC HALF-SPACES

5.1. Reflection and refraction at a plane interface

The presence of a discontinuity in the material properties generally produces a significant influence on systems of waves propagating through the medium. Consider, for example, the propagation of plane harmonic waves in an unbounded medium consisting of two joined elastic half-spaces of different material properties. In such a composite medium, systems of plane waves can be superposed to represent an incident wave in conjunction with reflections and refractions at the interface separating the two media. The wave which emanates from infinite depth in one of the media is called the *incident* wave. The question then is what combination of additional waves is required in order that the stresses and the displacements are continuous at the interface. These additional waves are called *reflected* and *refracted* waves. For the special case of an elastic half-space which adjoins a medium which does not transmit mechanical waves, the system of waves consists of course of incident and reflected waves only. Strictly speaking all media transmit waves, but for practical purposes refraction of elastic waves at an interface of a solid elastic body with air generally can be neglected. We will therefore examine in considerable detail the reflection of plane waves at a free surface.

The reflection and refraction pattern can be made unique by the requirement of causality. Although all the waves are steady-state traveling waves extending throughout the two joined half-spaces, the incident wave is taken to be the cause of the interface disturbance and the reflected and refracted waves are effects. This leads to the causality requirement that the reflected and refracted waves must propagate away from the interface.

Most of this chapter is concerned with plane waves representing disturbances that are uniform in planes of constant phase, i.e., in planes normal to the propagation vector. For bodies with a surface of material discontinuity there are, however, plane waves which are not uniform in planes of constant phase. These waves, which are called surface waves, propagate parallel to the surface of discontinuity. They have the property that the disturbance decays rapidly as the distance from the surface increases. For a free surface the surface waves are known as Rayleigh waves, and they are discussed in section 5.11. Surface waves at an interface of two media are called Stoneley waves.

5.2. Plane harmonic waves

For the discussion of this chapter a convenient representation of a plane harmonic displacement wave is given by

$$\boldsymbol{u} = A\boldsymbol{d}e^{i\eta}, \tag{5.1}$$

where

$$\eta = k(\boldsymbol{x}\cdot\boldsymbol{p}-ct). \tag{5.2}$$

As shown in section 4.2, eq. (5.1) describes a plane wave propagating with phase velocity c in a direction defined by the unit propagation vector $\boldsymbol{p}$. It is recalled that there are two types of plane harmonic waves:

(1) Longitudinal waves for which $\boldsymbol{d} = \pm\boldsymbol{p}$ and $c = c_L$.

(2) Transverse waves for which $\boldsymbol{d}\cdot\boldsymbol{p} = 0$ and $c = c_T$.

By substituting the components of (5.1) into Hooke's law, see eq. (2.40) of section 2.5, the components of the stress tensor are obtained as

$$\tau_{lm} = [\lambda\delta_{lm}(d_j p_j)+\mu(d_l p_m+d_m p_l)]ikAe^{i\eta}, \tag{5.3}$$

where the summation convention must be invoked.

5.3. Flux of energy in time-harmonic waves

Considering a surface element of unit area, the instantaneous rate of work of the surface traction is the scalar product of the surface traction and the particle velocity. This scalar product is called the power per unit area and it is denoted by $\mathscr{P}$,

$$\mathscr{P} = \boldsymbol{t}\cdot\dot{\boldsymbol{u}}, \tag{5.4}$$

where $\boldsymbol{t}$ is the traction vector. As discussed in chapter 1, the power per unit area defines the rate at which energy is communicated per unit area of the surface; clearly it represents the energy flux across the surface element. If the outer normal on the surface element is $\boldsymbol{n}$, we have $t_l = \tau_{lm}n_m$, and thus

$$\mathscr{P} = \tau_{lm}n_m\dot{u}_l, \tag{5.5}$$

where the summation convention is implied.

Let us examine a harmonic longitudinal wave with propagation vector $\boldsymbol{p}$, and let

$$d_1 = p_1 = \sin\theta, \quad d_2 = p_2 = \cos\theta, \quad d_3 = p_3 = 0.$$

For a surface element normal to the direction of propagation we have

$$n_1 = -p_1 = -\sin\theta, \quad n_2 = -p_2 = -\cos\theta, \quad n_3 = 0.$$

The components of the stress tensor appearing in (5.5) then are τ_{11}, τ_{12}, τ_{21}, and τ_{22}, which are obtained from (5.3) as

$$\begin{aligned}\tau_{11} &= (\lambda+2\mu\sin^2\theta)ikAe^{i\eta}\\ \tau_{12} &= \tau_{21} = (2\mu\sin\theta\cos\theta)ikAe^{i\eta}\\ \tau_{22} &= (\lambda+2\mu\cos^2\theta)ikAe^{i\eta},\end{aligned}$$

where $\eta = k(x_1\sin\theta+x_2\cos\theta-c_Lt)$.

Substituting these results into (5.5) and assuming that A is real-valued, we find

$$\mathscr{P}_L = (\lambda+2\mu)c_Lk^2A^2\mathscr{R}(ie^{i\eta})\mathscr{R}(ie^{i\eta}).$$

By employing the relation (1.87), the time-average over a period of the stress power is obtained as

$$\begin{aligned}\langle\mathscr{P}_L\rangle &= \tfrac{1}{2}(\lambda+2\mu)c_Lk^2A^2\\ &= \tfrac{1}{2}(\lambda+2\mu)\frac{\omega^2}{c_L}A^2.\end{aligned} \tag{5.6}$$

The time-average of the kinetic energy density is easily computed as

$$\langle\mathscr{K}\rangle = \tfrac{1}{4}\rho\omega^2A^2. \tag{5.7}$$

Since the time-average of the total energy is twice $\langle\mathscr{K}\rangle$, we have

$$\langle\mathscr{H}\rangle = \tfrac{1}{2}\rho\omega^2A^2. \tag{5.8}$$

By virtue of the relation $\langle\mathscr{P}\rangle = \langle\mathscr{H}\rangle c_e$, it is concluded that the velocity of energy flux is

$$c_e = c_L. \tag{5.9}$$

Similarly for a plane transverse wave, the time-average stress power is

$$\langle\mathscr{P}_T\rangle = \tfrac{1}{2}\mu\frac{\omega^2}{c_T}A^2. \tag{5.10}$$

Comparing eqs. (5.6) and (5.9) it is noted that for the same frequencies and amplitudes the average energy transmission is larger for a P-wave than for an S-wave.

5.4. Joined half-spaces

Considering in-plane motions, the system of incident, reflected and refracted waves must satisfy four conditions of continuity on the stresses and the displacements at the interface of the two half-spaces. It can therefore be expected that two reflected and two refracted waves generally will be required for each incident wave. The unit propagation vectors of the system of incident, reflected and refracted waves are shown in figure 5.1. The

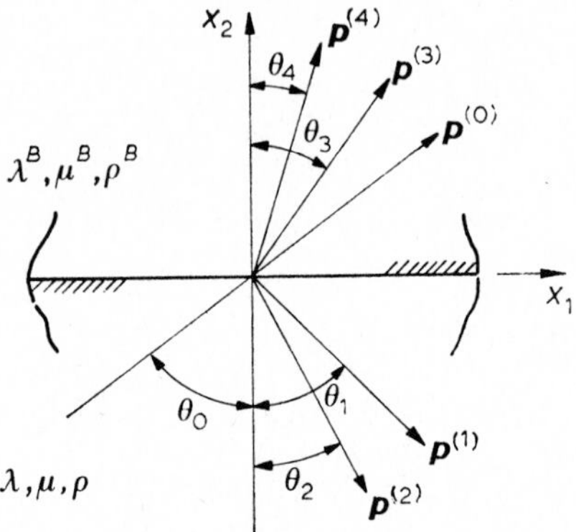

Fig. 5.1. Incident, reflected and refracted waves.

material properties of the medium carrying the incident and the reflected waves are the Lamé elastic constants λ and μ and the mass density ρ. The material constants of the medium into which refraction takes place are labeled by superscripts B, i.e., they are defined by λ^B, μ^B and ρ^B.

The incident as well as the reflected and refracted waves are represented by

$$\boldsymbol{u}^{(n)} = A_n \boldsymbol{d}^{(n)} \exp(i\eta_n), \tag{5.11}$$

where different values of the index n serve to label the various types of waves that occur, and where

$$\eta_n = k_n(x_1 p_1^{(n)} + x_2 p_2^{(n)} - c_n t). \tag{5.12}$$

The index n is assigned the value $n = 0$ for the incident wave, i.e.,

$$\boldsymbol{p}^{(0)} = \sin\theta_0 \, \boldsymbol{i}_1 + \cos\theta_0 \, \boldsymbol{i}_2 . \tag{5.13}$$

If the incident wave is a longitudinal wave we have

$$\boldsymbol{d}^{(0)} = \boldsymbol{p}^{(0)}, \qquad c_0 = c_L. \tag{5.14}$$

For an incident transverse wave we have

$$\boldsymbol{d}^{(0)} \cdot \boldsymbol{p}^{(0)} = 0, \qquad c_0 = c_T.$$

We distinguish between two types of transverse waves, the SV-waves and the SH-waves. An incident SV-wave is defined by

$$\boldsymbol{d}^{(0)} = \boldsymbol{i}_3 \wedge \boldsymbol{p}^{(0)}, \qquad c_0 = c_T. \tag{5.15}$$

For an incident SH-wave we have

$$\boldsymbol{d}^{(0)} = \boldsymbol{i}_3, \qquad c_0 = c_T. \tag{5.16}$$

The reflected and refracted waves may be both longitudinal and transverse waves. The reflected longitudinal waves and the reflected transverse waves are labeled by $n = 1$ and $n = 2$, respectively. The refracted longitudinal and transverse waves are labeled by $n = 3$ and $n = 4$, respectively. Thus we summarize:

Reflected longitudinal wave:

$$\boldsymbol{p}^{(1)} = \sin\theta_1 \boldsymbol{i}_1 - \cos\theta_1 \boldsymbol{i}_2, \qquad \boldsymbol{d}^{(1)} = \boldsymbol{p}^{(1)}, \qquad c_1 = c_L \tag{5.17}$$

Reflected transverse wave:

$$\boldsymbol{p}^{(2)} = \sin\theta_2 \boldsymbol{i}_1 - \cos\theta_2 \boldsymbol{i}_2, \qquad \boldsymbol{d}^{(2)} = \boldsymbol{i}_3 \wedge \boldsymbol{p}^{(2)} \qquad c_2 = c_T \tag{5.18}$$

Refracted longitudinal wave:

$$\boldsymbol{p}^{(3)} = \sin\theta_3 \boldsymbol{i}_1 + \cos\theta_3 \boldsymbol{i}_2, \qquad \boldsymbol{d}^{(3)} = \boldsymbol{p}^{(3)}, \qquad c_3 = c_L^B \tag{5.19}$$

Refracted transverse wave:

$$\boldsymbol{p}^{(4)} = \sin\theta_4 \boldsymbol{i}_1 + \cos\theta_4 \boldsymbol{i}_2, \qquad \boldsymbol{d}^{(4)} = \boldsymbol{i}_3 \wedge \boldsymbol{p}^{(4)}, \qquad c_4 = c_T^B. \tag{5.20}$$

For a given incident wave, the amplitudes, the unit propagation vectors and the wavenumbers of the reflected and refracted waves must be computed from the conditions on the displacements and the stresses at the interface between the two media. Incident waves are completely reflected in the special case when there is no upper medium, and the plane $x_2 = 0$ forms an external boundary. The reflection of harmonic waves at a plane boundary will be considered in the next four sections. Thereafter we will return to the joined half-spaces to examine the problem of reflection and refraction.

5.5. Reflection of SH-waves

Let us first consider the reflection of transverse waves which are horizontally polarized as defined by (5.16). Such SH-waves have displacement components in the x_3-direction only. An incident wave propagating in the half-space $x_2 < 0$ is represented by (see figure 5.2)

$$u_3^{(0)} = A_0 \exp\left[ik_0(x_1 \sin\theta_0 + x_2 \cos\theta_0 - c_T t)\right]. \tag{5.21}$$

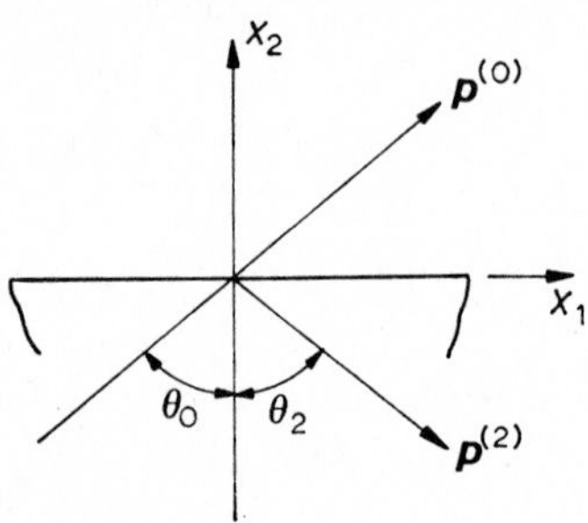

Fig. 5.2. Reflection of an SH-wave.

In the plane $x_2 = 0$ we have

$$u_3^{(0)} = A_0 \exp\left[ik_0(x_1 \sin\theta_0 - c_T t)\right].$$

It is noted that the displacement at $x_2 = 0$ can be viewed as a wave propagating in the x_1-direction with *apparent wavenumber* $k_0 \sin\theta_0$ and *apparent phase velocity* $c_T/\sin\theta_0$. Since $\sin\theta_0 \leqq 1$, the apparent phase velocity exceeds c_T, except for the case of grazing incidence, which corresponds to $\theta_0 = \pi/2$.

The relations governing the reflection depend on the boundary conditions at $x_2 = 0$. The following conditions may be considered:

(a) The displacements vanish at $x_2 = 0$.

(b) The plane $x_2 = 0$ is free of tractions.

The boundary conditions (a) and (b) correspond to a clamped and a free surface, respectively. Let us examine whether these two types of boundary conditions can be satisfied if it is assumed that an incident SH-wave is reflected as an SH-wave.

A reflected SH-wave is represented by

$$u_3^{(2)} = A_2 \exp\left[ik_2(x_1 \sin\theta_2 - x_2 \cos\theta_2 - c_T t)\right]. \tag{5.22}$$

If the displacements are to vanish at $x_2 = 0$ we have $u_3 = u_3^{(0)} + u_3^{(2)} \equiv 0$, i.e.,

$$A_0 \exp[ik_0(x_1 \sin\theta_0 - c_T t)] + A_2 \exp[ik_2(x_1 \sin\theta_2 - c_T t)] = 0. \quad (5.23)$$

Since this equation must be satisfied for all values of x_1 and t, the exponentials must be equal, which requires

$$k_0 \sin\theta_0 = k_2 \sin\theta_2, \qquad k_0 = k_2.$$

Substituting these results into (5.23) it also follows that the amplitudes must be equal but opposite in sign. We conclude that

$$k_2 = k_0, \qquad \theta_2 = \theta_0, \qquad A_2 = -A_0. \qquad (5.24\text{a, b, c})$$

It is noted that the wave undergoes a 180° phase shift in the displacement as it is reflected.

To examine the reflection at a plane that is free of tractions, we write out the components of the stress tensor τ_{2j}. By employing Hooke's law it is found that for an incident wave of the form (5.21) the one nontrivial stress component is τ_{23}, which equals $\mu\, u_{3,2}$. For the incident wave we find at $x_2 = 0$

$$\tau_{23}^{(0)} = ik_0 \mu A_0 \cos\theta_0 \exp[ik_0(x_1 \sin\theta_0 - c_T t)].$$

The reflected SH-wave yields

$$\tau_{23}^{(2)} = -ik_2 \mu A_2 \cos\theta_2 \exp[ik_2(x_1 \sin\theta_2 - c_T t)].$$

The requirement that $\tau_{23}^{(0)} + \tau_{23}^{(2)}$ vanishes at $x_2 = 0$ can be satisfied only if

$$k_2 = k_0, \qquad \theta_2 = \theta_0, \qquad A_2 = A_0. \qquad (5.25\text{a, b, c})$$

For a free boundary the reflected SH-wave thus is in phase with the incident wave.

For the clamped surface the superposition of the incident and the reflected wave is computed by employing (5.21), (5.22) and (5.24a, b, c). The result is

$$u_3 = 2iA_0 \sin(k_0 x_2 \cos\theta_0) \exp[ik_0(x_1 \sin\theta_0 - c_T t)]. \qquad (5.26)$$

Similarly for a free surface the superposition yields

$$u_3 = 2A_0 \cos(k_0 x_2 \cos\theta_0) \exp[ik_0(x_1 \sin\theta_0 - c_T t)]. \qquad (5.27)$$

Eqs. (5.26) and (5.27) represent wave motions behaving as standing waves in the x_2-direction and progressive waves in the x_1-direction. For normal incidence when $\theta_0 = 0$, the expressions (5.26) and (5.27) represent purely standing waves. If the surface is clamped, the case of grazing incidence, which corresponds to $\theta_0 = \frac{1}{2}\pi$, is not compatible with horizontally polarized wave motion.

5.6. Reflection of P-waves

In this section we examine the reflection of plane longitudinal waves whose displacement vectors and propagation vectors are situated in the $(x_1 x_2)$-plane. Employing the notation introduced in section 5.4 the incident as well as the reflected waves are denoted by

$$\boldsymbol{u}^{(n)} = A_n \boldsymbol{d}^{(n)} \exp(i\eta_n), \tag{5.28}$$

where different values of the index n serve to label the various types of waves that occur when a longitudinal wave is reflected, and where η_n is defined by (5.12). For reflection at a plane $x_2 = 0$ the relevant stresses are $\tau_{2j}^{(n)}$, where $j = 1, 2$. These components are readily computed by the use of Hooke's law, eq. (2.40), as

$$\tau_{22}^{(n)} = ik_n[(\lambda+2\mu)d_2^{(n)}p_2^{(n)}+\lambda d_1^{(n)}p_1^{(n)}]A_n \exp(i\eta_n) \tag{5.29}$$

$$\tau_{21}^{(n)} = ik_n \mu[d_2^{(n)}p_1^{(n)}+d_1^{(n)}p_2^{(n)}]A_n \exp(i\eta_n). \tag{5.30}$$

The displacements and the stresses at $x_2 = 0$ are obtained by replacing in (5.28)–(5.30) the term η_n by $\bar{\eta}_n$, where

$$\bar{\eta}_n = k_n(x_1 p_1^{(n)} - c_n t). \tag{5.31}$$

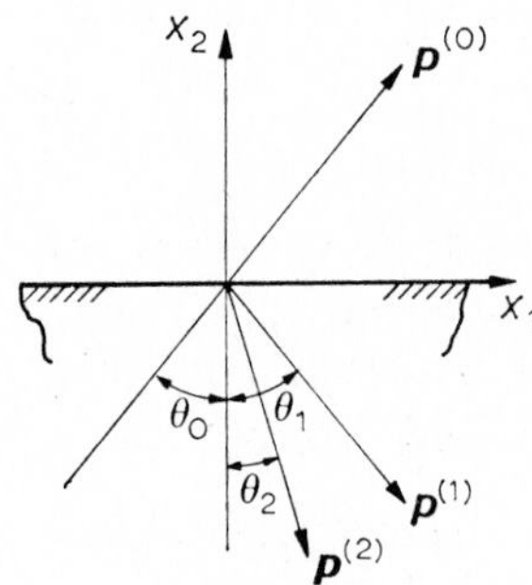

Fig. 5.3. Reflection of a P-wave.

The index n is assigned the value $n = 0$ for the incident P-wave (see figure 5.3). According to (5.13) and (5.14) we have

$$d_1^{(0)} = \sin\theta_0, \qquad d_2^{(0)} = \cos\theta_0 \tag{5.32a}$$

$$p_1^{(0)} = \sin\theta_0, \qquad p_2^{(0)} = \cos\theta_0 \tag{5.32b}$$

$$c_0 = c_L. \tag{5.32c}$$

In the plane $x_2 = 0$ the displacements and the stresses of the incident wave then become

$$u_1^{(0)} = A_0 \sin\theta_0 \exp(i\bar{\eta}_0) \tag{5.33}$$

$$u_2^{(0)} = A_0 \cos\theta_0 \exp(i\bar{\eta}_0) \tag{5.34}$$

$$\tau_{22}^{(0)} = ik_0(\lambda+2\mu\cos^2\theta_0)A_0 \exp(i\bar{\eta}_0) \tag{5.35}$$

$$\tau_{21}^{(0)} = 2ik_0\,\mu \sin\theta_0 \cos\theta_0\, A_0 \exp(i\bar{\eta}_0), \tag{5.36}$$

where

$$\bar{\eta}_0 = k_0(x_1 \sin\theta_0 - c_L t). \tag{5.37}$$

The reflected P-wave is labeled by $n = 1$. According to (5.17) we have

$$d_1^{(1)} = \sin\theta_1, \qquad d_2^{(1)} = -\cos\theta_1 \tag{5.38a}$$

$$p_1^{(1)} = \sin\theta_1, \qquad p_2^{(1)} = -\cos\theta_1 \tag{5.38b}$$

$$c_1 = c_L. \tag{5.38c}$$

It is anticipated that an incident P-wave gives rise to a reflected P-wave as well as to a reflected transverse wave with the displacement polarized in the $(x_1 x_2)$-plane. The latter type of transverse wave was earlier introduced as an SV-wave. The reflected SV-wave is labeled $n = 2$, and we have (see also figure 5.3)

$$d_1^{(2)} = \cos\theta_2, \qquad d_2^{(2)} = \sin\theta_2 \tag{5.39a}$$

$$p_1^{(2)} = \sin\theta_2, \qquad p_2^{(2)} = -\cos\theta_2 \tag{5.39b}$$

$$c_2 = c_T. \tag{5.39c}$$

The relations governing the reflection depend on the boundary conditions at the reflecting plane $x_2 = 0$. The following conditions may be considered:

(a) The plane $x_2 = 0$ is free of tractions: $\tau_{22} = \tau_{21} = 0$ (free boundary).

(b) The displacements vanish at $x_2 = 0$: $u_1 = u_2 = 0$ (clamped boundary).

(c) The normal displacement and the tangential stress vanish: $u_2 = 0$, $\tau_{21} = 0$ (smooth boundary).

(d) The tangential displacement and the normal stress vanish: $u_1 = 0$, $\tau_{22} = 0$.

Rather than consider these boundary conditions one by one, it is more efficient to first ask the question for which boundary conditions an incident

P-wave is reflected as a P-wave, under the angle $\theta_1 = \theta_0$, and with the same amplitude and wavenumber, i.e.,

$$A_2 \equiv 0, \qquad \theta_1 = \theta_0, \qquad A_1 = A_0, \qquad k_1 = k_0. \tag{5.40}$$

By employing (5.28)–(5.30), (5.38) and (5.40), we find at $x_2 = 0$ for the reflected P-wave

$$u_1^{(1)} = A_0 \sin\theta_0 \exp(i\bar{\eta}_0) \tag{5.41}$$

$$u_2^{(1)} = -A_0 \cos\theta_0 \exp(i\bar{\eta}_0) \tag{5.42}$$

$$\tau_{22}^{(1)} = ik_0(\lambda+2\mu\cos^2\theta_0)A_0 \exp(i\bar{\eta}_0) \tag{5.43}$$

$$\tau_{21}^{(1)} = -2ik_0\,\mu \sin\theta_0 \cos\theta_0\, A_0 \exp(i\bar{\eta}_0). \tag{5.44}$$

Superimposing (5.33)–(5.36) on (5.41)–(5.44) to obtain the total displacements and stresses at $x_2 = 0$, it is immediately seen that the normal displacement and the tangential stress vanish at $x_2 = 0$. Apparently (5.40) applies to the reflection of a longitudinal wave at a smooth surface. Thus, at a smooth boundary as defined under (c) a longitudinal wave is reflected as a longitudinal wave with the same amplitude and wavenumbers. Similarly we can consider

$$A_2 \equiv 0, \qquad \theta_1 = \theta_0, \qquad A_1 = -A_0,\; k_1 = k_0.$$

It is readily checked that these results apply if the tangential displacement and the normal stress vanish at $x_2 = 0$. Thus for boundary conditions (d) a longitudinal displacement wave is also reflected as a longitudinal wave only.

The two cases of "mixed" boundary conditions defined by (c) and (d) are unfortunately physically somewhat unrealistic. Of more practical significance are the reflections from a free or a clamped surface. In view of the foregoing results for mixed boundary conditions it is now expected that at a free or a clamped surface an incident longitudinal wave will generate not only a reflected P-wave but also a reflected SV-wave.

For a free surface the sum of the three tractions must vanish at $x_2 = 0$, and we obtain from eqs. (5.29), (5.30), (5.32), (5.38) and (5.39):

$$\tau_{22} = \tau_{22}^{(0)} + \tau_{22}^{(1)} + \tau_{22}^{(2)} \equiv 0:$$

$$ik_0(\lambda+2\mu\cos^2\theta_0)A_0\exp(i\bar{\eta}_0) + ik_1(\lambda+2\mu\cos^2\theta_1)A_1\exp(i\bar{\eta}_1)$$
$$-2ik_2\,\mu\sin\theta_2\cos\theta_2\,A_2\exp(i\bar{\eta}_2) = 0 \tag{5.45}$$

$$\tau_{21} = \tau_{21}^{(0)} + \tau_{21}^{(1)} + \tau_{21}^{(2)} \equiv 0:$$

$$2ik_0\,\mu\sin\theta_0\cos\theta_0\,A_0\exp(i\bar{\eta}_0) - 2ik_1\,\mu\sin\theta_1\cos\theta_1\,A_1\exp(i\bar{\eta}_1)$$
$$+ik_2\,\mu(\sin^2\theta_2 - \cos^2\theta_2)A_2\exp(i\bar{\eta}_2) = 0. \tag{5.46}$$

Eqs. (5.45) and (5.46) must be valid for all values of x_1 and t, and the exponentials must thus appear as factors in both equations, i.e.,

$$\bar{\eta}_0 = \bar{\eta}_1 = \bar{\eta}_2 .$$

Inspection of $\bar{\eta}_n$ from (5.31) then leads to the following conclusions:

$$k_0 \sin\theta_0 = k_1 \sin\theta_1 = k_2 \sin\theta_2 = k = \text{apparent wavenumber}$$
$$k_0 c_L = k_1 c_L = k_2 c_T = \omega = \text{circular frequency.}$$

These results provide, in turn, the simpler relations

$$k_1 = k_0 \tag{5.47}$$

$$k_2/k_0 = c_L/c_T = \kappa \tag{5.48}$$

$$\theta_1 = \theta_0 \tag{5.49}$$

$$\sin\theta_2 = \kappa^{-1}\sin\theta_0 . \tag{5.50}$$

The material constant κ was defined by eq. (4.8) as

$$\kappa = \left[\frac{2(1-\nu)}{1-2\nu}\right]^{\frac{1}{2}} . \tag{5.51}$$

Since $\kappa > 1$ and $\theta_2 \leqq \frac{1}{2}\pi$, it is apparent from (5.50) that $\theta_2 < \theta_1$. The wavenumber $k = k_0 \sin\theta_0$ is the wavenumber of the wave propagating along the surface $x_2 = 0$. The phase velocity along the surface $x_2 = 0$ is obtained as

$$c = \omega/k = c_L/\sin\theta_0 .$$

With the aid of (5.47)–(5.50) the algebraic equations for A_1/A_0 and A_2/A_0 can now be simplified to

$$(\lambda+2\mu\cos^2\theta_0)(A_1/A_0)-\kappa\mu\sin 2\theta_2(A_2/A_0) = -(\lambda+2\mu\cos^2\theta_0)$$
$$-\mu\sin 2\theta_0(A_1/A_0)-\kappa\mu\cos 2\theta_2(A_2/A_0) = -\mu\sin 2\theta_0 .$$

The solutions of this set of equations are

$$\frac{A_1}{A_0} = \frac{\sin 2\theta_0 \sin 2\theta_2 - \kappa^2\cos^2 2\theta_2}{\sin 2\theta_0 \sin 2\theta_2 + \kappa^2\cos^2 2\theta_2} \tag{5.52}$$

$$\frac{A_2}{A_0} = \frac{2\kappa\sin 2\theta_0\cos 2\theta_2}{\sin 2\theta_0 \sin 2\theta_2 + \kappa^2\cos^2 2\theta_2} . \tag{5.53}$$

Inspection of the amplitude ratios (5.52) and (5.53) leads to several observations:

(1) The amplitude ratios are independent of the wavelength of the incident wave, but depend only on the angle of incidence θ_0 and the material constant κ.

(2) For normal incidence, $\theta_0 = 0$, and thus $\theta_1 = 0$, $A_2/A_0 = 0$ and $A_1/A_0 = -1$. The incident P-wave is reflected as a P-wave. Since we have $A_0 d_2^{(0)} = A_1 d_2^{(1)}$, the reflected displacement wave is in phase with the incident wave. Superposition of the two waves produces a standing wave.

(3) For $\theta_0 = 90°$ (grazing incidence), the incident P-wave is again reflected as a P-wave, which is, however, 180°, out of phase with the incident wave. If the two waves are superposed the displacement vanishes altogether. An alternative way of approaching the limit $\theta_0 = 90°$, which was discussed by Goodier and Bishop[1], leads to displacements that increase linearly with the distance from the free surface.

(4) If $\sin 2\theta_0 \sin 2\theta_2 = \kappa^2 \cos^2 2\theta_2$, the incident P-wave is reflected as an SV-wave only. This phenomenon is called *mode conversion*. The amplitude of the reflected SV-wave is obtained as

$$A_2/A_0 = \kappa \cot 2\theta_2 .$$

As an example, we consider $\nu = 0.25$, which corresponds to $\lambda = \mu$, when (5.51) yields $\kappa = \sqrt{3}$, and mode conversion occurs for two angles of incidence, one of which is $\theta_0 = 60°$.

In figure 5.4, the amplitude ratios A_1/A_0 and A_2/A_0 are shown versus the angle of incidence θ_0, for Poisson's ratio $\nu = 0.25$. For various values of ν,

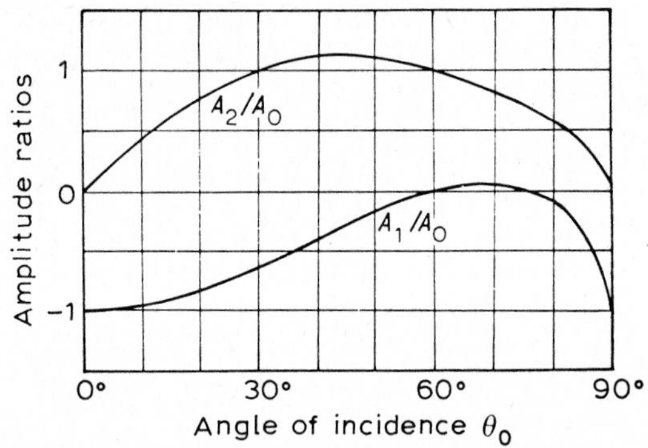

Fig. 5.4. Amplitude ratios for the reflection of a P-wave; $\nu = 0.25$. A_1/A_0 = relative amplitude of reflected P-wave. A_2/A_0 = relative amplitude of reflected SV-wave.

[1] J. N. Goodier and R. E. D. Bishop, *Journal of Applied Physics* **23** (1952), 124.

the ratio A_1/A_0 is shown versus θ_0 in figure 5.5. The curves in figure 5.5 are after results computed by Arenberg.[2]

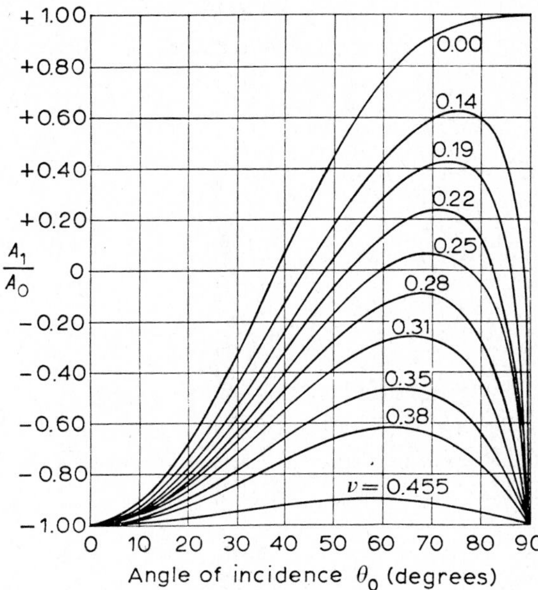

Fig. 5.5. Relative amplitude of reflected P-wave for various values of Poisson's ratio ν.

The reflection of a longitudinal wave at a fixed (clamped) boundary can be discussed in a completely analogous manner. We list just the amplitude ratios,

$$\frac{A_1}{A_0} = \frac{\cos(\theta_0 + \theta_2)}{\cos(\theta_0 - \theta_2)}$$

$$\frac{A_2}{A_0} = \frac{\sin 2\theta_0}{\cos(\theta_0 - \theta_2)},$$

where the relation between θ_2 and θ_0 is given by (5.50).

5.7. Reflection of SV-waves

To examine the reflection of incident SV-waves eqs. (5.28)–(5.30) are employed. We assign $n = 0$ to the incident SV-wave, so that

[2] D. L. Arenberg, *Journal of the Acoustical Society of America* **20** (1948), 1.

$$p_1^{(0)} = \sin\theta_0, \qquad p_2^{(0)} = \cos\theta_0 \tag{5.54a}$$

$$d_1^{(0)} = -\cos\theta_0, \qquad d_2^{(0)} = \sin\theta_0 \tag{5.54b}$$

$$c_0 = c_T. \tag{5.54c}$$

In the plane $x_2 = 0$, the displacements and the stresses of the incident wave then are of the forms

$$u_1^{(0)} = -A_0 \cos\theta_0 \exp(i\bar{\eta}_0) \tag{5.55}$$

$$u_2^{(0)} = A_0 \sin\theta_0 \exp(i\bar{\eta}_0) \tag{5.56}$$

$$\tau_{22}^{(0)} = 2ik_0\mu \sin\theta_0 \cos\theta_0 A_0 \exp(i\bar{\eta}_0) \tag{5.57}$$

$$\tau_{21}^{(0)} = ik_0\mu(\sin^2\theta_0 - \cos^2\theta_0)A_0 \exp(i\bar{\eta}_0), \tag{5.58}$$

where

$$\bar{\eta}_0 = k_0(x_1 \sin\theta_0 - c_T t). \tag{5.59}$$

We will consider only the case that the reflecting surface $x_2 = 0$ is free of tractions. For that case it is to be expected that an incident SV-wave generates a reflected P-wave as well as a reflected SV-wave. If the tractions vanish at the reflecting surface we have at $x_2 = 0$:

$$\tau_{22} = \tau_{22}^{(0)} + \tau_{22}^{(1)} + \tau_{22}^{(2)} \equiv 0:$$

$$2ik_0\mu \sin\theta_0 \cos\theta_0 A_0 \exp(i\bar{\eta}_0) + ik_1(\lambda + 2\mu\cos^2\theta_1)A_1 \exp(i\bar{\eta}_1)$$
$$-2ik_2\mu \sin\theta_2 \cos\theta_2 A_2 \exp(i\bar{\eta}_2) = 0 \tag{5.60}$$

$$\tau_{21} = \tau_{21}^{(0)} + \tau_{21}^{(1)} + \tau_{21}^{(2)} \equiv 0$$

$$ik_0\mu(\sin^2\theta_0 - \cos^2\theta_0)A_0 \exp(i\bar{\eta}_0) - 2ik_1\mu \sin\theta_1 \cos\theta_1 A_1 \exp(i\bar{\eta}_1)$$
$$+ik_2\mu(\sin^2\theta_2 - \cos^2\theta_2)A_2 \exp(i\bar{\eta}_2) = 0, \tag{5.61}$$

where, as in the previous section, the quantities associated with the reflected P- and SV-waves are denoted by $n = 1$ and $n = 2$, respectively. Since the exponentials must appear as factors, we conclude from the definition of $\bar{\eta}_n$

$$k_0 \sin\theta_0 = k_1 \sin\theta_1 = k_2 \sin\theta_2 = k$$

$$k_0 c_T = k_1 c_L = k_2 c_T = \omega.$$

These equations yield the simpler results

$$k_2 = k_0 \tag{5.62}$$

$$k_1/k_0 = c_T/c_L = \kappa^{-1} \tag{5.63}$$

$$\theta_2 = \theta_0 \tag{5.64}$$

$$\sin\theta_1 = \kappa \sin\theta_0. \tag{5.65}$$

By employing (5.62)–(5.65), the algebraic equations for A_1/A_0 and A_2/A_0 can be simplified to

$$(\lambda+2\mu\cos^2\theta_1)(A_1/A_0)-\kappa\mu\sin 2\theta_0(A_2/A_0) = -\kappa\mu\sin 2\theta_0$$
$$-\mu\sin 2\theta_1(A_1/A_0)-\kappa\mu\cos 2\theta_0(A_2/A_0) = +\kappa\mu\cos 2\theta_0.$$

The solutions to this set of equations are

$$\frac{A_1}{A_0} = -\frac{\kappa\sin 4\theta_0}{\sin 2\theta_0\sin 2\theta_1+\kappa^2\cos^2 2\theta_0} \tag{5.66}$$

$$\frac{A_2}{A_0} = \frac{\sin 2\theta_0\sin 2\theta_1-\kappa^2\cos^2 2\theta_0}{\sin 2\theta_0\sin 2\theta_1+\kappa^2\cos^2 2\theta_0}. \tag{5.67}$$

From the expression for A_1/A_0 it is observed that the reflected P-wave vanishes for $\theta_0 = 0$, $\theta_0 = \pi/4$, $\theta_0 = \pi/2$. For these particular values of θ_0, the incident SV-wave is reflected as an SV-wave. An incident SV-wave is reflected as a P-wave if the numerator of (5.67) vanishes. We must, however, also consider the relation (5.65), which shows that θ_1 is a real-valued angle only if θ_0 is smaller than the critical angle θ_{cr}, where

$$\theta_{cr} = \sin^{-1}(1/\kappa).$$

For example, if $\nu = 0.25$ ($\lambda = \mu$), we have $\kappa = \sqrt{3}$, and the angle of incidence must satisfy the restriction $\theta_0 < 35°\ 16'$ in order that θ_1 be real-valued. For $\theta_0 = \theta_{cr}$, (5.65)–(5.67) become

$$\theta_1 = \frac{\pi}{2}, \qquad \frac{A_1}{A_0} = \frac{4(\kappa^2-1)^{\frac{1}{2}}}{\kappa(2-\kappa^2)}, \qquad \frac{A_2}{A_0} = -1.$$

If $\theta_0 > \theta_{cr}$, the component $p_2^{(1)}$ becomes

$$p_2^{(1)} = -\cos\theta_1 = -i\kappa\beta,$$

where

$$\beta = (\sin^2\theta_0-\kappa^{-2})^{\frac{1}{2}}.$$

The reflected P-wave may then be written as

$$\boldsymbol{u}^{(1)} = S\boldsymbol{d}^{(1)}\exp(k_0\beta x_2)\exp[ik_0\sin\theta_0(x_1-c_L t/\kappa\sin\theta_0)-i\alpha], \tag{5.68}$$

where

$$S = \frac{A_0\sin 4\theta_0}{[\kappa^2\cos^4 2\theta_0+4(\kappa^2\sin^2\theta_0-1)\sin^2 2\theta_0\sin^2\theta_0]^{\frac{1}{2}}} \tag{5.69}$$

$$\tan\alpha = \frac{2(\kappa^2\sin^2\theta_0-1)^{\frac{1}{2}}\sin 2\theta_0\sin\theta_0}{\kappa\cos^2 2\theta_0}. \tag{5.70}$$

Eq. (5.68) is an example of the type of inhomogeneous plane waves that were discussed in section 4.2. Referring to (4.14), the vectors $\boldsymbol{p}'$ and $\boldsymbol{p}''$, the wavenumber k and the phase velocity c are

$$\boldsymbol{p}' = p_1^{(1)}\boldsymbol{i}_1 = \kappa \sin\theta_0\, \boldsymbol{i}_1$$

$$\boldsymbol{p}'' = p_2^{(1)}\boldsymbol{i}_2 = -\kappa(\sin^2\theta_0 - \kappa^{-2})\boldsymbol{i}_2$$

$$k = k_1 = k_0/\kappa, \qquad c = c_L.$$

Since (5.67) is the ratio of two complex conjugates, we have $|A_2/A_0| = 1$, and

$$\boldsymbol{u}^{(2)} = -A_0\, \boldsymbol{d}^{(2)} \exp\left[ik_0(x_1 \sin\theta_0 - x_2 \cos\theta_0 - c_T t) - 2i\alpha\right],$$

where α is defined by (5.70). The reflected P-wave is a wave propagating in the x_1-direction with wavenumber $k_0 \sin\theta_0$ and phase velocity c_L/κ $\sin\theta_0$. The amplitude of the reflected P-wave decays with the depth into the material (decreasing x_2). This type of wave is called a surface wave.

The amplitude ratio A_2/A_0 is plotted in figure 5.6. The critical angle θ_c, as well as the angle at which an incident SV-wave is reflected as a P-wave, is plotted in figure 5.7. The latter figure also shows the angle at which an incident P-wave is completely converted into a reflected SV-wave.

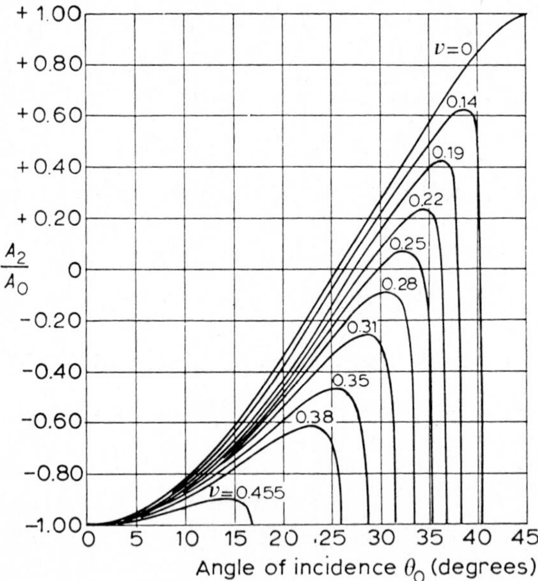

Fig. 5.6. Relative amplitude of the reflected SV-wave for various values of Poisson's ratio ν.

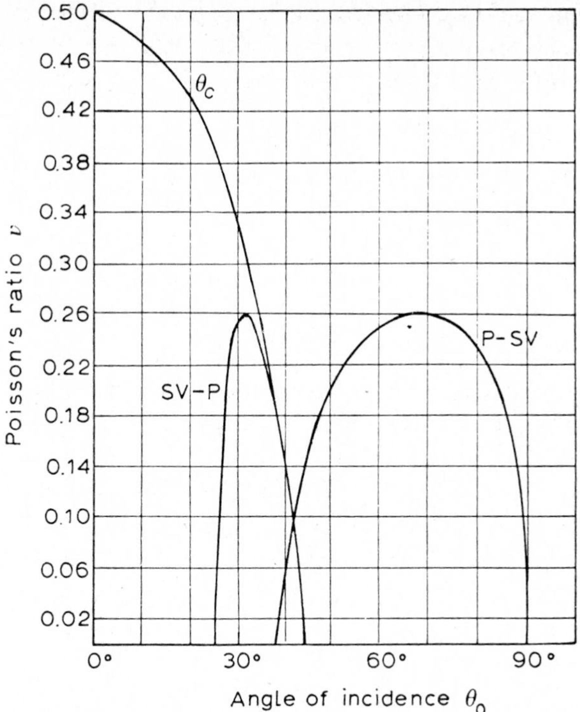

Fig. 5.7. Special angles of incidence for various values of Poisson's ratio ν. Total reflection of an SV-wave: θ_c. Reflection of an SV-wave as a P-wave: SV-P. Reflection of a P-wave as an SV-wave: P-SV.

Figures 5.6 and 5.7 are after results from the previously cited paper by Arenberg.

5.8. Reflection and partition of energy at a free surface

As shown in section 5.3, the average energy transmissions per unit area for longitudinal and transverse waves may be expressed as

$$\langle \mathscr{P}_L \rangle = \tfrac{1}{2}(\lambda + 2\mu)\frac{\omega^2}{c_L}A^2$$

and

$$\langle \mathscr{P}_T \rangle = \tfrac{1}{2}\mu \frac{\omega^2}{c_T}A^2,$$

respectively. Now let us consider (see figure 5.8) a beam of incident P-waves of cross-sectional area ΔS_0. The corresponding beams of reflected P-waves and SV-waves are of cross-sectional areas ΔS_1 and ΔS_2, respectively. Since the surface area ΔS is free of tractions and since no energy is dissipated, the

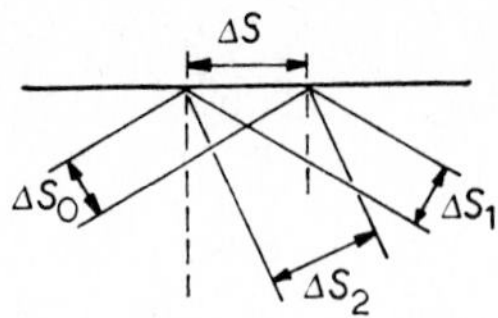

Fig. 5.8. Beams of incident and reflected waves.

average energy transmission across ΔS_0 must equal the sum of the average energy transmissions across ΔS_1 and ΔS_2. Thus

$$\tfrac{1}{2}(\lambda+2\mu)\frac{\omega^2}{c_L}(A_0)^2\Delta S_0 = \tfrac{1}{2}(\lambda+2\mu)\frac{\omega^2}{c_L}(A_1)^2\Delta S_1+\tfrac{1}{2}\mu\frac{\omega^2}{c_T}(A_2)^2\Delta S_2. \qquad (5.71)$$

By using

$$\Delta S_0 = \Delta S_1 = \Delta S\cos\theta_0, \qquad \Delta S_2 = \Delta S\cos\theta_2,$$

we find

$$\left(\frac{A_1}{A_0}\right)^2+\left(\frac{A_2}{A_0}\right)^2\frac{c_T}{c_L}\frac{\cos\theta_2}{\cos\theta_0} = 1. \qquad (5.72)$$

If θ_2 is eliminated from (5.72) by means of (5.50), we obtain

$$\left(\frac{A_1}{A_0}\right)^2+\left(\frac{A_2}{A_0}\right)^2\frac{1}{\kappa\cos\theta_0}\left(1-\frac{\sin^2\theta_0}{\kappa^2}\right)^{\frac{1}{2}} = 1. \qquad (5.73)$$

It can be checked that the previously derived amplitude ratios (5.52) and (5.53) satisfy (5.73). From (5.71) we can also determine how the average energy transmission is partitioned over the reflected P-wave and the reflected SV-wave.

5.9. Reflection and refraction of SH-waves

If the space $x_2 > 0$ is filled by another medium, either gaseous, liquid or solid, waves are transmitted across the interface $x_2 = 0$. Some of the salient aspects of reflection and refraction can be exhibited by analyzing the problem of an SH-wave incident on the interface of two solids. For this case both the reflected and the transmitted waves are SH-waves. The wave

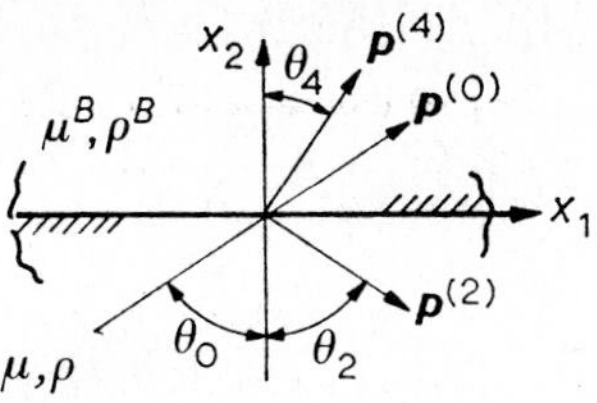

Fig. 5.9. Reflection and refraction of SH-waves.

normals are indicated in figure 5.9, using the notation which was introduced in section 5.4. The incident wave is represented by

$$u_3^{(0)} = A_0 \exp\left[ik_0(x_1 \sin\theta_0 + x_2 \cos\theta_0 - c_T t)\right].$$

On the plane $x_2 = 0$ we find

$$u_3^{(0)} = A_0 \exp\left[ik_0(x_1 \sin\theta_0 - c_T t)\right],$$

and

$$\tau_{23}^{(0)} = ik_0\,\mu \cos\theta_0\, A_0 \exp\left[ik_0(x_1 \sin\theta_0 - c_T t)\right].$$

There are no other stress and displacement components associated with the incident SH-wave. The reflected SH-wave yields at $x_2 = 0$

$$u_3^{(2)} = A_2 \exp\left[ik_2(x_1 \sin\theta_2 - c_T t)\right]$$

$$\tau_{23}^{(2)} = -ik_2\,\mu \cos\theta_2\, A_2 \exp\left[ik_2(x_1 \sin\theta_2 - c_T t)\right].$$

Identifying the material constants in the half-space $x_2 > 0$ by superscripts B, we find for the refracted SH-wave at $x_2 = 0$

$$u_3^{(4)} = A_4 \exp\left[ik_4(x_1 \sin\theta_4 - c_T^B t)\right]$$

$$\tau_{23}^{(4)} = ik_4\,\mu^B \cos\theta_4\, A_4 \exp\left[ik_4(x_1 \sin\theta_4 - c_T^B t)\right].$$

At the interface $x_2 = 0$ the displacement u_3 and the stress τ_{23} are continuous, which leads to two equations for A_2 and A_4 in terms of A_0. Since the exponentials must appear as factors, we conclude that the following relations hold true

$$k_0 \sin\theta_0 = k_2 \sin\theta_2 = k_4 \sin\theta_4$$

$$k_0\, c_T = k_2\, c_T = k_4\, c_T^B,$$

and thus

$$k_2 = k_0, \qquad \theta_2 = \theta_0 \tag{5.74}$$

$$k_4 = (c_T/c_T^B)k_0 \tag{5.75}$$

$$\sin\theta_4 = (c_T^B/c_T)\sin\theta_0. \tag{5.76}$$

The amplitude ratios are subsequently computed as

$$\frac{A_2}{A_0} = \frac{\mu \cos \theta_0 - \mu^B(c_T/c_T^B) \cos \theta_4}{\mu \cos \theta_0 + \mu^B(c_T/c_T^B) \cos \theta_4} \tag{5.77}$$

$$\frac{A_4}{A_0} = \frac{2\mu \cos \theta_0}{\mu \cos \theta_0 + \mu^B(c_T/c_T^B) \cos \theta_4}. \tag{5.78}$$

The amplitude ratios are plotted in figure 5.10 as functions of θ_0. Eqs. (5.74)–(5.78) prompt the following observations.

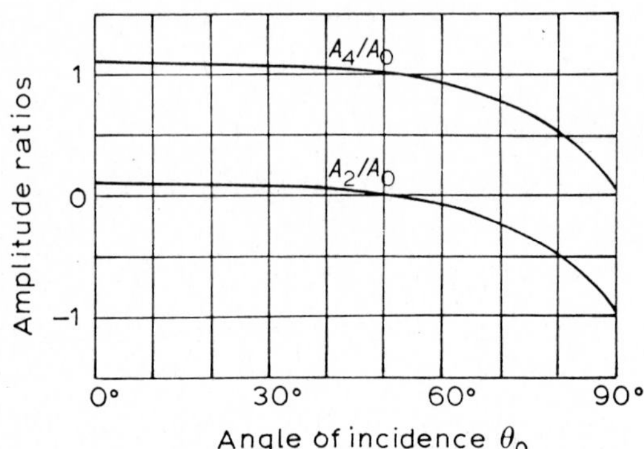

Fig. 5.10. Reflection and refraction of an incident SH-wave. $\mu^B/\mu = 0.64$, $c_T^B/c_T = 0.8$. A_2/A_0 = relative amplitude of reflected SH-wave; A_4/A_0 = relative amplitude of refracted SH-wave.

(1) The wave is completely transmitted if

$$\mu \cos \theta_0 - \mu^B(c_T/c_T^B) \cos \theta_4 = 0.$$

Thus, a combination of angle of incidence and material properties is possible for which no SH-wave is reflected.

(2) If $(c_T^B/c_T) \sin \theta_0 > 1$, $\sin \theta_4$ exceeds unity. In this case $\cos \theta_4$ is pure imaginary, and the transmitted wave takes the form

$$u_3^{(4)} = A_4 \exp(-bx_2) \exp[ik_4(x_1 \sin \theta_4 - c_T^B t)],$$

where $b = k_0[(c_T^B/c_T)^2 \sin^2 \theta_0 - 1]^{\frac{1}{2}}$. Instead of a refracted wave this gives a surface movement in the upper half-space whose amplitude diminishes exponentially with increasing distance from the interface. When $\cos \theta_4$ is imaginary, A_2/A_0 equals the ratio of two conjugate complex numbers. Hence there is only a change of phase and the amplitude of the reflected wave equals the amplitude of the incident wave. Thus, if $\sin \theta_0 > c_T/c_T^B$, we have total reflection with a shift in phase. If sin $\theta_0 = c_T/c_T^B$, i.e., if sin $\theta_4 = 1$, there is total reflection without change of phase.

(3) If $x_2 > 0$ is vacuum, there is, of course, a reflected wave only, and we find

$$k_2 = k_0, \qquad \theta_2 = \theta_0, \qquad A_2 = A_0,$$

in agreement with eq. (5.25).

5.10. Reflection and refraction of P-waves

In general it should be anticipated that a P-wave incident on the interface of two elastic solids will give rise to reflection and transmission of both P-waves and SV-waves, as shown in figure 5.1. The incident P-wave is represented by (5.33)–(5.36). From (5.28)–(5.30) we find the displacements and the stresses for the reflected P-wave and the reflected SV-wave by employing (5.28)–(5.30) with (5.38) and (5.39), respectively. For the transmitted P-wave ($n = 3$), we have

$$d_1^{(3)} = \sin\theta_3, \qquad d_2^{(3)} = \cos\theta_3 \tag{5.79a}$$

$$p_1^{(3)} = \sin\theta_3, \qquad p_2^{(3)} = \cos\theta_3 \tag{5.79b}$$

$$c_4 = c_L^B. \tag{5.79c}$$

The transmitted SV-wave ($n = 4$) is defined by

$$d_1^{(4)} = -\cos\theta_4, \qquad d_2^{(4)} = \sin\theta_4 \tag{5.80a}$$

$$p_1^{(4)} = \sin\theta_4, \qquad p_2^{(4)} = \cos\theta_4 \tag{5.80b}$$

$$c_4 = c_T^B. \tag{5.80c}$$

The wave speeds c_L^B and c_T^B are defined as

$$c_L^B = [(\lambda^B + 2\mu^B)/\rho^B]^{\frac{1}{2}}, \qquad c_T^B = (\mu^B/\rho^B)^{\frac{1}{2}}.$$

For the physically most significant case of *perfect contact* the displacements and the stresses are continuous at $x_2 = 0$:

$$u_j^{(0)} + u_j^{(1)} + u_j^{(2)} = u_j^{(3)} + u_j^{(4)} \tag{5.81}$$

$$\tau_{2j}^{(0)} + \tau_{2j}^{(1)} + \tau_{2j}^{(2)} = \tau_{2j}^{(3)} + \tau_{2j}^{(4)}, \tag{5.82}$$

where $j = 1, 2$. Eqs. (5.81) and (5.82) must be valid for all values of x_1 and t, which implies

$$\bar{\eta}_0 = \bar{\eta}_1 = \bar{\eta}_2 = \bar{\eta}_3 = \bar{\eta}_4. \tag{5.83}$$

By employing (5.31) we conclude that (5.83) implies

$$k_0 \sin\theta_0 = k_1 \sin\theta_1 = k_2 \sin\theta_2 = k_3 \sin\theta_3 = k_4 \sin\theta_4 \tag{5.84}$$

$$k_0 c_L = k_1 c_L = k_2 c_T = k_3 c_L^B = k_4 c_T^B. \tag{5.85}$$

For a given θ_0 and k_0, the other angles and wavenumbers can thus be obtained from (5.84) and (5.85). Eqs. (5.81) and (5.82) form a system of four equations for the amplitudes A_1, A_2, A_3 and A_4 in terms of A_0. In matrix notation the system can be written as

$$\begin{bmatrix} -\sin\theta_1 & -\cos\theta_2 & \sin\theta_3 & -\cos\theta_4 \\ \cos\theta_1 & -\sin\theta_2 & \cos\theta_3 & \sin\theta_4 \\ \sin 2\theta_1 & \kappa\cos 2\theta_2 & \dfrac{\mu^B}{\mu}\dfrac{c_L}{c_L^B}\sin 2\theta_3 & -\dfrac{\mu^B}{\mu}\dfrac{c_L}{c_T^B}\cos 2\theta_4 \\ -\kappa^2\cos 2\theta_2 & \kappa\sin 2\theta_2 & \dfrac{\mu^B}{\mu}\dfrac{c_L}{c_L^B}(\kappa^B)^2\cos 2\theta_4 & \dfrac{\mu^B}{\mu}\dfrac{c_L}{c_T^B}\sin 2\theta_4 \end{bmatrix} \begin{bmatrix} A_1 \\ A_2 \\ A_3 \\ A_4 \end{bmatrix}$$

$$= A_0 \begin{bmatrix} \sin\theta_0 \\ \cos\theta_0 \\ \sin 2\theta_0 \\ \kappa^2\cos 2\theta_2 \end{bmatrix} \tag{5.86}$$

In (5.86) we have used relations of the form

$$\frac{\lambda + 2\mu\cos^2\theta_0}{\mu} = \kappa^2 \cos 2\theta_2 ,$$

where $\kappa = c_L/c_T$ is defined by (5.51).

Explicit expressions for the amplitude ratios A_1/A_0, etc. are given in the book by Ewing et al.[3], where additional references are also listed. Here we will just briefly examine the special case of normal incidence which is defined by

$$\theta_0 = 0.$$

In view of (5.84), the other angles then also vanish, and from the first and the third of (5.86) it is found that

$$A_2 = A_4 \equiv 0.$$

[3] W. M. Ewing, W. S. Jardetzky and F. Press, *Elastic waves in layered media*. New York, McGraw-Hill Book Co., (1957), p. 87.

The two remaining equations yield

$$\frac{A_1}{A_0} = \frac{\rho^B c_L^B - \rho c_L}{\rho^B c_L^B + \rho c_L} \tag{5.87}$$

$$\frac{A_3}{A_0} = \frac{2\rho c_L}{\rho^B c_L^B + \rho c_L} . \tag{5.88}$$

Eq. (5.87) shows that no wave will be reflected at normal incidence when the product of the mass density and the velocity of longitudinal waves is the same for the two media. A product of the form ρc_L is known as a mechanical impedance of the medium. If $\rho^B c_L^B > \rho c_L$, the signs of A_1 and A_3 are the same. Since the direction of propagation is, however, reversed upon reflection, the wave undergoes a change of phase of 180°.

Eqs. (5.87) and (5.88) are, respectively, the reflection and the transmission coefficients for a harmonic displacement wave under normal incidence. In chapter 1, the coefficients were obtained for *stress* pulses of arbitrary shape.

5.11. Rayleigh surface waves

The possibility of a wave traveling along the free surface of an elastic half-space such that the disturbance is largely confined to the neighborhood of the boundary was considered by Rayleigh.[4]

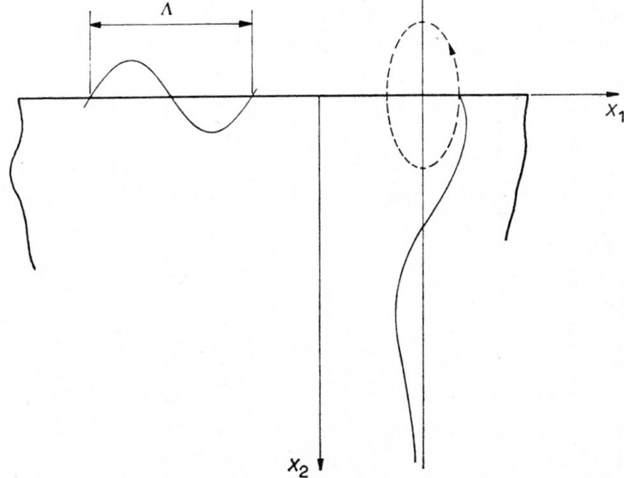

Fig. 5.11. Rayleigh waves

[4] Lord Rayleigh, *Proceedings London Mathematical Society* **17** (1887), 4.

The criterion for surface waves or Rayleigh waves is that the displacement decays exponentially with distance from the free surface. Here we investigate the existence of Rayleigh waves for the two-dimensional case of plane waves propagating in the x_1-direction (see figure 5.11).

We consider displacement components of the form,

$$u_1 = Ae^{-bx_2} \exp [ik(x_1 - ct)] \tag{5.89}$$

$$u_2 = Be^{-bx_2} \exp [ik(x_1 - ct)] \tag{5.90}$$

$$u_3 \equiv 0. \tag{5.91}$$

The real part of b is supposed to be positive, so that the displacements decrease with increasing x_2 and tend to zero as x_2 increases beyond bounds.

Substitution of eqs. (5.89)–(5.91) into the displacement equations of motion yields two homogeneous equations for the constants A and B. A nontrival solution of this system of equations exists if the determinant of the coefficients vanishes, which leads to the equation

$$[c_L^2 b^2 - (c_L^2 - c^2)k^2][c_T^2 b^2 - (c_T^2 - c^2)k^2] = 0. \tag{5.92}$$

The roots of (5.92) are

$$b_1 = k\left(1 - \frac{c^2}{c_L^2}\right)^{\frac{1}{2}}, \qquad b_2 = k\left(1 - \frac{c^2}{c_T^2}\right)^{\frac{1}{2}}.$$

It is noted that b_1 and b_2 are real and positive if $c < c_T < c_L$ and if positive roots are taken.

The ratios (B/A) corresponding to b_1 and b_2 can now be computed as

$$\left(\frac{B}{A}\right)_1 = -\frac{b_1}{ik}, \qquad \left(\frac{B}{A}\right)_2 = \frac{ik}{b_2}.$$

Returning to eqs. (5.89)–(5.91), a general solution of the displacement equations of motion may thus be written in the form

$$u_1 = [A_1 e^{-b_1 x_2} + A_2 e^{-b_2 x_2}] \exp [ik(x_1 - ct)] \tag{5.93}$$

$$u_2 = \left[-\frac{b_1}{ik} A_1 e^{-b_1 x_2} + \frac{ik}{b_2} A_2 e^{-b_2 x_2}\right] \exp [ik(x_1 - ct)]. \tag{5.94}$$

The constants A_1 and A_2 and the phase velocity c have to be chosen such that the stresses τ_{22} and τ_{21} vanish at $x_2 = 0$. By substituting eqs. (5.93), (5.94) and (5.91) into the expressions for τ_{22} and τ_{21} at $x_2 = 0$, we obtain after some manipulation

$$2b_1 A_1 + \left(2 - \frac{c^2}{c_T^2}\right) k^2 \frac{A_2}{b_2} = 0$$

$$\left(2 - \frac{c^2}{c_T^2}\right) A_1 + 2b_2 \frac{A_2}{b_2} = 0.$$

For a nontrival solution the determinant of the coefficients of A_1, A_2 must vanish, which yields the following well-known equation for the phase velocity of Rayleigh waves:

$$\left(2 - \frac{c^2}{c_T^2}\right)^2 - 4\left(1 - \frac{c^2}{c_L^2}\right)^{\frac{1}{2}} \left(1 - \frac{c^2}{c_T^2}\right)^{\frac{1}{2}} = 0. \tag{5.95}$$

It is noted that the wavenumber does not enter in (5.95), and surface waves at a free surface of an elastic half-space are thus nondispersive.

The computations leading to eq. (5.95) simplify somewhat if the displacement potentials are employed, which were introduced in section 2.10. For the two-dimensional case of plane strain the displacement components u_1 and u_2 can be expressed in terms of $\varphi(x_1, x_2, t)$ and $\psi_3(x_1, x_2, t)$. By considering expressions for the potentials of the general forms (5.89) and (5.90) it follows from the wave equations $\nabla^2\varphi = (1/c_L^2)\ddot{\varphi}$ and $\nabla^2\psi_3 = (1/c_T^2)\ddot{\psi}_3$ that

$$\varphi = Ce^{-b_1 x_2} e^{ik(x_1 - ct)}$$

and

$$\psi_3 = De^{-b_2 x_2} e^{ik(x_1 - ct)},$$

respectively. Substituting these expressions into τ_{22} and τ_{21}, and invoking the boundary conditions at $x_2 = 0$ yields eq. (5.95).

If we substitute $c = c_T$ into the left-hand side of (5.95) we obtain unity. Substitution of $c = \varepsilon c_T$, where ε is a very small number, yields $-2[1-(c_T/c_L)^2]\varepsilon^2$, which is always negative. Hence (5.95) has at least one real root lying between $c = 0$ and $c = c_T$.

The questions now arise whether there is only one real root in the interval $0 < c < c_T$, and whether there are possibly roots elsewhere along the real axis or in the complex plane. A convenient method to check on the number of roots of (5.95) is by means of the principle of the argument. This principle may be stated as follows as a theorem of the theory of complex variables: Let $G(z)$ be analytic everywhere inside and on a simple closed curve C, except for a finite number of poles inside C, and let $G(z)$ have no zeros on C. Then

$$\frac{1}{2\pi i}\int_C \frac{dG}{dz}\frac{dz}{G(z)} = Z - P,$$

where Z is the number of zeros inside C, and P is the number of poles. The numbers Z and P include the orders of poles and zeros, i.e., a pole of order three counts as three in the number P. The formula can be checked by replacing C by a sum of contours surrounding each zero or pole of $G(z)$, since these are the only singularities of the integrand. By the use of Laurent-series expansions these individual integrals are easily evaluated and their sum yields the number $Z-P$.

To apply the principle of the argument to the Rayleigh equation it is convenient to rewrite eq. (5.95) in the form

$$R(s) = (2s^2 - s_T^2)^2 + 4s^2(s_L^2 - s^2)^{\frac{1}{2}}(s_T^2 - s^2)^{\frac{1}{2}} = 0, \tag{5.96}$$

where $s = 1/c$ is the slowness of surface waves, and s_L and s_T are the slownesses of longitudinal and transverse waves, respectively:

$$s_L = \frac{1}{c_L}, \qquad s_T = \frac{1}{c_T}. \tag{5.97a, b}$$

In the complex s-plane the function $R(s)$ is rendered single-valued by introducing branch cuts along $(1/c_L) \leqq |\mathscr{R}(s)| \leqq (1/c_T)$, $\mathscr{I}(s) = 0$. Now consider the contour C consisting of Γ, and Γ_l and Γ_r as indicated in figure 5.12a. Since the function $R(s)$ clearly does not have poles in the complex

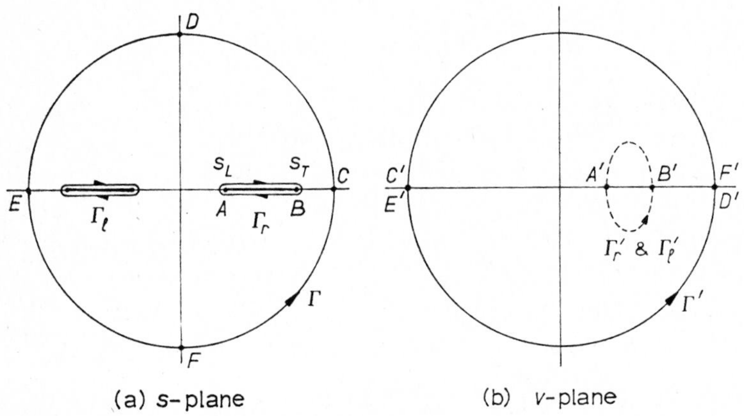

Fig. 5.12. Mapping from s- to v-plane.

s-plane, we find that within the contour $C = \Gamma + \Gamma_l + \Gamma_r$ the number of zeros is given by

$$Z = \frac{1}{2\pi i}\int_C \frac{\mathrm{d}R}{\mathrm{d}s}\,\frac{\mathrm{d}s}{R(s)}. \tag{5.98}$$

The counting of the number of zeros is carried out by mapping the s-plane on the v-plane through the relation

$$v = R(s).$$

If C_v is the mapping of C in the v-plane, the integral (5.98) in the v-plane becomes

$$\frac{1}{2\pi i}\int_{C_v} \frac{dv}{v} = Z.$$

The latter integral has a simple pole at $v = 0$, and thus Z is simply the number of times the image contour C_v encircles the origin in the v-plane in the counter-clockwise direction. To determine the number of zeros in the s-plane we thus carefully trace the mapping of the contour C into the v-plane.

Since $R(s) = R(-s)$ the images of Γ_r and Γ_l are the same, and one of them, say Γ_r, needs to be considered. We have (see figure 5.12)

at A: $$R(s_L) = (2s_L^2 - s_T^2)^2$$

along AB: $$R(s) = (2s^2 - s_T^2)^2 \mp i4s^2(s^2 - s_L^2)^{\frac{1}{2}}(s_T^2 - s^2)^{\frac{1}{2}},$$

where the minus sign applies above the cut, and the plus sign applies below the cut. Also,

at B: $$R(s_T) = s_T^4.$$

Note that along Γ_r we have $\mathscr{R}[R(s)] > 0$, and the mapping into the v-plane is thus qualitatively as indicated in figure 4.12b. For $|s|$ large, we find

$$R(s) = 2s^2(s_L^2 - s_T^2) + O(1). \tag{5.99}$$

The mappings of Γ_r and Γ_l do not encircle the origin in the v-plane but, as is seen from eq. (5.99), the mapping of Γ encircles the origin twice. In the s-plane there are thus two zeros of the equation $R(s) = 0$, and consequently eq. (5.95) also possesses two roots.

Since (5.95) is an equation for c^2 the two roots are each other's opposite. As noted earlier, eq. (5.95) shows that the roots may be expected along the real axis for $-c_T < c < c_T$. Obviously only the positive real root is of interest. The root for c^2 is usually computed by rationalizing (5.95), whereupon a cubic equation emerges, which may, however, yield three real roots for c^2. Two of these roots are extraneous; they arise from the rationalization process of squaring.

Denoting the phase velocity of Rayleigh waves by c_R, eq. (5.95) can be considered as an equation for c_R/c_T, with Poisson's ratio $\nu(0 \leqq \nu \leqq 0.5)$ as independent parameter.

TABLE 5.1

Velocity of Rayleigh waves for various values of Poisson's ratio

ν	c_R/c_T
0	0.862
0.25	0.919
0.333	0.932
0.5	0.955

For various values of Poisson's ratio the phase velocity is tabulated in table 5.1. A good approximation of c_R can be written as

$$c_R = \frac{0.862 + 1.14\nu}{1+\nu} c_T. \tag{5.100}$$

As ν varies from 0 to 0.5, the Rayleigh wave phase velocity increases monotonically from $0.862c_T$ to $0.955\,c_T$.

For two values of Poisson's ratio, figure 5.13 shows the variations of the displacements with depth. The displacements are referred to the normal displacement u_2 at the surface, and they are plotted versus the ratio of x_2

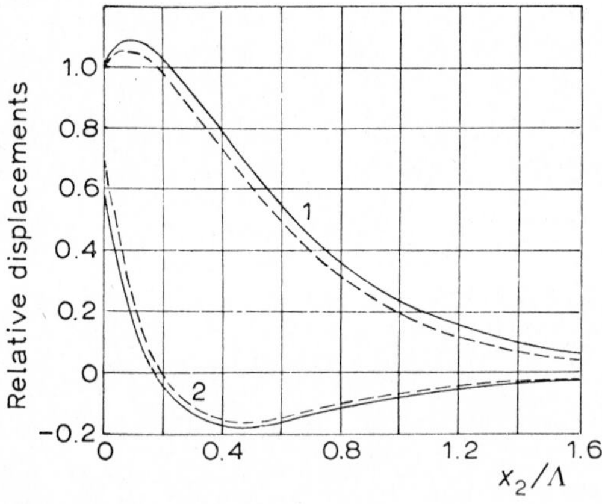

Fig. 5.13. Displacement amplitudes $u_1/u_2(x_2 = 0)$, curves 2, and $u_2/u_2(x_2 = 0)$, curves 1, for $\nu = 0.25$ (- - - -) and $\nu = 0.34$ (——).

and the wavelength. The variations of the stresses with depth are shown in figure 5.14, where the stresses are referred to τ_{11} at $x_2 = 0$.

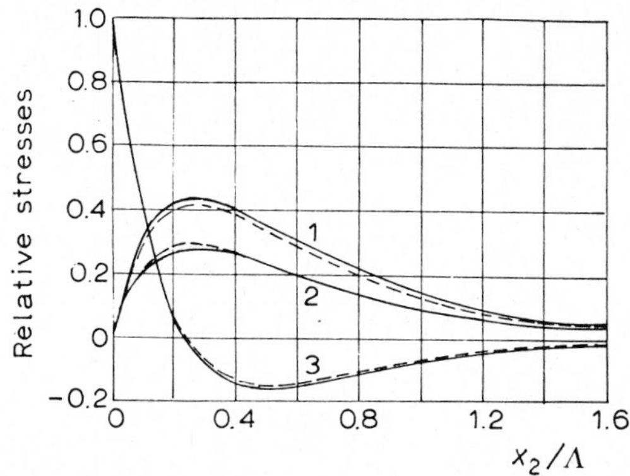

Fig. 5.14. Stress amplitudes: curves 3, $\tau_{11}/\tau_{11}(x_2 = 0)$; curves 1, $\tau_{12}/\tau_{11}(x_2 = 0)$; and, curves 2, $\tau_{22}/\tau_{11}(x_2 = 0)$ for $\nu = 0.25$ (- - - -) and $\nu = 0.34$ (——).

The figures show the localization of the wave motion in a thin layer near the surface, of a thickness which is about twice the wavelength of the surface waves.

Since the displacement components u_1 and u_2 are 90° out of phase, the trajectories of the particles are ellipses. For the coordinate axes of figure 5.11 the motion is counterclockwise at the free surface.

At a depth of $x_2 \sim 0.2\ \Lambda$ the direction of rotation reverses, since u_1 changes sign. The semimajor axes of the ellipses are normal to the free surface; the semiminor axes are parallel to the free surface. At the free surface the normal displacement is about 1.5 times the tangential displacement.

Rayleigh waves have been studied in great detail, and they have found several applications. For additional information we refer to the book by Viktorov.[5]

The attractive features are the absence of dispersion and the localization of the motion in the vicinity of the surface.

Given suitable generating conditions, surface waves as well as body waves are generated at a bounding surface. For a two-dimensional geometry the surface waves are essentially one-dimensional, but the body waves are cylindrical and undergo geometrical attenuation. Thus at some distance from the source the disturbance due to the surface wave becomes predominant.

[5] I. A. Viktorov, *Rayleigh and Lamb waves*. New York, Plenum Press, (1967).

Surface motions of a more general nature than discussed in this section were investigated by Knowles[6], who considered all possible motions of the half-space for which the scalar potential φ and the components of the vector potential ψ_j have the forms

$$\varphi = \Phi(x_1, x_3)e^{-ax_2+ikct} \tag{5.101}$$

$$\psi_j = \Psi_j(x_1, x_3)e^{-bx_2+ikct}. \tag{5.102}$$

From the wave equations for φ and ψ_j and from the condition that the shear stresses $\tau_{2\alpha}$ vanish at $x_2 = 0$, where $\alpha = 1, 3$, it can be shown that φ and ψ_2 can be expressed in terms of ψ_1, ψ_3, k and c as

$$\varphi = -\frac{1}{k}\left(1-\frac{c^2}{c_L^2}\right)^{-\frac{1}{2}}\left(1-\frac{1}{2}\frac{c^2}{c_T^2}\right) e_{\alpha\beta}\psi_{\alpha,\beta} \tag{5.103}$$

$$\psi_2 = \frac{1}{k}\left(1-\frac{c^2}{c_T^2}\right)^{\frac{1}{2}}\psi_{\beta,\beta}, \tag{5.104}$$

where α, $\beta = 1, 3$. It also follows that

$$a = k\left(1-\frac{c^2}{c_L^2}\right)^{\frac{1}{2}} \qquad b = k\left(1-\frac{c^2}{c_T^2}\right)^{\frac{1}{2}}. \tag{5.105a, b}$$

In eq. (5.103), $e_{\alpha\beta}$ is the two-dimensional alternator defined by $e_{11} = e_{33} = 0$, $e_{13} = -e_{31} = 1$. Furthermore ψ_1 and ψ_3 must satisfy the reduced wave equation

$$\nabla^2\psi_\alpha + k^2\psi_\alpha = 0, \tag{5.106}$$

where $\nabla^2 = \partial^2/\partial x_1^2 + \partial^2/\partial x_3^2$. From the remaining condition of vanishing normal stress, $\tau_{22} = 0$ at $x_2 = 0$, it then follows that c must satisfy eq. (5.95).

5.12. Stoneley waves

Propagating disturbances confined to the neighborhood of a surface occur not only in the vicinity of a free surface but also at the interface of two half-spaces filled with different materials. Thus, there can be surface waves at the interface of a solid and a fluid and also at the interface of two solids. The latter waves, which are called Stoneley waves, will be briefly discussed in this section.

[6] J. K. Knowles, *Journal of Geophysical Research* **71** (1966) 5480.

Let the material constants in the half-space $x_2 < 0$ be denoted by subscripts and superscripts B. The material constants in the half-space $x_2 > 0$ do not carry subscripts and superscripts. For $x_2 > 0$ the displacement components u_1 and u_2 are given by (5.93) and (5.94). In the half-space $x_2 < 0$ we have

$$u_1 = [A_3 e^{b_3 x_2} + A_4 e^{b_4 x_2}] \exp[ik(x_1 - ct)] \tag{5.107}$$

$$u_2 = \left[+\frac{b_3}{ik} A_3 e^{b_3 x_2} - \frac{ik}{b_4} A_4 e^{b_4 x_2}\right] \exp[ik(x_1 - ct)], \tag{5.108}$$

where

$$b_3 = k\left[1 - \left(\frac{c}{c_L^B}\right)^2\right]^{\frac{1}{2}}, \qquad b_4 = k\left[1 - \left(\frac{c}{c_T^B}\right)^2\right]^{\frac{1}{2}}, \tag{5.109}$$

where c_L^B and c_T^B denote the phase velocities of longitudinal and transverse waves, respectively, in the half-space $x_2 < 0$.

The condition that the displacements and the stresses are continuous at $x_2 = 0$ yields four homogeneous equations for the four constants A_1, A_2, A_3 and A_4. The determinant of the coefficients must vanish which yields the following equation

$$\begin{vmatrix} 1 & 1 & -1 & -1 \\ \dfrac{b_1}{k} & \dfrac{k}{b_2} & \dfrac{b_3}{k} & \dfrac{k}{b_4} \\ 2\dfrac{b_1}{k} & \left(2 - \dfrac{c^2}{c_T^2}\right)\dfrac{k}{b_2} & 2\dfrac{\mu^B}{\mu}\dfrac{b_3}{k} & \dfrac{\mu^B}{\mu}\left[2 - \left(\dfrac{c}{c_T^B}\right)^2\right]\dfrac{k}{b_4} \\ 2 - \dfrac{c^2}{c_T^2} & 2 & -\dfrac{\mu^B}{\mu}\left[2 - \left(\dfrac{c}{c_T^B}\right)^2\right] & -2\dfrac{\mu^B}{\mu} \end{vmatrix} = 0 \tag{5.110}$$

This equation is, of course, much more difficult to analyze than the much simpler Rayleigh equation (5.95).

Taking note of the definitions of b_1, b_2, b_3 and b_4, we observe that the wavenumber k does not appear in eq. (5.110). Thus Stoneley waves are not dispersive. The number and the nature of the roots of (5.110) can again be determined by the principle of the argument. This analysis is presented in the book by Cagniard.[7] It turns out that there are always two roots. The roots are, however, not necessarily real or on the proper Riemann surface. Only over a certain range of the ratios ρ^B/ρ and μ^B/μ are real roots of (5.110) obtained.

[7] L. Cagniard, *Reflection and refraction of progressive seismic waves*, translated and revised by E. A. Flinn and C. H. Dix, New York, McGraw-Hill Book Co. (1962), p. 47.

5.13. Slowness diagrams

Some of the results on the reflection and refraction of waves can be illustrated graphically by means of slowness diagrams. In particular, the relations between the angles of incidence, reflection and refraction can be shown, the critical angles can be identified, and the decay rates normal to the direction of propagation for plane waves with complex-valued unit propagation vectors can be determined.

In this section, we provide a few examples of the use of slowness diagrams by reproducing some results from a paper by Crandall.[8] Slowness diagrams were introduced in section 4.2, where it was shown that the components of the slowness vector $\boldsymbol{q}$ (which appears in the representation of a plane wave) are related by a circle or an equilateral hyperbola. Given $q_1 < q$, the corresponding value of q_2 is on the circle of radius q. When $q < q_1$, the corresponding value of β is on the hyperbola, as indicated in figure 4.1. In a homogeneous isotropic elastic medium, $q = q_L = 1/c_L$ for longitudinal waves, and $q = q_T = 1/c_T$ for transverse waves.

Considering reflections at a free surface, we draw two slowness diagrams, each consisting of a circle and a hyperbola, one pair with $q = q_L = 1/c_L$ and one with $q = q_T = 1/c_T$. In figure 5.15, the incident longitudinal wave is given. The slowness vector of the incident wave is $\boldsymbol{q}^{(0)}$. The slowness vectors of the reflected P- and SV-waves are $\boldsymbol{q}^{(1)}$ and $\boldsymbol{q}^{(2)}$, respectively.

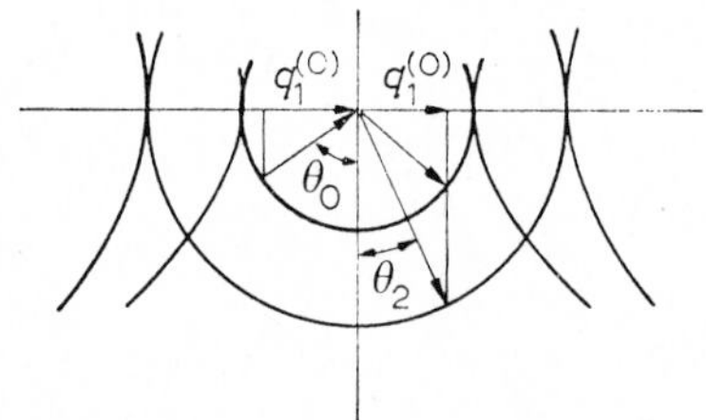

Fig. 5.15. Slowness diagram for incident P-wave.

Since the space-time distribution of the incident wave, see eqs. (5.33)–(5.36), on the surface $x_2 = 0$ is a sinusoidal wave with frequency ω and slowness $q_1^{(0)}$, the reflected wave must have this same slowness and frequency on $x_2 = 0$, if two independent boundary conditions are to be satisfied at *all* times and *all* positions on the surface. In figure 5.15, the directions of $\boldsymbol{q}^{(1)}$ and $\boldsymbol{q}^{(2)}$ are fixed by the requirement that their component in the x_1-direction be $q_1^{(0)}$. Note that

[8] S. H. Crandall, *Journal of the Acoustical Society of America* **47** (1970), 1338.

$$\theta_1 = \theta_0, \qquad \sin\theta_2 = (q_L/q_T)\sin\theta_0$$

in agreement with (5.49) and (5.50).

Figure 5.16 illustrates the similar configuration for an incident transverse wave. In both figures 5.15 and 5.16, the slowness component q_1^0 along the

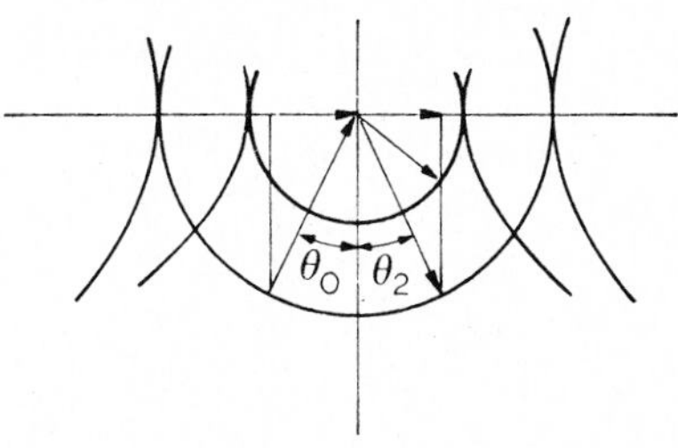

Fig. 5.16. Slowness diagram for incident SV-wave, $0 < q_1^{(0)} < q_L$.

free surface is less than either q_L or q_T, so all waves have real slowness components. In figure 5.17, the incident transverse wave has a slowness $q_1^{(0)}$ such that $q_L < q_1^{(0)} < q_T$, so that the reflected P-wave has a complex slowness. The amplitude decay rate is fixed by β_L, see eq. (4.20). This case corresponds to the results presented by eqs. (5.68)–(5.70).

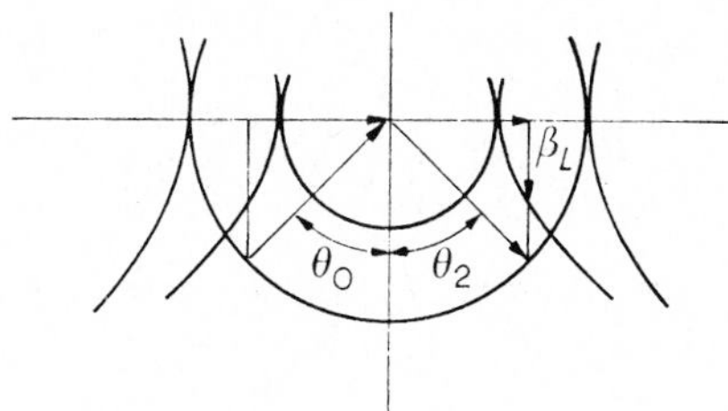

Fig. 5.17. Slowness diagram for incident SV-wave, $q_L < q_1^{(0)} < q_T$.

The configurations in figures 5.15–5.17 represent reflections of incident waves with entirely real slowness components. Various other cases are examined in the previously cited paper by Crandall.

Slowness diagrams can also be used to examine refraction and reflection at an interface, as discussed in some detail in a paper by McNiven and Mengi.[9]

[9] H. D. McNiven and Y. Mengi, *Journal of the Acoustical Society of America* **44** (1968), 1658.

5.14. Problems

5.1. A thin layer of a substance whose elastic constants are very small and can be neglected is adhered to the surface of an elastic half-space. The mass density of the substance is ρ_s. A plane harmonic longitudinal wave is incident on the covered surface.

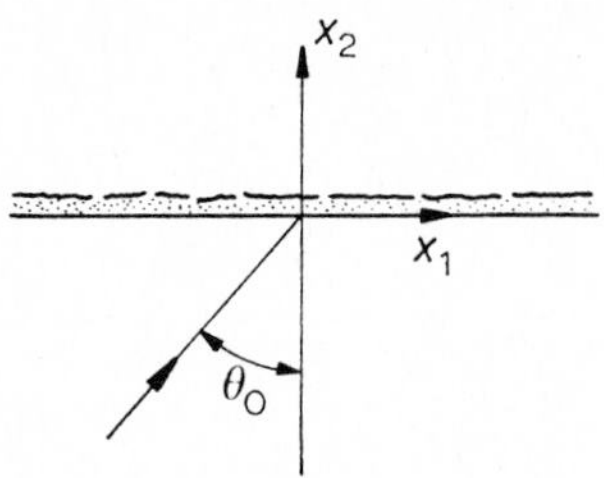

(a) Determine the influence of the mass-layer on the amplitudes of the reflected waves.

(b) Consider beams of incident and reflected waves and investigate the averaged energy transmission over the cross sections of the beams.

5.2. An elastic half-space is covered by a stretched thin membrane, as shown in the figure. The equation governing the motion of a membrane subjected to a load $q(x_1, x_3, t)$ per unit area is

$$\frac{\partial^2 u_2}{\partial x_1^2} + \frac{\partial^2 u_2}{\partial x_3^2} + \frac{q}{T} = \frac{\rho}{T}\frac{\partial^2 u_2}{\partial t^2}.$$

Now let a plane harmonic longitudinal wave be incident on the covered surface, as shown in the figure. Assuming that there is perfect contact

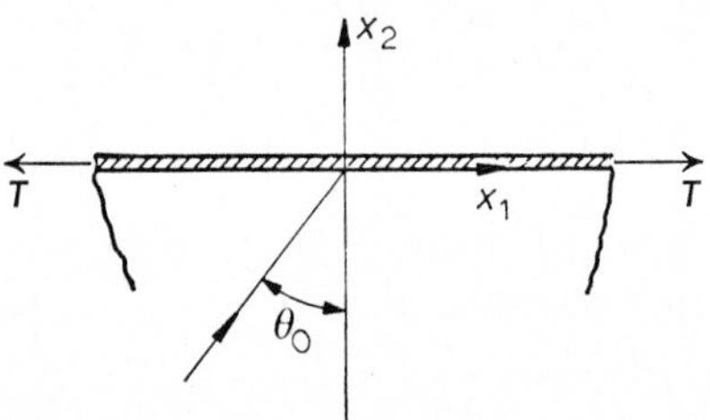

between the membrane and the half-space, and assuming that the membrane is infinitely rigid in its own plane, determine the amplitudes of the reflected wave(s).

5.3. Consider the reflection and refraction of SH-waves at an interface of two materials. This problem is discussed in section 5.9. For a specific

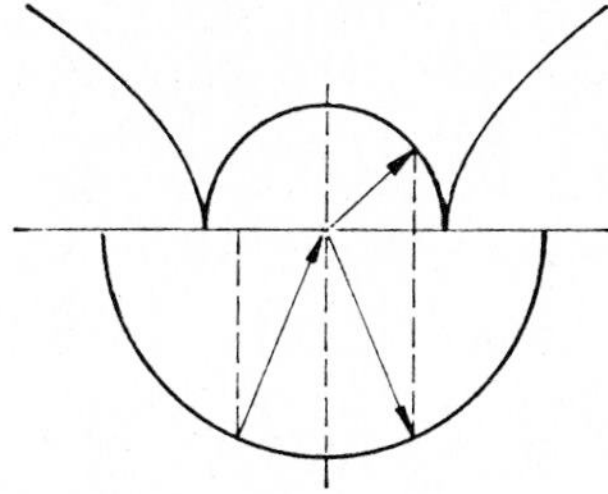

set of material properties a slowness diagram is shown in the figure. Obtain the results (5.74)–(5.76) and the observations 1–3 (p. 184) from an examination of slowness diagrams.

5.4. The reflection and refraction of SH-waves is discussed in section 5.9. Consider beams of incident, reflected and refracted waves and examine the averaged transmission of energy, in particular for the case $(c_T^B/c_T)\sin\theta_0 > 1$.

5.5. An elastic half-space is covered with an ideal elastic fluid. Equations governing the motion of an ideal fluid are given in section 2.15. A harmonic longitudinal wave propagating in the solid is incident on the interface.

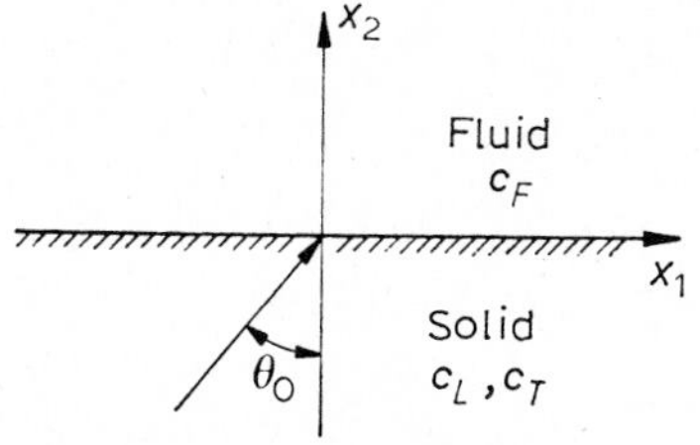

(a) Draw slowness diagrams for the case $c_F < c_T < c_L$, $c_T < c_F < c_L$ and $c_T < c_L < c_F$, and determine the critical angles.

(b) Determine the amplitudes of the reflected and transmitted waves.

5.6. The origin of a system of cylindrical coordinates is placed at a point on the surface of an elastic half-space. The z-axis is normal to the surface, with the positive z-axis pointing into the half-space.

To investigate the propagation of axially symmetric surface waves it is convenient to use displacement potentials. The pertinent relations can be found in section 2.13. Start off with expressions for $\varphi(r, z, t)$ and $\psi_\theta(r, z, t)$ of the forms

$$\varphi = \Phi(r)e^{-b_1 z}e^{ikct}$$

and

$$\psi_\theta = \Psi_\theta(r)e^{-b_2 z}e^{ikct},$$

where b_1 and b_2 are as stated in section 5.11, and determine $\Phi(r)$ and $\Psi_\theta(r)$. Show that the phase velocity c must satisfy eq. (5.95).

5.7. An elastic half-space is covered by a thin layer of fluid, as shown in the figure. Suppose that the layer is subjected to surface tension, so that it acts as a stretched thin membrane. We assume that the fluid adheres perfectly to the half-space, but that it is nonviscous, so that shear interaction with the half-space can be ignored.

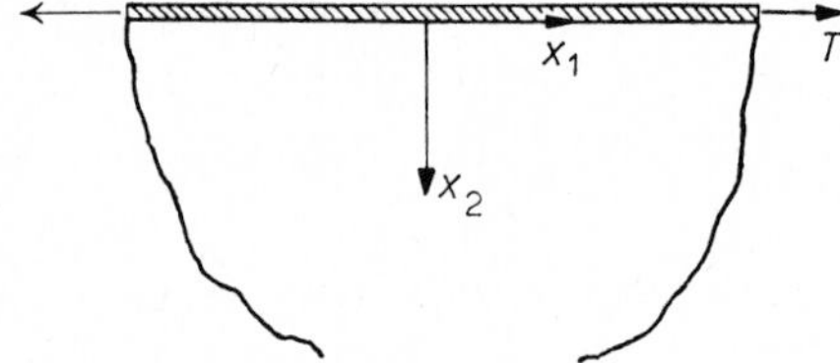

Consider the propagation of surface waves in the half-space:

(1) Determine the equation for the phase velocity.

(2) Are the waves dispersive?

(3) Check that the classical equation for Rayleigh waves along a free surface is a proper limit case.

(4) Examine the roots of the equation.

5.8. Examine the propagation of surface waves at an interface of a solid and an ideal elastic fluid.

5.9. Consider two semi-infinite elastic solids in smooth contact, i.e., the shear stresses vanish as the interface while the normal displacement is continuous. Examine the propagation of Stoneley waves along the interface.

(a) Show that the equation for the phase velocity can be expressed in the form

$$S(\eta) = \alpha^2 \left[1 - \left(\frac{\alpha\eta}{\kappa^B}\right)^2\right]^{\frac{1}{2}} R_1 + \left[1 - \left(\frac{\eta}{\kappa}\right)^2\right]^{\frac{1}{2}} \frac{\mu^B}{\mu} R_2 = 0,$$

where

$$\alpha = \frac{c_T}{c_T^B} \qquad \kappa = \frac{c_L}{c_T}$$

$$R_1 = (2-\eta^2)^2 - 4\left[1-\left(\frac{\eta}{\kappa}\right)^2\right]^{\frac{1}{2}}(1-\eta^2)^{\frac{1}{2}}$$

$$R_2 = (2-\alpha^2\eta^2)^2 - 4\left[1-\left(\frac{\alpha\eta}{\kappa^B}\right)^2\right]^{\frac{1}{2}}(1-\alpha^2\eta^2)^{\frac{1}{2}}$$

and

$$\eta = \frac{c}{c_T}.$$

(b) Does $S(\eta) = 0$ always have real roots? If not, what are the conditions for real roots?

5.10. Two semi-infinite elastic solids are in perfect contact. Draw a slowness diagram for the case

$$c_T^B < c_L^B < c_T < c_L,$$

and determine the critical angle(s) for an incident longitudinal wave. Do the same for the other cases that are conceivable.

CHAPTER 6

HARMONIC WAVES IN WAVEGUIDES

6.1. Introduction

The analysis of plane time-harmonic waves in a half-space does not meet with many difficulties. Indeed, the rules governing the reflections at the surface can be derived in a straightforward manner, as was shown in some detail in chapter 5. In the present chapter we will examine the complications that enter if a body has a finite cross-sectional dimension.

Let us first consider harmonic wave motion in a homogeneous elastic layer of thickness $2h$ and of infinite in-plane dimensions. The results of chapter 5 suggest that harmonic waves can propagate in a layer by being reflected back and forth between the two plane surfaces. In a steady-state situation, which is assumed for harmonic waves, the systems of incident and reflected waves form, however, a standing wave across the thickness of the layer, so that the propagation is essentially in the direction of the layer. This motivates the term waveguide for the layer or, for that matter, for any extended body with a cross section of finite dimensions. As another example of wave motion in a waveguide we will in a later section of this chapter examine waves in a circular cylinder of infinite length.

For motion in the $(x_1 x_2)$-plane the picture of reflecting waves in a layer becomes complicated because, as depicted in figure 6.1, both P- and SV-waves are generally reflected at the incidence of either one. In a circular

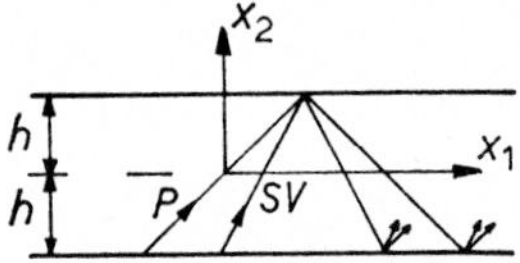

Fig. 6.1. Waves in a layer.

cylinder it becomes rather impractical to trace the reflection of waves. The idea of reflecting waves is, therefore, usually discarded and the analysis is approached by *a priori* considering a wave motion of the form

$$f(x_2, x_3) \exp [ik(x_1 - ct)]. \tag{6.1}$$

Eq. (6.1) represents a standing wave across the cross section and a traveling wave in the direction of the waveguide. For the much simpler case of horizontally polarized shear motion of an elastic layer (SH-waves) some interesting insight is, however, gained by indeed considering the reflections at the two surfaces.

The function $f(x_2, x_3)$ is the interference pattern formed by the reflecting waves. In an unbounded layer the interference pattern is a function of the thickness coordinate only, and it is represented by trigonometric functions. The analysis of motions of a circular cylinder is more complicated because the distribution of the motion across the cross section involves Bessel functions which are somewhat more difficult to deal with.

The investigation of progressive waves in waveguides leads to the introduction of several new concepts such as modes of wave propagation, the frequency spectrum, dispersion and group velocity.

In the last sections of this chapter we examine some approximate theories which considerably simplify the analysis of waves in layers and cylinders. The classical approximate theories are generally applicable for wavelengths that are large compared to the cross-sectional dimension of the waveguide. The higher order approximations are good up to frequencies that are somewhat higher than the frequency of the highest thickness mode contained in the equations.

6.2. Horizontally polarized shear waves in an elastic layer

The reflection at a free surface of an incident SH-wave was studied in section 5.5, where it was found that there is total reflection without change of phase, i.e., $\theta_2 = \theta_0$, $A_2 = A_0$ and $k_2 = k_0$, see eqs. (5.25a, b, c). If we consider a layer it is therefore conceivable that two systems of plane waves will propagate in the layer. One system has unit propagation vector $\sin \theta_0 \boldsymbol{i}_1 + \cos \theta_0 \boldsymbol{i}_2$ and can be considered as an "incident" wave on the free surface $x_2 = +h$, and a "reflected" wave from the surface $x_2 = -h$. The second system has unit propagation vector $\sin \theta_0 \boldsymbol{i}_1 - \cos \theta_0 \boldsymbol{i}_2$ and can be considered as a reflected wave from the surface $x_2 = +h$, and an incident wave on the surface $x_2 = -h$. Since the bounding surfaces of the layer are free of tractions, the system of incident and reflected waves must sustain itself if it is to form a steady-state pattern. For free waves the reflections must thus constructively interfere with each other.

Referring to figure 6.2, we consider a ray *ADEF* which has been reflected once from each boundary. If the disturbance in the plane *PBE* is to interfere

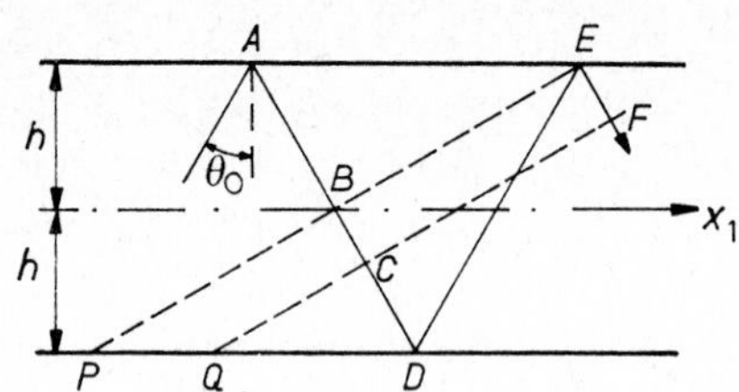

Fig. 6.2. SH-waves in a layer.

constructively with the coincident disturbance which has traversed the additional path BDE, it is required that

$$\frac{1}{\Lambda'} \times \text{length } BDE = n, \tag{6.2}$$

where Λ' is the wavelength and n is an integer. From the geometry of figure 6.2 we obtain

$$\text{length } BDE = \frac{2h}{\cos\theta_0} + \frac{2h\cos 2\theta_0}{\cos\theta_0} = 4h\cos\theta_0 . \tag{6.3}$$

As the disturbance in the plane PBE moves a distance BC it traces the distance PQ in the horizontal direction. If c is the phase velocity of wave motion in the x_1-direction, we find from figure 6.2

$$\sin\theta_0 = BC/PQ = c_T/c. \tag{6.4}$$

Denoting the wavenumbers in the x_1-direction and the direction of the ray AD by k and k', respectively, we have

$$\omega = kc = k'c_T,$$

and thus

$$k = \frac{2\pi}{\Lambda} = k'\sin\theta_0 = \frac{2\pi\sin\theta_0}{\Lambda'} . \tag{6.5}$$

In (6.5), Λ and Λ' are the wavelengths in the x_1-direction and in the direction of AD, respectively. Eq. (6.5) implies

$$\Lambda = \Lambda'/\sin\theta_0 . \tag{6.6}$$

Substitution of (6.3) into (6.2), with the aid of (6.5), yields

$$k\cot\theta_0 = \frac{n\pi}{2h} . \tag{6.7}$$

Also,

$$\cot \theta_0 = \left(\frac{1}{\sin^2 \theta_0} - 1\right)^{\frac{1}{2}} = \left[\left(\frac{c}{c_T}\right)^2 - 1\right]^{\frac{1}{2}},$$

where we have used eq. (6.4). Eq. (6.7) can thus be expressed as

$$\left(\frac{c}{c_T}\right)^2 = 1 + \left(\frac{n\pi}{2kh}\right)^2. \tag{6.8}$$

This result shows that except for $n = 0$ the phase velocity depends on the wavenumber k. Harmonic SH-waves in an elastic layer thus are dispersive. Since $\omega = kc$ the result (6.8) can also be written in the form

$$\Omega^2 = n^2 + \xi^2, \tag{6.9}$$

where the dimensionless frequency Ω and the dimensionless wavenumber ξ are defined as

$$\Omega = \frac{2h\omega}{\pi c_T}, \qquad \xi = \frac{2kh}{\pi}, \tag{6.10a, b}$$

respectively. Eq. (6.9) is called the frequency equation.

The treatment based on the principle of constructive interference is instructive. It is, however, mathematically more straightforward to derive the frequency equation (6.9) by postulating at the outset solutions of the form

$$u_3 = f(x_2) \exp\left[i(kx_1 - \omega t)\right]. \tag{6.11}$$

This solution must satisfy the equation

$$\frac{\partial^2 u_3}{\partial x_1^2} + \frac{\partial^2 u_3}{\partial x_2^2} = \frac{1}{c_T^2}\frac{\partial^2 u_3}{\partial t^2}, \tag{6.12}$$

and the boundary conditions at $x_2 = \pm h$

$$\mu \frac{\partial u_3}{\partial x_2} = 0. \tag{6.13}$$

Substituting (6.11) into (6.12), and solving for $f(x_2)$ we find

$$f(x_2) = B_1 \sin(qx_2) + B_2 \cos(qx_2), \tag{6.14}$$

where

$$q^2 = \frac{\omega^2}{c_T^2} - k^2. \tag{6.15}$$

The boundary conditions (6.13) yield

$$B_1 \cos(qh) \pm B_2 \sin(qh) = 0. \tag{6.16}$$

Eq. (6.16) can be satisfied in two ways:

either

$$B_1 = 0 \quad \text{and} \quad \sin(qh) = 0 \tag{6.17}$$

or

$$B_2 = 0 \quad \text{and} \quad \cos(qh) = 0. \tag{6.18}$$

For an arbitrarily specified value of the wavenumber k, eqs. (6.17) and (6.18) yield an infinite number of solutions for the frequency ω. A specific wave motion of the layer, called a mode of propagation, corresponds to each frequency satisfying (6.17) or (6.18). If $B_1 = 0$, the expression for $f(x_2)$ shows that the displacement is symmetric with respect to the midplane of the layer. The displacement is antisymmetric if $B_2 = 0$. In both cases the frequencies follow from

$$qh = \frac{n\pi}{2}, \tag{6.19}$$

where, however, $n = 0, 2, 4, \ldots$ for symmetric modes, and $n = 1, 3, 5, \ldots$ for antisymmetric modes. By using the definition of q, (6.15), and the definitions of the dimensionless frequency and the dimensionless wavenumber, (6.10a, b), eq. (6.19) can also be written as $\Omega^2 = n^2 + \xi^2$, which was derived earlier on the basis of constructive interference as eq. (6.9).

6.3. The frequency spectrum of SH-modes

In the Ω–ξ-plane the frequency equation

$$\Omega^2 = n^2 + \xi^2, \tag{6.20}$$

yields an infinite number of continuous curves, called *branches*, each corresponding to an integer value of n. A branch displays the relationship between the dimensionless frequency Ω and the dimensionless wavenumber ξ for a particular mode of propagation. The collection of branches constitutes the *frequency spectrum*.

To identify the modes we call the symmetric SH-modes ($n = 0, 2, 4, \ldots$) the SS-modes. Similarly, the antisymmetric SH-modes ($n = 1, 3, 5, \ldots$) are called the AS-modes. Moreover, we introduce the index r in such a manner that SS(r) and AS(r) identify the rth symmetric and antisymmetric SH-modes, respectively. For symmetric modes, r is related to n by $r = \frac{1}{2}n$.

For antisymmetric modes we have $r = \frac{1}{2}n+\frac{1}{2}$. In figure 6.3 the displacement distributions for several modes are sketched. For the lowest symmetric mode ($n = 0$) the displacement is constant across the layer, for the lowest antisymmetric mode ($n = 1$) there is one node, etc.

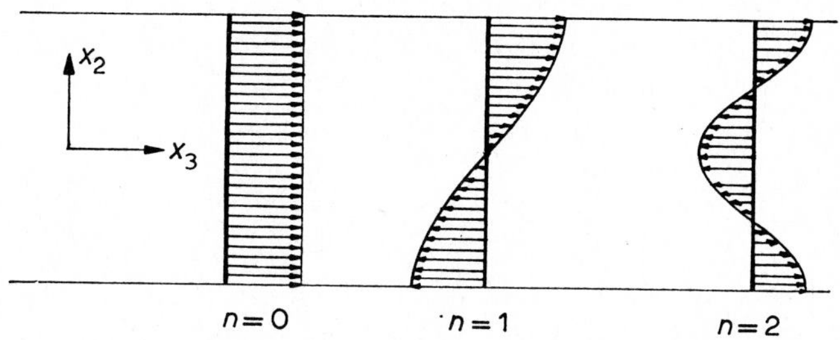

Fig. 6.3. Displacement distributions in the x_2x_3-plane.

By eliminating Ω through the relation $\omega = kc$ or $\Omega = \xi c/c_T$, the following relation between ξ and the dimensionless phase velocity c/c_T can be written

$$\frac{c}{c_T} = \pm\left(1+\frac{n^2}{\xi^2}\right)^{\frac{1}{2}}.$$

If ξ is eliminated, c/c_T may also be expressed in the form

$$\frac{c}{c_T} = \pm\left(1-\frac{n^2}{\Omega^2}\right)^{-\frac{1}{2}}. \qquad (6.21)$$

In these equations the sign before the radical indicates the propagation sense of the wave. It is clear from both equations that except for SS(0) the phase velocity depends on the wavenumber (or on the frequency), and the higher modes are thus dispersive.

In free motions the frequency should be taken as real-valued. It is now seen from the frequency equation (6.20) that for $\Omega < n$ this equation can be satisfied only if ξ is purely imaginary. As is evident from eq. (6.11), displacements associated with positive imaginary wave numbers decay exponentially with x_1. Such displacements do not represent progressive waves, but rather localized standing wave motions. For a particular mode the frequency at which the wavenumber changes from real to imaginary (or complex) values is called the *cut-off frequency*. It is noted that for horizontally polarized shear waves the cut-off frequencies are given by the frequencies at vanishing wavenumbers.

The frequency spectrum is shown in figure 6.4. If ξ is real, the branches are hyperbolas with asymptotes $\Omega = \xi$. For imaginary wavenumbers eq. (6.20) represents circles with radii n. Since the frequency spectrum is symmetric in ξ, the second quadrant can be used to plot Ω versus positive

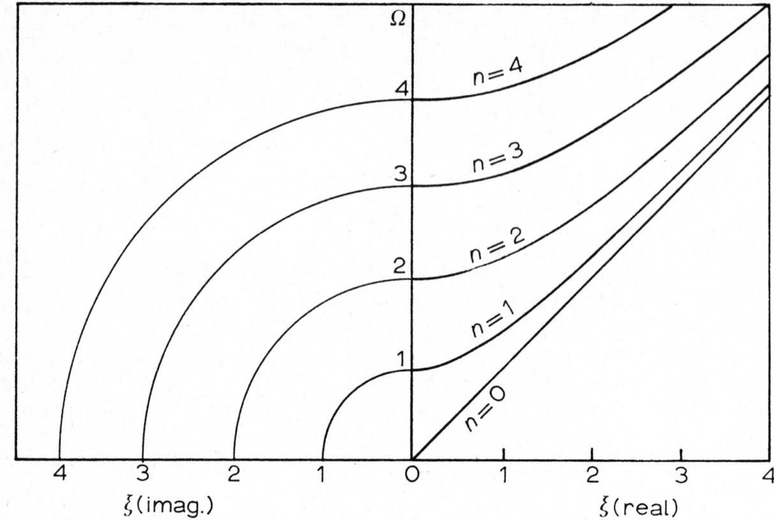

Fig. 6.4. Frequency spectrum for SH-modes in a layer.

imaginary values of ξ. Negative values of Ω need not be considered. Since the two relevant material constants μ and ρ and the one relevant geometrical parameter h appear in Ω and ξ only, the frequency spectrum can be employed for any homogeneous isotropic layer.

6.4. Energy transport by SH-waves in a layer

In section 5.3 it was shown that the propagation of time-harmonic waves is concomitant with a flux of energy. The time average energy transmission per unit area for transverse waves was computed as eq. (5.10) of chapter 5. Let us now examine the transport of energy for the case of SH-waves propagating in a layer. We are particularly interested in the velocity of energy flux along the layer.

The time average over a period of the power per unit area integrated over the thickness of the layer is denoted by $\langle P \rangle$,

$$\langle P \rangle = -\frac{1}{T}\int_0^T \mathrm{d}t \int_{-h}^{h} \tau_{13}\,\dot{u}_3\,\mathrm{d}x_2. \tag{6.22}$$

The time averages of the integrals over the thickness of the total energy density, the kinetic energy density and the strain energy density are denoted by $\langle H\rangle$, $\langle K\rangle$ and $\langle U\rangle$, respectively, where

$$\langle H\rangle = \langle K\rangle + \langle U\rangle = 2\langle K\rangle. \tag{6.23}$$

In eq. (6.23),

$$\langle K\rangle = \frac{1}{T}\int_0^T \mathrm{d}t \int_{-h}^{h} \tfrac{1}{2}\rho(\dot{u}_3)^2 \mathrm{d}x_2 . \tag{6.24}$$

To compute the velocity of energy flux we use the relation between the time average power transmission and the time average total energy

$$\langle P\rangle = \langle H\rangle c_e . \tag{6.25}$$

Suppose we consider an antisymmetrical mode, i.e.,

$$u_3 = B_1 \sin(qx_2)\cos[k(x_1 - ct)], \tag{6.26}$$

where q and c are defined by eqs. (6.15) and (6.8). We can now determine τ_{13} and $\dot{u}_3$ and easily evaluate the integrals in eqs. (6.22) and (6.24). The results are

$$\langle P\rangle = \tfrac{1}{2}\mu c h B_1^2 k^2$$

$$\langle K\rangle = \tfrac{1}{4}\rho c^2 h B_1^2 k^2 .$$

By means of (6.25) and (6.23) the velocity of energy transmission is then computed as

$$c_e = \frac{c_T^2}{c}$$

or

$$\frac{c_e}{c_T} = \frac{c_T}{c} = \frac{\xi}{(n^2 + \xi^2)^{\frac{1}{2}}}, \tag{6.27}$$

where we have used eq. (6.8). From the frequency equation (6.20) we derive

$$\frac{\mathrm{d}\Omega}{\mathrm{d}\xi} = \frac{\xi}{(n^2 + \xi^2)^{\frac{1}{2}}} . \tag{6.28}$$

Comparing (6.27) and (6.28) it is concluded that

$$\frac{c_e}{c_T} = \frac{\mathrm{d}\Omega}{\mathrm{d}\xi} \tag{6.29}$$

or

$$c_e = \frac{\mathrm{d}\omega}{\mathrm{d}k} . \tag{6.30}$$

It has thus been shown that for the propagation of SH-waves in a layer, the velocity of energy flux equals the derivative of the frequency with respect to the wavenumber. Using the relation $\omega = kc$, we can also write this result

$$c_e = c + k\,\frac{\mathrm{d}c}{\mathrm{d}k}. \tag{6.31}$$

For the lowest symmetric mode ($n = 0$) c is constant ($c = c_T$) and then c_e equals the phase velocity. If there is dispersion, i.e., $c = c(k)$, it follows from (6.31) that the velocity of energy transport is different from the phase velocity.

Historically the velocity defined by $\mathrm{d}\omega/\mathrm{d}k$ entered the literature under the name *group velocity*, independent of the idea of energy propagation, but rather through a kinematic argument. The kinematic argument which motivated the terminology "group" velocity involves superposition of waves of various frequencies and will be presented in the next section. From now on we will however refer to $c_g = \mathrm{d}\omega/\mathrm{d}k$ as the group velocity.

The dimensionless group velocity of SH-waves can be computed as

$$\frac{c_g}{c_T} = \frac{\mathrm{d}\Omega}{\mathrm{d}\xi} = \frac{\pm\xi}{(n^2+\xi^2)^{\frac{1}{2}}}. \tag{6.32}$$

An alternative expression can be obtained by eliminating ξ,

$$\frac{c_g}{c_T} = \pm\left(1-\frac{n^2}{\Omega^2}\right)^{\frac{1}{2}}.$$

For a particular mode the group velocity is zero when $\mathrm{d}\Omega/\mathrm{d}\xi$ vanishes, which implies that the group velocity is zero at the cut-off frequencies. The dimensionless group velocity is plotted for several modes in figure 6.5. We note that for the lowest symmetrical mode, which is not dispersive, $c_g/c_T = 1$. For the higher modes $c_g/c_T < 1$, but c_g/c_T approaches unity as ξ increases, i.e., as the wavelength decreases. The asymptotic limit $c = c_T$ as ξ increases, of both the phase velocity and the group velocity, is easily understood on physical grounds. As ξ increases the wavelength decreases, which means that the thickness of the layer becomes relatively large and waves propagate as in an unbounded medium.

For imaginary wavenumbers the motion is nonpropagating. Indeed for imaginary ξ the stress is 90° out of phase with the particle velocity, and the energy flux through a plane normal to the x_1-axis is zero. It is thus seen that for the modes with imaginary wavenumbers the associated energy

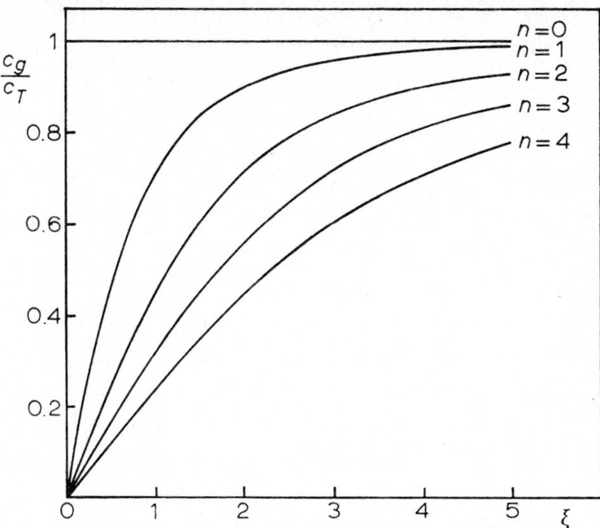

Fig. 6.5. Group velocity for SH-modes.

is stored in the vicinity of $x_1 = 0$. The modes with imaginary wavenumbers play an important function in the forced horizontally polarized shear motions of a layer. In general an external forcing agency will generate motion in a large number of modes, consisting of both propagating and nonpropagating modes. Only the inclusion of nonpropagating modes generally makes it possible to satisfy prescribed boundary conditions on planes $x_1 = \text{const.}$ In a later chapter we shall return to forced motions of layers.

6.5. Energy propagation velocity and group velocity

The result that in a perfectly harmonic motion of fixed wavenumber and fixed frequency, energy is propagated at a velocity which can be expressed as a ratio of changes of frequency and wavenumber in going to a neighboring wave solution, is rather surprising. For the case of SH-waves in a layer the result was verified rather than proven in the previous section. For several other examples similar verifications were presented by Biot.[1]

As we shall see in this section it can in fact be proven for general periodic wave motion that energy propagates with the velocity $\mathrm{d}\omega/\mathrm{d}k$.

For a one-dimensional case the relation between $\mathrm{d}\omega/\mathrm{d}k$ and energy

[1] M. A. Biot, *The Physical Review* **105** (1957) 1129.

transport was already discussed by Rayleigh.[2] In Rayleigh's discussion it is supposed that in a general dynamical system a pure imaginary change in wavenumber is made. It is then shown that the corresponding imaginary change in the frequency would be replaced by a zero frequency change if the motion of every particle in the system were resisted by an additional small force proportional to its momentum. The energy flow across a plane in this steady state is then calculated by balancing it against the dissipative action of those forces throughout the region beyond that plane.

A general proof applicable to periodic waves in rather *general* systems is due to Lighthill.[3] The proof is based on an argument that utilizes real rather than imaginary changes in frequency and wavenumber. Here we present a simpler proof that is suitable for waves in a waveguide.

A time-harmonic wave propagating in an elastic waveguide of constant cross-sectional area A is represented by

$$u_i = \mathrm{f}_i(x_2, x_3)g(\eta), \tag{6.33}$$

where

$$g(\eta) = B_1 \sin(\eta) \text{ or } g(\eta) = B_2 \cos(\eta), \tag{6.34a}$$

and

$$\eta = kx_1 - \omega t. \tag{6.34b}$$

For example, a symmetric mode of horizontally polarized shear waves in a layer is represented by

$$u_3 = B_2 \cos(qx_2)\cos(\eta).$$

In this expression, q is a constant which depends on the boundary conditions and on the mode that is considered. Thus, q does not change along a particular branch of the frequency spectrum. Similarly, the function $\mathrm{f}_i(x_2, x_3)$ in eq. (6.33) is independent of k and ω along a branch of the corresponding frequency spectrum.

A mechanical system can be completely specified by the Lagrangian density $\mathscr{L}$. For a linearly elastic, homogeneous and isotropic body the Lagrangian density is

$$\mathscr{L}(\dot{u}_i, u_{i,j}) = \tfrac{1}{2}\rho\dot{u}_i\dot{u}_i - [\tfrac{1}{2}\lambda(\varepsilon_{kk})^2 + \mu\varepsilon_{ij}\varepsilon_{ij}],$$

where the summation convention must be invoked, and the components of the small strain tensor ε_{ij} are

$$\varepsilon_{ij} = \tfrac{1}{2}(u_{i,j} + u_{j,i}).$$

[2] Lord Rayleigh, *The theory of sound*, Vol. I. New York, Dover Publications (1945), p. 475.

[3] M. J. Lighthill, *Journal of the Institute of Mathematics and Applications* **1** (1965) 1.

Possible motions of the body must obey a set of governing equations that can be obtained from Hamilton's principle. As noted in section 2.9, the usual form of the principle states that

$$\delta \int_{t_1}^{t_2} \mathrm{d}t \int_V \mathscr{L}\,\mathrm{d}V = 0, \tag{6.35}$$

for any changes δu_i of the function $u_i(x_1, x_2, x_3, t)$ which vanish at $t = t_1$ and $t = t_2$, and on the boundary of the arbitrary volume of integration V. For a Lagrangian density which depends on $\dot{u}_i$ and $u_{i,j}$, the equations of motion implied by (6.35) are

$$\frac{\partial}{\partial t}\left(\frac{\partial \mathscr{L}}{\partial \dot{u}_i}\right) + \sum_{j=1}^{3} \frac{\partial}{\partial x_j}\left(\frac{\partial \mathscr{L}}{\partial u_{i,j}}\right) = 0. \tag{6.36}$$

For a time-harmonic wave of the general form (6.33), propagating in a waveguide with free or clamped cylindrical surfaces, Hamilton's principle can be used in a slightly special form to obtain an expression for the velocity of energy propagation. This special form states that

$$\delta I = \delta \int_0^{2\pi} \mathrm{d}\eta \int_A \mathscr{L}\,\mathrm{d}x_2\,\mathrm{d}x_3 = 0, \tag{6.37}$$

for all δu_i that are harmonic functions of the form (6.33), with the same frequency and wavenumber as the u_i themselves. In eq. (6.37), A is the cross-sectional area. To prove (6.37), we write

$$\delta I = \int_0^{2\pi} \mathrm{d}\eta \int_A \sum_{i=1}^{3} \left(\frac{\partial \mathscr{L}}{\partial \dot{u}_i}\,\delta \dot{u}_i + \sum_{j=1}^{3} \frac{\partial \mathscr{L}}{\partial u_{i,j}}\,\delta u_{i,j}\right) \mathrm{d}x_2\,\mathrm{d}x_3, \tag{6.38}$$

which can also be written as

$$\begin{aligned}\delta I = &-\int_0^{2\pi} \mathrm{d}\eta \int_A \sum_{i=1}^{3} \left[\frac{\partial}{\partial t}\left(\frac{\partial \mathscr{L}}{\partial \dot{u}_i}\right) + \sum_{j=1}^{3} \frac{\partial}{\partial x_j}\left(\frac{\partial \mathscr{L}}{\partial u_{i,j}}\right)\right] \delta u_i\,\mathrm{d}x_2\,\mathrm{d}x_3 \\ &+ \int_0^{2\pi} \mathrm{d}\eta \int_A \frac{\partial}{\partial t}\left(\sum_{i=1}^{3} \frac{\partial \mathscr{L}}{\partial \dot{u}_i}\,\delta u_i\right) \mathrm{d}x_2\,\mathrm{d}x_3 \\ &+ \int_0^{2\pi} \mathrm{d}\eta \int_A \sum_{i=1}^{3} \sum_{j=1}^{3} \frac{\partial}{\partial x_j}\left(\frac{\partial \mathscr{L}}{\partial u_{i,j}}\,\delta u_i\right) \mathrm{d}x_2\,\mathrm{d}x_3.\end{aligned}$$

The first of these integrals vanishes by (6.36). The second integral can be rewritten as

$$-\omega \int_A \mathrm{d}x_2\,\mathrm{d}x_3 \int_0^{2\pi} \frac{\partial}{\partial \eta}\left(\sum_{i=1}^{3} \frac{\partial \mathscr{L}}{\partial \dot{u}_i}\,\delta u_i\right) \mathrm{d}\eta,$$

whereupon it is immediately seen that the inner integral vanishes because of the periodicity of δu_i. The third integral can be expressed in the form

$$\int_0^{2\pi} \mathrm{d}\eta \int_A \left\{ \sum_{j=2}^{3} \frac{\partial}{\partial x_j} \left(\sum_{i=1}^{3} \frac{\partial \mathscr{L}}{\partial u_{i,j}} \delta u_i \right) + k \frac{\partial}{\partial \eta} \left(\sum_{i=1}^{3} \frac{\partial \mathscr{L}}{\partial u_{i,1}} \delta u_i \right) \right\} \mathrm{d}x_2 \, \mathrm{d}x_3 .$$

The first of these vanishes in view of the conditions on the cylindrical surfaces of the waveguide (note that $\tau_{ij} = -\partial \mathscr{L}/\partial u_{i,j}$), and the second is zero because of the periodicity of δu_i with respect to η.

By a simpler argument eq. (6.37) follows from the observation that for a linearly elastic solid the time average over a period of the kinetic energy equals the time average of the strain energy, and the time average over a period of the Lagrangian thus vanishes.

In the usual manner the velocity of energy transport is defined as the ratio of the time average of the power per cross section and the time average of the total energy per unit length of the waveguide

$$c_e = \frac{\langle P \rangle}{\langle H \rangle}$$

where

$$\langle P \rangle = \frac{1}{T} \int_0^T P \, \mathrm{d}t,$$

and

$$\langle H \rangle = \frac{1}{T} \int_0^T \mathrm{d}t \int_A (\mathscr{U} + \mathscr{K}) \mathrm{d}x_2 \, \mathrm{d}x_3 .$$

In the latter equation, $\mathscr{U}$ and $\mathscr{K}$ are the strain energy density and the kinetic energy density, respectively. We can write

$$\langle P \rangle = - \left\langle \int_A \tau_{i1} \dot{u}_i \mathrm{d}x_2 \, \mathrm{d}x_3 \right\rangle = \left\langle \int_A \sum_{i=1}^{3} \frac{\partial \mathscr{L}}{\partial u_{i,1}} \dot{u}_i \mathrm{d}x_2 \, \mathrm{d}x_3 \right\rangle$$

$$\langle H \rangle = \frac{2}{T} \int_0^T \mathrm{d}t \int_A \mathscr{K} \, \mathrm{d}x_2 \, \mathrm{d}x_3 = \left\langle \int_A \sum_{i=1}^{3} \frac{\partial \mathscr{L}}{\partial \dot{u}_i} \dot{u}_i \mathrm{d}x_2 \, \mathrm{d}x_3 \right\rangle$$

Thus

$$c_e = \frac{\left\langle \int_A \sum_{i=1}^{3} \frac{\partial \mathscr{L}}{\partial u_{i,1}} \dot{u}_i \mathrm{d}x_2 \, \mathrm{d}x_3 \right\rangle}{\left\langle \int_A \sum_{i=1}^{3} \frac{\partial \mathscr{L}}{\partial \dot{u}_i} \dot{u}_i \mathrm{d}x_2 \, \mathrm{d}x_3 \right\rangle} . \tag{6.39}$$

To show that c_e also equals $\mathrm{d}\omega/\mathrm{d}k$, we consider a perturbation of the displacement field in which the wavenumber and the frequency change,

$$u_i+\delta u_i = \mathrm{f}_i(x_2, x_3)g^*(k^*x_1-\omega^* t),$$

where

$$k^* = k+\delta k, \qquad \omega^* = \omega+\delta\omega. \tag{6.40a, b}$$

We find by employing (6.38),

$$\delta I = \int_0^{2\pi} \mathrm{d}\eta \int_A \sum_{i=1}^{3} \left[\frac{\partial \mathscr{L}}{\partial \dot{u}_i}\,\delta(-\omega^*\mathrm{f}_i g')+\frac{\partial \mathscr{L}}{\partial u_{i,1}}\,\delta(k^*\mathrm{f}_i g')+\sum_{j=2}^{3}\frac{\partial \mathscr{L}}{\partial u_{i,j}}\,\delta\left(\frac{\partial \mathrm{f}_i}{\partial x_j}g\right)\right]$$

where

$$g' = \frac{\mathrm{d}g}{\mathrm{d}\eta}.$$

In view of (6.40a, b), this expression can be rewritten as

$$\begin{aligned}\delta I = \int_0^{2\pi} \mathrm{d}\eta \int_A \sum_{i=1}^{3} &\left[\frac{\partial \mathscr{L}}{\partial \dot{u}_i}\,\delta(-\omega\mathrm{f}_i g')+\frac{\partial \mathscr{L}}{\partial u_{i,1}}\,\delta(k\mathrm{f}_i g')\right.\\ &\left.+\sum_{j=2}^{3}\frac{\partial \mathscr{L}}{\partial u_{i,j}}\,\delta\left(\frac{\partial \mathrm{f}_i}{\partial x_j}g\right)\right]\mathrm{d}x_2\,\mathrm{d}x_3\\ -\frac{\mathrm{d}\omega}{\mathrm{d}k}\,\delta k\int_0^{2\pi}\mathrm{d}\eta\int_A\sum_{i=1}^{3}\frac{\partial \mathscr{L}}{\partial \dot{u}_i}\,&\mathrm{f}_i g'\mathrm{d}x_2\,\mathrm{d}x_3+\delta k\int_0^{2\pi}\mathrm{d}\eta\int_A\sum_{i=1}^{3}\frac{\partial \mathscr{L}}{\partial u_{i,1}}\,\mathrm{f}_i g'\mathrm{d}x_2\,\mathrm{d}x_3.\end{aligned}$$

Of the three terms on the right-hand side, the first ıepresents the changes resulting from changes in the function $g(\eta)$ without changes in ω and k. By virtue of the version of Hamilton's principle for changes in u_i which maintain frequency and wavenumber, i.e., eq. (6.37), this integral must vanish. In the remaining two integrals the integrations can be changed from η to t. By identifying

$$\dot{u} = -\mathrm{f}_i(x_2, x_3)\,\omega\,\frac{\mathrm{d}g}{\mathrm{d}\eta},$$

we find

$$\frac{\mathrm{d}\omega}{\mathrm{d}k} = \frac{\displaystyle\int_0^T \mathrm{d}t\int_A\sum_{i=1}^{3}\frac{\partial \mathscr{L}}{\partial u_{i,1}}\,\dot{u}_i\,\mathrm{d}x_2\,\mathrm{d}x_3}{\displaystyle\int_0^T \mathrm{d}t\int_A\sum_{i=1}^{3}\frac{\partial \mathscr{L}}{\partial \dot{u}_i}\,\dot{u}_i\,\mathrm{d}x_2\,\mathrm{d}x_3}. \tag{6.41}$$

Comparison of (6.39) and (6.41) yields the desired result

$$c_e = \frac{\mathrm{d}\omega}{\mathrm{d}k}.$$

The treatment of energy propagation velocity as it is presented in this section also provides a link with the concept of group velocity as it was originally introduced by means of a kinematic argument. The first definition and derivation of the group velocity is apparently due to Stokes, whose treatment was discussed by Sommerfeld.[4]

In Stokes' treatment the group velocity appears when two plane waves that advance in the positive x-direction, with the same amplitude, but slightly different wavenumber, hence slightly different frequency, are superposed

$$\begin{aligned} u &= A\{\sin(k_1 x - \omega_1 t) + \sin(k_2 x - \omega_2 t)\} \\ &= 2A \cos\left(\frac{k_1 - k_2}{2} x - \frac{\omega_1 - \omega_2}{2} t\right) \sin\left(\frac{k_1 + k_2}{2} x - \frac{\omega_1 + \omega_2}{2} t\right). \end{aligned} \tag{6.42}$$

Introducing

$$\frac{k_1 + k_2}{2} = k_0; \qquad \frac{\omega_1 + \omega_2}{2} = \omega_0$$

$$\frac{k_1 - k_2}{2} = \Delta k; \qquad \frac{\omega_1 - \omega_2}{2} = \Delta\omega,$$

Eq. (6.42) can be written as

$$u = C \sin(k_0 x - \omega_0 t), \tag{6.43}$$

where

$$C = 2A \cos(\Delta k x - \Delta\omega t). \tag{6.44}$$

It is seen that a modulation represented by C is impressed on the carrier. The situation is depicted in figure 6.6.

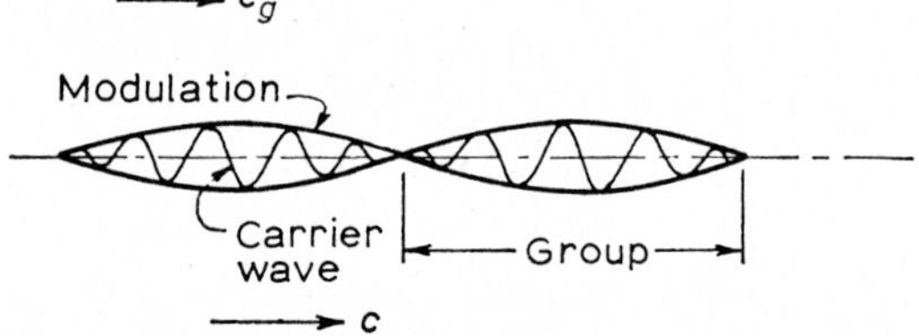

Fig. 6.6. Propagation of a group.

The introduction of the "amplitude factor" $C(x, t)$, which is responsible for the modulation, suggests that the cosine in eq. (6.44) is a slowly variable

[4] A. Sommerfeld, *Mechanics of deformable bodies*. New York, Academic Press, Inc. (1964), p. 184.

quantity. The modulation results in the building up of groups with amplitude C which move along with the *group velocity* c_g. The succession of wavelets represented by $\sin (k_0 x - \omega_0 t)$ moves along with the *phase velocity* $c_0 = \omega_0/k_0$, which is not very different from the phase velocities of the component waves, $c_1 = \omega_1/k_1$ and $c_2 = \omega_2/k_2$, respectively. On the other hand, the velocity of propagation of the modulation, or of the group, is found by setting

$$\Delta k x - \Delta \omega t = \text{constant},$$

which when differentiated yields

$$\frac{\mathrm{d}x}{\mathrm{d}t} = \frac{\Delta\omega}{\Delta k}.$$

In the limit of $\Delta k \to 0$ we obtain the group velocity as

$$c_g = \frac{\mathrm{d}\omega}{\mathrm{d}k},$$

which is just the same as the velocity of energy transport of a monochromatic wave, as discussed earlier in this section.

Both the dynamic and the kinematic arguments take as points of departure supposed small changes of the wavenumber and the associated small changes of the frequency. In the kinematic argument one can infer that the rate of transfer of energy is identical with the group velocity from the fact that no energy can travel past the nodes which move with velocity c_g. Thus, even in the limit as Δk approaches zero and when the wavelength of the modulation increases beyond bounds, the energy still propagates with velocity c_g. This limitcase just yields the superposition of two waves of the same amplitude, wavenumber and frequency, i.e., a monochromatic wave. Thus in a monochromatic wave the energy also propagates with the group velocity.

There are several ways of expressing the group velocity:

$$c_g = \frac{\mathrm{d}\omega}{\mathrm{d}k},$$

or

$$c_g = c + k\frac{\mathrm{d}c}{\mathrm{d}k},$$

or

$$c_g = c - \Lambda\frac{\mathrm{d}c}{\mathrm{d}\Lambda}$$

The group velocity and the phase velocity are shown in figure 6.7 in the (ω–k)-plane.

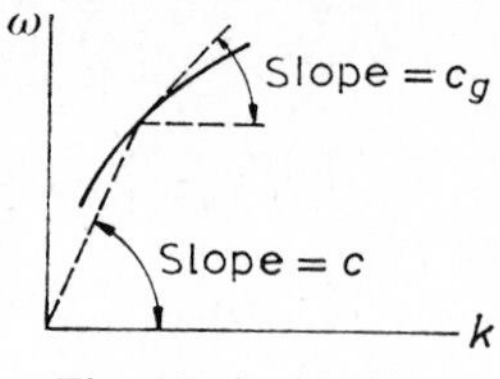

Fig. 6.7. (ω–k)-plane.

Another useful relation is

$$\frac{1}{c_g} = \frac{1}{c} - \frac{\omega}{c^2}\frac{\mathrm{d}c}{\mathrm{d}\omega}.$$

In general, $\mathrm{d}c/\mathrm{d}k < 0$ and thus $c_g < c$. The wavelets of the carrier wave are then building up at the back of the group, progressing through the group and disappearing in the front. If $\mathrm{d}c/\mathrm{d}k < 0$ we speak of normal dispersion. The converse case when $\mathrm{d}c/\mathrm{d}k > 0$ and $c_g > c$ is referred to as anomalous dispersion.

6.6. Love waves

The criterion for surface waves is that the propagating disturbance decays exponentially with distance from the surface. In chapter 5 Rayleigh surface waves propagating along the free surface of a half-space were examined. For Rayleigh waves the material particles move in the plane of propagation. Thus, for propagation in the x_1-direction along the surface of the half-space $x_2 \geqq 0$, the displacement $u_3(x_1, x_2, x)$ vanishes for classical Rayleigh waves.

The question may now be raised whether surface waves with displacements perpendicular to the plane of propagation, the $(x_1 x_2)$-plane, are possible in a homogeneous isotropic linearly elastic half-space. We recall that SH-waves are governed by the equation

$$\frac{\partial^2 u_3}{\partial x_1^2} + \frac{\partial^2 u_3}{\partial x_2^2} = \frac{1}{c_T^2}\frac{\partial^2 u_3}{\partial t^2}. \tag{6.45}$$

A solution of (6.45) representing a surface wave would be of the form

$$u_3 = Ae^{-bx_2}\exp[ik(x_1 - ct)], \tag{6.46}$$

where the real part of b must be positive. By substituting (6.46) into (6.45) we find

$$b = k\left[1-\left(\frac{c}{c_T}\right)^2\right]^{\frac{1}{2}}. \tag{6.47}$$

For a free surface the boundary condition at $x_2 = 0$ is

$$\frac{\partial u_3}{\partial x_2} = 0. \tag{6.48}$$

The boundary condition (6.48) can, however, be satisfied only if either $A = 0$ or $b = 0$. Neither case represents a surface wave.

Experimental data, particularly as gathered from seismological observations, have, however, shown that SH surface waves may occur along free surfaces. An analytical resolution of this question was provided by Love, who showed that SH-waves are possible if the half-space is covered by a layer of a different material, as shown in figure 6.8.

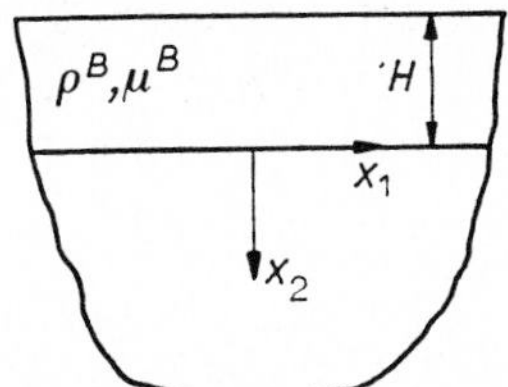

Fig. 6.8. Layered half-space.

The wave motion in the layer can be represented by eq. (6.11),

$$u_3^B = [B_1 \sin(q_B x_2)+B_2 \cos(q_B x_2)] \exp[ik(x_1-ct)], \tag{6.49}$$

where

$$q_B = k\left[\left(\frac{c}{c_T^B}\right)^2-1\right]^{\frac{1}{2}}. \tag{6.50}$$

The motion of the half-space is given by eq. (6.46). The condition of vanishing shear stress at the free surface, $x_2 = -H$, yields

$$B_1 \cos(q_B H)+B_2 \sin(q_B H) = 0.$$

Continuity of the shear stress and the displacement at the interface $x_2 = 0$ is satisfied if

$$\mu^B q_B B_1 = -\mu b A$$
$$B_2 = A.$$

By combining these three results we find the following equation for the phase velocity c,

$$\tan\left\{\left[\left(\frac{c}{c_T^B}\right)^2-1\right]^{\frac{1}{2}}kH\right\}-\frac{\mu}{\mu^B}\frac{[1-(c/c_T)^2]^{\frac{1}{2}}}{[(c/c_T^B)^2-1]^{\frac{1}{2}}}=0. \tag{6.51}$$

The left-hand side of (6.51) is positive for $c = c_T$, while it is negative for $c = c_T^B$. Apparently we can thus find a real root in the interval $c_T^B < c \leqq c_T$. No real root exists if $c_T < c_T^B$.

Eq. (6.51) shows that Love waves are dispersive, as opposed to Rayleigh waves which are not dispersive. If we consider kH as the independent variable, we have $c = c_T$ for $kH = 0$. As the wavelength decreases (kH increases), the phase velocity decreases. The phase velocity also approaches c_T as $[(c/c_T^B)^2-1]^{\frac{1}{2}}kH$ approaches π, 2π, etc. The latter limits are for the higher modes. For the lowest mode, c/c_T is shown versus the dimensionless wavenumber in figure 6.9. More extensive quantitative information can be found in the book by Ewing et al.[5]

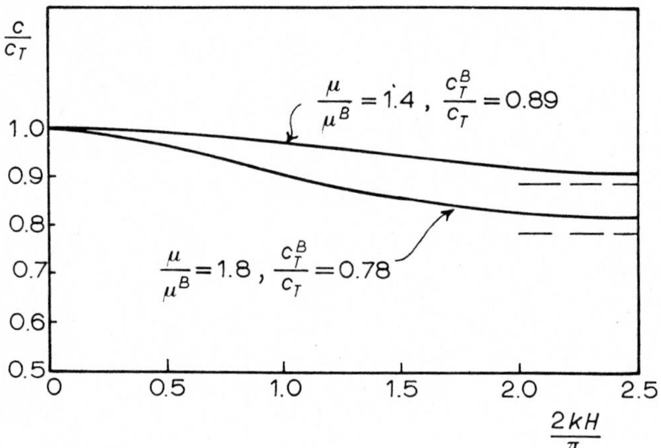

Fig. 6.9. Phase velocity for the lowest mode of Love waves.

6.7. Waves in plane strain in an elastic layer

For time-harmonic wave motion in plane strain of an elastic layer the equation relating frequency or phase velocity to the wavenumber can also

[5] W. M. Ewing, W. S. Jardetzky and F. Press, *Elastic waves in layered media.* New York, McGraw-Hill Book Company (1957), p. 210.

be derived on the basis of the principle of constructive interference. This approach was pursued by Tolstoy and Usdin.[6]

In most treatments, however, the alternative approach is followed of employing expressions for the field variables representing a standing wave in the x_2-direction and a propagating wave in the x_1-direction. The expressions are then substituted into the boundary conditions to derive the frequency equation. This more straightforward approach will be followed here.

It is convenient to decompose the displacement field by the use of scalar and vector potentials, as discussed in chapter 2. For motion in plane strain in the $(x_1 x_2)$-plane, we have

$$u_3 \equiv 0, \qquad \frac{\partial}{\partial x_3}(\;) \equiv 0. \tag{6.52}$$

Eq. (2.87) then reduces to

$$u_1 = \frac{\partial \varphi}{\partial x_1} + \frac{\partial \psi}{\partial x_2}, \tag{6.53}$$

$$u_2 = \frac{\partial \varphi}{\partial x_2} - \frac{\partial \psi}{\partial x_1}. \tag{6.54}$$

For simplicity of notation the subscript 3 has been omitted from ψ in (6.53) and (6.54). The relevant components of the stress tensor follow from Hooke's law as

$$\tau_{21} = \mu\left(\frac{\partial u_2}{\partial x_1} + \frac{\partial u_1}{\partial x_2}\right) = \mu\left(2\frac{\partial^2 \varphi}{\partial x_1 \partial x_2} - \frac{\partial^2 \psi}{\partial x_1^2} + \frac{\partial^2 \psi}{\partial x_2^2}\right), \tag{6.55}$$

$$\tau_{22} = \lambda\left(\frac{\partial u_1}{\partial x_1} + \frac{\partial u_2}{\partial x_2}\right) + 2\mu\frac{\partial u_2}{\partial x_2} = \lambda\left(\frac{\partial^2 \varphi}{\partial x_1^2} + \frac{\partial^2 \varphi}{\partial x_2^2}\right) + 2\mu\left(\frac{\partial^2 \varphi}{\partial x_2^2} - \frac{\partial^2 \psi}{\partial x_1 \partial x_2}\right). \tag{6.56}$$

As discussed in chapter 2, the potentials φ and ψ satisfy wave equations, which for plane strain are two-dimensional,

$$\frac{\partial^2 \varphi}{\partial x_1^2} + \frac{\partial^2 \varphi}{\partial x_2^2} = \frac{1}{c_L^2}\frac{\partial^2 \varphi}{\partial t^2}, \tag{6.57}$$

$$\frac{\partial^2 \psi}{\partial x_1^2} + \frac{\partial^2 \psi}{\partial x_2^2} = \frac{1}{c_T^2}\frac{\partial^2 \psi}{\partial t^2}. \tag{6.58}$$

[6] I. Tolstoy and E. Usdin, *Geophysics* **18** (1953) 844.

To investigate wave motion in the elastic layer, we consider solutions of (6.57) and (6.58) of the form

$$\varphi = \Phi(x_2) \exp\,[i(kx_1 - \omega t)], \tag{6.59}$$

$$\psi = \Psi(x_2) \exp\,[i(kx_1 - \omega t)]. \tag{6.60}$$

Substituting (6.59) and (6.60) into (6.57) and (6.58), respectively, the solutions of the resulting equations are obtained as

$$\Phi(x_2) = A_1 \sin(px_2) + A_2 \cos(px_2) \tag{6.61}$$

$$\Psi(x_2) = B_1 \sin(qx_2) + B_2 \cos(qx_2), \tag{6.62}$$

wherein

$$p^2 = \frac{\omega^2}{c_L^2} - k^2, \qquad q^2 = \frac{\omega^2}{c_T^2} - k^2. \tag{6.63a, b}$$

In the expressions for the displacement and the stress components, which are obtained from (6.53)–(6.56), the term $\exp\,[i(kx_1 - \omega t)]$ appears as a multiplier. Since the exponential appears in all of the expressions it does not play a further role in the determination of the frequency equation and it is therefore omitted in the sequel. Thus we write

$$u_1 = ik\Phi + \frac{\mathrm{d}\Psi}{\mathrm{d}x_2} \tag{6.64}$$

$$u_2 = \frac{\mathrm{d}\Phi}{\mathrm{d}x_2} - ik\Psi \tag{6.65}$$

$$\tau_{21} = \mu\left(2ik\frac{\mathrm{d}\Phi}{\mathrm{d}x_2} + k^2\Psi + \frac{\mathrm{d}^2\Psi}{\mathrm{d}x_2^2}\right) \tag{6.66}$$

$$\tau_{22} = \lambda\left(-k^2\Phi + \frac{\mathrm{d}^2\Phi}{\mathrm{d}x_2^2}\right) + 2\mu\left(\frac{\mathrm{d}^2\Phi}{\mathrm{d}x_2^2} - ik\frac{\mathrm{d}\Psi}{\mathrm{d}x_2}\right). \tag{6.67}$$

Inspection of (6.64) and (6.65) shows that the displacement components can be written in terms of elementary functions. For the displacement in the x_1-direction the motion is symmetric (antisymmetric) with regard to $x_2 = 0$, if u_1 contains cosines (sines). The displacement in the x_2-direction is symmetric (antisymmetric) if u_2 contains sines (cosines). The modes of wave propagation in the elastic layer may thus be split up into two systems of symmetric and antisymmetric modes, respectively:

Symmetric modes:

$$\begin{aligned}
\Phi &= A_2 \cos(px_2)\\
\Psi &= B_1 \sin(qx_2)\\
u_1 &= ikA_2 \cos(px_2)+qB_1 \cos(qx_2)\\
u_2 &= -pA_2 \sin(px_2)-ikB_1 \sin(qx_2)\\
\tau_{21} &= \mu[-2ikpA_2 \sin(px_2)+(k^2-q^2)B_1 \sin(qx_2)]\\
\tau_{22} &= -\lambda(k^2+p^2)A_2 \cos(px_2)-2\mu[p^2A_2 \cos(px_2)+ikqB_1 \cos(qx_2)].
\end{aligned}$$

Antisymmetric modes:

$$\begin{aligned}
\Phi &= A_1 \sin(px_2)\\
\Psi &= B_2 \cos(qx_2)\\
u_1 &= ikA_1 \sin(px_2)-qB_2 \sin(qx_2)\\
u_2 &= pA_1 \cos(px_2)-ikB_2 \cos(qx_2)\\
\tau_{21} &= \mu[2ikpA_1 \cos(px_2)+(k^2-q^2)B_2 \cos(qx_2)]\\
\tau_{22} &= -\lambda(k^2+p^2)A_1 \sin(px_2)-2\mu[p^2A_1 \sin(px_2)-ikqB_2 \sin(qx_2)].
\end{aligned}$$

The frequency relation, i.e. the expression relating ω to the wave number k is now obtained from the boundary conditions. If the boundaries are free, we have at $x_2 = \pm h$:

$$\tau_{21} = \tau_{22} \equiv 0.$$

For the symmetric modes the boundary conditions yield a system of two homogeneous equations for the constants A_2 and B_1. Similarly, for the antisymmetric modes two homogeneous equations for the constants A_1 and B_2 are obtained. Since the systems are homogeneous, the determinant of the coefficients must vanish, which yields the frequency equation. Thus, for the symmetric modes we find

$$\frac{(k^2-q^2)\sin(qh)}{2ikp\sin(ph)} = -\frac{2\mu ikq\cos(qh)}{(\lambda k^2+\lambda p^2+2\mu p^2)\cos(ph)}$$

This equation can be rewritten as

$$\frac{\tan(qh)}{\tan(ph)} = -\frac{4k^2pq}{(q^2-k^2)^2}. \tag{6.68}$$

For the antisymmetric modes the boundary conditions yield

$$-\frac{(k^2-q^2)\cos(qh)}{2ikp\cos(ph)} = \frac{2\mu ikq\sin(qh)}{(\lambda k^2+\lambda p^2+2\mu p^2)\sin(ph)}.$$

or

$$\frac{\tan(qh)}{\tan(ph)} = -\frac{(q^2-k^2)^2}{4k^2pq}. \tag{6.69}$$

Eqs. (6.68) and (6.69) are the well-known Rayleigh-Lamb frequency equations. These transcendental equations look deceptively simple. Although the frequency equations were derived at the end of the 19th century it was not until quite recently that the frequency spectrum was unraveled in complete detail by Mindlin.[7]

The frequency equations for horizontally polarized shear waves, eq. (6.9) and for symmetric and antisymmetric motions in plane strain, eqs. (6.68) and (6.69), respectively, can also be obtained as the result of a somewhat more general approach if we take as point of departure expressions representing propagating waves for the scalar potential φ and for *all* three components of $\boldsymbol{\psi}$:

$$\begin{aligned}
\varphi &= (A\cos px_2 + B\sin px_2)\exp[i(kx_1-\omega t)] \\
\psi_1 &= (C\cos qx_2 + D\sin qx_2)\exp[i(kx_1-\omega t)] \\
\psi_2 &= (E\cos qx_2 + F\sin qx_2)\exp[i(kx_1-\omega t)] \\
\psi_3 &= (G\cos qx_2 + H\sin qx_2)\exp[i(kx_1-\omega t)].
\end{aligned}$$

The boundary conditions of vanishing stresses τ_{22}, τ_{21} and τ_{23} at $x_2 = \pm h$ yield six homogeneous equations for the eight constants. Two additional equations are obtained, however, by evaluating the condition $\nabla\cdot\boldsymbol{\psi} = 0$ at $x_2 = \pm h$. A necessary and sufficient condition that the eight equations for the eight constants A, B, C, D, E, F, G and H possess solutions is that the determinant of the coefficients be zero. It can subsequently be shown that the eight by eight determinant can be reduced to the product of four subdeterminants.[8] Wave motions are thus possible if for given material parameters the frequency ω and the wave number k are related in such a manner that one of these four subdeterminants vanishes. The equations that are obtained by equating the subdeterminants individually to zero are just the frequency equations (6.9), (6.68) and (6.69).

Much simpler frequency equations than the Rayleigh-Lamb equations are obtained if we consider (the unfortunately rather unrealistic) mixed

[7] R. D. Mindlin, "Waves and vibrations in isotropic elastic plates", in: *Structural mechanics*, ed. by J. N. Goodier and N. J. Hoff. New York, Pergamon Press (1960).

[8] See T. R. Meeker and A. H. Meitzler, "Guided wave propagation in elongated cylinders and plates", in: *Physical acoustics*, ed. by W. P. Mason. New York, Academic Press (1964), p. 115.

boundary conditions. These are described by the following conditions at $x_2 = \pm h$: either $u_2 = 0$, $\tau_{21} = 0$, or $u_1 = 0$, $\tau_{22} = 0$. As conceivable approximations to these conditions one may think in the first case of a thin layer of lubricant, and in the second case of a thin layer of felt separating the elastic layer from a rigid boundary.

We have seen earlier in the discussion of the reflection of time-harmonic waves that the mixed boundary conditions do not couple the equivoluminal and dilatational waves. Thus, if the boundary conditions $u_2 = 0$, $\tau_{21} = 0$ are examined for symmetric motions, we note that both boundary conditions can be satisfied by either $A_2 = 0$ and $\sin qh = 0$, or $B_1 = 0$ and $\sin ph = 0$. These two cases correspond to uncoupled equivoluminal and dilatational modes, respectively. The condition $\sin qh = 0$ yields the following frequency equation

$$q = \frac{n\pi}{2h}, \qquad n = 2, 4, 6.$$

By employing the definition of q, eq. (6.63b), we rewrite the frequency equation as

$$\Omega^2 = n^2 + \xi^2,$$

where Ω and ξ are the dimensionless frequency and the dimensionless wave number, respectively, which were earlier defined as

$$\Omega = \frac{2h\omega}{\pi c_T}, \qquad \xi = \frac{2kh}{\pi}.$$

The frequency equations corresponding to the boundary conditions $u_2 = 0$, $\tau_{21} = 0$ at $x_2 = \pm h$ can be summarized as follows:

Symmetric equivoluminal modes:

$$A_2 = 0; \qquad \Omega^2 = n^2 + \xi^2, \qquad n = 0, 2, 4, 6, \ldots \tag{6.70}$$

Antisymmetric equivoluminal modes:

$$A_1 = 0; \qquad \Omega^2 = n^2 + \xi^2, \qquad n = 1, 3, 5, \ldots \tag{6.71}$$

Symmetric dilatational modes:

$$B_1 = 0; \qquad \Omega^2 = \kappa^2(m^2 + \xi^2), \qquad m = 0, 2, 4, 6, \ldots \tag{6.72}$$

Antisymmetric dilatational modes:

$$B_2 = 0; \qquad \Omega^2 = \kappa^2(m^2 + \xi^2), \qquad m = 1, 3, 5, \ldots \tag{6.73}$$

where $\kappa = c_L/c_T$, see eq. (4.8).

For real-valued wave numbers, (6.70)–(6.73) represent hyperbolas in the Ω–ξ-plane, with asymptotes $\Omega = \xi$ and $\Omega = \kappa\xi$ for (6.70), (6.71) and (6.72), (6.73), respectively. For imaginary wave numbers, eqs. (6.70) and (6.71) represent circles with radius n, and eqs. (6.72) and (6.73) represent ellipses with semi-axes m and κm. It is clear that the frequency spectrum is very similar to that of horizontally polarized shear waves.

For completeness, we also summarize the frequency equations for the boundary conditions $u_1 = 0$ and $\tau_{22} = 0$ at $x_2 = \pm h$:

Symmetric equivoluminal modes:

$$A_2 = 0; \qquad \Omega^2 = n^2+\xi^2, \qquad n = 1, 3, 5, \ldots \tag{6.74}$$

Antisymmetric equivoluminal modes:

$$A_1 = 0; \qquad \Omega^2 = n^2+\xi^2, \qquad n = 0, 2, 4, 6, \ldots \tag{6.75}$$

Symmetric dilatational modes:

$$B_1 = 0; \qquad \Omega^2 = \kappa^2(m^2+\xi^2), \qquad m = 1, 3, 5, \ldots \tag{6.76}$$

Antisymmetric dilatational modes:

$$B_2 = 0; \qquad \Omega^2 = \kappa^2(m^2+\xi^2), \qquad m = 0, 2, 4, 6, \ldots \tag{6.77}$$

6.8. The Rayleigh-Lamb frequency spectrum

The relations between the frequencies and the wave numbers expressed by the Rayleigh-Lamb frequency equations yield an infinite number of branches for an infinite number of symmetric and antisymmetric modes. The symmetric modes are usually termed the *longitudinal* modes because the average displacement over the thickness is in the longitudinal direction. For the antisymmetric modes, the average displacement is in the transverse direction, and these modes are generally termed the *flexural* modes.

It is again convenient to introduce the dimensionless frequency Ω and the dimensionless wave number ξ by

$$\Omega = \frac{2h\omega}{\pi c_T}, \qquad \xi = \frac{2kh}{\pi}.$$

The frequency equation for the longitudinal modes may then be rewritten as

$$\frac{\tan\left[\frac{1}{2}\pi(\Omega^2-\xi^2)^{\frac{1}{2}}\right]}{\tan\left[\frac{1}{2}\pi(\Omega^2/\kappa^2-\xi^2)^{\frac{1}{2}}\right]} = -\frac{4\xi^2(\Omega^2/\kappa^2-\xi^2)^{\frac{1}{2}}(\Omega^2-\xi^2)^{\frac{1}{2}}}{(\Omega^2-2\xi^2)^2}, \tag{6.78}$$

while for the flexural modes we have

$$\frac{\tan\left[\frac{1}{2}\pi(\Omega^2-\xi^2)^{\frac{1}{2}}\right]}{\tan\left[\frac{1}{2}\pi(\Omega^2/\kappa^2-\xi^2)^{\frac{1}{2}}\right]} = -\frac{(\Omega^2-2\xi^2)^2}{4\xi^2(\Omega^2/\kappa^2-\xi^2)^{\frac{1}{2}}(\Omega^2-\xi^2)^{\frac{1}{2}}}. \tag{6.79}$$

If the frequency equations are written in this form it is apparent that only one material parameter represented by κ needs to be specified to compute a system of curves showing Ω as functions of ξ. The material constant $\kappa = c_L/c_T$ may be expressed in terms of Poisson's ratio ν by

$$\kappa = \left[\frac{2(1-\nu)}{1-2\nu}\right]^{\frac{1}{2}}.$$

A frequency spectrum for in-plane motions of a layer thus displays the branches of the longitudinal and flexural modes for a specific value of Poisson's ratio.

Although the Rayleigh-Lamb equations look rather simple, it is not possible to write analytical expressions for the branches. To obtain detailed and precise numerical information, the roots of eqs. (6.78) and (6.79) must be computed on a digital computer. In rough outline the most commonly used numerical technique is to choose a value of the frequency Ω, and then scan in the domain of wave numbers for values of ξ at which the expressions (6.78) and (6.79) change signs. For real-valued ξ, this is quite straightforward and extensive numerical results are available. The wave number ξ may, however, also be imaginary or complex, and for these cases numerical information is less readily available. Without actual numerical computation of the roots it is, however, quite possible to examine the frequency spectrum and to discuss its most important features.

In analyzing the Rayleigh-Lamb frequency equations the dimensionless frequency Ω is taken as real and positive. The dimensionless wave number ξ may be real, but analogously to the frequency spectrum of SH-waves in a layer, imaginary wave numbers should also be expected. It will become evident later in this section that for motion in plane strain we should also expect complex wave numbers. Real-valued wave numbers correspond to time-harmonic waves that are not attenuated while propagating in the x_1-direction. Imaginary values of ξ give displacements in the form of sums of exponentials. Complex wave numbers result in products of exponentials and trigonometric functions. In physical terms, imaginary and complex wave numbers correspond to standing waves with decaying amplitudes as x_1 increases.

It is informative to investigate the limiting values of Ω for very small and very large values of ξ. In particular, an examination of $\xi \ll 1$, i.e., $kh \ll \frac{1}{2}\pi$ is of interest. Since $k = 2\pi/\Lambda$, the condition $\xi \ll 1$ corresponds to the, for practical applications, most common case of a layer thickness $2h$ which is much smaller than half the wavelength.

For real-valued ξ, there are three ranges in the Ω–ξ-plane in which the radicals in eqs. (6.78) and (6.79) are of different character: $\xi > \Omega$, $\Omega > \xi > \Omega/\kappa$, and $\Omega/\kappa > \xi > 0$. We first consider the range $\xi > \Omega$, where the two radicals in (6.78) and (6.79) are both purely imaginary and the trigonometric functions become hyperbolic tangents. For small values of ξ, we now assume the following expansion for Ω:

$$\Omega = \Omega_1 \xi + \tfrac{1}{2}\Omega_2 \xi^2 + \ldots . \tag{6.80}$$

The expansion does not include a constant term because the expansion is within the region $\xi > \Omega$. Upon substituting (6.80) into the frequency equation for flexural waves, eq. (6.79), we find, after expanding the hyperbolic tangents as well as the radicals, whereby a sufficient number of terms must be retained,

$$\Omega_1 = 0, \qquad \Omega_2 = 2\pi \left[\frac{1}{3}\left(1 - \frac{1}{\kappa^2}\right)\right]^{\frac{1}{2}}.$$

Thus, for $\xi \ll 1$ we may write

$$\Omega = \left(\frac{\kappa^2 - 1}{3\kappa^2}\right)^{\frac{1}{2}} \pi \xi^2. \tag{6.81}$$

It will be shown in a later section that (6.81) is the frequency found from the Lagrange-Germain plate theory for flexural motions, which indicates that the classical plate theory can be used to describe wave motions provided that the wavelength is much larger than the thickness of the plate. If the expansion (6.80) is substituted into the frequency equation for longitudinal motions, eq. (6.78), no solutions can be obtained in the range $\xi > \Omega$.

In the region $\xi > \Omega$, the left-hand sides of both (6.78) and (6.79) approach unity as ξ increases beyond bounds. For very large values of ξ, these equations then reduce to

$$(\Omega^2 - 2\xi^2)^2 - 4\xi^2(\xi^2 - \Omega^2/\kappa^2)^{\frac{1}{2}}(\xi^2 - \Omega^2)^{\frac{1}{2}} = 0.$$

This equation is recognized as the frequency equation for Rayleigh surface waves, which was derived earlier in the previous chapter. The asymptotic behavior of the frequency as the wave number increases is intuitively very acceptable because for very short waves the frequency spectrum of the elastic layer should include the frequency of surface motions in a half-space.

We next consider the range $\Omega > \xi > \Omega/\kappa$, where one of the radicals is imaginary and the other one is real. It can be checked that for large values of ξ the solutions of the two frequency equations asymptotically approach the line $\Omega = \xi$.

For $\xi \ll 1$, we obtain from (6.78), by a limiting process,

$$\Omega = 2\left(\frac{\kappa^2-1}{\kappa^2}\right)^{\frac{1}{2}}\xi. \tag{6.82}$$

The frequency equation (6.79) does not yield a solution near $\xi = 0$ in the range $\Omega/\kappa < \xi < \Omega$. It will be shown in the sequel that (6.82) is also the frequency according to the elementary Poisson theory for extensional motions of a plate. The exact curve relating the frequency and the wave number for this lowest longitudinal mode later crosses the line $\Omega = \xi$ and asymptotically approaches $\Omega = (c_R/c_T)\xi$ as ξ increases, where c_R is the velocity of Rayleigh surface waves.

In the region $0 < \xi < \Omega/\kappa$, there are no asymptotic limits for large ξ. For small values of ξ, we find that the frequency equation is satisfied for symmetric motions if

$$\sin\left(\tfrac{1}{2}\pi\Omega\right) = 0 \quad \text{or} \quad \cos\left(\tfrac{1}{2}\pi\Omega/\kappa\right) = 0, \tag{6.83a, b}$$

and for antisymmetric motions if

$$\sin\left(\tfrac{1}{2}\pi\Omega/\kappa\right) = 0 \quad \text{or} \quad \cos\left(\tfrac{1}{2}\pi\Omega\right) = 0. \tag{6.84a, b}$$

These equations yield the frequencies for waves of infinitely long wavelengths, i.e., for motions that are independent of the x_1-coordinate. We can, of course, study these motions directly by writing out the expressions for the displacements and the stresses. It is then found that the displacements in the x_1- and x_2-directions are not coupled. We have either dilatational motions ($u_1 \equiv 0$, $u_2 \neq 0$), or equivoluminal motions ($u_1 \neq 0$, $u_2 \equiv 0$). The frequency equations (6.83a, b) and (6.84a, b) then immediately follow from the expressions for the stresses and the conditions that the boundaries are free of tractions. In the limit of vanishing ξ the dimensionless frequencies Ω may thus be written as follows:

Symmetric dilatational modes:

$$\Omega = \kappa m, \qquad m = 1, 3, 5, \ldots \tag{6.83}$$

Antisymmetric dilatational modes:

$$\Omega = \kappa m, \qquad m = 0, 2, 4, 6, \ldots \tag{6.84}$$

Symmetric equivoluminal modes:

$$\Omega = n, \qquad n = 0, 2, 4, 6, \ldots \tag{6.85}$$

Antisymmetric equivoluminal modes:

$$\Omega = n, \qquad n = 1, 3, 5, \ldots. \tag{6.86}$$

The dilatational motions for $m \neq 0$ and the equivoluminal motions for $n \neq 0$ are also termed thickness stretch and thickness shear motions, respectively. It is noted that the limiting frequencies coincide with the corresponding limiting frequencies for the layer with mixed boundary conditions, as obtained from (6.76), (6.77), (6.70) and (6.71), respectively.

Much more can be said about the frequencies for real and imaginary values of the wave numbers. We shall just summarize some of the salient results, which are due mostly to Mindlin.[9] It is of interest that it can be shown that the branches for the layer with mixed boundary conditions, as defined by eqs. (6.76), (6.77), (6.70) and (6.71), form bounds for the curves of the layer with free boundaries, for real as well as for imaginary wave numbers. Over a finite interval a curve of the Rayleigh-Lamb frequency spectrum is confined between two of these bounds, but then it must cross one of them. The crossings take place at successive intersections of bounds m even with n even, and m odd with n odd. At these intersections the corresponding modes satisfy both mixed and traction-free boundary conditions. Furthermore, the frequencies at $\xi = 0$ are known, and information on the slopes and the curvatures at $\xi = 0$ can easily be obtained. It is also known that the two lowest modes, whose behavior for $\xi \ll 1$ is given by (6.81) and (6.82), approach $\Omega = (c_R/c_T)\xi$ as ξ is real and increases. All the other modes approach $\Omega = \xi$ as ξ is real and increasing. In addition, the slopes and the curvatures at the points where the bounds are crossed can be computed. Altogether, the foregoing information is sufficient to sketch the branches for real and imaginary values of ξ on the gridwork of bounds.

For real-valued wavenumbers the branches for the four lowest longitudinal modes governed by eq. (6.78) have been plotted in figure 6.10. Analogous branches computed for the flexural modes from eq. (6.79) have been plotted

[9] R. D. Mindlin, see fn. 7, p. 224.

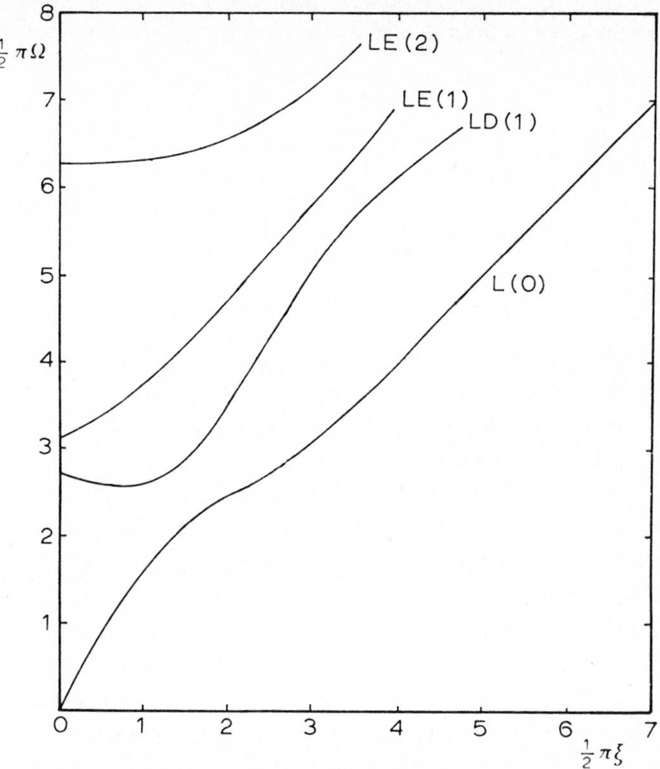

Fig. 6.10. Branches for the four lowest longitudinal modes; $\nu = 0.25$.

in figure 6.11. The curves were taken from a report by Potter and Leedham.[10] An observation which would be particularly obvious if a large number of branches were shown is that a line Ω = const intersects only a finite number of curves for real and imaginary ξ. Since we would expect an infinite number of wavenumbers for any value of the frequency, it is to be expected that there is an infinite number of branches with complex wavenumbers. The existence of these branches can easily be checked for very small values of Ω. Let us consider longitudinal modes and rewrite eq. (6.78) as

$$F(\Omega, \xi) = 0, \tag{6.87}$$

where

$$F(\Omega, \xi) = \frac{\tan\left[\frac{1}{2}\pi(\Omega^2-\xi^2)^{\frac{1}{2}}\right]}{\tan\left[\frac{1}{2}\pi(\Omega^2/\kappa^2-\xi^2)^{\frac{1}{2}}\right]} + \frac{4\xi^2(\Omega^2/\kappa^2-\xi^2)^{\frac{1}{2}}(\Omega^2-\xi^2)^{\frac{1}{2}}}{(\Omega^2-2\xi^2)^2}. \tag{6.88}$$

[10] D. S. Potter and C. D. Leedham, *Normalized numerical solutions for Rayleigh frequency equation.* Santa Barbara, Calif., GM Defense Research Laboratories, TR 66–57.

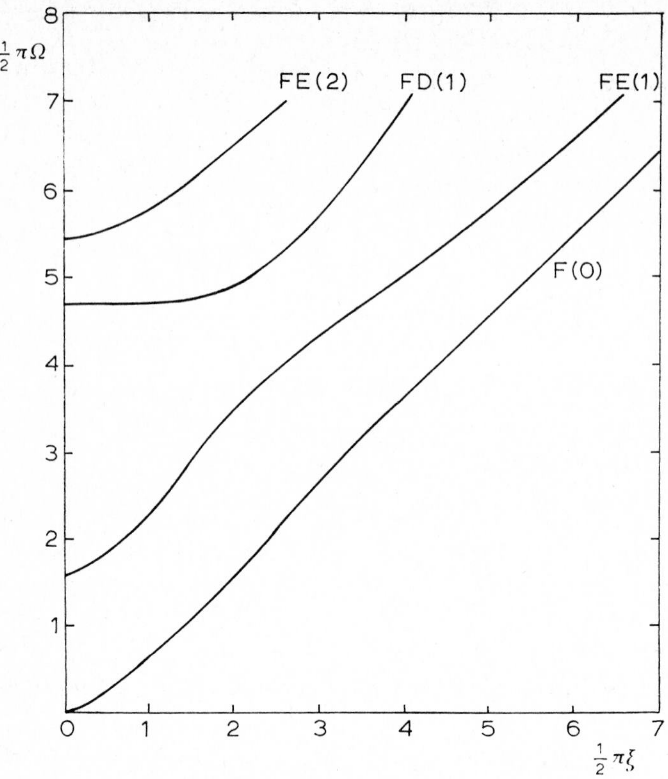

Fig. 6.11. Branches for the four lowest flexural modes; $\nu = 0.25$.

For small values of Ω we write

$$F(\Omega, \xi) = F_0(\xi) + F_1(\xi)\Omega + F_2(\xi)\Omega^2 + \dots . \tag{6.89}$$

In view of (6.87), we have $F_0(\xi) \equiv 0$. Also, since $F(\Omega, \xi)$ is a function of Ω^2, a solution of (6.87) intersects the plane $\Omega = 0$ at a right angle, which implies $F_1(\xi) \equiv 0$. For very small values of Ω the first term of the expansion (6.89) is thus quadratic. The points at which a curve satisfying eq. (6.87) intersects the plane $\Omega = 0$ then follows from the equation

$$F_2(\xi) = 0. \tag{6.90}$$

According to eq. (6.89), the function $F_2(\xi)$ can be computed as

$$F_2(\xi) = \lim_{\Omega \to 0} \frac{F(\Omega, \xi)}{\Omega^2},$$

where $F(\Omega, \xi)$ is defined by eq. (6.88). By applying l'Hôpital's rule we can obtain $F_2(\xi)$, whereupon (6.90) yields

$$\sinh(\pi\xi)+\pi\xi = 0.$$

The solutions of this equation are the intersections of the curves defined by (6.87) with the plane $\Omega = 0$. In a completely analogous manner we find for the flexural modes

$$\sinh(\pi\xi)-\pi\xi = 0.$$

If we write $\xi = \xi_1+i\xi_2$, we obtain the following system of simultaneous equations for ξ_1 and ξ_2:

$$\cos(\pi\xi_2) = \mp\pi\xi_1/\sin(\pi\xi_1)$$
$$\cosh(\pi\xi_1) = \mp\pi\xi_2/\sinh(\pi\xi_2),$$

where the minus and the plus signs apply to longitudinal and flexural modes, respectively. Since this set of equations possesses an infinite number of solutions, it is now clear that the frequency spectrum will contain an infinite number of modes with complex wave numbers.

It is evident that detailed information on the behavior of the curves for complex ξ and $\Omega > 0$ can be gathered only by actual numerical computations. Analytically it is also possible, however, to determine the intersection points with the planes $\xi_1 = 0$ and $\xi_2 = 0$. This can be done on the basis of the observation that ξ occurs in the frequency equation in ξ to the second power. The negative of a solution for ξ, as well as its complex conjugate, thus also satisfies the frequency equation, which implies that at the points of intersections the curves must be normal to the planes $\xi_1 = 0$ and $\xi_2 = 0$, respectively. As a consequence, the derivative of $F(\Omega, \xi)$ with respect to ξ must vanish, i.e., $\partial F/\partial\xi_2 = 0$ and $\partial F/\partial\xi_1 = 0$ at $\xi_1 = 0$ and $\xi_2 = 0$, respectively. We have also

$$\frac{\partial F}{\partial\Omega}\frac{\partial\Omega}{\partial\xi}+\frac{\partial F}{\partial\xi} = 0,$$

and thus in both cases the intersection points are located at points where $\partial\Omega/\partial\xi \equiv 0$. Since the additional condition is still $F(\Omega, \xi) = 0$, the intersection points of the curves for complex values of ξ with the planes $\xi_1 = 0$ and $\xi_2 = 0$ are defined by

$$F(\Omega, \xi) = 0, \qquad \frac{\partial\Omega}{\partial\xi} = 0. \tag{6.91}$$

Eqs. (6.91) define the extreme values of the frequency curves. For real-valued wave numbers the minimums defined by eq. (6.91) were earlier termed the cut-off frequencies. For symmetric motions a few curves, including one with complex wave numbers, have been sketched in figure 6.12. Numerical information on the frequencies for imaginary and complex-valued wave-numbers can be found in the previously cited report by Potter and Leedham.

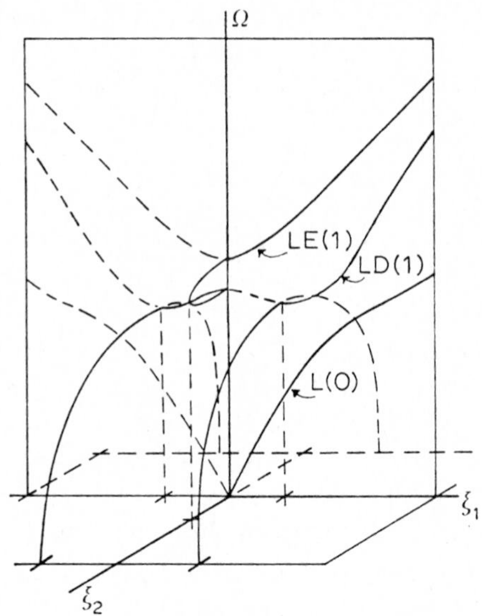

Fig. 6.12. Symmetric modes with complex wavenumbers.

To trace a particular branch we start at the point where the frequency Ω is zero. Except for the two lowest modes, whose behavior at small values of Ω is given by eqs. (6.81) and (6.82), respectively, this means that we start with a complex wave number. We choose $\xi_1 > 0$ and $\xi_2 > 0$, so that the amplitude decays as x_1 increases. For the branch shown in figure 6.12, the mode is nonpropagating at small frequencies and decays more slowly as Ω increases. At the point defined by the cut-off frequency, where the wave number becomes real-valued, the mode is converted into a propagating mode as we proceed along the curve for which real ξ increases as Ω increases. The group velocity vanishes for complex wave numbers and, in choosing the branch as described above, the group velocity is always positive for the part of the branch with real wave numbers. In figure 6.12, the branch is shown by a heavy line and indicated by LD(1). The terminology LD(1) indicates that we consider a symmetric mode which, for real-valued wave numbers larger than the wave number corresponding to the cut-off frequency, follows the curve which intersects $\xi = 0$ at the frequency of the first dilatational mode as given by eq. (6.83). It should be emphasized that a solution curve in the plane of Ω and real ξ is thus not necessarily part of a single branch. For example, for real ξ, the dashed part in figure 6.12 does not belong to the LD(1) branch. In general, the LD(r) branch is the branch which in the plane of real wave numbers runs completely or in part along the curve that intersects $\xi = 0$ at the frequency of the rth purely dilatational mode. In a completely analogous manner the mode defined by LE(r) is associated with the rth purely equivoluminal mode. The antisymmetric or flexural modes are labeled FD(r) and FE(r), and they are defined in a completely equivalent manner as the LD(r) and LE(r) modes. This labeling system does not hold for the two lowest modes whose branches come out of the origin of the Ω–ξ-plane, and which are completely defined by real-valued wave numbers. For small values of ξ and Ω the branches of these modes are represented by (6.82) and (6.81) for the longitudinal and the flexural modes, respectively. Although some authors call these modes the first longitudinal and flexural modes, we will denote them by the more consistent terminology of zeroth modes, L(0) and F(0). Often these two modes are referred to as *the* flexural and *the* longitudinal mode.

Consistent with the scheme of selecting the branches such that the group velocity does not change sign along a branch, the dashed part which would appear to be the lower end of the LD(1) branch (see figure 6.12) is to be considered part of the branch which emanates from the point $\Omega = 0$, $\xi_1 > 0$, but $\xi_2 < 0$. This branch decays toward decreasing x_1, and following the dashed line, via a brief detour (not shown) in the plane of purely imaginary (negative) ξ, it becomes a mode propagating in the negative x_1-direction. Over a short range of frequencies in between the cut-off frequency and the frequency of the first purely dilatational mode, the energy propagates in the negative x_1-direction but the phase velocity is positive. A similar situation develops when we start at $\Omega = 0$, $\xi_1 < 0$ and $\xi_2 > 0$. In consistently following a curve such that the group velocity is either zero or positive, we define the branch LE(1), which is indicated by a solid line in figure 6.12. For real-valued but negative wave numbers this branch includes a range of frequencies where the group velocity is positive but the phase velocity is negative. Such wave motions carry energy in one direction but appear to propagate in the other direction. The wave troughs and crests appear to move against the energy flux. This "backward-wave" transmission was investigated experimentally and theoretically by Meitzler.[11]

We will return to the Rayleigh-Lamb frequency spectrum in our discussion of the forced motions of an elastic layer.

6.9. Waves in a rod of circular cross section

In cylindrical coordinates the equations governing the motions of a homogeneous, isotropic, linearly elastic medium are given in section 2.13. To these equations we now seek solutions which represent time-harmonic wave motions propagating in the axial direction along a circular cylindrical rod.

The displacement equations of motion may be written as

$$\nabla^2 u - \frac{u}{r^2} - \frac{2}{r^2}\frac{\partial v}{\partial \theta} + \frac{1}{1-2\nu}\frac{\partial \Delta}{\partial r} = \frac{1}{c_T^2}\frac{\partial^2 u}{\partial t^2} \tag{6.92}$$

$$\nabla^2 v - \frac{v}{r^2} + \frac{2}{r^2}\frac{\partial u}{\partial \theta} + \frac{1}{1-2\nu}\frac{1}{r}\frac{\partial \Delta}{\partial \theta} = \frac{1}{c_T^2}\frac{\partial^2 v}{\partial t^2} \tag{6.93}$$

$$\nabla^2 w + \frac{1}{1-2\nu}\frac{\partial \Delta}{\partial z} = \frac{1}{c_T^2}\frac{\partial^2 w}{\partial t^2}, \tag{6.94}$$

where ∇^2 is the Laplacian

$$\nabla^2 = \frac{\partial^2}{\partial r^2} + \frac{1}{r}\frac{\partial}{\partial r} + \frac{1}{r^2}\frac{\partial^2}{\partial \theta^2} + \frac{\partial^2}{\partial z^2}, \tag{6.95}$$

and the dilatation Δ is defined as

$$\Delta = \frac{\partial u}{\partial r} + \frac{1}{r}\left(\frac{\partial v}{\partial \theta} + u\right) + \frac{\partial w}{\partial z}. \tag{6.96}$$

The pertinent stress-strain relations are

$$\tau_r = \lambda\Delta + 2\mu\frac{\partial u}{\partial r} \tag{6.97}$$

$$\tau_{r\theta} = \mu\left[\frac{1}{r}\left(\frac{\partial u}{\partial \theta} - v\right) + \frac{\partial v}{\partial r}\right] \tag{6.98}$$

$$\tau_{rz} = \mu\left(\frac{\partial u}{\partial z} + \frac{\partial w}{\partial r}\right). \tag{6.99}$$

[11] A. H. Meitzler, "Backward-wave transmission of stress pulses in elastic cylinders and plates", *J. Acoust. Soc. Am.* **38** (1965) 835.

As discussed in chapter 2, the displacements u, v and w may be expressed in terms of potentials φ and ψ. In cylindrical coordinates the expressions are stated by eqs. (2.132)–(2.134). The scalar potential φ and the component ψ_z of the vector potential satisfy the following uncoupled wave equations

$$\nabla^2\varphi = \frac{1}{c_L^2}\frac{\partial^2\varphi}{\partial t^2} \tag{6.100}$$

$$\nabla^2\psi_z = \frac{1}{c_T^2}\frac{\partial^2\psi_z}{\partial t^2}. \tag{6.101}$$

The equations for ψ_r and ψ_θ are coupled

$$\nabla^2\psi_r - \frac{\psi_r}{r^2} - \frac{2}{r^2}\frac{\partial\psi_\theta}{\partial\theta} = \frac{1}{c_T^2}\frac{\partial^2\psi_r}{\partial t^2} \tag{6.102}$$

$$\nabla^2\psi_\theta - \frac{\psi_\theta}{r^2} + \frac{2}{r^2}\frac{\partial\psi_r}{\partial\theta} = \frac{1}{c_T^2}\frac{\partial^2\psi_\theta}{\partial t^2}. \tag{6.103}$$

In addition, the components of the vector potential must satisfy a constraint condition, for example $\nabla \cdot \psi = 0$.

We consider an infinitely long cylinder with a solid circular cross section of radius a. If the cylindrical surface is free of tractions the conditions at $r = a$ are

$$\tau_r = 0, \qquad \tau_{r\theta} = 0, \qquad \tau_{rz} = 0. \tag{6.104a, b, c}$$

Let us first consider the scalar potential $\varphi(r, \theta, z, t)$. A wave propagating in the positive z-direction is of the form

$$\varphi = \Phi(r)\Theta(\theta)\exp[i(kz-\omega t)],$$

where, in view of the finite cross-sectional dimensions of the rod, we have assumed a separation-of-variables type of solution in r and θ. Eq. (6.100) yields the following two equations for $\Phi(r)$ and $\Theta(\theta)$:

$$\frac{d^2\Phi}{dr^2} + \frac{1}{r}\frac{d\Phi}{dr} + \left(\frac{\omega^2}{c_L^2} - k^2\right)\Phi - \frac{n^2}{r^2}\Phi = 0 \tag{6.105}$$

$$\frac{d^2\Theta}{d\theta^2} + n^2\Theta = 0. \tag{6.106}$$

The solutions of the equation for $\Theta(\theta)$ are sines and cosines of argument $n\theta$. Since the solutions should be continuous functions of θ, with continuous derivatives, n can only be zero or an integer. The equation for $\Phi(r)$ yields ordinary Bessel functions as solutions. For a solid cylinder the field quantities should be finite at the center of the cylinder, and only Bessel functions of the first kind thus should be retained. The expression for the scalar potential then emerges as

$$\varphi = [A_1 \cos(n\theta) + A_2 \sin(n\theta)] J_n(pr) \exp[i(kz - \omega t)], \tag{6.107}$$

where $J_n(\ )$ is the Bessel function of the first kind of order n, and where p was defined in eq. (6.63a) as

$$p^2 = \frac{\omega^2}{c_L^2} - k^2.$$

The wave equation governing ψ_z can be treated in the same manner. The solution can be written as

$$\psi_z = [B_1 \cos(n\theta) + B_2 \sin(n\theta)] J_n(qr) \exp[i(kz - \omega t)], \tag{6.108}$$

where q is defined as

$$q^2 = \frac{\omega^2}{c_T^2} - k^2.$$

We now turn to the equations governing ψ_r and ψ_θ. These equations are somewhat more difficult to deal with because they are coupled. It is, however, evident that ψ_r and ψ_θ also contain trigonometric functions of θ. Moreover, the form of the coupling in eqs. (6.102) and (6.103) indicates that a sine-dependence on θ in ψ_r is consistent with a cosine-dependence on θ in ψ_θ, and vice-versa. Thus we can consider the pair

$$\psi_r = \Psi_r(r) \sin(n\theta) \exp[i(kz - \omega t)] \tag{6.109}$$

$$\psi_\theta = \Psi_\theta(r) \cos(n\theta) \exp[i(kz - \omega t)]. \tag{6.110}$$

The equations for $\Psi_r(r)$ and $\Psi_\theta(r)$ are then obtained from (6.102) and (6.103) as

$$\frac{\mathrm{d}^2\Psi_r}{\mathrm{d}r^2} + \frac{1}{r}\frac{\mathrm{d}\Psi_r}{\mathrm{d}r} + \frac{1}{r^2}(-n^2\Psi_r + 2n\Psi_\theta - \Psi_r) - k^2\Psi_r + \frac{\omega^2}{c_T^2}\Psi_r = 0$$

$$\frac{\mathrm{d}^2\Psi_\theta}{\mathrm{d}r^2} + \frac{1}{r}\frac{\mathrm{d}\Psi_\theta}{\mathrm{d}r} + \frac{1}{r^2}(-n^2\Psi_\theta + 2n\Psi_r - \Psi_\theta) - k^2\Psi_\theta + \frac{\omega^2}{c_T^2}\Psi_\theta = 0.$$

In a convenient manner these two equations can be solved simultaneously.

First we subtract to obtain an equation for $\Psi_r - \Psi_\theta$, which can easily be solved as

$$\Psi_r - \Psi_\theta = 2C_2 J_{n+1}(qr). \tag{6.111}$$

Upon adding the two equations, we can solve for $\Psi_r + \Psi_\theta$ as

$$\Psi_r + \Psi_\theta = 2C_1 J_{n-1}(qr). \tag{6.112}$$

The corresponding expressions for Ψ_r and Ψ_θ are

$$\Psi_r = C_1 J_{n-1}(qr) + C_2 J_{n+1}(qr) \tag{6.113}$$

$$\Psi_\theta = C_1 J_{n-1}(qr) - C_2 J_{n+1}(qr). \tag{6.114}$$

The scalar potential and the three components of the vector potential have now been determined in terms of four arbitrary constants.

The displacement vector is, however, specified in terms of three constants and there are, moreover, only three boundary conditions to provide us with three homogeneous equations. The required additional condition is provided by the condition $\nabla \cdot \boldsymbol{\psi} = 0$. The latter condition is, however, somewhat arbitrary, and since it yields an awkward equation it is often replaced by the simpler condition[12],

$$\Psi_r = -\Psi_\theta, \tag{6.115}$$

which implies $C_1 = 0$.

On the basis of the foregoing discussion we may then consider the following set of potentials

$$\varphi = A_1 J_n(pr) \cos(n\theta) \exp[i(kz - \omega t)] \tag{6.116}$$

$$\psi_z = B_1 J_n(qr) \sin(n\theta) \exp[i(kz - \omega t)] \tag{6.117}$$

$$\psi_r = C_2 J_{n+1}(qr) \sin(n\theta) \exp[i(kz - \omega t)] \tag{6.118}$$

$$\psi_\theta = -C_2 J_{n+1}(qr) \cos(n\theta) \exp[i(kz - \omega t)]. \tag{6.119}$$

These expressions can be employed to compute the stresses in terms of the potentials. The boundary conditions (6.104a, b, c) then yield three homogeneous equations for the three constants A_1, B_1 and C_2. The requirement that the determinant of the coefficients vanishes provides us with the frequency equation which relates ω, n and k. The frequency equation is stated in the article by Meeker and Meitzler.[13]

[12] T. R. Meeker and A. H. Meitzler, see fn. 8, p. 224.
[13] See fn. 12 above.

As an alternative to eqs. (6.116)–(6.119) the following analogous set of potentials may be considered

$$\varphi = A_2 J_n(pr) \sin(n\theta) \exp[i(kz-\omega t)] \tag{6.120}$$

$$\psi_z = B_2 J_n(qr) \cos(n\theta) \exp[i(kz-\omega t)] \tag{6.121}$$

$$\psi_r = C J_{n+1}(qr) \cos(n\theta) \exp[i(kz-\omega t)] \tag{6.122}$$

$$\psi_\theta = -C J_{n+1}(qr) \sin(n\theta) \exp[i(kz-\omega t)]. \tag{6.123}$$

The analysis of wave motion in a hollow circular cylindrical rod can be carried out in a completely analogous manner. In the solutions of the Bessel equations we must now retain the Bessel functions of the second kind. Thus instead of (6.116) we have

$$\varphi = [A_1 J_n(pr) + A_2 Y_n(pr)] \cos(n\theta) \exp[i(kz-\omega t)].$$

The expressions for ψ_z, ψ_r and ψ_θ are modified in an analogous manner. The corresponding expressions for the stresses are now in terms of six arbitrary constants. If the stresses vanish at the inner as well as the outer radius of the rod, the boundary conditions yield six homogeneous equations for the six constants. The requirement that the determinant of the coefficients must vanish yields the frequency equation. For details of the analysis and for numerical information we refer to the work of Armenàkas et al.[14]

6.10. The frequency spectrum of the circular rod of solid cross section

A reasonably complete examination of the transcendental equation relating the frequency ω, the axial wavenumber k and the circumferential order number n requires a rather extensive effort of numerical computation. For every choice of the integer n and the real-valued wavenumber k an infinite number of roots of the frequency equation can be found, which represent the frequencies of an infinite number of modes of wave propagation in the rod. Moreover, for a complete investigation of the frequency spectrum it is necessary to consider imaginary and complex-valued wavenumbers as well.

Some insight in the structure of the frequency spectrum can be gained by examining the motions in the limit of vanishing wavenumber. The wavenumber becomes smaller as the wavelength increases and the limit $k \to 0$ thus corresponds to infinite wavelength, i.e., to motions which are indepen-

[14] A. E. Armenàkas, D. C. Gazis and G. Herrmann, *Free vibrations of circular cylindrical shells*. New York, Pergamon Press (1969).

dent of the axial coordinate z. Inspection of (6.92)–(6.99) reveals that for motions that are independent of z the displacements $u(r, \theta, t)$ and $v(r, \theta, t)$ remain coupled. The axial displacement $w(r, \theta, t)$ uncouples, however, from the other displacement components. For infinite wavelength we thus have uncoupled motions in plane strain and axial shear. If the motions are not only independent of z but also independent of θ, the displacements $u(r, t)$ and $v(r, t)$ become uncoupled, to describe radial dilatational and circumferential shear motions, respectively. The results for vanishing wavenumber can be extended to small values of k, as shown by Achenbach and Fang.[15] By means of an asymptotic analysis these authors determined higher-order derivatives $d\omega_n/dk$ at $k = 0$ for all branches in the real ω–k-plane.

Motions which are independent of θ, but do depend on z may be separated in torsional motions involving $v(r, z, t)$ only, and longitudinal motions involving $u(r, z, t)$ and $w(r, z, t)$. For motions which depend on both z and θ we will examine in some detail the case $n = 1$ which corresponds to flexural waves.

6.10.1. *Torsional waves*

Torsional waves involve a circumferential displacement only which is independent of θ. The governing equation follows from (6.93) as

$$\frac{\partial^2 v}{\partial r^2} + \frac{1}{r}\frac{\partial v}{\partial r} - \frac{v}{r^2} + \frac{\partial^2 v}{\partial z^2} = \frac{1}{c_T^2}\frac{\partial^2 v}{\partial t^2}. \tag{6.124}$$

The displacement may be written as

$$v(r, z, t) = \frac{1}{q} B_2 J_1(qr) \exp\left[i(kz-\omega t)\right]. \tag{6.125}$$

From the three boundary conditions (6.104a, b, c) only the condition $\tau_{r\theta}$ at $r = a$ is nontrivial. This condition yields the frequency equation in the form

$$(qa)J_0(qa) - 2J_1(qa) = 0. \tag{6.126}$$

Eq. (6.126) is a transcendental equation whose roots have been tabulated.[16] The first three roots are $q_1 a = 5.136$, $q_2 a = 8.417$ and $q_3 a = 11.62$. It is

[15] J. D. Achenbach and S. J. Fang, *Journal of the Acoustical Society of America* **47** (1970) 1282.

[16] Cf. *Handbook of mathematical functions*, ed. by M. Abramowitz and I. A. Stegun. Washington, National Bureau of Standards (1964), table 9.7, p. 414.

noted that $q = 0$ is also a solution of the frequency equation. By taking the limit $q \to 0$ of eq. (6.125) we find

$$v = \tfrac{1}{2}B_2 r \exp\,[i(kz-\omega t)]. \tag{6.127}$$

This displacement represents the well-known lowest torsional mode. In the lowest mode the displacement is proportional to the radius, and the motion is thus a rotation of each cross-section of the cylinder as a whole about its center.

Since $q = 0$ implies that the phase velocity equals c_T, the lowest torsional mode is not dispersive. The higher torsional modes are dispersive with frequencies which follow from the definition of q as

$$\left(\frac{\omega a}{c_T}\right)^2 = (q_n a)^2+(ka)^2, \tag{6.128}$$

where $q_n a$ are the solutions of (6.126). It is noted that given a real-valued frequency the wavenumber may be real-valued or imaginary. Just as for the case of SH-waves in a layer, the branches are hyperboles for real-valued k and circles for imaginary values of the wavenumbers.

As pointed out by, among others, Redwood[17], the lowest torsional mode can be used in delay lines when undistorted pulse propagation is required. This mode requires a special sort of excitation, with the amplitude of the displacement proportional to the radius, but such an excitation is practicable. Even if the excitation takes some other form, the additional modes will be evanescent if the frequency and the radius are adjusted so that the waveguide is operating below the cut-off frequency of the second mode.

6.10.2. *Longitudinal waves*

Longitudinal waves are axially symmetric waves characterized by the presence of displacement components in the radial and axial directions. The governing equations follow from eqs. (6.92)–(6.94). It is, however, convenient to employ the displacement potentials, which follow from (6.116) and (6.119) as

$$\varphi = AJ_0(pr)\exp\,[i(kz-\omega t)]$$

$$\psi_\theta = CJ_1(qr)\exp\,[i(kz-\omega t)].$$

The corresponding radial and axial displacements are obtained from (2.132) and (2.134) as

[17] M. Redwood, *Mechanical waveguides*. New York, Pergamon Press (1960), p. 148.

$$u = \{-pAJ_1(pr) - ikCJ_1(qr)\}\exp[i(kz-\omega t)] \tag{6.129}$$

$$w = \{ikAJ_0(pr) + qCJ_0(qr)\}\exp[i(kz-\omega t)]. \tag{6.130}$$

At the cylindrical surface ($r = a$) the stresses must be zero. Substituting eqs. (6.129) and (6.130) into τ_r, and setting the resulting expression equal to zero at $r = a$, we find

$$\left[-\tfrac{1}{2}(q^2-k^2)J_0(pa) + \frac{p}{a}J_1(pa)\right]A + \left[-ikqJ_0(qa) + \frac{ik}{a}J_1(qa)\right]C = 0.$$

A second equation comes from the condition that τ_{rz} vanishes at $r = a$,

$$[-2ikpJ_1(pa)]A - (q^2-k^2)J_1(qa)C = 0.$$

The requirement that the determinant of the coefficients must vanish yields the frequency equation as

$$\frac{2p}{a}(q^2+k^2)J_1(pa)J_1(qa) - (q^2-k^2)^2J_0(pa)J_1(qa) - 4k^2pqJ_1(pa)J_0(qa) = 0, \tag{6.131}$$

which is known as the Pochhammer frequency equation.

Eq. (6.131) appears to relate five quantities ω (or c), k, a, c_L and c_T. Just

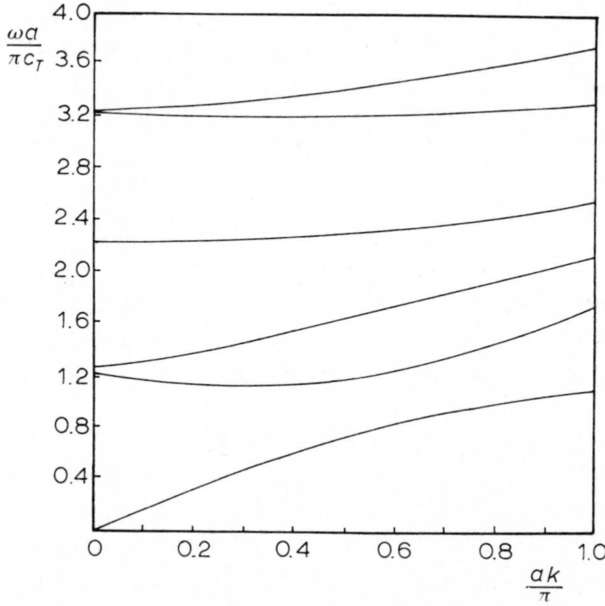

Fig. 6.13. Dimensionless frequencies for longitudinal modes; $\nu = 0.30$.

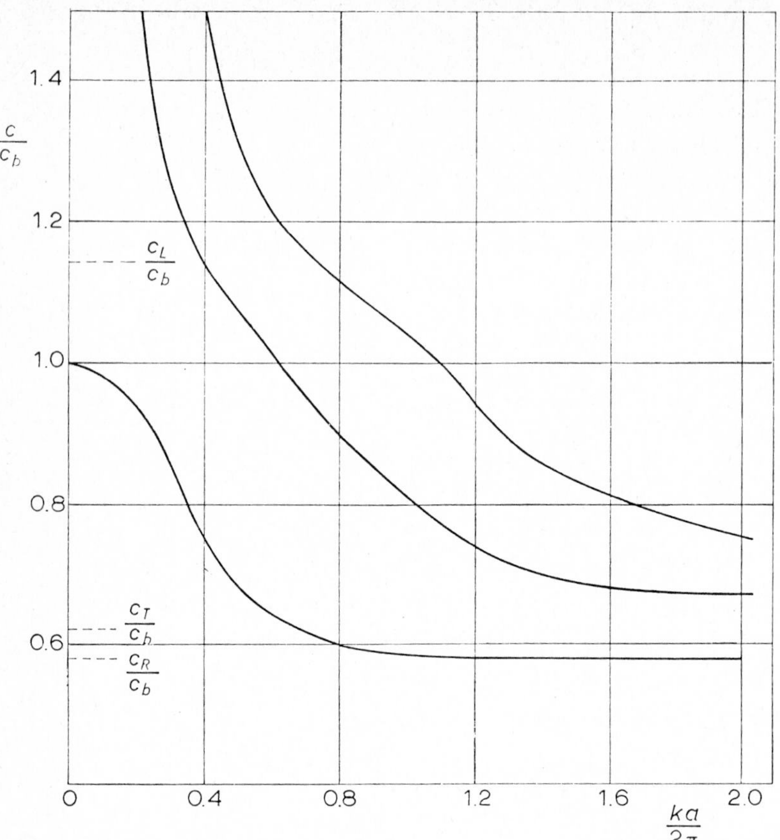

Fig. 6.14. Dimensionless phase velocity for longitudinal modes; $\nu = 0.29$.

as in the case of the Rayleigh-Lamb frequency equation for the layer, a choice of appropriate dimensionless quantities reduces the number, however, to three. The three variables in dimensionless form are Poisson's ratio ν (or the constant κ), the dimensionless frequency ω/ac_T (or the dimensionless phase velocity c/c_T) and the dimensionless wavenumber ka. A rather detailed discussion of the frequency spectrum of longitudinal modes, including real, imaginary and complex branches was given by Onoe et al.[18]

For real-valued wavenumbers numerical results for the frequency spectrum are included in the work of Armenàkas et al.[19]

[18] M. Onoe, H. D. McNiven and R. D. Mindlin, *Journal of Applied Mechanics* **28** (1962) 729.

[19] A. E. Armenàkas, D. C. Gazis and G. Herrmann, see fn. 14, p. 240.

Some numerical values for real-valued wavenumbers are displayed in figures 6.13–6.15. For Poisson's ratio $\nu = 0.31$, the dimensionless frequencies are shown in figure 6.13, after the results of Armenàkas et al. For $\nu = 0.29$ (steel), figure 6.14 shows the dimensionless phase velocity. The numerical data plotted in figure 6.14 are after a paper by Davies.[20] Davies also presented curves showing the variation of the group velocity with frequency, and these curves are reproduced in figure 6.15.

The lowest longitudinal mode is the most important mode from the practical point of view. At very small and very large values of ka the frequencies or the phase velocities can be computed by taking appropriate expansions of the Bessel functions in eq. (6.131) in the same manner as was discussed in section 6.8 for the layer. For small values of ka ($ka \ll 1$) we find

$$\frac{\omega a}{c_T} = \omega_1(ka) + \omega_3(ka)^3 + O[(ka)^5], \tag{6.132}$$

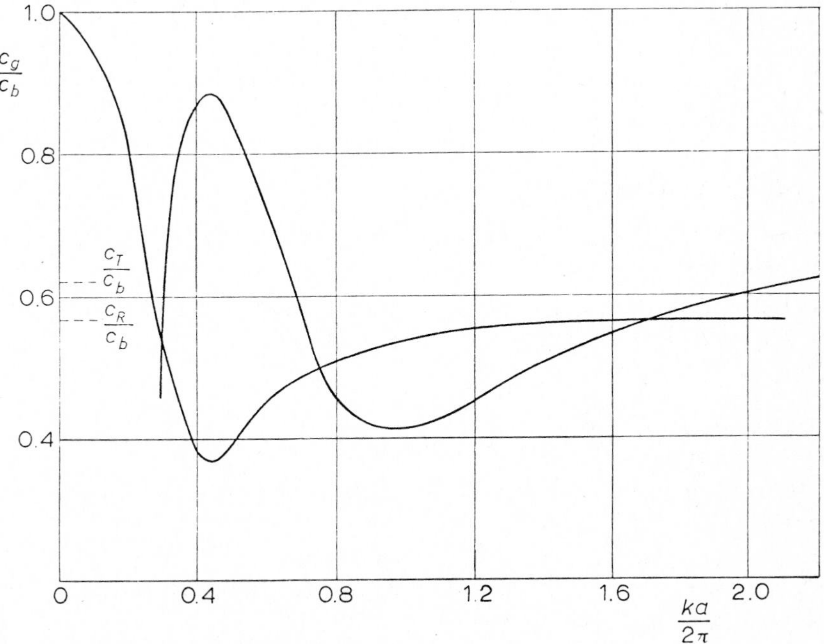

Fig. 6.15. Dimensionless group velocity for longitudinal modes; $\nu = 0.29$.

[20] R. M. Davies, *Philosophical Transactions of the Royal Society* **A240** (1948) 375.

where

$$\omega_1 = \left(\frac{3\lambda+2\mu}{\lambda+\mu}\right)^{\frac{1}{2}} = \left(\frac{E}{\mu}\right)^{\frac{1}{2}} \tag{6.133}$$

$$\omega_3 = -\frac{1}{4}\left(\frac{E}{\mu}\right)^{\frac{1}{2}} \nu^2. \tag{6.134}$$

In these expressions E is Young's modulus. The expression for the phase velocity corresponding to eq. (6.132) is

$$c = \left(\frac{E}{\rho}\right)^{\frac{1}{2}} [1-\tfrac{1}{4}\nu^2(ka)^2] + O[(ka)^4]. \tag{6.135}$$

In the limit as $(ka) \to 0$ the phase velocity thus becomes equal to $(E/\rho)^{\frac{1}{2}}$, which is called the bar velocity and which is the value found from the simplest theory of rods.

As $(ka) \to \infty$ the phase velocity approaches from below the velocity of Rayleigh waves. At some intermediate wavenumber the phase velocity has a minimum value slightly less than c_R. As can be seen from figure 6.15, the group velocity of the lowest mode shows a pronounced minimum for an intermediate value of ka.

6.10.3. *Flexural waves*

It remains to examine wave motions that do depend on the circumferential angle θ through the trigonometric functions shown in eqs. (6.116)–(6.123). Of the circumferential modes the family defined by $n = 1$ is the most important. Let us thus examine potentials of the form

$$\begin{aligned}
\varphi &= AJ_1(pr)\cos(\theta)\exp[i(kz-\omega t)] \\
\psi_z &= BJ_1(qr)\sin(\theta)\exp[i(kz-\omega t)] \\
\psi_r &= CJ_2(qr)\sin(\theta)\exp[i(kz-\omega t)] \\
\psi_\theta &= -CJ_2(qr)\cos(\theta)\exp[i(kz-\omega t)].
\end{aligned}$$

By the use of (2.132)–(2.134) the displacements are obtained as

$$u = U(r)\cos(\theta)\exp[i(kz-\omega t)] \tag{6.136}$$

$$v = V(r)\sin(\theta)\exp[i(kz-\omega t)] \tag{6.137}$$

$$w = W(r)\cos(\theta)\exp[i(kz-\omega t)], \tag{6.138}$$

where

$$U(r) = A\frac{\partial}{\partial r}J_1(pr)+\frac{B}{r}J_1(qr)+ikCJ_2(qr)$$

$$V(r) = -\frac{A}{r}J_1(pr)+ikCJ_2(qr)-B\frac{\partial}{\partial r}J_1(qr)$$

$$W(r) = ikAJ_1(pr)-\frac{C}{r}\frac{\partial}{\partial r}[rJ_2(qr)]-\frac{C}{r}J_2(qr).$$

To illustrate the motions represented by these displacement distributions we choose the (yz)-plane (the vertical plane) as the plane from which θ is measured, as shown in figure 6.16. The (xz)-plane is termed the horizontal plane. It now follows from (6.137) that for points in the vertical plane the

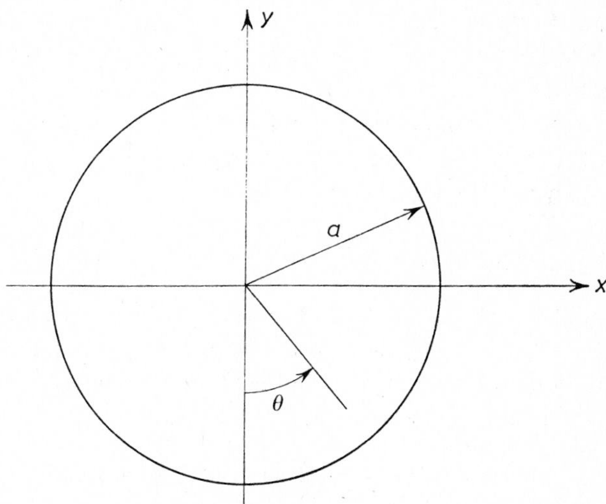

Fig. 6.16. Cross section of rod.

displacement component v vanishes, so that these points remain in the vertical plane. In the horizontal plane ($\theta = \pm\pi/2$) the displacement components u and w vanish. Points in the horizontal plane thus perform purely vertical oscillations, since in the (xz)-plane v points in the y-direction. These observations suggest the terminology *flexural waves* for the motions defined by eqs. (6.136)–(6.138). Indeed, it can further be checked that w is odd in y, and the displacement component in the y-direction is even in y.

To determine the frequency equation the displacement (6.136)–(6.138) must be substituted into the expressions for the stresses, and τ_r, τ_{rz} and $\tau_{r\theta}$ must subsequently be set equal to zero at $r = a$. This leads to a system of three homogeneous equations for A, B and C. The requirement that the determinant of the coefficients must vanish yields the frequency equation. This frequency equation was examined in considerable detail by Pao and Mindlin.[21] Numerical computations on the frequency spectrum were carried out by Armenàkas et al.[22] Some typical curves showing the dimensionless frequency versus the dimensionless wavenumber are shown in figure 6.17.

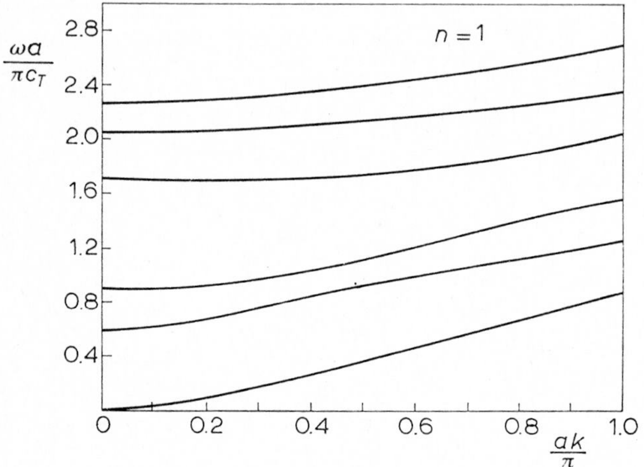

Fig. 6.17. Dimensionless frequencies for flexural modes; $\nu = 0.30$.

Of particular interest is the lowest flexural mode. Either by employing expansions of the Bessel functions in the frequency equation, or more efficiently by the asymptotic method of the previously cited paper by Achenbach and Fang, it can be shown that for small values of $ka(ka \ll 1)$ we may write

$$\frac{\omega a}{c_T} = \frac{1}{2}\left(\frac{3\lambda+2\mu}{\lambda+\mu}\right)^{\frac{1}{2}}(ka)^2 + O[(ka)^4]. \tag{6.139}$$

As ka increases the phase velocity of the lowest mode approaches the velocity of Rayleigh waves. For the lowest flexural mode the phase velocity,

[21] Y. H. Pao and R. D. Mindlin, *Journal of Applied Mechanics* **27** (1960) 513.
[22] A. E. Armenàkas, D. C. Gazis and G. Herrmann, see fn. 14, p. 240.

as well as the group velocity, is shown in figure 6.18. The numerical results displayed in figure 6.18 are after the work of Davies.[23]

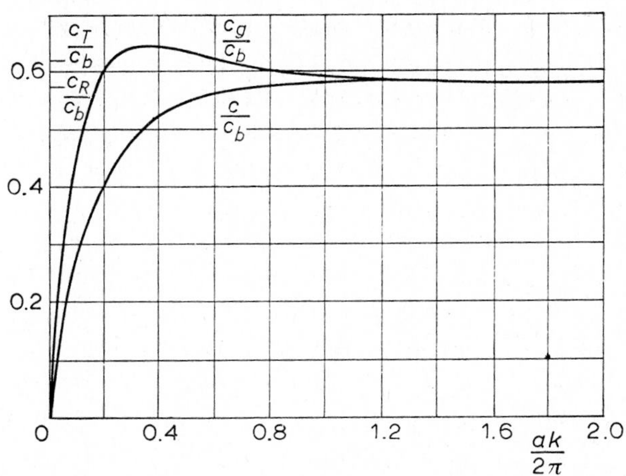

Fig. 6.18. Phase velocity and group velocity for the lowest flexural mode; $\nu = 0.29$.

Families of flexural modes of higher circumferential orders have also been investigated. For specific information and references we refer to the previously cited article by Meeker and Meitzler and to the book by Armenàkas et al.

6.11. Approximate theories for rods

The exact treatment of harmonic wave motions in an elastic circular cylinder is already rather complicated. For a cylinder with other than a circular or an elliptical cross section it becomes impossible to carry out an exact analysis. Even for a strip of rectangular cross section whose lateral surfaces are free of tractions it is not possible to analyze general harmonic wave motions rigorously within the context of the linear theory of elasticity.[24] It is for that reason that several models have been proposed which provide an approximate description of wave motions in rods of rather arbitrary cross section.[25] In this section we review the models that are commonly

[23] R. M. Davies, see fn. 20, p. 245.

[24] Some special cases which are amenable to a rigorous analysis are reviewed in the paper by Meeker and Meitzler, see fn. 8, p. 224.

[25] For a survey see: W. A. Green, in: *Progress in solid mechanics*, Vol. 1, ed. by R. A. Hill and I. N. Sneddon. Amsterdam, North-Holland Publishing Company (1960), p. 225.

used. These models are based on *a priori* assumptions with regard to the deformation of the cross-sectional area of the rod. The assumptions simplify the description of the kinematics to such an extent that the wave motions can be described by one-dimensional approximate theories. For the propagation of time-harmonic waves it was found that the approximate theories can adequately account for the dispersive behavior of the lowest axisymmetric and flexural modes over a limited but significant range of wavenumbers and frequencies. In chapter 8 we will comment on the applicability of the approximate theories for transient motions.

The governing equations can be obtained either by using variational methods or by straightforward momentum considerations of an element of the rod. The latter have the advantage that the physical concepts are conveyed more clearly. For the more complicated theories it is, however, easier to employ the assumed displacement distributions to compute the corresponding kinetic and strain energies for an element of the rod, whereupon Hamilton's principle can be applied to obtain the governing equations. Since an abundance of literature on the derivation of the approximate theories is already available we will present only a brief derivation of the equations for the Timoshenko model, and we will state the governing equations for some other models.

6.11.1. Extensional motions

In extensional wave motions the dominant component of the displacement is in the longitudinal direction. Based on the assumption that the (arbitrary) cross-sectional area of the rod remains plane, it was shown in section 1.5 that consideration of the forces acting on an element leads to the equation

$$\frac{\partial^2 u}{\partial x^2} = \frac{1}{c_b} \frac{\partial^2 u}{\partial t^2}, \tag{6.140}$$

where

$$c_b^2 = \frac{E}{\rho}. \tag{6.141}$$

Eq. (6.140) predicts that harmonic waves are not dispersive.

For a rod of circular cross section Mindlin and McNiven[26] derived a system of one-dimensional equations which takes into account the coupling between longitudinal axial shear and radial modes. The spectrum of frequencies for real, imaginary and complex wavenumbers was explored in

[26] R. D. Mindlin and H. D. McNiven, *Journal of Applied Mechanics* **27** (1960) 145.

detail and compared with the corresponding branches from the exact spectrum.

Several other approximate theories which can describe the effect of dispersion are discussed in the previously cited paper by W. A. Green.

6.11.2. *Torsional motions*

In the approximate theory it is assumed that transverse sections remain plane and that the motion consists of a rotation of the sections about the axis. This leads to a wave equation for the angle of rotation with a propagation velocity

$$c = \left(\frac{C}{\rho A}\right)^{\frac{1}{2}} \frac{1}{K}. \tag{6.142}$$

Here K is the radius of gyration of a cross section of the rod about its axis, A is the cross-sectional area and C is the torsional rigidity of the rod. For a circular cylindrical rod c reduces to $(\mu/\rho)^{\frac{1}{2}}$. As shown in section 6.10, this is the exact result for the lowest mode of torsional wave propagation in a circular cylinder.

An approximate dynamical theory of torsion for rods of noncircular cross section which includes the effects of both the warping and the in-plane motions was developed by Bleustein and Stanley.[27] The approximation is based on expansions of the displacements together with a truncation procedure which retains only the torsional, contour-shear and warping motions.

6.11.3. *Flexural motions – Bernoulli-Euler model*

In the simplest theory of flexural motions of rods of arbitrary but uniform cross section with a plane of symmetry it is assumed that the dominant displacement component is parallel to the plane of symmetry. It is also assumed that the deflections are small and that cross-sectional areas remain plane and normal to the neutral axis. For a beam which is free of lateral loading the equation of motion becomes

$$\frac{\partial^2 w}{\partial t^2} + \frac{EI}{\rho A}\frac{\partial^4 w}{\partial x^4} = 0, \tag{6.143}$$

where w is the deflection, I is the second moment of the cross-sectional area about the neutral axis and A is the cross-sectional area. Substituting a

[27] J. L. Bleustein and R. M. Stanley, *International Journal of Solids and Structures* **6** (1970) 569.

harmonic wave, we find for the phase velocity

$$c = \left(\frac{E}{\rho}\right)^{\frac{1}{2}} \left(\frac{I}{A}\right)^{\frac{1}{2}} k. \tag{6.144}$$

Thus the phase velocity is proportional to the wavenumber, which suggests that (6.144) cannot be correct for large wavenumbers (short waves). For a circular cylindrical rod, (6.144) becomes

$$c = \frac{1}{2}\left(\frac{E}{\rho}\right)^{\frac{1}{2}} ka. \tag{6.145}$$

The frequency according to (6.145) is

$$\omega = \frac{1}{2}\left(\frac{E}{\rho}\right)^{\frac{1}{2}} k^2 a. \tag{6.146}$$

It can be checked that (6.146) agrees with the asymptotic expression (6.139), which was obtained as a limitcase from the exact frequency equation.

6.11.4. *Flexural motions – Timoshenko model*

By taking into account shear deformation in the description of the flexural motion of a rod we obtain a model which yields more satisfactory results for shorter wavelengths. In this model it is still assumed that plane sections remain plane; it is, however, not assumed that plane sections remain normal to the neutral plane. After deformation the neutral axis has

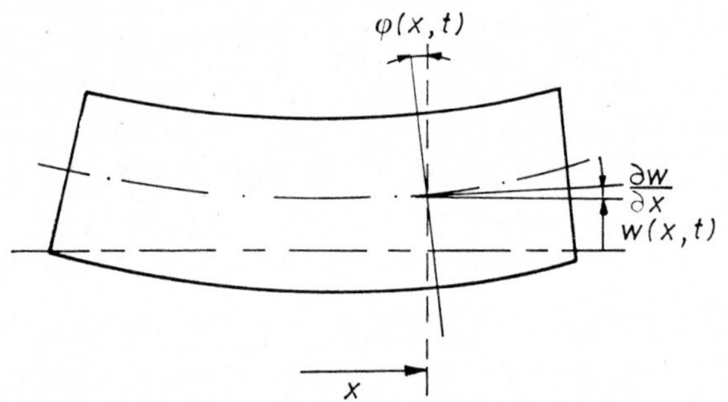

Fig. 6.19. Timoshenko beam.

been rotated through the small angle $\partial w/\partial x$, while the cross section has been rotated through the angle φ, as depicted in figure 6.19. The shearing angle γ is the net decrease in angle

$$\gamma = \frac{\partial w}{\partial x} - \varphi. \tag{6.145}$$

The bending moment M acting over the cross section is related to φ by the well-known relation

$$M = -EI\frac{\partial \varphi}{\partial x}. \tag{6.146}$$

The relation between the shear force Q and the angle γ is

$$Q = \kappa\mu A\gamma, \tag{6.147}$$

where κ is a numerical factor which reflects the fact that the beam is not in a state of uniform shear, but that (6.146) represents a relation between the resultant shear force and some kind of average shear angle. The factor κ depends on the cross-sectional shape and on the rationale adopted in the averaging process. Fortunately there is not very much spread in the values of κ obtained by different averaging processes. The factor does, however, depend noticeably on the shape of the cross section. A table of shear coefficients is included in a paper by Mindlin and Deresiewicz.[28]

By employing (6.146) and (6.147) the strain energy of a finite segment of a Timoshenko beam can be computed as

$$U = \int_{x_1}^{x_2} \left[\tfrac{1}{2}EI\left(\frac{\partial \varphi}{\partial x}\right)^2 + \tfrac{1}{2}\kappa\mu A\left(\frac{\partial w}{\partial x} - \varphi\right)^2\right] \mathrm{d}x. \tag{6.148}$$

The corresponding kinetic energy is

$$K = \int_{x_1}^{x_2} \left[\tfrac{1}{2}\rho A\left(\frac{\partial w}{\partial t}\right)^2 + \tfrac{1}{2}\rho I\left(\frac{\partial \varphi}{\partial t}\right)^2\right] \mathrm{d}x. \tag{6.149}$$

Subsequently Hamilton's principle and the Euler equations (2.74) can be employed to obtain the following set of governing equations for a homogeneous beam:

$$\kappa\mu\frac{\partial}{\partial x}\left(\frac{\partial w}{\partial x} - \varphi\right) = \rho\frac{\partial^2 w}{\partial t^2} \tag{6.150}$$

$$EI\frac{\partial^2 \varphi}{\partial x^2} + \kappa\mu A\left(\frac{\partial w}{\partial x} - \varphi\right) = \rho I\frac{\partial^2 \varphi}{\partial t^2}. \tag{6.151}$$

[28] R. D. Mindlin and H. Deresiewicz, *Proceedings Second National Congress of Applied Mechanics*, ASME (1954).

Since there are two degrees of freedom this set of equations describes two modes of motion. The corresponding frequencies, phase velocities and group velocities, as well as the character of the motions, are discussed in the book by Crandall et al.[29]

For a circular rod it was found that the lowest mode generally provides a very good approximation to the phase velocity. The phase velocity of the second mode agrees less well with the exact result. Often the second mode is discarded by ignoring rotatory inertia. This simplifies the equation and it does not appreciably affect the results for the lowest mode. Curves comparing the branches according to the Bernoulli-Euler model, the Timoshenko model and the exact theory can be found in many places in the literature.[30]

6.12. Approximate theories for plates

The analysis of free and forced harmonic motions of elastic layers of infinite extent does not present essential difficulties. As was shown in sections 6.7 and 6.8, the free motions of an elastic layer can take place in an infinite number of modes whose frequencies are governed by the Rayleigh-Lamb frequency equation. Forced motions will be considered in chapter 8. In a plate of finite dimensions, with free or clamped edges, each of these modes (or its overtones) couples, however, with all the others (or their overtones), leading to an extraordinarily complex spectrum. These complications have motivated the formulation of approximate theories to describe the motions of plates. In these theories the system of field equations governing the three-dimensional elastic continuum of the layer is reduced to a set of two-dimensional equations for field quantities defined in the mid-plane of the layer. The reduction from three- to two-dimensional equations is achieved by making certain kinematical assumptions with regard to the deformation of the cross-sectional area, such as the Kirchhoff assumption. In a general approach which was developed by Mindlin[31], the displacement components

[29] S. H. Crandall, D. C. Karnopp, E. F. Kurtz and D. C. Pridmore-Brown, *Dynamics of mechanical and electromechanical systems*. New York, McGraw-Hill Book Co. (1968), p. 360.

[30] Cf. Y. C. Fung, *Foundations of solid mechanics*. Englewood Cliffs, N.J., Prentice-Hall (1965), p. 325.

[31] R. D. Mindlin, *An introduction to the mathematical theory of the vibrations of elastic plates*. Fort Monmouth, N.J., U.S. Army Signal Corps Eng. Lab. (1955). It appears that this monograph is not readily available. The approach is, however, also discussed in R. D. Mindlin, *Quarterly of Applied Mathematics* **19** (1961) 51; while the general techniques are also displayed in the book by Tiersten: H. Tiersten, *Linear piezoelectric plate vibrations*. New York, Plenum Press (1969), p. 141.

are expanded in an infinite series of powers of the thickness coordinate. This series is substituted in the equations of motion from the theory of elasticity. The resulting equations are subsequently integrated across the thickness of the layer. Upon introduction of the boundary conditions on the tractions, the three-dimensional equations of elasticity are in this manner converted into an infinite series of two-dimensional equations in the in-plane coordinates. The system is then truncated to produce the approximate equations.

The truncation of the series expansions implies in physical terms that only a finite number of modes will be described. The approximate theories yield, nevertheless, very good results for the frequencies and the phase velocities over a substantial range of wavenumbers. This is due to the fact that the higher modes do not greatly affect the spectrum at lower frequencies.

6.12.1. *Flexural motions – classical theory*

The classical theory of the bending of plates is based on the hypothesis that every straight line in the plate which was originally perpendicular to the middle surface of the plate remains straight after deformation and perpendicular to the deflected middle surface. In terms of an (x, y, z)-coordinate system, where the x- and y-axes are in the plane of the plate, the equation governing bending of a plate of uniform thickness $2h$ is

$$\frac{\partial^4 w}{\partial x^4} + 2\frac{\partial^4 w}{\partial x^2 \partial y^2} + \frac{\partial^4 w}{\partial y^2} + \frac{2\rho h}{D}\frac{\partial^2 w}{\partial t^2} = 0, \tag{6.152}$$

where $w(x, y, t)$ is the transverse displacement and D is the flexural rigidity

$$D = \frac{8Eh^3}{12(1-\nu^2)}. \tag{6.153}$$

For a derivation of (6.152) we refer to the book by Fung.[32] Eq. (6.152) can also be obtained as a limitcase of Mindlin's more general approach.[33]

By considering a harmonic wave of the form

$$u_1 = Ae^{i(kx-\omega t)}, \tag{6.154}$$

we find

$$\omega = k^2 h\left[\frac{E}{3\rho(1-\nu^2)}\right]^{\frac{1}{2}}. \tag{6.155}$$

[32] Y. C. Fung, *Foundations of solid mechanics*. Englewood Cliffs, N.J., Prentice-Hall (1965), p. 456.

[33] See p. 6.14 of the previously cited monograph by Mindlin.

This expression agrees with eq. (6.81) which was derived from the Rayleigh-Lamb frequency equation by considering long waves (small values of k). The classical theory thus is applicable for long waves.

6.12.2. *Effects of transverse shear and rotary inertia*

These effects can be accounted for by extending the classical plate theory in a manner which is completely analogous to the extension of the Bernoulli-Euler model to the Timoshenko model for beams. A plate theory for elastic isotropic plates which includes transverse shear and rotary inertia was presented by Mindlin.[34] For flexural motions the displacement components are expressed in the forms

$$u = -z\psi_x(x, y, t) \tag{6.156}$$

$$v = -z\psi_y(x, y, t) \tag{6.157}$$

$$w = w(x, y, t), \tag{6.158}$$

where ψ_x and ψ_y are the local rotations in the x- and y-directions, respectively, of lines normal to the mid-plane before deformation. These rotations are analogous to the rotation φ shown in figure 6.19. For a plate of thickness $2h$, these displacement expressions lead to the following system of governing equations:

$$\tfrac{1}{2}D\left[(1-\nu)\nabla^2\psi_x+(1+\nu)\frac{\partial}{\partial x}\left(\frac{\partial\psi_x}{\partial x}+\frac{\partial\psi_y}{\partial y}\right)\right] -2\kappa\mu h\left(\psi_x-\frac{\partial w}{\partial x}\right) = 8\,\frac{\rho h^3}{12}\,\frac{\partial^2\psi_x}{\partial t^2} \tag{6.159}$$

$$\tfrac{1}{2}D\left[(1-\nu)\nabla^2\psi_y+(1+\nu)\frac{\partial}{\partial y}\left(\frac{\partial\psi_x}{\partial x}+\frac{\partial\psi_y}{\partial y}\right)\right] -2\kappa\mu h\left(\psi_y-\frac{\partial w}{\partial y}\right) = 8\,\frac{\rho h^3}{12}\,\frac{\partial^2\psi_y}{\partial t^2} \tag{6.160}$$

$$2\kappa\mu h\left(\nabla^2 w-\frac{\partial\psi_x}{\partial x}-\frac{\partial\psi_y}{\partial y}\right) = 2\rho h\,\frac{\partial^2 w}{\partial t^2}, \tag{6.161}$$

where

$$\nabla^2 = \frac{\partial^2}{\partial x^2}+\frac{\partial^2}{\partial y^2}.$$

[34] R. D. Mindlin, *Journal of Applied Mechanics* **18** (1951) 31.

In eqs. (6.159)–(6.161), κ is a numerical factor (correction coefficient) which is introduced to account for the fact that the shear stresses are not constant over the thickness as the simple kinematic relations would really imply. The correction coefficient is chosen so that the frequency of the lowest thickness shear mode computed from (6.159)–(6.161) yields the same result as given by (6.86) for the exact theory. Setting

$$\psi_y = w \equiv 0, \qquad \psi_x = e^{i\omega t}, \tag{6.162}$$

we obtain from (6.159)

$$\omega = \left(\frac{12\kappa}{4h^2}\right)^{\frac{1}{2}} c_T. \tag{6.163}$$

According to (6.86) the exact result is

$$\omega = \frac{\pi}{2h} c_T. \tag{6.164}$$

Eqs. (6.163) and (6.164) will agree if

$$\kappa = \frac{\pi^2}{12}. \tag{6.165}$$

Plots showing a comparison of the phase velocities for the lowest mode of harmonic waves according to the classical theory, eqs. (6.159)–(6.161), and the exact theory of sections 6.7 and 6.8 are also shown in Mindlin's paper. It is found that eqs. (6.159)–(6.161) yield very good results for frequencies up to about 20% higher than the frequency of the thickness shear mode, eq. (6.164).

6.12.3. *Extensional motions*

In the elementary Poisson theory for the stretching of a plate of uniform thickness a state of generalized plane stress is assumed. By integration the governing equations follow from eqs. (2.60) and (2.53) as

$$\frac{\partial^2 u}{\partial x^2} + \frac{1-\nu}{2}\frac{\partial^2 u}{\partial y^2} + \frac{1+\nu}{2}\frac{\partial^2 v}{\partial x\,\partial y} = \frac{(1-\nu^2)\rho}{E}\frac{\partial^2 u}{\partial t^2} \tag{6.166}$$

$$\frac{\partial^2 v}{\partial y^2} + \frac{1-\nu}{2}\frac{\partial^2 v}{\partial x^2} + \frac{1+\nu}{2}\frac{\partial^2 u}{\partial x\,\partial y} = \frac{(1-\nu^2)\rho}{E}\frac{\partial^2 v}{\partial t^2}. \tag{6.167}$$

Considering a harmonic wave of the form

$$u = Ae^{i(kx-\omega t)}, \qquad v = 0, \tag{6.168}$$

we obtain

$$\omega^2 = \frac{E}{\rho} \frac{k^2}{(1-\nu^2)}. \tag{6.169}$$

Thus, this theory cannot describe dispersion. Eq. (6.169) agrees however with eq. (6.82) which was obtained for long waves from the Rayleigh-Lamb frequency equation.

A theory which goes well beyond the elementary theory was presented by Mindlin and Medick.[35] Their theory takes into account the coupling between extensional symmetric thickness-stretch and thickness-shear modes. The spectrum of frequencies for real, imaginary, and complex wavenumbers in an infinite plate was explored in detail and compared with the corresponding branches of the Rayleigh-Lamb spectrum.

6.13. Problems

6.1. An elastic layer of thickness $2h$ is referred to a rectangular coordinate system. The x_1 and x_3 axes are placed in the midplane of the layer. Examine the wave motion which is described by

$$\varphi = \psi_1 = \psi_3 \equiv 0$$

$$\psi_2 = A \cos qx_2 \sin lx_3 \, e^{ik(x_1 - ct)}.$$

Determine q from the condition that the surfaces $x_2 = \pm h$ are free of tractions. What is the relation between the phase velocity c and the wavenumbers k and l? Sketch the displacement distributions for the lowest three modes.

6.2. In section 5.7 it was noted that an SV-wave incident under an angle of incidence $\theta_0 = 45°$ on a traction-free surface is reflected as an SV-wave only. This observation, in conjunction with the idea of constructive interference which was discussed in section 6.2, can be used to construct simple modes of motion of an elastic layer over a discrete spectrum of wavelengths.

(1) What is the phase velocity of these modes?
(2) What is the velocity of energy transmission?
(3) Show that the frequencies are given by

$$\frac{\omega h}{c_L} = \sqrt{2}\,\frac{n\pi}{2}, \qquad \text{where} \quad n = 1, 2, 3, \ldots$$

[35] R. D. Mindlin and M. A. Medick, *Journal of Applied Mechanics* **26** (1959) 561.

(4) Do these motions belong to the families of longitudinal or flexural waves, or both?

(5) Sketch the displacement distributions for a few of the modes.

These modes are known as the Lamé modes. Do the Lamé modes apply to a strip ($-a \leqq x_3 \leqq a$, $-h \leqq x_2 \leqq h$, $-\infty < x_1 < \infty$) which is free of surface tractions?

6.3. The analysis of motions in a layer can easily be extended beyond the case of plane strain. Consider, for example, wave motions that are described by

$$\varphi = B \cos px_2 \cos lx_3 \sin k(x_1 - ct)$$

$$\psi_1 = Cl \sin qx_2 \sin lx_3 \sin k(x_1 - ct)$$

$$\psi_3 = Ck \sin qx_2 \cos lx_3 \cos k(x_1 - ct).$$

What are the planes of symmetry for these motions? Determine the stresses τ_{22}, τ_{21} and τ_{23}, and show that the frequency equation for a layer free of surface tractions at $x_2 = \pm h$ is given by

$$\frac{\tan (ph)}{\tan (qh)} = -\frac{(k^2+l^2-q^2)^2}{4pq(k^2+l^2)},$$

where

$$p^2+l^2+k^2 = \frac{\omega^2}{c_L^2}$$

$$q^2+l^2+k^2 = \frac{\omega^2}{c_T^2}.$$

6.4. We consider standing waves in an elastic layer bounded by traction-free surfaces at $x_2 = \pm h$. Let

$$\varphi = -A\zeta \cos \xi x_1 \cos \alpha x_2 \cos \zeta x_3 \sin \omega t$$

$$\psi_1 = B\zeta \cos \xi x_1 \sin \beta x_2 \sin \zeta x_3 \sin \omega t$$

$$\psi_3 = -B\xi \sin \xi x_1 \cos \beta x_2 \cos \zeta x_3 \sin \omega t.$$

These give modes which are symmetric with respect to the three coordinate planes. Show that the conditions of vanishing tractions at $x_2 = \pm h$ require

$$\frac{B}{A} = \frac{\zeta(\xi^2+\zeta^2-\beta^2)\cos \alpha h}{2\beta(\xi^2+\zeta^2)\cos \beta h} = \frac{2\alpha\zeta \sin \alpha h}{(\beta^2-\xi^2-\zeta^2)\sin \beta h}.$$

Choose α and β as

$$\alpha = \frac{m\pi}{2h} \quad \text{and} \quad \beta = \frac{n\pi}{2h},$$

respectively, and compute the corresponding frequencies.

Consider in addition the horizontally polarized shear motions defined by

$$\psi_2 = [C_1 \zeta \sin \eta x_1 \cos \alpha x_2 \sin \zeta x_3 + C_2 \zeta \sin \xi x_1 \cos \beta x_2 \sin \zeta x_3] \sin \omega t.$$

Check that the surfaces $x_2 = \pm h$ are free of tractions if α and β are chosen as stated previously.

Now compute the tractions on $x_1 = \pm a$. First determine the specific values of ξ and η for which the shear stresses vanish at $x_1 = \pm a$. Proceed to compute the specific ratios of a/h for which the normal stresses vanish at $x_1 = \pm a$.

The results of this problem give an exact solution of the equations of the theory of elasticity for a family of modes in a strip of infinite length and certain specific ratios of width to depth.

6.5. A sandwich construction consists of three layers: a core and two cover-sheets, as shown in the figure. Investigate the propagation of horizontally

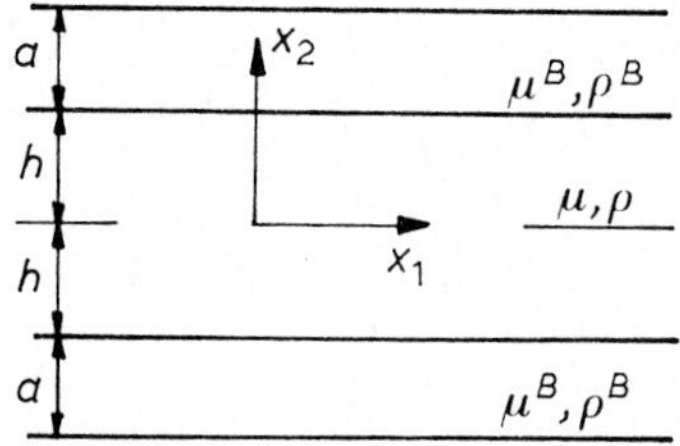

polarized shear waves in the sandwich, i.e., consider displacement solutions of the forms

$$u_3 = f(x_2)e^{ik(x_1 - ct)}.$$

Determine the frequency equation. What are the limiting frequencies for $kh \ll 1$ and $kh \gg 1$?

6.6. Determine the frequency equation for the propagation of torsional waves in a hollow circular cylinder of inner radius a and outer radius b.

6.7. Axial shear motions of a circular cylinder are defined by the following displacement distribution:

$$w = W(r)e^{in\theta}e^{i\omega t}.$$

Consider a solid circular cylinder of radius a and determine the frequency equation.

6.8. The Pochhammer frequency equation for longitudinal waves of a rod is given by eq. (6.131). Show that for the lowest mode the phase velocity approaches from below the velocity of Rayleigh waves as ka increases beyond bounds.

6.9. Derive the equations of motion for the homogeneous Timoshenko beam, eqs. (6.150) and (6.151), by considerations based on the balances of linear and angular momentums of an element of the beam.

6.10. Employ the Mindlin plate equations, eqs. (6.159)–(6.161), to examine the propagation of straight crested waves. Neglect rotatory inertia and compute the dimensionless frequency $\Omega = 2h\omega/\pi c_T$ as a function of the dimensionless wavenumber $\xi = 2kh/\pi$. For $\nu = 0.25$ the corresponding branch of the exact frequency spectrum is shown in figure 6.11. Plot the frequencies according to the Mindlin theory, the classical plate theory and the theory of elasticity in one graph, and estimate the wavelengths for which the approximate frequencies differ by about 5 % from the exact frequencies.

CHAPTER 7

FORCED MOTIONS OF A HALF-SPACE

7.1. Integral transform techniques

The dynamic response of elastic bodies to time-varying external loads can be investigated in an efficient manner by the use of integral transform techniques.

The integral transform $f^T(\xi)$ of a function $f(x)$ defined in an interval (x_1, ∞) is an expression of the form

$$f^T(\xi) = \int_{x_1}^{\infty} f(x)K(x, \xi)dx, \qquad \xi_1 \leqq \xi < \infty, \tag{7.1}$$

where x_1 and ξ_1 are real numbers, and $K(x, \xi)$ is called the kernel of the transformation. Provided that the function $f(x)$ satisfies appropriate conditions, we can express $f(x)$ in terms of its integral transform by using an inversion formula of the general form

$$f(x) = \int_{\xi_1}^{\infty} f^T(\xi)M(x, \xi)d\xi. \tag{7.2}$$

Here, $M(x, \xi)$ is a suitable function defined in the region $x_1 < x < \infty$, $\xi_1 < \xi < \infty$, and is called the kernel of the inverse transform.

The definition of an integral transform can be made more general by letting the kernel K depend on a complex parameter ζ varying over some region D of the complex plane. Eq. (7.1) is then replaced by

$$f^T(\zeta) = \int_{x_1}^{\infty} f(x)K(x, \zeta)dx \qquad \zeta \in D. \tag{7.3}$$

In this case, the inversion formula takes the form

$$f(x) = \frac{1}{2\pi i}\int_{\Gamma} f^T(\zeta)M(x, \zeta)d\zeta, \tag{7.4}$$

where $M(x, \zeta)$ is the kernel of the inverse transform, which is defined for all x in the interval (x_1, ∞). The complex variable ζ is in the region D, while Γ is a suitable path of integration contained in D.

Integral transform techniques are very useful in solving partial differential equations. The basic idea is to select an appropriate kernel K so that application of the transform to an equation for f yields a simpler equation for f^T, with one less independent variable. If the reduced equation for f^T can be solved, the solution f can be expressed in terms of the inversion integral, which must then be evaluated.

To operate effectively with integral transforms some knowledge of complex variable theory is required. We will summarize the definitions and theorems that are needed in the sequel. Generally Greek symbols will be used to denote complex variables, e.g., $\zeta = \zeta_1 + i\zeta_2$. If to each point ζ in a certain region R there correspond one or more complex numbers, denoted by χ, then we write $\chi = \mathrm{f}(\zeta)$, and we say that χ is a function of the complex variable ζ. The function $\chi = \mathrm{f}(\zeta)$ is *analytic* at the point ζ when it is single-valued and differentiable at this point. The function $\mathrm{f}(\zeta)$ is said to be *regular in a region D* if it is analytic at *every* point of D. We say that $\mathrm{f}(\zeta)$ is an analytic function in a region D if $\mathrm{f}(\zeta)$ is analytic at every point of a region *except* for a certain number of exceptional points, which are called singularities. An analytic function which is regular in every finite region of the ζ-plane is called an *entire* function.

Of great importance is Cauchy's theorem, which leads directly to the residue theorem which was stated in section 1.9. We also recall Jordan's lemma, which was also stated in section 1.9.

An integral transform as defined by eq. (7.3) is essentially an integral of the form

$$\mathrm{f}^T(\zeta) = \int_a^b g(x, \zeta)\mathrm{d}x, \tag{7.5}$$

where $g(x, \zeta)$ is a function of the complex variable ζ and the real variable x. The variable ζ will be assumed to lie inside a region D, i.e., the boundary of D, if any, is excluded.

We now state a theorem for the conditions under which $\mathrm{f}^T(\zeta)$ as defined by (7.5) is regular. This theorem is a simplification to integrals over a real-valued variable of a theorem stated by Noble[1]:

Theorem 7.1: Let $g(x, \zeta) = \mathrm{f}(x)K(x, \zeta)$ satisfy the conditions

(1) $K(x, \zeta)$ is a continuous function of the complex variable ζ and the real variable x, where ζ lies inside the region D, and x lies in the closed interval $[a, b]$,

[1] B. Noble, *Methods based on the Wiener-Hopf technique.* New York, Pergamon Press (1958), p. 11.

(2) $K(x, \zeta)$ is a regular function of ζ in D for every x in $[a, b]$,

(3) $\mathrm{f}(x)$ has only a finite number of finite discontinuities in $[a, b]$ and a finite number of maxima and minima on any finite subinterval of $[a, b]$,

(4) $\mathrm{f}(x)$ is bounded except at a finite number of points. If x_0 is such a point, so that $g(x, \zeta) \to \infty$ as $x \to x_0$, then

$$\int_a^b g(x, \zeta)\mathrm{d}x = \lim_{\delta \to 0} \int_{[a,\, b-\delta]} g(x, \zeta)\mathrm{d}x$$

exists, where the notation $[a, b-\delta]$ denotes the interval $[a, b]$ apart from a small length δ on both sides of x_0, and lim $(\delta \to 0)$ denotes the limit as this excluded length tends to zero. The limit must be approached uniformly when ζ lies in any closed domain D' within D.

(5) If the range of integration goes to (∞, ∞) then conditions (1) and (2) must be satisfied for any bounded part of the range of integration. The infinite integral $\mathrm{f}^T(\zeta)$ must be uniformly convergent when ζ lies in any closed domain D' within D.

Then $\mathrm{f}^T(\zeta)$ defined by (7.5) is a regular function of ζ in D.

In the next two sections we briefly summarize the most commonly used integral transforms.

7.2. Exponential transforms

Integral transforms are based on the Fourier integral theorem. For a real function $\mathrm{f}(x)$ defined in the interval $(-\infty, \infty)$ of the real variable x, the Fourier integral theorem may be stated as

$$\mathrm{f}(x) = \frac{1}{2\pi} \int_{-\infty}^{\infty} \mathrm{d}\xi \int_{-\infty}^{\infty} \mathrm{f}(u) e^{i\xi(u-x)} \mathrm{d}u. \tag{7.6}$$

If $\mathrm{f}(x)$ has a jump discontinuity at the point $x = x_1$, the left-hand side should at $x = x_1$ be replaced by the sum

$$\tfrac{1}{2}[\mathrm{f}(x_1+0)+\mathrm{f}(x_1-0)].$$

A heuristic derivation of (7.6) can be found in section 1.8. The Fourier integral theorem is valid provided that the function $\mathrm{f}(x)$ satisfies certain conditions. The theorem is easiest to prove if $\mathrm{f}(x)$ is piecewise smooth and if $|\mathrm{f}(x)|$ is integrable from $-\infty$ to $+\infty$.[2] Proofs of the Fourier integral

[2] For a proof see G. F. Carrier, M. Krook and C. E. Pearson, *Functions of a complex variable*. New York, McGraw-Hill Book Co. (1966), p.305.

theorem for functions satisfying less restrictive conditions were presented by Titchmarsh.[3]

7.2.1. *Exponential Fourier transform*

The Fourier integral theorem suggests the definition of $\mathscr{F}_E[\mathrm{f}(x)]$, the exponential Fourier transform of $\mathrm{f}(x)$, as

$$\mathrm{f}^*(\xi) = \mathscr{F}_E[\mathrm{f}(x)] = \int_{-\infty}^{\infty} e^{i\xi x}\mathrm{f}(x)\mathrm{d}x, \tag{7.7}$$

with inverse transform

$$\mathrm{f}(x) = \mathscr{F}_E^{-1}[\mathrm{f}^*(\xi)] = \frac{1}{2\pi}\int_{-\infty}^{\infty} e^{-i\xi x}\mathrm{f}^*(\xi)\mathrm{d}\xi. \tag{7.8}$$

It is actually immaterial which of the transform operations is conducted with the negative exponent and where the factor $1/2\pi$ is placed. As an alternative to (7.7) and (7.8) we can thus also define the exponential Fourier transform as

$$\mathrm{f}^*(\xi) = \int_{-\infty}^{\infty} e^{-i\xi x}\mathrm{f}(x)\mathrm{d}x, \tag{7.9}$$

with the inversion

$$\mathrm{f}(x) = \frac{1}{2\pi}\int_{-\infty}^{\infty} e^{i\xi x}\mathrm{f}^*(\xi)\mathrm{d}\xi. \tag{7.10}$$

The transform pairs (7.7), (7.8) or (7.9), (7.10) play an important role in solving a wide variety of problems in elastic wave propagation.

The transform of a derivative is related in a simple manner to the transform of the function itself. Let us confine our attention to functions $\mathrm{f}(x)$ which vanish as $|x| \to \infty$. By employing the definition (7.7), the Fourier transform of $\mathrm{df}/\mathrm{d}x$ is then given by

$$\mathscr{F}_E\left[\frac{\mathrm{df}}{\mathrm{d}x}\right] = \int_{-\infty}^{\infty} e^{i\xi x}\frac{\mathrm{df}}{\mathrm{d}x}\,\mathrm{d}x.$$

By an integration by parts,

$$\mathscr{F}_E\left[\frac{\mathrm{df}}{\mathrm{d}x}\right] = [e^{i\xi x}\mathrm{f}(x)]_{x=-\infty}^{x=\infty} - i\xi\int_{-\infty}^{\infty} e^{i\xi x}\mathrm{f}(x)\mathrm{d}x.$$

[3] E. C. Titchmarsh, *Introduction to the theory of Fourier transforms*, 2nd ed. London, Oxford University Press (1948), p. 16.

Thus,

$$\mathscr{F}_E\left[\frac{\mathrm{df}}{\mathrm{d}x}\right] = -i\xi \mathrm{f}^*(\xi).$$

Similarly, assuming that all derivatives of $\mathrm{f}(x)$ up to the $(n-1)$st vanish at $\pm\infty$, we find

$$\mathscr{F}_E\left[\frac{\mathrm{d}^n\mathrm{f}}{\mathrm{d}x^n}\right] = (-i\xi)^n \mathrm{f}^*(\xi). \tag{7.11}$$

As an example we consider the exponential Fourier transform of the Gaussian function

$$\mathrm{f}(x) = \mathrm{f}_0 \exp\left(-\frac{x^2}{\sigma^2}\right).$$

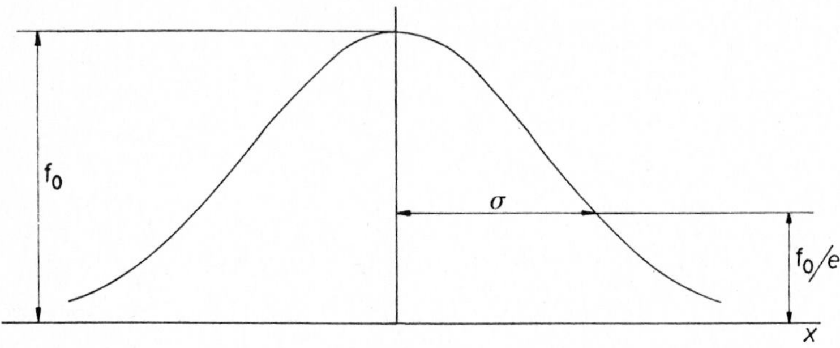

Fig. 7.1. The Gaussian function.

This function has the form shown in figure 7.1. Its shape is characterized by the central height f_0 and the width parameter σ. Its transform is

$$\begin{aligned}\mathrm{f}^*(\xi) &= \mathrm{f}_0 \int_{-\infty}^{\infty} e^{i\xi x} e^{-x^2/\sigma^2} \mathrm{d}x \\ &= \mathrm{f}_0 \exp\left(-\frac{\xi^2\sigma^2}{4}\right) \int_{-\infty}^{\infty} \exp\left[-\left(\frac{x}{\sigma} - \tfrac{1}{2}i\xi\sigma\right)^2\right] \mathrm{d}x.\end{aligned}$$

This integral can be evaluated by employing the standard integral

$$\int_{-\infty}^{\infty} \exp(-s^2)\mathrm{d}s = \sqrt{\pi}, \tag{7.12}$$

with $s = x/\sigma - \frac{1}{2}i\xi\sigma$. We obtain

$$\mathrm{f}^*(\xi) = \mathrm{f}_0\sigma\sqrt{\pi} \exp(-\xi^2\sigma^2/4). \tag{7.13}$$

The transform is again a Gaussian function, but now with width parameter $2/\sigma$, which is inversely related to the width parameter of the transformed function. We see that a tall and narrow pulse gives rise to a low and wide pulse in the transformation variable.

The Gaussian function can very conveniently be employed to introduce the Dirac delta function. The delta function was introduced in section 3.3. According to (7.12), the area under the curve defined by the Gaussian function is $f_0\sigma\sqrt{\pi}$. If we choose $f_0 = 1/\sigma\sqrt{\pi}$, the area becomes unity. Suppose we consider a sequence of functions

$$f_n(x) = \frac{1}{\sigma_n\sqrt{\pi}} \exp\left(-\frac{x^2}{\sigma_n^2}\right),$$

with decreasing values of σ_n, and thus increasing values of the central height. The limit of this sequence as $\sigma_n \to 0$ defines the Dirac delta function. This limit has the value zero, except when $x = 0$, but its integral over an interval which includes $x = 0$ is unity. If $f_0 = 1/\sigma\sqrt{\pi}$, the Fourier transform of the delta function is unity, as can be seen from (7.13).

7.2.2. *Two-sided Laplace transform*

The two-sided Laplace transform, which is a slight modification of the exponential Fourier transform, will often be used in the sequel. Let us return to the Fourier integral theorem (7.6) and let us assume that the theorem holds for a function of the form $g(x) = f(x) \exp(-\xi x)$. Upon substitution of $g(x)$ into (7.6), the integral theorem may be expressed in the form

$$f(x) = \frac{1}{2\pi i}\int_{\zeta_1 - i\infty}^{\zeta_1 + i\infty} e^{\zeta x} d\zeta \int_{-\infty}^{\infty} e^{-\zeta u} f(u) du, \tag{7.14}$$

where $\zeta = \zeta_1 + i\zeta_2$.

Eq. (7.14) suggests the following definition of the two-sided Laplace transform of $f(x)$:

$$f^*(\zeta) = \int_{-\infty}^{\infty} e^{-\zeta x} f(x) dx \tag{7.15}$$

with the inverse transform

$$f(x) = \frac{1}{2\pi i}\int_{\zeta_1 - i\infty}^{\zeta_1 + i\infty} e^{\zeta x} f^*(\zeta) d\zeta. \tag{7.16}$$

For $x = x_1$ the left-hand side of (7.16) should be replaced by

$$\tfrac{1}{2}[f(x_1+0)+f(x_1-0)],$$

if $f(x)$ suffers a discontinuity at the point $x = x_1$.

Note that we use the same notation $f^*(\zeta)$ to indicate the two-sided Laplace transform and the exponential Fourier transform. This will not give rise to confusion because it will always be clearly stated which transform is actually used.

To determine the domain of regularity of $f^*(\zeta)$, we consider a function $f(x)$ which satisfies the conditions (3) and (4) of theorem 7.1. The function $\exp(-\zeta x)$ obviously satisfies the conditions (1) and (2). Now suppose that $|f(x)| < A \exp(\zeta_{1-}x)$ for $x \to +\infty$, and $|f(x)| < B \exp(\zeta_{1+}x)$ for $x \to -\infty$, then $f^*(\zeta)$ as given by (7.15) is uniformly convergent for $\zeta_{1-} < \mathscr{R}(\zeta) < \zeta_{1+}$, which implies that condition (5) is satisfied. In view of these observations it follows from the theorem that $f^*(\zeta)$ is a regular function of ζ provided $\zeta_{1-} < \mathscr{R}(\zeta) < \zeta_{1+}$.

A very useful property of the two-sided Laplace transform is the convolution theorem. An integral of the form

$$h(x) = \int_{-\infty}^{\infty} k(x-u)f(u)du \tag{7.17}$$

is called the convolution of $k(x)$ and $f(x)$. It can be shown that

$$h(x) = \int_{-\infty}^{\infty} k(u)f(x-u)du. \tag{7.18}$$

The convolution theorem states

$$\int_{-\infty}^{\infty} e^{-\zeta x}h(x)dx = k^*(\zeta)f^*(\zeta), \tag{7.19}$$

where $k^*(\zeta)$ and $f^*(\zeta)$ are the two-sided Laplace transforms of $k(x)$ and $f(x)$, respectively. Conversely, the inverse two-sided Laplace transform of k^*f^* is the function $h(x)$ as given by (7.17) or (7.18).

7.2.3. One-sided Laplace transform

The one-sided Laplace transform which is frequently used for initial value problems with time t as the independent variable is defined as

$$\bar{f}(p) = \mathscr{L}[f(t)] = \int_0^{\infty} f(t)e^{-pt}dt. \tag{7.20}$$

It is customary to use p as the transform parameter for the one-sided Laplace transform. If the integral of eq. (7.20) converges for $p = p_1$, then it also converges for any value of p satisfying $\mathscr{R}(p) > \mathscr{R}(p_1)$. In general, the function $\bar{f}(p)$ is a regular function of the complex variable p for $\mathscr{R}(p) > \mathscr{R}(p_1)$. The inverse Laplace transform follows from (7.16) as

$$f(t) = \mathscr{L}^{-1}[\bar{f}(p)] = \frac{1}{2\pi i}\int_{\gamma-i\infty}^{\gamma+i\infty} \bar{f}(p)e^{pt}\mathrm{d}p, \tag{7.21}$$

where $\mathscr{R}(\gamma) > \mathscr{R}(p_1)$. Thus the path of integration in eq. (7.21) can be any vertical line to the right of all singularities of $\bar{f}(p)$. As in the case of the exponential Fourier transform, the left-hand side of (7.21) should be replaced by

$$\tfrac{1}{2}[f(t+0)+f(t-0)]$$

at a point of discontinuity of $f(t)$.

In later applications we will employ the one-sided Laplace transform without ever having to involve the inversion integral (7.21). In those cases p may be considered as real.

The Laplace transforms of the derivatives of a function can be obtained by integrations by part

$$\mathscr{L}\left[\frac{\mathrm{d}f}{\mathrm{d}t}\right] = p\bar{f}(p) - f(0+) \tag{7.22}$$

$$\mathscr{L}\left[\frac{\mathrm{d}^2f}{\mathrm{d}t^2}\right] = p^2\bar{f}(p) - pf(0+) - f'(0+) \tag{7.23}$$

etc.,

where by $f'(0+)$ and $f(0+)$ we mean the limits of $\mathrm{d}f/\mathrm{d}t$ and $f(t)$, respectively, as $t \to 0$ with $t > 0$.

We also quote from certain well-known results concerning asymptotic relations, known as Abelian theorems, between functions and their Laplace transforms. If $\bar{f}(p)$ is related to $f(t)$ by (7.20), then, if for $-1 < \alpha < 0$

$$f(t) \sim At^{\alpha} \quad \text{for} \quad t \to 0+, \tag{7.24}$$

then

$$\bar{f}(p) \sim A\Gamma(\alpha+1)p^{-\alpha-1} \quad \text{for} \quad p \to \infty, \tag{7.25}$$

where p tends to infinity along paths in the right half-plane, $\mathscr{R}(p) > 0$.

7.3. Other integral transforms

In this section we summarize some of the integral transforms whose kernel are *not* exponential functions.

7.3.1. *Fourier sine transform*

For conditions on $f(x)$ under which the Fourier integral theorem is valid, the sine transform is defined as

$$f^S(\xi) = \mathscr{F}_S[f(x)] = \int_0^\infty f(x) \sin \xi x \, dx, \tag{7.26}$$

with inverse transform

$$f(x) = \mathscr{F}_S^{-1}[f^S(\xi)] = \frac{2}{\pi}\int_0^\infty f^S(\xi) \sin \xi x \, d\xi. \tag{7.27}$$

7.3.2. *Fourier cosine transform*

Analogously to (7.26) and (7.27), we have

$$f^C(\xi) = \mathscr{F}_C[f(x)] = \int_0^\infty f(x) \cos \xi x \, dx \tag{7.28}$$

$$f(x) = \mathscr{F}_C^{-1}[f^C(\xi)] = \frac{2}{\pi}\int_0^\infty f^C(\xi) \cos \xi x \, d\xi. \tag{7.29}$$

7.3.3. *Hankel transform*

Let the real function $f(r)$ be defined in the interval $(0, \infty)$. Under rather weak restrictions on the behavior of $f(r)$ we can, analogously to the Fourier integral theorem, state Hankel's integral theorem

$$f(r) = \int_0^\infty J_\nu(\xi r)\xi \, d\xi \int_0^\infty f(r) J_\nu(\xi r) r \, dr, \tag{7.30}$$

where $J_\nu(\xi r)$ is the Bessel function of the first kind of order $\nu > -\frac{1}{2}$. If $f(r)$ has a jump discontinuity at the point $r = c$, the left-hand side should be replaced by

$$\tfrac{1}{2}[f(c+0)+f(c-0)].$$

The Hankel transform of a function is defined as

$$f^{H\nu}(\xi) = \mathscr{H}_\nu[f(r)] = \int_0^\infty f(r) J_\nu(\xi r) r \, dr. \tag{7.31}$$

By virtue of eq. (7.30), the inverse is given by

$$f(r) = \mathscr{H}_\nu^{-1}[f^{H\nu}(\xi)] = \int_0^\infty f^{H\nu}(\xi) J_\nu(\xi r)\xi \, d\xi. \tag{7.32}$$

7.3.4. *Mellin transform*

Let $f(r)$ be a real function defined in the interval $(0, \infty)$ such that both integrals

$$\int_0^1 r^{\sigma_1 - 1}|f(r)|dr \quad \text{and} \quad \int_1^\infty r^{\sigma_2 - 1}|f(r)|dr$$

are finite for suitably chosen real numbers σ_1 and σ_2. Then, the Mellin transform of $f(r)$ is defined as

$$f^M(p) = \int_0^\infty f(r) r^{p-1} dr,$$

where $p = \sigma + i\tau$ is any complex number in the strip $\sigma_1 < Re\, p < \sigma_2$. The inversion is given by the formula

$$f(r) = \frac{1}{2\pi i} \int_{\gamma - i\infty}^{\gamma + i\infty} f^M(p) r^{-p} dp, \tag{7.33}$$

where $\sigma_1 < \gamma < \sigma_2$.

7.4. Asymptotic expansions of integrals

7.4.1. *General considerations*

The application of integral transform techniques to wave propagation problems yields expressions for the field variables that are of the general form

$$I(\lambda) = \int_\Gamma F(\zeta) e^{-\lambda f(\zeta)} d\zeta, \tag{7.34}$$

where Γ is a path in the complex ζ-plane and the parameter λ may be considered as real-valued and positive. Expressed in its real and imaginary parts, the function $f(\zeta)$ is

$$f(\zeta) = g(\xi, \eta) + ih(\xi, \eta), \tag{7.35}$$

where

$$\zeta = \xi + i\eta. \tag{7.36}$$

It is often impossible to reduce these integrals to closed-form expressions by such analytical methods as contour integration. For that reason we will discuss ways to develop asymptotic expansions for integrals of the type (7.34).

To define the idea of an asymptotic expansion we consider a function for large values of a parameter; say we consider $I(\lambda)$ for large values of λ.

The function $I(\lambda)$ is now said to have the asymptotic expansion

$$I(\lambda) \sim S_N(\lambda) = \sum_{n=0}^{N} \frac{a_n}{\lambda^n}, \qquad \lambda \to \infty \tag{7.37}$$

if for fixed N

$$\lim_{\lambda\to\infty} \lambda^N[I(\lambda)-S_N(\lambda)] = 0. \tag{7.38}$$

The asymptotic character of the expansion is indicated by the use of the $\sim$ symbol. An alternative way of writing (7.37) is

$$I(\lambda) = \sum_{n=0}^{N} \frac{a_n}{\lambda^n} + O\left(\frac{1}{\lambda^{N+1}}\right), \qquad \lambda \to \infty. \tag{7.39}$$

It should be noted that for fixed λ the summation $S_N(\lambda)$ usually diverges as N increases. Nevertheless, if the limit (7.38) holds, the difference between $I(\lambda)$ and $S_N(\lambda)$ can be made arbitrarily small by taking λ large enough, and $S_N(\lambda)$ can be used as an approximation to $I(\lambda)$.

We will first consider two special cases of (7.34).

7.4.2. *Watson's lemma*

An asymptotic expansion of the type defined by (7.39) can easily be found for an integral of the form

$$I(\lambda) = \int_0^a e^{-\lambda\xi}\xi^\mu g(\xi)\mathrm{d}\xi. \tag{7.40}$$

The result is known as Watson's lemma. It may be stated as follows:

Suppose that in some interval $(0, \xi_1)$ the function $g(\xi)$ can be written as

$$g(\xi) = g_0+g_1\xi+g_2\xi^2+\ldots g_N\xi^N+R_{N+1}(\xi),$$

where N is some nonnegative integer and where some constant C exists such that

$$|R_{N+1}(\xi)| < C\xi^{N+1}$$

for ξ in $(0, \xi_1)$. Also, μ is real and $\mu > -1$, and constants K and b exist so that $|g(\xi)| < K \exp(b\xi)$ in $(0, a)$. Then as $\lambda \to \infty$,

$$I(\lambda) = \sum_{n=0}^{N} g_n \frac{\Gamma(\mu+n+1)}{\lambda^{\mu+n+1}} + O\left(\frac{1}{\lambda^{\mu+n+2}}\right), \tag{7.41}$$

where $\Gamma(\ )$ is the gamma function. The proof of this statement of Watson's lemma can be found in the book by Carrier et al.[5]

[5] G. F. Carrier, M. Krook and C. E. Pearson, *Functions of a complex variable*. New York, McGraw-Hill Book Co. (1966), p. 253.

7.4.3. *Fourier integrals*

Some useful asymptotic expansions of integrals of the form

$$\int_a^b e^{ix\xi} g(\xi)\mathrm{d}\xi \tag{7.42}$$

are given by Erdélyi.[6] It is assumed that (a, b) is a real interval and $g(\xi)$ is an integrable function so that (7.42) exists for all real x.

Of particular interest are the expansions for the case that the integrand has a singularity of a simple type at one end point of the interval. Let us consider the following integral:

$$I(x) = \int_a^b e^{ix\xi}(\xi - a)^{\mu-1} g(\xi)\mathrm{d}\xi. \tag{7.43}$$

It can now be stated that if $g(\xi)$ is N times continuously differentiable for $a \leqq \xi \leqq b$, $g^{(n)}(b) = 0$ for $n = 0, 1, \ldots, N-1$, and $0 < \mu < 1$, then

$$I(x) = -A_N(x) + O(x^{-N}), \quad \text{as} \quad x \to \infty, \tag{7.44}$$

where

$$A_N(x) = \sum_{n=0}^{N-1} \frac{\Gamma(n+\mu)}{n!} e^{\frac{1}{2}\pi i(n+\mu-2)} g^{(n)}(a) x^{-n-\mu} e^{ixa}. \tag{7.45}$$

The proof is given in the book by Erdélyi (p. 48).

Asymptotic expansions of Fourier-type integrals are also discussed in the book by Carrier et al.[7]

In the preceding special cases the path of integration was along the real axis. Let us now return to the form of the integral as it is stated by (7.34), and let us consider a rather general method of constructing asymptotic expansions for integrals in the complex plane.

7.4.4. *The saddle point method*

In the classical presentation of the saddle point method a path of "steepest descent" passing over the saddle point(s) of the function f(ζ) is found, and contour integration is employed to effect the change of contour from Γ to the path of steepest descent. The classical approach will be discussed in the next section.

An alternative and perhaps simpler version of the saddle point method

[6] A. Erdélyi, *Asymptotic expansions*. New York, Dover Publications, Inc. (1956), p. 46.
[7] *Loc. cit.*, p. 255.

was presented by van der Waerden.[8] In van der Waerden's version an asymptotic expansion of (7.34) is obtained by carrying out the following operations:

(1) Introduce $\lambda \mathrm{f}(\zeta)$ or (if λ is real) $\mathrm{f}(\zeta)$ as a new variable.

(2) Draw the contour C in the f-plane. This is easy since the function f is given on the contour.

(3) Determine the branch points and the poles of the functions occurring in the integral in the f-plane.

(4) Expand the integrand in a power series in the neighborhood of every branch point and every pole.

(5) Integrate term by term.

The branch points correspond to the saddle points of the classical presentation. Only those branch points and poles lying to the right of the contour C in the f-plane need be considered. As a further simplification, those branch points or poles which lie more to the right than others may be neglected.

The method of van der Waerden is particularly useful if a pole is located close to the saddle point. For details we refer to the cited paper.

7.5. The methods of stationary phase and steepest descent

These two methods have been used in numerous wave propagation problems. The two methods are related in that they both involve concentrating the integration in the vicinity of the stationary point of the exponent appearing in integrals of the form (7.34).

7.5.1. Stationary-phase approximation

The stationary-phase approximation is usually employed if the path of integration is along the real axis and if the exponent is imaginary, i.e., if (7.34) is of the form

$$I(t) = \int_a^b F(\xi) e^{ith(\xi)} \mathrm{d}\xi, \tag{7.46}$$

where t is the large parameter and $h(\xi)$ is taken in the form

$$h(\xi) = \omega(\xi) - \frac{\xi x}{t}. \tag{7.47}$$

[8] B. L. van der Waerden, *Applied Scientific Research* **B2** (1950) 33.

Eq. (7.46) is the type of representation that is obtained for a traveling pulse by Fourier transform methods. If the medium is not dispersive, $\omega = c\xi$, where c is a constant, and the integral is a function of $ct-x$, so that the pulse does not change shape as it propagates. If there is dispersion, i.e., if $c = c(\xi)$, it is usually impossible to evaluate (7.46) exactly. The alternatives then are a numerical evaluation or an approximate analytical evaluation. With the availability of electronic computers it has become feasible to evaluate integrals of the form (7.46) by numerical methods.[9] It is, however, also possible to employ a quite accurate analytical approximation which is known as the stationary phase approximation and which is due to Stokes and Kelvin. The approximation can be used for eq. (7.46) when t is large.

The integral (7.46) may also be written in the form

$$I = \int_a^b F(\xi)\{\cos[th(\xi)] + i\sin[th(\xi)]\}\mathrm{d}\xi. \tag{7.48}$$

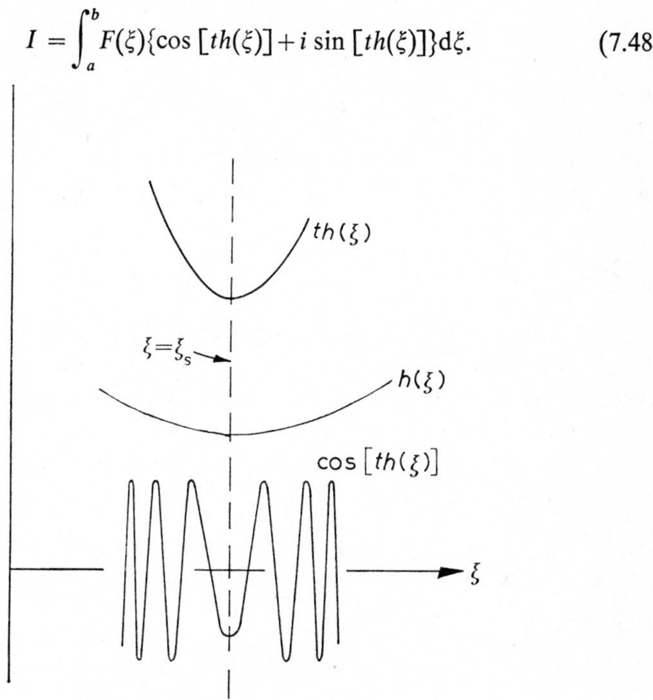

Fig. 7.2. Point of stationary phase.

Assuming that $F(\xi)$ is a relatively slowly varying function, the approximate evaluation of (7.48) rests on the observation that for large t the trigonometric

[9] Numerical methods were discussed by I. M. Longman, *Proceedings Cambridge Philosophical Society* **52** (1956) 764.

functions are very rapidly oscillating, with a self-cancelling effect on the integral. This is illustrated in figure 7.2. Only when the phase $h(\xi)$ varies slowly will there be an appreciable contribution to the integral. In particular, if there is a point of stationary phase defined by a root of $\mathrm{d}h(\xi)/\mathrm{d}\xi = 0$, it is reasonable to argue that the positive and negative loops tend to be mutually canceling except near the point(s) of stationary phase.

For eq. (7.46), the point of stationary phase ξ_s is obtained from $\mathrm{d}h/\mathrm{d}\xi = 0$, i.e.,

$$\frac{\mathrm{d}\omega}{\mathrm{d}\xi} - \frac{x}{t} = 0. \tag{7.49}$$

Thus ξ_s depends on x and t. At the point of stationary phase we have $x/t = \mathrm{d}\omega/\mathrm{d}\xi$, i.e., the point propagates with the group velocity c_g. Since t is positive, the integrand of (7.46) will have points of stationary phase for positive x only if $\mathrm{d}\omega/\mathrm{d}\xi > 0$.

By virtue of the foregoing arguments eq. (7.46) may be approximated by

$$I = F(\xi_s)\int_{\xi_s-\varepsilon}^{\xi_s+\varepsilon} e^{ith(\xi)}\mathrm{d}\xi, \tag{7.50}$$

where $\xi = \xi_s$ is the point of stationary phase and ε is a small number. To approximate (7.50) further, we expand $h(\xi)$ about $\xi = \xi_s$

$$h(\xi) = h_0+(\xi-\xi_s)h_1+\tfrac{1}{2}(\xi-\xi_s)^2h_2+\ldots, \tag{7.51}$$

where

$$h_0 = h(\xi)|_{\xi=\xi_s} \tag{7.52}$$

$$h_n = \left.\frac{\mathrm{d}^n h}{\mathrm{d}\xi^n}\right|_{\xi=\xi_s}. \tag{7.53}$$

Since $h_1 = 0$, eq. (7.50) can be written in the form

$$I = e^{ih_0t}F(\xi_s)\int_{\xi_s-\varepsilon}^{\xi_s+\varepsilon} e^{\frac{1}{2}i(\xi-\xi_s)^2h_2t}\mathrm{d}\xi, \tag{7.54}$$

where we limit ourselves for the present time to second-order terms. Once again using the argument that the contributions from values of ξ not in the immediate vicinity of ξ_s are mutually cancelling, the limits of integration are extended to $\pm\infty$, so that (7.54) becomes

$$I = e^{ih_0t}F(\xi_s)\int_{-\infty}^{\infty} e^{\frac{1}{2}ih_2t\xi^2}\mathrm{d}\xi. \tag{7.55}$$

To evaluate (7.55) we use the following formulas:

$$\int_0^\infty \sin\left(\tfrac{1}{2}a\xi^2\right)\mathrm{d}\xi = \int_0^\infty \cos\left(\tfrac{1}{2}a\xi^2\right)\mathrm{d}\xi = \frac{1}{2}\left(\frac{\pi}{a}\right)^{\frac{1}{2}},$$

and thus

$$\int_{-\infty}^{\infty} e^{\pm\frac{1}{2}ia\xi^2}\mathrm{d}\xi = (1\pm i)\left(\frac{\pi}{a}\right)^{\frac{1}{2}}. \tag{7.56}$$

Applying this result to (7.55), we find

$$I = F(\xi_s)\left[\frac{2\pi}{t|h_2|}\right]^{\frac{1}{2}} e^{i(h_0 t\pm\pi/4)}, \tag{7.57}$$

where $+$ and $-$ apply for $h_2 > 0$ and $h_2 < 0$, respectively. If (7.49) can be solved explicitly for ξ_s, eq. (7.57) yields an asymptotic representation of the integral (7.46), valid for large values of t.

Near points where $h_2 = 0$, that is near a stationary point of the group velocity, the approximation provided by (7.54) breaks down. A sufficient degree of accuracy can usually be secured by following the same procedure but including third-order terms in (7.51). Rather than expanding around the point of stationary phase, we compute the value $\xi = \xi_{sg}$ from the condition

$$\frac{\mathrm{d}^2 h}{\mathrm{d}\xi^2} = \frac{\mathrm{d}^2\omega}{\mathrm{d}\xi^2} = \frac{\mathrm{d}c_g}{\mathrm{d}\xi} = 0, \tag{7.58}$$

and we employ the expansion

$$th(\xi) = th(\xi_{sg})+(\xi-\xi_{sg})b+(\xi-\xi_{sg})^3 a,$$

where

$$b = t\left.\frac{\mathrm{d}h}{\mathrm{d}\xi}\right|_{\xi=\xi_{sg}}$$

$$a = \tfrac{1}{6}t\left.\frac{\mathrm{d}^3 h}{\mathrm{d}\xi^3}\right|_{\xi=\xi_{sg}}.$$

Instead of (7.54) we then have

$$I = e^{ih(\xi_{sg})t}F(\xi_{sg})\int_{-\infty}^{\infty} e^{i(b\xi+a\xi^3)}\mathrm{d}\xi. \tag{7.59}$$

This integral is recognized as the Airy function. Using the definition[10]

[10] *Handbook of mathematical functions*, ed. by M. Abramowitz and I. A. Stegun. Washington, National Bureau of Standards (1964), p. 446.

$$(3a)^{-\frac{1}{3}}\pi Ai[\pm(3a)^{-\frac{1}{3}}b] = \int_0^\infty \cos\,[a\xi^3 \pm b\xi]\mathrm{d}\xi, \tag{7.60}$$

we see that for $ab > 0$ we may write

$$I = e^{ih(\xi_{sg})t}F(\xi_{sg})\,\frac{2\pi}{(3|a|)^{\frac{1}{3}}}\,Ai[3^{-\frac{1}{3}}|a|^{-\frac{1}{3}}|b|], \tag{7.61}$$

while for $ab < 0$ we have

$$I = e^{ih(\xi_{sg})t}F(\xi_{sg})\,\frac{2\pi}{(3|a|)^{\frac{1}{3}}}\,Ai[-3^{-\frac{1}{3}}|a|^{-\frac{1}{3}}|b|]. \tag{7.62}$$

Graphs as well as tables of the Airy function can be found elsewhere.[10] For positive argument the Airy function is monotonically decreasing. For negative values of the argument the Airy function is oscillatory with a pronounced maximum at $3^{-\frac{1}{3}}|a|^{-\frac{1}{3}}|b| = 1$. Thus for values of x and t corresponding to stationary values of the group velocity a conspicuously large amplitude wave is propagated. Since the Airy function is of the order of magnitude of unity, the amplitude is proportional to $|a|^{-\frac{1}{3}}$, that is to $x^{-\frac{1}{3}}$ approximately, since $x \sim tc_g(\xi_{sg})$. On the other hand, it follows from (7.57) that wave groups corresponding to the point of stationary phase are proportional to $x^{-\frac{1}{2}}$. Thus, as the dispersive wave travels outward the relative importance of solutions corresponding to $\mathrm{d}c_g/\mathrm{d}\xi = 0$ increases.

Some other interesting points related to the solutions (7.61) and (7.62) are brought out in the book by Tolstoy and Clay.[11]

If $F(\xi)$ has a pole on the real axis, an additional contribution to the integral may occur. Another special case arises if a pole of the integrand and a point of stationary phase coincide. This case comes up in section 8.5.

For a formal discussion of other special cases we refer to the book by Erdélyi[12] and to the paper by Eckart.[13]

7.5.2. *Steepest-descent approximation*

For the purpose of introducing the steepest-descent approximation let us first consider the case that the integration in eq. (7.34) is along the real axis

$$I(\lambda) = \int_a^b F(\xi)e^{-\lambda g(\xi)}\mathrm{d}\xi. \tag{7.63}$$

[11] I. Tolstoy and C. S. Clay, *Ocean acoustics*. New York, McGraw-Hill Book Co. (1966), p. 42.

[12] A. Erdélyi, *Asymptotic expansions*. New York, Dover Publications, Inc. (1956), p. 51.

[13] C. Eckart, *Reviews of Modern Physics* **20** (1948) 399.

If the function $g(\xi)$ is real valued and positive, an approximate evaluation of the integral is based on the observation that for large λ the exponential $\exp[-\lambda g(\xi)]$ is relatively small except in the vicinity of the point where $g(\xi)$ shows a minimum (see figure 7.3). The major contribution then comes from the part of the path where the integrand shows a steep peak.

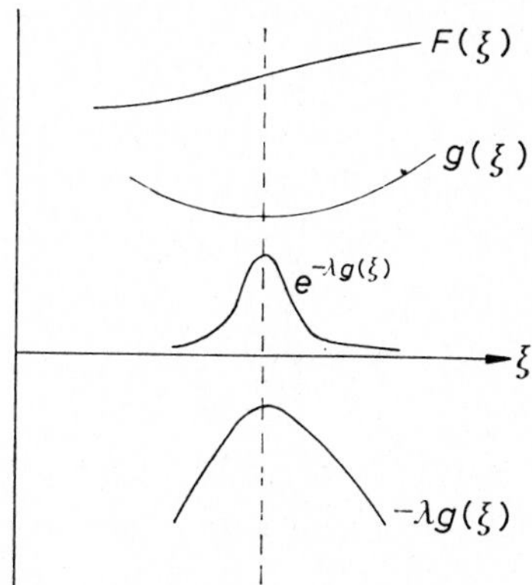

Fig. 7.3. Behavior of the integrand.

Let us now examine the integral of eq. (7.34) in the complex plane,

$$I(\lambda) = \int_{\Gamma} F(\zeta) e^{-\lambda f(\zeta)} d\zeta. \tag{7.64}$$

It is assumed that $f(\zeta)$ and $F(\zeta)$ are analytic functions of ζ in a domain containing the path of integration. It is also assumed that these functions are independent of λ. We will summarize the classical presentation of the saddle point approximation as it can be found in several treatises.[14]

In terms of its real and imaginary parts the function $f(\zeta)$ is written

$$f(\zeta) = g(\xi, \eta) + ih(\xi, \eta). \tag{7.65}$$

The integrand in (7.64) then assumes the form

$$e^{-\lambda g(\xi, \eta)} e^{-i\lambda h(\xi, \eta)} F(\xi + i\eta).$$

When λ is large a small change of $h(\xi, \eta)$ due to a small change of ζ will produce rapid oscillations of the trigonometric functions. If a path is now chosen along which $h(\xi, \eta)$ is constant, the rapid oscillations disappear and

[14] E.g., H. Jeffreys and B. C. Jeffreys, *Methods of mathematical physics*, 2nd ed. Cambridge, University Press (1950), p. 503.

the most quickly varying part of the integrand is $\exp[-\lambda g(\xi, \eta)]$. Then, just as was discussed in the context of eq. (7.63), the main contribution to the integral will come from the neighborhood of the point(s) $\zeta = \zeta_s$, where $g(\xi, \eta)$ is smallest. It will be shown in the sequel that the path through $\zeta = \zeta_s$ defined by $h =$ constant is in fact the path along which $g(\xi, \eta)$ changes most rapidly (the path of steepest descent), so that the requirement of concentrating the largest values of $g(\xi, \eta)$ to the shortest possible segment of the integration path is optimized along the path $h =$ constant. The idea of the steepest descent approximation thus is to deform the path of integration in the ζ-plane into a contour on which $h(\xi, \eta)$ is constant and which passes through the point $\zeta = \zeta_s$, where $g(\xi, \eta)$ is stationary.

The point $\zeta = \zeta_s$ is determined from the condition that

$$\frac{\partial g}{\partial \xi} = \frac{\partial g}{\partial \eta} \equiv 0, \tag{7.66}$$

which shows that the tangent plane at the surface $g(\xi, \eta)$ is horizontal at $\zeta = \zeta_s$. If this point were an absolute minimum (maximum), we would have

$$\frac{\partial^2 g}{\partial \xi^2} > 0\,(< 0) \quad \text{and} \quad \frac{\partial^2 g}{\partial \eta^2} > 0\,(< 0).$$

The function $g(\xi, \eta)$ is, however, harmonic, i.e.,

$$\frac{\partial^2 g}{\partial \xi^2} + \frac{\partial^2 g}{\partial \eta^2} = 0,$$

and the point $\zeta = \zeta_s$ thus is neither an absolute maximum nor an absolute minimum, but must be a saddle point, as shown in figure 7.4.

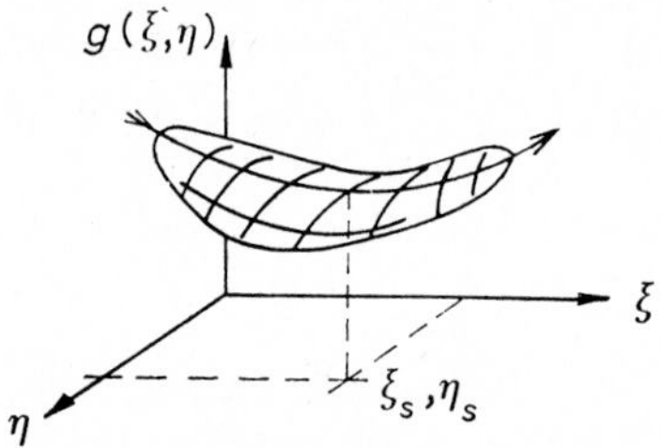

Fig. 7.4. The saddle point.

By the Cauchy-Riemann conditions we have

$$\frac{\partial g}{\partial \xi} = \frac{\partial h}{\partial \eta}, \qquad \frac{\partial g}{\partial \eta} = -\frac{\partial h}{\partial \xi}. \tag{7.67a, b}$$

These results, together with (7.66), show that the position of the saddle point can be determined from the condition

$$\frac{\mathrm{d}f}{\mathrm{d}\zeta} = 0 \quad \text{at} \quad \zeta = \zeta_s. \tag{7.68}$$

It will now be verified that the path along which h = constant is the path of steepest descent. At a position defined by $\zeta = \bar{\zeta}$ we consider a local coordinate s in a direction defined by the angle θ with the positive ζ-axis. Then

$$\left.\frac{\partial g}{\partial s}\right|_{\zeta=\bar{\zeta}} = g_\xi \cos\theta + g_\eta \sin\theta,$$

where

$$g_\xi = \left.\frac{\partial g}{\partial \xi}\right|_{\zeta=\bar{\zeta}}, \qquad g_\eta = \left.\frac{\partial g}{\partial \eta}\right|_{\zeta=\bar{\zeta}}.$$

If $\partial g/\partial s$ is to be a maximum for variable θ, we must have

$$-g_\xi \sin\theta + g_\eta \cos\theta = 0.$$

By using the Cauchy-Riemann conditions, (7.67), this expression is rewritten as

$$-h_\eta \sin\theta - h_\xi \cos\theta = -\frac{\partial h}{\partial s} = 0.$$

Since $\bar{\zeta}$ is arbitrary, h must be constant along the path of steepest descent.

The saddle points link the "valleys" and the "ridges" on the surface $g(\xi, \eta)$. The curves h = constant will go either up a ridge or down a valley, since these are the directions of greatest change. The ones of most interest

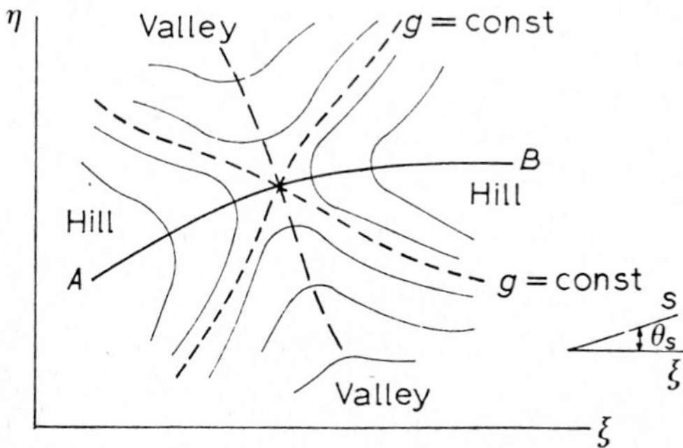

Fig. 7.5. Saddle point and path of steepest descent AB.

to us are those going along a ridge. These are the paths of steepest descent for which the neighborhood of the saddle point produces the most significant contribution.

A sketch of the saddle point and the path of steepest descent is shown in figure 7.5. In employing the method of steepest descent it is assumed that the path of integration Γ is deformed into paths of steepest descent running along the rims of the hills.

Along the path of steepest descent we expand $\mathrm{f}(\zeta)$ in a Taylor series

$$\mathrm{f}(\zeta) = \mathrm{f}_0+(\zeta-\zeta_s)\mathrm{f}_1+\tfrac{1}{2}(\zeta-\zeta_s)^2\mathrm{f}_2+\ldots,$$

where

$$f_n = \left.\frac{\mathrm{d}^n\mathrm{f}}{\mathrm{d}\zeta^n}\right|_{\zeta=\zeta_s}.$$

Keeping only second-order terms, we have, since $\mathrm{f}_1 = 0$ by eq. (7.68),

$$\mathrm{f}(\zeta)-\mathrm{f}_0 = \tfrac{1}{2}(\zeta-\zeta_s)^2\mathrm{f}_2, \tag{7.69}$$

where the right-hand side is real because $h(\xi, \eta)$ is constant. The right-hand side is also positive because we are on the path of steepest descent. Thus we introduce the real variable t by

$$\tfrac{1}{2}(\zeta-\zeta_s)^2\mathrm{f}_2 = t^2, \tag{7.70}$$

whereupon (7.64) becomes

$$I \sim e^{-\lambda \mathrm{f}_0}\int_{\Gamma_s} F(\zeta)e^{-\lambda t^2}\frac{\mathrm{d}\zeta}{\mathrm{d}t}\,\mathrm{d}t.$$

Since the exponential decays very rapidly, this integral may be rewritten as

$$I = e^{-\lambda \mathrm{f}_0}F(\zeta_s)\int_{-\varepsilon}^{+\varepsilon} e^{-\lambda t^2}\frac{\mathrm{d}\zeta}{\mathrm{d}t}\,\mathrm{d}t, \tag{7.71}$$

provided that $F(\zeta)$ is not singular in the vicinity of $\zeta = \zeta_s$. Writing

$$\zeta-\zeta_s = re^{i\theta_s}, \tag{7.72}$$

we conclude from (7.70) that

$$\arg\left[\tfrac{1}{2}\mathrm{f}_2\, e^{2i\theta_s}\right] = 0, \tag{7.73}$$

which yields θ_s. It also follows that

$$r = t|\tfrac{1}{2}\mathrm{f}_2|^{-\frac{1}{2}}. \tag{7.74}$$

Substituting (7.74) into (7.72), $d\zeta/dt$ can be computed and (7.71) may be written as

$$I \sim e^{-\lambda f_0} F(\zeta_s) \frac{e^{i\theta_s}}{|\frac{1}{2}f_2|^{\frac{1}{2}}} \int_{-\varepsilon}^{+\varepsilon} e^{-\lambda t^2} dt.$$

The limits of integration are now extended to $\pm\infty$, and in first approximation we find

$$I \sim \left(\frac{2\pi}{\lambda}\right)^{\frac{1}{2}} \frac{F(\zeta_s)}{|f_2|^{\frac{1}{2}}} e^{-\lambda f_0 + i\theta_s}. \tag{7.75}$$

The steepest descent approximation is valid for large values of t.

In the complex plane, integrals of the type (7.64) can also be represented by the stationary-phase approximation. In fact, the stationary-phase and steepest-descent approximations are nearly equivalent. If one wishes to use the stationary-phase idea, the integration through the saddle point should be taken along a path for which $g(\xi, \eta)$ is constant.

In section 7.6 we will present an application of the steepest-descent approximation to a wave propagation problem. An application to the problem of wave motion generated in a half-space by a buried line-source is given by Newlands.[15]

7.6. Half-space subjected to antiplane surface disturbances

The exponential Fourier transform can conveniently be used to determine the dynamic response of a half-space $y \geqq 0$ to surface disturbances of the form (at $y = 0$)

$$\tau_{yz} = \mu T(x) e^{i\omega t}.$$

Since the external excitation is independent of the z-coordinate, the problem is two-dimensional in the coordinates x and y. The distribution of shear tractions generates an antiplane shear motion in the half-space, which is governed by

$$\frac{\partial^2 w}{\partial x^2} + \frac{\partial^2 w}{\partial y^2} = \frac{1}{c_T^2} \frac{\partial^2 w}{\partial t^2}, \tag{7.76}$$

where $w(x, y, t)$ is the displacement in the z-direction.

If the surface tractions have been operating for a long time it is reasonable to assume that a steady state has been reached. Under that assumption we

[15] M. Newlands, *Philosophical Transactions of the Royal Society* (*London*) **A245** (1952) 213.

may consider displacements of the form

$$w(x, y, t) = w(x, y)e^{i\omega t}. \tag{7.77}$$

The governing equation for $w(x, y)$ follows from (7.76) as

$$\frac{\partial^2 w}{\partial x^2} + \frac{\partial^2 w}{\partial y^2} + k_T^2 w = 0, \tag{7.78}$$

where

$$k_T = \frac{\omega}{c_T}. \tag{7.79}$$

The boundary condition at $y = 0$ reduces, to

$$\frac{\partial w}{\partial y} = T(x). \tag{7.80}$$

7.6.1. *Exact solution*

By applying the exponential Fourier transform according to the definition given by eq. (7.7), eq. (7.78) becomes an ordinary differential equation,

$$\frac{\mathrm{d}^2 w^*}{\mathrm{d}y^2} - (\xi^2 - k_T^2)w^* = 0, \tag{7.81}$$

while the boundary condition assumes the form

$$\frac{\mathrm{d}w^*}{\mathrm{d}y} = T^*(\xi). \tag{7.82}$$

In eqs. (7.81) and (7.82), ξ is the Fourier transform parameter.

The general solution of (7.81) is

$$w^*(\xi, y) = A(\xi)\exp[(\xi^2 - k_T^2)^{\frac{1}{2}}y] + B(\xi)\exp[-(\xi^2 - k_T^2)^{\frac{1}{2}}y],$$

where $A(\xi)$ and $B(\xi)$ are as yet unknown functions of ξ. The condition at $y = 0$ is not enough to determine both $A(\xi)$ and $B(\xi)$. The additional requirement is that $w(x, y)$ be bounded as $y \to \infty$, which implies that $w^*(\xi, y)$ cannot grow exponentially as $y \to \infty$. Since $A(\xi)$ and $B(\xi)$ do not depend on y, the only way to avoid such an exponential growth is to set the coefficient of the growing exponential term equal to zero.

The radicals in the exponents are made single-valued by choosing branch cuts emanating from the branch points $\xi = \pm k_T$. We shall choose the branches such that $\mathcal{R}(\xi^2 - k_T^2)^{\frac{1}{2}}$ is nonnegative on the real ξ-axis. The desired behavior at the limit $y \to \infty$ then requires that we set $A(\xi) \equiv 0$. The function $B(\xi)$ can then be determined from the boundary condition (7.82). We find

$$w^*(\xi, y) = -\frac{T^*(\xi)}{(\xi^2-k_T^2)^{\frac{1}{2}}}\, e^{-(\xi^2-k_T^2)^{\frac{1}{2}}y}. \tag{7.83}$$

According to (7.8), the inverse transform is

$$w(x, y) = -\frac{1}{2\pi}\int_{-\infty}^{\infty}\frac{T^*(\xi)}{(\xi^2-k_T^2)^{\frac{1}{2}}}\, e^{-i\xi x-(\xi^2-k_T^2)^{\frac{1}{2}}y}\mathrm{d}\xi. \tag{7.84}$$

The major task of the analysis consists of the evaluation of the integral in eq. (7.84). Since the branch points are located on the path of integration it is necessary to describe in more detail the choice of the branch cuts. For

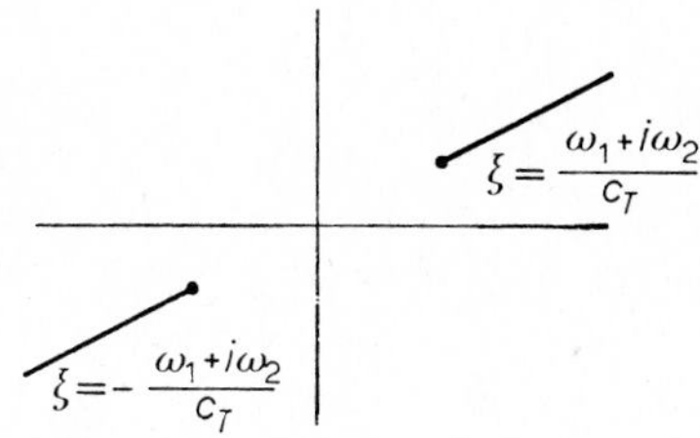

Fig. 7.6. Branch points for a complex frequency.

that purpose it is simpler to consider first a forcing term with a complex frequency

$$\omega = \omega_1 + i\omega_2.$$

For positive ω_2, the motion exponentially decays with increasing time. The branch points now are located at

$$\xi = \pm\frac{\omega_1 + i\omega_2}{c_T},$$

as shown in figure 7.6. The location of the branch points for $\omega_2 \neq 0$ indicates that in the limit of vanishing ω_2 the path of integration should be

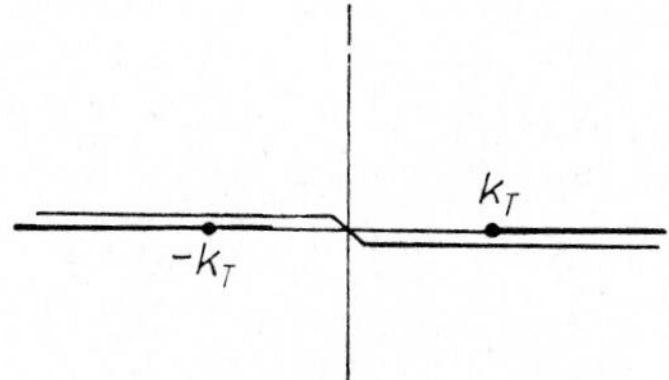

Fig. 7.7. Path in the ξ-plane.

chosen as depicted in figure 7.7, where the appropriate branch cuts are also shown. We choose $(\xi^2-k_T^2)^{\frac{1}{2}}$ positive for $\mathscr{R}(\xi) > k_T$, $\mathscr{I}(\xi) = 0-$.

To carry out an exact evaluation of (7.84) a specific choice must be made for $T(x)$. The simplest choice is $T(x) = \delta(x)$, which yields $T^*(\xi) = 1$. The integral then reduces to

$$w(x, y) = -\frac{1}{2\pi}\int_{-\infty}^{\infty} \frac{e^{-i\xi x-(\xi^2-k_T{}^2)^{\frac{1}{2}}y}}{(\xi^2-k_T^2)^{\frac{1}{2}}}\,d\xi. \tag{7.85}$$

As a first step we introduce

$$x = r\cos\theta, \qquad y = r\sin\theta, \tag{7.86a, b}$$

whereupon (7.85) assumes the form

$$w(x, y) = -\frac{1}{2\pi}\int_{-\infty}^{\infty} \frac{e^{-ir[\xi\cos\theta-i(\xi^2-k_T{}^2)^{\frac{1}{2}}\sin\theta]}}{(\xi^2-k_T^2)^{\frac{1}{2}}}\,d\xi. \tag{7.87}$$

An elegant way of evaluating this integral is to seek a curve in the complex ξ-plane along which the exponent assumes a simple form, for example, the simple form $\exp(-irs)$. To achieve this simplification we seek a contour in the ξ-plane defined by

$$s = \xi\cos\theta - i(\xi^2-k_T^2)^{\frac{1}{2}}\sin\theta. \tag{7.88}$$

Eq. (7.88) can be solved for ξ to yield

$$\xi_\pm = s\cos\theta \pm i(s^2-k_T^2)^{\frac{1}{2}}\sin\theta. \tag{7.89}$$

In the ξ-plane eq. (7.89) describes a hyperbola as shown in figure 7.8 by C_1 and C_2. If $\mathscr{I}(\xi_\pm) = 0$ we have $s = k_T$, and the vertex of the hyperbola is thus defined by $\xi = k_T\cos\theta$. The modulus of ξ increases as s increases.

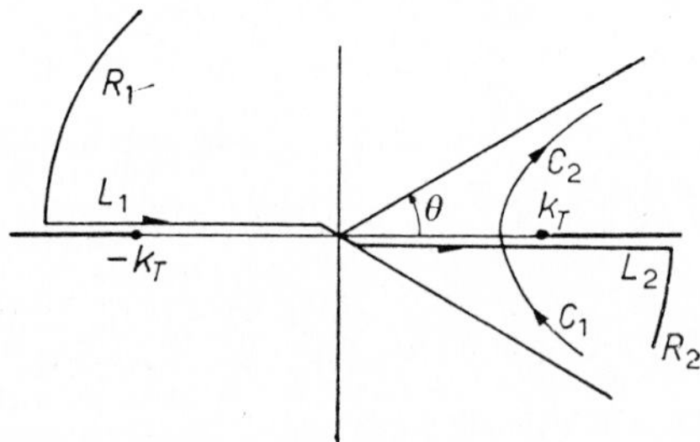

Fig. 7.8. Change of contour in the ξ-plane.

The asymptotes of the hyperbola are evidently defined by

$$\lim_{s\to\infty} \frac{\mathscr{I}(\xi_\pm)}{\mathscr{R}(\xi_\pm)} = \pm \frac{\sin\theta}{\cos\theta} = \pm \tan\theta.$$

The real variable s increases from $s = k_T$ to ∞ as we move out along the branches of the hyperbola. For $\theta \neq 0$, the vertex of the hyperbola is located in between the origin and the branch point at k_T; the vertex and the branch point coincide for $\theta = 0$.

The change of path of integration from the real axis to the hyperbola is very simple. In changing from L_1 to C_1 and L_2 to C_2, respectively (see figure 7.8) no poles or branch cuts are crossed. The complementary integrations along the circular contours R_1 and R_2 vanish as $|\xi| \to \infty$. Thus

$$w = \frac{1}{2\pi}\int_{k_T}^{\infty} (\xi_+^2 - k_T^2)^{-\frac{1}{2}} \frac{\partial \xi_+}{\partial s} e^{-irs} ds + \frac{1}{2\pi}\int_{\infty}^{k_T} (\xi_-^2 - k_T^2)^{-\frac{1}{2}} \frac{\partial \xi_-}{\partial s} e^{-irs} ds. \tag{7.90}$$

By employing (7.89) we find

$$\frac{\partial \xi_\pm}{\partial s} = \cos\theta \pm \frac{is\sin\theta}{(s^2 - k_T^2)^{\frac{1}{2}}}. \tag{7.91}$$

Also

$$(\xi_+^2 - k_T^2)^{\frac{1}{2}} = -(s^2 - k_T^2)^{\frac{1}{2}}\cos\theta - is\sin\theta \tag{7.92}$$

$$(\xi_-^2 - k_T^2)^{\frac{1}{2}} = (s^2 - k_T^2)^{\frac{1}{2}}\cos\theta - is\sin\theta, \tag{7.93}$$

and thus

$$\frac{\partial \xi_\pm}{\partial s}(\xi_\pm^2 - k_T^2)^{-\frac{1}{2}} = \mp \frac{1}{(s^2 - k_T^2)^{\frac{1}{2}}}. \tag{7.94}$$

Substituting the result (7.94) into the integrals in eq. (7.90), we obtain

$$w(x, y) = -\frac{1}{\pi}\int_{k_T}^{\infty} \frac{e^{-irs}}{(s^2 - k_T^2)^{\frac{1}{2}}} ds. \tag{7.95}$$

By introducing $s = k_T u$ the integral becomes

$$w(x, y) = -\frac{1}{\pi}\int_{1}^{\infty} \frac{e^{-ik_T ur}}{(u^2 - 1)^{\frac{1}{2}}} du. \tag{7.96}$$

Apart from a multiplying constant the integral in (7.96) is a well-known representation of the Hankel function of the second kind, see eq. (4.71). It can be verified that $w(x, y)$ may be written as

$$w(x, y) = \tfrac{1}{2} i H_0^{(2)}(k_T r). \tag{7.97}$$

7.6.2. *Asymptotic representation*

An asymptotic representation of $w(x, y)$ can be obtained in a convenient manner by taking the first term of the expansion (7.44). Thus we rewrite (7.95) in the form

$$w = -\frac{1}{\pi}\int_{k_T}^{\infty} e^{-irs}(s-k_T)^{-\frac{1}{2}}g(s)\mathrm{d}s,$$

where

$$g(s) = \frac{1}{(s+k_T)^{\frac{1}{2}}}.$$

According to (7.44), we can write

$$w \sim \frac{A_1}{\pi}, \quad \text{as} \quad r \to \infty,$$

where, according to (7.45),

$$A_1 = \Gamma(\tfrac{1}{2})e^{\frac{1}{2}\pi i(\frac{1}{2}-2)}(2k_T)^{-\frac{1}{2}}(-r)^{-\frac{1}{2}}e^{-ik_T r}.$$

Thus

$$w \sim \frac{i}{(2\pi k_T r)^{\frac{1}{2}}}\, e^{-i(k_T r-\pi/4)}. \tag{7.98}$$

7.6.3. *Steepest-descent approximation*

An approximate evaluation of (7.84) provides an instructive exercise in the application of the method of steepest descent. First we introduce the polar coordinates

$$x = r\cos\theta, \qquad y = r\sin\theta.$$

Eq. (7.84) may then be written as

$$w(r, \theta) = -\frac{1}{2\pi}\int_{-\infty}^{\infty}\frac{T^*(\zeta)}{(\zeta^2-k_T^2)^{\frac{1}{2}}}\, e^{-r\mathrm{f}(\zeta)}\mathrm{d}\zeta, \tag{7.99}$$

where

$$\mathrm{f}(\zeta) = i\zeta\cos\theta+(\zeta^2-k_T^2)^{\frac{1}{2}}\sin\theta. \tag{7.100}$$

By differentiation of (7.100) it is found that the position of the saddle point follows from

$$i\cos\theta+\frac{\zeta\sin\theta}{(\zeta^2-k_T^2)^{\frac{1}{2}}} = 0. \tag{7.101}$$

Eq. (7.101) can hold for a point on the real axis in the range $-k_T < \zeta < k_T$, where, consistent with the branch cuts shown in figure 7.7, we have

$$(\zeta^2 - k_T^2) = i(k_T^2 - \zeta^2)^{\frac{1}{2}}. \tag{7.102}$$

After substitution of (7.102) into (7.101) it can easily be verified that the solution of (7.101) is

$$\zeta_s = k_T \cos\theta \qquad (0 \leqq \theta \leqq \pi).$$

At the saddle point we subsequently find

$$\mathfrak{f}_0 = ik, \qquad \mathfrak{f}_2 = -\frac{i}{k_T \sin^2\theta}.$$

From (7.73) it follows that

$$2\theta_s - \frac{\pi}{2} = 0, \qquad \text{i.e.,} \quad \theta_s = \frac{\pi}{4}.$$

By the use of (7.75) the approximation for $w(r, \theta)$ can now be written as

$$w(r, \theta) = \frac{i}{(2\pi k_T r)^{\frac{1}{2}}} T^*(k_T \cos\theta) e^{-i(k_T r - \pi/4)}. \tag{7.103}$$

For the special case $T^*(\zeta) = 1$, eq. (7.103) agrees with (7.98).

It is an instructive exercise to determine at least roughly the course of the path of steepest descent elsewhere in the ζ-plane.

7.7. Lamb's problem for a time-harmonic line load

One of the contributions of lasting significance in the area of wave propagation in elastic solids is the article entitled "On the propagation of tremors over the surface of an elastic solid", by H. Lamb.[16] In this work, Lamb investigated the wave motion generated at the surface of an elastic half-space by the application of concentrated loads at the surface or inside the half-space. Most fully discussed were the surface motions generated by a line load and a point load applied normally to the surface. Both loads of harmonic time dependence and impulsive loads were considered. In recent years the methods and solutions in Lamb's paper have been cast in a somewhat more elegant form and more detailed computations have been carried out, particularly for loads of arbitrary time dependence. These modifications,

[16] H. Lamb, *Philosophical Transactions of the Royal Society* (*London*) **A203** (1904) 1.

which mainly concern the transient dynamic response of an elastic half-space, will be discussed in section 7.11. It is, however, of interest to discuss the surface motion generated by a time-harmonic line load applied normal to the surface. This problem, which was solved by Lamb in almost complete detail, is one of the classical problems in elastic wave propagation. The boundary conditions for the problem are (at $y = 0$)

$$\tau_y = -Q\delta(x)e^{i\omega t} \tag{7.104}$$

$$\tau_{yx} = 0. \tag{7.105}$$

7.7.1. Equations governing a state of plane strain

A line load which is applied normal to the free surface of an elastic half-space along a line coincident with the z-axis generates a deformation in plane strain. It is expedient to employ the decomposition of the displacement vector discussed in chapter 2. For the case of plane strain the displacement vector and the ∇-operator are in the (xy)-plane and consequently the vector potential $\boldsymbol{\psi}$ is directed normal to the (xy)-plane; we write $\boldsymbol{\psi} = \psi\boldsymbol{k}$. The displacement components $u(x, y, t)$ and $v(x, y, t)$ may then be written as

$$u = \frac{\partial \varphi}{\partial x} + \frac{\partial \psi}{\partial y} \tag{7.106}$$

$$v = \frac{\partial \varphi}{\partial y} - \frac{\partial \psi}{\partial x}, \tag{7.107}$$

provided that φ and ψ satisfy the following two-dimensional wave equations

$$\frac{\partial^2 \varphi}{\partial x^2} + \frac{\partial^2 \varphi}{\partial y^2} = s_L^2 \frac{\partial^2 \varphi}{\partial t^2} \tag{7.108}$$

$$\frac{\partial^2 \psi}{\partial x^2} + \frac{\partial^2 \psi}{\partial y^2} = s_T^2 \frac{\partial^2 \psi}{\partial t^2}, \tag{7.109}$$

where s_L and s_T are wave slownesses defined by

$$s_L = \frac{1}{c_L}, \qquad s_T = \frac{1}{c_T}. \tag{7.110a, b}$$

The relevant components of the stress tensor are

$$\tau_x = \lambda \left(\frac{\partial^2 \varphi}{\partial x^2} + \frac{\partial^2 \varphi}{\partial y^2}\right) + 2\mu \left(\frac{\partial^2 \varphi}{\partial x^2} + \frac{\partial^2 \psi}{\partial x\, \partial y}\right) \tag{7.111}$$

$$\tau_y = \lambda\left(\frac{\partial^2\varphi}{\partial x^2}+\frac{\partial^2\varphi}{\partial y^2}\right)+2\mu\left(\frac{\partial^2\varphi}{\partial y^2}-\frac{\partial^2\psi}{\partial x\,\partial y}\right) \tag{7.112}$$

$$\tau_{xy} = \mu\left(2\frac{\partial^2\varphi}{\partial x\,\partial y}-\frac{\partial^2\psi}{\partial x^2}+\frac{\partial^2\psi}{\partial y^2}\right). \tag{7.113}$$

7.7.2. *Steady-state solution*

Assuming that a steady state has been reached, the potentials may be expressed in the forms

$$\varphi(x, y, t) = \varphi(x, y)e^{i\omega t} \tag{7.114}$$

$$\psi(x, y, t) = \psi(x, y)e^{i\omega t}, \tag{7.115}$$

where $\varphi(x, y)$ and $\psi(x, y)$ satisfy the equations

$$\frac{\partial^2\varphi}{\partial x^2}+\frac{\partial^2\varphi}{\partial y^2}+k_L^2\varphi = 0 \tag{7.116}$$

$$\frac{\partial^2\psi}{\partial x^2}+\frac{\partial^2\psi}{\partial y^2}+k_T^2\psi = 0, \tag{7.117}$$

and k_L and k_T are defined as

$$k_L = s_L\omega, \qquad k_T = s_T\omega. \tag{7.118a, b}$$

In the expressions for the displacements and the stresses the term exp $(i\omega t)$ appears as a multiplier. In the sequel this exponential term is omitted. The formulae (7.111)–(7.113) then give the stresses

$$\tau_x = \mu\left[-k_T^2\varphi-2\frac{\partial^2\varphi}{\partial y^2}+2\frac{\partial^2\psi}{\partial x\,\partial y}\right] \tag{7.119}$$

$$\tau_y = \mu\left[-k_T^2\varphi-2\frac{\partial^2\varphi}{\partial x^2}-2\frac{\partial^2\psi}{\partial x\,\partial y}\right] \tag{7.120}$$

$$\tau_{xy} = \mu\left[2\frac{\partial^2\varphi}{\partial x\,\partial y}-k_T^2\psi-2\frac{\partial^2\psi}{\partial x^2}\right]. \tag{7.121}$$

The problem is solved by applying the exponential Fourier transform over x. The appropriate expressions for φ^* and ψ^* then follow from (7.116) and (7.117) as

$$\varphi^* = \Phi(\xi)\exp\left[-(\xi^2-k_L^2)^{\frac{1}{2}}y\right] \tag{7.122}$$

$$\psi^* = \Psi(\xi)\exp\left[-(\xi^2-k_T^2)^{\frac{1}{2}}y\right]. \tag{7.123}$$

The radicals in the exponents are made single-valued by choosing branch cuts emanating from the branch points $\xi = \pm k_L$ and $\xi = \pm k_T$, respectively. The choice of the branch cuts relative to the path of integration is dictated by the same arguments that were used in the previous section. Thus we first take a complex valued frequency $\omega = \omega_1 + i\omega_2$, with positive imaginary part ω_2. In the limit $\omega_2 \to 0$ we arrive at the branch cuts shown in figure 7.9. The radicals $(\xi^2 - k_L^2)^{\frac{1}{2}}$ and $(\xi^2 - k_T^2)^{\frac{1}{2}}$ are taken as real and positive for $\mathscr{I}(\xi) = 0^-$ and $\mathscr{R}(\xi) > k_L$ and $\mathscr{R}(\xi) > k_T$, respectively.

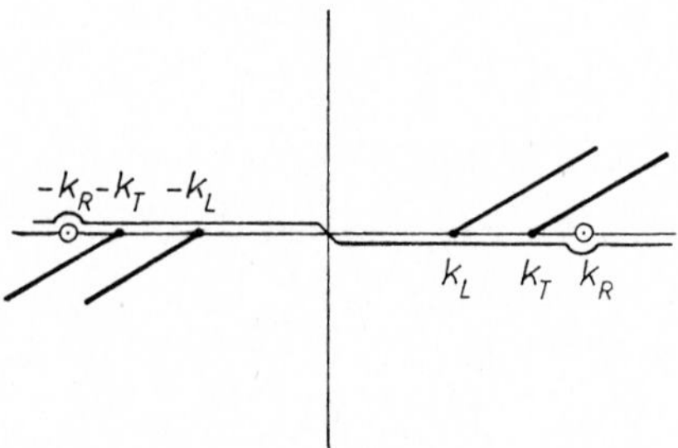

Fig. 7.9. Branch cuts for Lamb's problem.

Substituting (7.122) and (7.123) into the Fourier transforms of (7.120) and (7.121), and employing the boundary conditions (7.104) and (7.105), we arrive at the following equations for $\Phi(\xi)$ and $\Psi(\xi)$:

$$(2\xi^2 - k_T^2)\Phi - 2i\xi(\xi^2 - k_T^2)^{\frac{1}{2}}\Psi = -\frac{Q}{\mu} \tag{7.124}$$

$$2i\xi(\xi^2 - k_L^2)^{\frac{1}{2}}\Phi + (2\xi^2 - k_T^2)\Psi = 0. \tag{7.125}$$

The solutions are

$$\Phi = -\frac{2\xi^2 - k_T^2}{F(\xi)}\frac{Q}{\mu} \tag{7.126}$$

$$\Psi = \frac{2i\xi(\xi^2 - k_L^2)^{\frac{1}{2}}}{F(\xi)}\frac{Q}{\mu}, \tag{7.127}$$

where

$$F(\xi) = (2\xi^2 - k_T^2)^2 - 4\xi^2(\xi^2 - k_L^2)^{\frac{1}{2}}(\xi^2 - k_T^2)^{\frac{1}{2}}. \tag{7.128}$$

At this stage, exponential Fourier transforms of the displacements and the stresses have been obtained. The next step is the evaluation of the inversion integrals.

The exponential Fourier transforms of the displacements can be obtained from (7.106) and (7.107) as

$$u^* = -i\xi\varphi^* - (\xi^2 - k_T^2)^{\frac{1}{2}}\psi^* \tag{7.129}$$

$$v^* = -(\xi^2 - k_L^2)^{\frac{1}{2}}\varphi^* + i\xi\psi^*. \tag{7.130}$$

By application of the inversion integral of the exponential Fourier transform the displacements then follow as

$$u = (I_{uL} + I_{uT})\frac{Q}{\mu} \tag{7.131}$$

$$v = (I_{vL} + I_{vT})\frac{Q}{\mu}, \tag{7.132}$$

where

$$I_{uL} = \frac{i}{2\pi}\int_{-\infty}^{\infty} \frac{\xi(2\xi^2 - k_T^2)}{F(\xi)} e^{-i\xi x - (\xi^2 - k_L{}^2)^{\frac{1}{2}} y} d\xi \tag{7.133}$$

$$I_{uT} = -\frac{i}{\pi}\int_{-\infty}^{\infty} \frac{\xi(\xi^2 - k_T^2)^{\frac{1}{2}}(\xi^2 - k_L^2)^{\frac{1}{2}}}{F(\xi)} e^{-i\xi x - (\xi^2 - k_T{}^2)^{\frac{1}{2}} y} d\xi \tag{7.134}$$

$$I_{vL} = \frac{1}{2\pi}\int_{-\infty}^{\infty} \frac{(\xi^2 - k_L^2)^{\frac{1}{2}}(2\xi^2 - k_T^2)}{F(\xi)} e^{-i\xi x - (\xi^2 - k_L{}^2)^{\frac{1}{2}} y} d\xi \tag{7.135}$$

$$I_{vT} = -\frac{1}{\pi}\int_{-\infty}^{\infty} \frac{\xi^2(\xi^2 - k_L^2)^{\frac{1}{2}}}{F(\xi)} e^{-i\xi x - (\xi^2 - k_T{}^2)^{\frac{1}{2}} y} d\xi. \tag{7.136}$$

The path of integration in the ξ-plane is indicated in figure 7.9. The poles are located at the roots of the equation

$$F(\xi) = 0.$$

This equation was examined in our study of Rayleigh waves along the free surface of an elastic half-space. It was found in section 5.11 that there are two real roots which may be written as $\xi = \pm k_R$, where $k_R = \omega/c_R$, and c_R is the velocity of Rayleigh waves. Since the poles are located on the path of integration the path must be indented as shown in figure 7.9.

Attempts to evaluate the integrals in eqs. (7.133)–(7.136) by suitable contour integrations in the complex ξ-plane have met with only moderate success. The best that apparently can be achieved is to replace the integration along the real axis by integrals around the branch cuts plus contributions from the poles. For the special case of $y \equiv 0$, i.e., for the displacements at the surface of the half-space, such computations were carried out by

Lamb[17], who also presented asymptotic representations for the branch-line integrals. These computations require subsequent changes of the contours of integration as well as the introduction of new variables. Lamb's computations were reproduced in the book by Ewing et al.[18], and in the book by Båth.[19] Here we state the results from Lamb's original paper (at $y = 0$):

$$u = -\frac{Q}{\mu} He^{i(\omega t - k_R x)} + \frac{Q}{\mu}\left(\frac{2}{\pi}\right)^{\frac{1}{2}}\left(1-\frac{k_L^2}{k_T^2}\right)^{\frac{1}{2}} \frac{e^{i(\omega t - k_T x - \pi/4)}}{(k_T x)^{\frac{3}{2}}} - \frac{Q}{\mu}\left(\frac{2}{\pi}\right)^{\frac{1}{2}} \frac{k_L^3 k_T^2 (k_T^2 - k_L^2)^{\frac{1}{2}}}{(k_T^2 - 2k_L^2)^3} \frac{ie^{i(\omega t - k_L x - \pi/4)}}{(k_L x)^{\frac{3}{2}}} \tag{7.137}$$

$$v = -\frac{iQ}{\mu} Ke^{i(\omega t - k_R x)} + \frac{2Q}{\mu}\left(\frac{2}{\pi}\right)^{\frac{1}{2}}\left(1-\frac{k_L^2}{k_T^2}\right) \frac{ie^{i(\omega t - k_T x - \pi/4)}}{(k_T x)^{\frac{3}{2}}} + \frac{Q}{2\mu}\left(\frac{2}{\pi}\right)^{\frac{1}{2}} \frac{k_L^2 k_T^2}{(k_T^2 - 2k_L^2)^2} \frac{ie^{i(\omega t - k_L x - \pi/4)}}{(k_L x)^{\frac{3}{2}}}, \tag{7.138}$$

where

$$H = -\frac{k_R[2k_R^2 - k_T^2 - 2(k_R^2 - k_L^2)^{\frac{1}{2}}(k_R^2 - k_T^2)^{\frac{1}{2}}]}{F'(k_R)} \tag{7.139}$$

and

$$K = -\frac{k_T^2 (k_R^2 - k_L^2)^{\frac{1}{2}}}{F'(k_R)}. \tag{7.140}$$

In (7.139) and (7.140),

$$F'(k_R) = \left.\frac{\mathrm{d}F(\xi)}{\mathrm{d}\xi}\right|_{\xi = k_R}.$$

Clearly the first terms in (7.137) and (7.138) represent surface waves. In the two-dimensional case these waves do not suffer geometrical attenuation as the distance x from the disturbance increases. The remaining terms in (7.137) and (7.138) are cylindrical body waves which decay with increasing x as $x^{-\frac{3}{2}}$. Some additional remarks on the solutions may be found in Lamb's original paper or in the cited books by Ewing et al. and Båth.

It would seem that it should also be possible to write out approximate expressions for the displacements for $y \neq 0$ by the methods that were

[17] H. Lamb, *loc. cit.*

[18] W. M. Ewing, W. S. Jardetzky and F. Press, *Elastic waves in layered media.* New York, McGraw-Hill Book Co. (1957), p. 44.

[19] M. Båth, *Mathematical aspects of seismology.* Amsterdam, Elsevier Publishing Co. (1968), p. 343.

employed in the previous section. On the other hand this may not be worth the effort since an exact closed form solution for *arbitrary* time dependence of the surface excitation will be presented in section 7.11.

7.8. Suddenly applied line load in an unbounded medium

Conceptually the transient wave motion generated by a load of arbitrary time dependence can be expressed as a superposition integral over the response to a corresponding time-harmonic load. It is, however, generally more efficient to analyze the transient problem separately by employing Laplace transform techniques. This will be illustrated by the example of a suddenly applied line load in an unbounded medium.

Employing a right-handed (x, y, z)-coordinate system, the concentrated line load may be represented by the following distribution of body forces:

$$\boldsymbol{F} = \boldsymbol{j} Q\delta(x)\delta(y)\mathrm{f}(t), \tag{7.141}$$

where $\mathrm{f}(t)$ vanishes for $t < 0$. Since the applied load is independent of z, and since there are no boundaries, the displacement in the z-direction as well as the derivatives of all field variables with respect to z vanish identically.

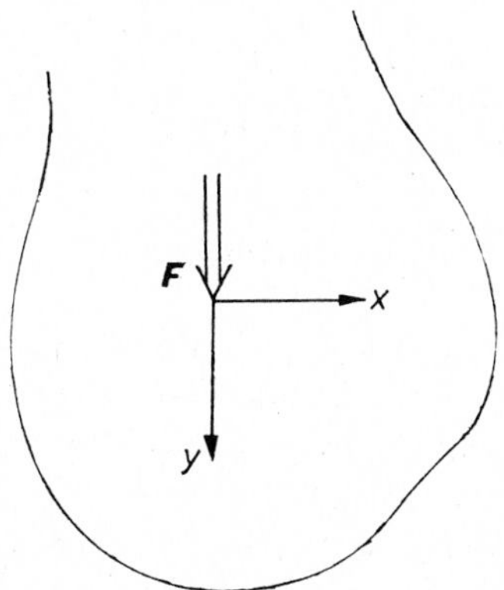

Fig. 7.10. Line load in an unbounded medium.

The response of the unbounded medium to the line load is antisymmetric with respect to the plane $y = 0$. As a consequence the normal stress τ_y vanishes in the plane $y = 0$, at least for $x \neq 0$. For the same reason the displacement in the x-direction also vanishes in the plane $y = 0$. In view of these observations the problem of the suddenly applied line load in an unbounded medium can be reformulated as a boundary value problem for a half-space $y \geqq 0$, with the following boundary conditions at $y = 0$:

$$\tau_y(x, 0, t) = -\tfrac{1}{2}Q\delta(x)\mathrm{f}(t) \tag{7.142}$$

$$u(x, 0, t) = 0. \tag{7.143}$$

Since the medium is initially undisturbed, the initial conditions are

$$u(x, y, 0) = v(x, y, 0) \equiv 0 \tag{7.144}$$

$$\dot{u}(x, y, 0) = \dot{v}(x, y, 0) \equiv 0. \tag{7.145}$$

The governing equations are stated by eqs. (7.106)–(7.113). The solution of the problem at hand is obtained by applying the one-sided Laplace transform over time and the two-sided Laplace transform over the spatial coordinate x. These transforms are defined in section 7.2. Upon application of the integral transforms eqs. (7.108) and (7.109) reduce to the following ordinary differential equations:

$$\frac{\mathrm{d}^2\bar{\varphi}^*}{\mathrm{d}y^2} - (s_L^2 p^2 - \zeta^2)\bar{\varphi}^* = 0 \tag{7.146}$$

$$\frac{\mathrm{d}^2\bar{\psi}^*}{\mathrm{d}y^2} - (s_T^2 p^2 - \zeta^2)\bar{\psi}^* = 0, \tag{7.147}$$

where p and ζ are the variables of the one-sided and the two-sided Laplace transforms, respectively. Solutions with the proper behavior for large positive values of y are

$$\bar{\varphi}^* = \Phi(p, \zeta)e^{-(s_L{}^2p^2-\zeta^2)^{\frac{1}{2}}y} \tag{7.148}$$

$$\bar{\psi}^* = \Psi(p, \zeta)e^{-(s_T{}^2p^2-\zeta^2)^{\frac{1}{2}}y}, \tag{7.149}$$

provided that the branches are chosen such that the radicals have positive real parts.

Upon applying the Laplace transforms to eqs. (7.106) and (7.112) we obtain

$$\bar{u}^* = \zeta\bar{\varphi}^* - (s_T^2 p^2 - \zeta^2)^{\frac{1}{2}}\bar{\psi}^* \tag{7.150}$$

$$\bar{\tau}_y^* = \mu[s_T^2 p^2\bar{\varphi}^* - 2\zeta^2\bar{\varphi}^* + 2\zeta(s_T^2 p^2 - \zeta^2)^{\frac{1}{2}}\bar{\psi}^*]. \tag{7.151}$$

Substituting (7.148) and (7.149) into (7.150) and (7.151), and employing the boundary conditions (7.142) and (7.143) yields

$$\Phi = -\frac{1}{2}\frac{Q}{\mu s_T^2}\frac{\bar{\mathrm{f}}(p)}{p^2} \tag{7.152}$$

$$\Psi = -\frac{1}{2}\frac{Q}{\mu s_T^2}\frac{\bar{f}(p)}{p^2}\frac{\zeta}{(s_T^2 p^2-\zeta^2)^{\frac{1}{2}}}. \tag{7.153}$$

Let us consider the displacement in the x-direction in some detail. From (7.150) we obtain

$$\frac{2\mu s_T^2}{Q}\bar{u}^* = -\frac{\zeta\bar{f}(p)}{p^2}\{e^{-(s_L^2p^2-\zeta^2)^{\frac{1}{2}}y}-e^{-(s_T^2p^2-\zeta^2)^{\frac{1}{2}}y}\}. \tag{7.154}$$

The remaining task is to evaluate the inverse transforms of this expression. For the first term of (7.154), inversion of the two-sided Laplace transform requires evaluation of the following integral:

$$\bar{I}_L = \frac{1}{2\pi i}\frac{1}{p^2}\int_{\zeta_1-i\infty}^{\zeta_1+i\infty} e^{\zeta x}e^{-(s_L^2p^2-\zeta^2)^{\frac{1}{2}}y}\mathrm{d}\zeta, \tag{7.155}$$

where $-s_L p < \zeta_1 < s_L p$. By taking the path of integration along the imaginary axis and by introducing the substitution $\zeta = i\xi$, eq. (7.155) can be rewritten as

$$\bar{I}_L = \frac{1}{\pi p^2}\int_0^\infty \cos(\xi x)e^{-(\xi^2+s_L^2p^2)^{\frac{1}{2}}y}\mathrm{d}\xi. \tag{7.156}$$

The integral in eq. (7.156) can be looked up in a table of cosine transforms[20], and we find

$$\bar{I}_L = \frac{s_L}{\pi p}\frac{y}{r}K_1(s_L pr), \tag{7.157}$$

where

$$r^2 = x^2+y^2, \tag{7.158}$$

and $K_1(\ )$ is the modified Bessel function of the second kind. The inverse Laplace transform of (7.157) can be obtained from a table of Laplace transforms.[21] We find

$$I_L = \frac{1}{\pi}\frac{y}{r^2}(t^2-s_L^2r^2)^{\frac{1}{2}}H(t-s_L r). \tag{7.159}$$

The second term in (7.154) can be treated in the same manner. The integral corresponding to (7.155) is

$$\bar{I}_T = \frac{1}{2\pi i}\frac{1}{p^2}\int_{\zeta_1-i\infty}^{\zeta_1+i\infty} e^{\zeta x}e^{-(s_T^2p^2-\zeta^2)^{\frac{1}{2}}y}\mathrm{d}\zeta. \tag{7.160}$$

[20] *Tables of integral transforms*, ed. by A. Erdélyi et al., Vol. 1. New York, McGraw-Hill Book Co. (1953), p. 16, No. 26.

[21] *Tables of integral transforms*, *loc. cit.*, p. 277, No. 11.

Analogously to (7.159), the inverse Laplace transform of $\bar{I}_T$ is

$$I_T = \frac{1}{\pi}\frac{y}{r^2}(t^2 - s_T^2 r^2)^{\frac{1}{2}} H(t - s_T r). \tag{7.161}$$

If $\mathrm{f}(t)$ is a Dirac delta function the one-sided Laplace transform $\bar{\mathrm{f}}(p)$ is unity. The displacement $u(x, y, t)$ then follows from (7.154) as

$$\frac{2\pi\mu s_T^2}{Q} u(x, y, t)$$

$$= -\frac{\partial}{\partial x}\left[\frac{y}{r^2}(t^2 - s_L^2 r^2)^{\frac{1}{2}} H(t - s_L r) - \frac{y}{r^2}(t^2 - s_T^2 r^2)^{\frac{1}{2}} H(t - s_T r)\right]. \tag{7.162}$$

For a $\mathrm{f}(t)$ of a more general form the convolution theorem of one-sided Laplace transforms can be used to determine the corresponding displacement $u(x, y, t)$.

7.9. The Cagniard-de Hoop method

In the previous section the inverse transforms were obtained by evaluating the two inversion integrals one at a time. This process was aided considerably by the availability of the results in tables of integral transforms. By inspecting eq. (7.155), the intermediate result (7.157) and the expression (7.159), it becomes apparent that the intermediate result (7.157) is more complicated than the final expression for I_L. This suggests that there might be a more efficient method of obtaining I_L directly from its double transform by inverting the two transforms in one operation, and not one at a time. Lamb was aware of this possibility, but the method was formally proposed within the context of the application of Laplace transforms by Cagniard, and in modified form by de Hoop.[22] The Cagniard-de Hoop method is based on the following elementary property of the one-sided Laplace transform:

$$\mathscr{L}^{-1}\left\{\int_{t_1}^{\infty} e^{-pt}\mathrm{f}(t)\mathrm{d}t\right\} = \mathrm{f}(t)H(t - t_1). \tag{7.163}$$

To display the Cagniard-de Hoop scheme we return to the problem discussed in the previous section. Let us introduce the following substitution in the expression for $\bar{I}_L$, eq. (7.155),

$$\zeta = p\eta, \tag{7.164}$$

[22] A. T. de Hoop, *Applied Scientific Research* **B8** (1960) 349.

where p, which is the variable of the one-sided Laplace transform, is assumed real and positive. The result can be written as

$$\bar{I}_L = \frac{1}{2\pi i}\frac{1}{p}\int_{\eta_1-i\infty}^{\eta_1+i\infty} e^{-p[(s_L{}^2-\eta^2)^{\frac{1}{2}}y-\eta x]}\mathrm{d}\eta, \tag{7.165}$$

where $-s_L < \eta_1 < s_L$. The idea of the Cagniard-de Hoop method is to deform the path of integration in the η-plane in such a manner that the inverse Laplace transform of the integral along the new path of integration can be obtained by inspection, analogously to (7.163).

The desired path of integration in the η-plane is obviously defined by the equation

$$(s_L^2-\eta^2)^{\frac{1}{2}}y-\eta x = t. \tag{7.166}$$

Eq. (7.166) can be solved for η to yield

$$\eta_{L\pm}(r, \theta, t) = -\frac{t}{r}\cos\theta \pm i\left(\frac{t^2}{r^2}-s_L^2\right)^{\frac{1}{2}}\sin\theta, \tag{7.167}$$

where the positive square root is taken. In eq. (7.167),

$$r^2 = x^2+y^2, \tag{7.168}$$

and

$$\tan\theta = \frac{y}{x}, \tag{7.169}$$

where $0 \leqq \theta \leqq \pi$. In the η-plane, eq. (7.167) describes a hyperbola, as shown in figure 7.11. If $\mathscr{I}(\eta_{L\pm}) = 0$, we have $t = s_L r$, and the vertex of

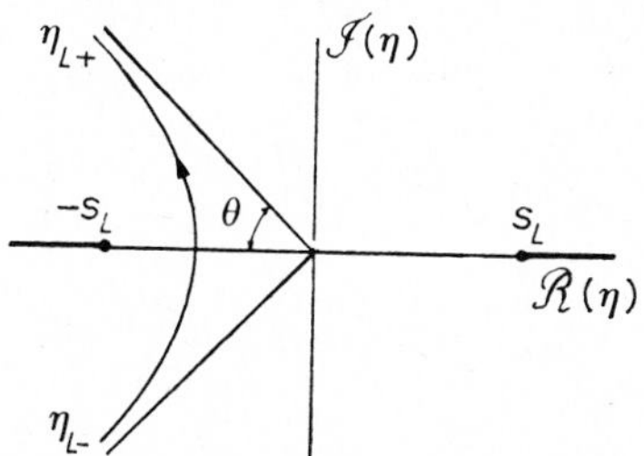

Fig. 7.11. Cagniard-de Hoop contour.

the hyperbola is thus defined by $\eta = -s_L\cos\theta$. Since the modulus $|\eta_{L\pm}|$ becomes larger as t increases, the asymptote of the hyperbola is evidently defined by

$$\frac{\mathscr{I}(\eta_{L\pm})}{\mathscr{R}(\eta_{L\pm})} = \mp \frac{\sin\theta}{\cos\theta} = \mp \tan\theta. \tag{7.170}$$

The variable t apparently increases from $s_L r$ to ∞ as we move out along the hyperbola. We note that for $0 < \theta < \pi$ the vertex of the hyperbola is located on the real axis between the branch points at $-s_L$ and $+s_L$.

It is noted that the change of contour is very similar to the one introduced in section 7.6 for the simplification of certain integrals. It should also be mentioned that for the special case of $y = 0$ and within the context of exponential Fourier transforms the basic idea of the approach, which is based on the substitution $\zeta = p\eta$, eq. (7.164), was in fact included in Lamb's paper.[23]

The transition to the Cagniard contour defined by $\eta_{L\pm}$ is very simple in this case because no poles and branch points are crossed. Along the Cagniard contour we now introduce t as the new variable, which leads to

$$\bar{I}_L = \frac{1}{p}\frac{1}{2\pi i}\left[\int_{\infty}^{s_L r}\left(\frac{\partial\eta_{L-}}{\partial t}\right)e^{-pt}\mathrm{d}t + \int_{s_L r}^{\infty}\left(\frac{\partial\eta_{L+}}{\partial t}\right)e^{-pt}\mathrm{d}t\right]. \tag{7.171}$$

From (7.167) we find

$$\frac{\partial\eta_{L\pm}}{\partial t} = -\frac{\cos\theta}{r} \pm i\left(\frac{t^2}{r^2} - s_L^2\right)^{-\frac{1}{2}}\frac{t}{r^2}\sin\theta, \tag{7.172}$$

and eq. (7.171) may thus be written

$$\bar{I}_L = \frac{1}{p}\frac{\sin\theta}{\pi r^2}\int_{s_L r}^{\infty} e^{-pt}\left(\frac{t^2}{r^2} - s_L^2\right)^{-\frac{1}{2}} t\,\mathrm{d}t. \tag{7.173}$$

Since

$$\mathscr{L}^{-1}\int_{s_L r}^{\infty} e^{-pt}\left(\frac{t^2}{r^2} - s_L^2\right)^{-\frac{1}{2}} t\,\mathrm{d}t = \left(\frac{t^2}{r^2} - s_L^2\right)^{-\frac{1}{2}} tH(t - s_L r),$$

the inverse Laplace transform of $\bar{I}_L$ may be expressed as

$$I_L = \frac{\sin\theta}{\pi r^2}\int_{s_L r}^{t}\left(\frac{t^2}{r^2} - s_L^2\right)^{-\frac{1}{2}} t\,\mathrm{d}t = \frac{\sin\theta}{\pi}\left(\frac{t^2}{r^2} - s_L^2\right)^{\frac{1}{2}} H(t - s_L r). \tag{7.174}$$

According to eq. (7.169), $\sin\theta = y/r$, and (7.174) can thus also be written as

$$I_L = \frac{1}{\pi}\frac{y}{r^2}(t^2 - s_L^2 r^2)^{\frac{1}{2}} H(t - s_L r). \tag{7.175}$$

[23] H. Lamb, *loc. cit.*, p. 22.

In a similar manner we can determine

$$I_T = \mathscr{L}^{-1}\left\{\frac{1}{2\pi i}\frac{1}{p^2}\int_{\zeta_1-i\infty}^{\zeta_1+i\infty} e^{\zeta x} e^{-(s_T{}^2p^2-\zeta^2)^{\frac{1}{2}}y}\mathrm{d}\zeta\right\}.$$

The appropriate Cagniard contour in the complex η-plane is now defined by

$$\eta_{T\pm}(r,\theta,t) = -\frac{t}{r}\cos\theta \pm i\left(\frac{t^2}{r^2}-s_T^2\right)^{\frac{1}{2}}\sin\theta. \tag{7.176}$$

The result is

$$I_T = \frac{1}{\pi}\frac{y}{r^2}(t^2-s_T^2r^2)^{\frac{1}{2}}H(t-s_Tr). \tag{7.177}$$

For the special case $\bar{f}(p) = 1$, the displacement $u(x, y, t)$ then follows from (7.154) as

$$\frac{2\pi\mu s_T^2}{Q}u(x,y,t) = -\frac{\partial}{\partial x}(I_L-I_T). \tag{7.178}$$

By substituting (7.175) and (7.177) we recover eq. (7.162).

7.10. Some observations on the solution for the line load

By carrying out the differentiations in eq. (7.162) the displacement in the horizontal direction is obtained as

$$\frac{2\pi\mu s_T^2}{Q}u(x,y,t) = \frac{xy}{r^2}\left[\frac{2}{r^2}(t^2-s_L^2r^2)^{\frac{1}{2}}+\frac{s_L^2}{(t^2-s_L^2r^2)^{\frac{1}{2}}}\right]H(t-s_Lr)$$
$$-\frac{xy}{r^2}\left[\frac{2}{r^2}(t^2-s_T^2r^2)^{\frac{1}{2}}+\frac{s_T^2}{(t^2-s_T^2r^2)^{\frac{1}{2}}}\right]H(t-s_Tr). \tag{7.179}$$

Eq. (7.179) agrees with the solution worked out by different methods by Eason et al.[24] Expressions for the displacement in the y-direction and for the stresses can be found in the paper by Eason et al.

The expression (7.179) shows that disturbances are propagated outward with the velocities c_L and c_T. The wave fronts are circles centered at the point of application of the load, and radii c_Lt and c_Tt, respectively. At the wave fronts the displacement shows algebraic singularities. This is due to the representation of the impulsive force by the idealized Dirac

[24] G. Eason, J. Fulton and I. N. Sneddon, *Philosophical Transactions of the Royal Society* (*London*) **A248** (1956) 575.

delta function. Near the point of application of the load we find for large-valued t

$$\frac{2\pi\mu s_T^2}{Q}\,u(x,y,t) \sim \frac{xy}{r^2}\,\frac{s_T^2-s_L^2}{t}.$$

Thus the displacement decreases with time. This is intuitively very acceptable since the load varies with time as a Dirac delta function.

If the concentrated line load varies with time as a Heaviside step function, the displacement $u(x, y, t)$ can be obtained by integration of (7.179). The result is

$$\frac{2\pi\mu\kappa^2 r^4}{Qxy}\,u = \begin{cases} 0 & r \geqq \dfrac{t}{s_L} \\[2ex] \dfrac{t}{s_L}\left(\dfrac{t^2}{s_L^2}-r^2\right)^{\frac{1}{2}} & \dfrac{t}{s_T} \leqq r \leqq \dfrac{t}{s_L} \\[2ex] \dfrac{t}{s_L}\left(\dfrac{t^2}{s_L^2}-r^2\right)^{\frac{1}{2}} - \dfrac{\kappa t}{s_L}\left(\dfrac{t^2}{s_T^2}-r^2\right)^{\frac{1}{2}} & r \leqq \dfrac{t}{s_T} \end{cases}$$

where

$$\kappa = \frac{c_L}{c_T} = \frac{s_T}{s_L}.$$

According to the previously cited paper by Eason et al., the displacement in the y-direction is

$$\frac{4\pi\mu\kappa^2 r^2}{Q}\,v$$

$$= \begin{cases} 0 & r \geqq \dfrac{t}{s_L} \\[2ex] \left(1-\dfrac{2x^2}{r^2}\right)\dfrac{t}{s_L}\left(\dfrac{t^2}{s_L^2}-r^2\right)^{\frac{1}{2}} + r^2 \ln\left[\dfrac{t}{s_L r} + \left(\dfrac{t^2}{s_L^2 r^2}-1\right)^{\frac{1}{2}}\right] & \dfrac{t}{s_T} \leqq r \leqq \dfrac{t}{s_L} \\[2ex] \left(1-\dfrac{2x^2}{r^2}\right)\dfrac{t}{s_L}\left\{\left(\dfrac{t^2}{s_L^2}-r^2\right)^{\frac{1}{2}} - \kappa\left(\dfrac{t^2}{s_T^2}-r^2\right)^{\frac{1}{2}}\right\} + r^2 \ln\left[\dfrac{t}{s_L r} + \left(\dfrac{t^2}{s_L^2 r^2}-1\right)^{\frac{1}{2}}\right] & \\ \qquad + \kappa^2 r^2 \ln\left[\dfrac{t}{s_T r} + \left(\dfrac{t^2}{s_T^2 r^2}-1\right)^{\frac{1}{2}}\right] & r \leqq \dfrac{t}{s_T}. \end{cases}$$

Substitution of these results into the stress-strain relation yields

$$-\frac{2\pi\kappa^2 r^2}{Qx}\tau_{xy}$$

$$=\begin{cases} 0 & r > \dfrac{t}{s_L} \\ \dfrac{2y^2}{r^2}\dfrac{t}{s_L}\left(\dfrac{t^2}{s_L^2}-r^2\right)^{-\frac{1}{2}} - \dfrac{2}{r^2}\dfrac{t}{s_L}\left(1-\dfrac{4y^2}{r^2}\right)\left(\dfrac{t^2}{s_L^2}-r^2\right)^{\frac{1}{2}} & \dfrac{t}{s_T} < r < \dfrac{t}{s_L} \\ \dfrac{\kappa t}{s_L}\left(\dfrac{t^2}{s_T^2}-r^2\right)^{-\frac{1}{2}} - \dfrac{2y^2}{r^2}\dfrac{t}{s_L}\left[\kappa\left(\dfrac{t^2}{s_T^2}-r^2\right)^{-\frac{1}{2}} - \left(\dfrac{t^2}{s_L^2}-r^2\right)^{-\frac{1}{2}}\right] & \\ \quad + \dfrac{2}{r^2}\dfrac{t}{s_L}\left(1-\dfrac{4y^2}{r^2}\right)\left[\kappa\left(\dfrac{t^2}{s_T^2}-r^2\right)^{\frac{1}{2}} - \left(\dfrac{t^2}{s_L^2}-r^2\right)^{\frac{1}{2}}\right] & r < \dfrac{t}{s_T}. \end{cases}$$

It is noted that the displacements now are continuous at the wavefronts. If we let t tend to infinity, we find that

$$\tau_{xy} = -\frac{xQ}{2\pi r^2\kappa^2}\left[1+\frac{2(\kappa^2-1)y^2}{r^2}\right],$$

in agreement with the expressions obtained from equilibrium theory.

7.11. Transient waves in a half-space

A study of transient waves in a half-space generated by a normal line load was included in Lamb's paper.[25] By applying Fourier superposition Lamb presented explicit expressions for the field variables at the surface of the half-space for a line load of arbitrary time dependence.

In this section we will employ the Cagniard-de Hoop scheme to determine the stresses at any point in the medium due to a surface excitation at $y = 0$ of the form

$$\tau_y = -Q\delta(x)\mathrm{f}(t) \tag{7.180}$$

$$\tau_{yx} = 0. \tag{7.181}$$

The equations governing the dynamic response of the elastic half-space are stated by eqs. (7.106)–(7.113). The half-space is initially at rest, and the initial conditions thus are given by eqs. (7.144) and (7.145). We will seek expressions for the field variables by applying the one-sided Laplace transform over time and the two-sided Laplace transform over x. In anticipation

[25] H. Lamb, see fn. 16, p. 289.

of the use of the Cagniard-de Hoop method the following form of the two-sided Laplace transform is used

$$f^*(p\eta) = \int_{-\infty}^{\infty} e^{-p\eta x} f(x) \mathrm{d}x. \tag{7.182}$$

The inverse transform is

$$f(x) = \frac{p}{2\pi i} \int_{\eta_1 - i\infty}^{\eta_1 + i\infty} e^{p\eta x} f^*(p\eta) \mathrm{d}\eta. \tag{7.183}$$

These definitions imply that the substitution $\zeta = p\eta$, see eq. (7.164), is introduced within the definition of the two-sided Laplace transform. This leads to an appreciable simplification of the algebra.

In the half-space $y \geqq 0$ the appropriate expressions for $\bar{\varphi}^*$ and $\bar{\psi}^*$ now are of the forms

$$\bar{\varphi}^* = \Phi(p, \eta) e^{-p\gamma_L y} \tag{7.184}$$

$$\bar{\psi}^* = \Psi(p, \eta) e^{-p\gamma_T y}, \tag{7.185}$$

where

$$\gamma_L = (s_L^2 - \eta^2)^{\frac{1}{2}}, \qquad \gamma_T = (s_T^2 - \eta^2)^{\frac{1}{2}}. \tag{7.186}$$

The corresponding transforms of the displacements and the stresses are

$$\bar{u}^* = p\eta\bar{\varphi}^* - p\gamma_T\bar{\psi}^* \tag{7.187}$$

$$\bar{v}^* = -p\gamma_L\bar{\varphi}^* - p\eta\bar{\psi}^* \tag{7.188}$$

$$\bar{\tau}_x^* = \mu p^2[(s_T^2 - 2s_L^2 + 2\eta^2)\bar{\varphi}^* - 2\eta\gamma_T\bar{\psi}^*] \tag{7.189}$$

$$\bar{\tau}_y^* = \mu p^2[(s_T^2 - 2\eta^2)\bar{\varphi}^* + 2\eta\gamma_T\bar{\psi}^*] \tag{7.190}$$

$$\bar{\tau}_{yx}^* = \mu p^2[-2\eta\gamma_L\bar{\varphi}^* + (s_T^2 - 2\eta^2)\bar{\psi}^*]. \tag{7.191}$$

The boundary conditions (7.180) and (7.181) yield the following expressions for $\Phi(p, \eta)$ and $\Psi(p, \eta)$:

$$\Phi(p, \eta) = -\frac{Q}{\mu} \frac{\bar{\mathrm{f}}(p)}{p^2} \frac{s_T^2 - 2\eta^2}{R(\eta)} \tag{7.192}$$

$$\Psi(p, \eta) = -\frac{Q}{\mu} \frac{\bar{\mathrm{f}}(p)}{p^2} \frac{2\eta\gamma_L}{R(\eta)}, \tag{7.193}$$

where

$$R(\eta) = (s_T^2 - 2\eta^2)^2 + 4\eta^2\gamma_L\gamma_T. \tag{7.194}$$

To exemplify the remaining manipulations we will consider the inversion of $\bar{\tau}_y^*$ in some detail. The inversion formula for the two-sided Laplace transform, eq. (7.183), leads to the expression

$$\bar{\tau}_y = -Qp\bar{f}(p)(\bar{I}_L+\bar{I}_T), \tag{7.195}$$

where

$$\bar{I}_L = \frac{1}{2\pi i}\int_{\eta_1-i\infty}^{\eta_1+i\infty}\frac{(s_T^2-2\eta^2)^2}{R(\eta)}e^{p(\eta x-\gamma_L y)} \tag{7.196}$$

$$\bar{I}_T = \frac{1}{2\pi i}\int_{\eta_1-i\infty}^{\eta_1+i\infty}\frac{4\eta^2\gamma_L\gamma_T}{R(\eta)}e^{p(\eta x-\gamma_T y)} \tag{7.197}$$

In both (7.196) and (7.197) we have $-s_L < \eta_1 < s_L$. The equation $R(\eta) = 0$ is recognized as the equation for the slowness of Rayleigh surface waves along the free surface of an elastic half-space, see eq. (5.96) of section 5.11. Thus we know that $R(\eta) = 0$ has two real roots at $\eta = \pm s_R$, where $s_R = 1/c_R$.

For the integrals (7.196) and (7.197) there is now a slight complication in the application of the Cagniard-de Hoop method because the radicals γ_L and γ_T are not appearing in separate terms. Nevertheless the method still works very well. It is noted that the integrands have the following singularities in the complex η-plane: branch points at $\eta = \pm s_L$ and $\eta = \pm s_T$, and simple poles at $\eta = \pm s_R$.

It is clear that the path defined by (7.167) casts the integrand of (7.196) in the desired form

$$\bar{I}_L = \int_{s_L r}^{\infty} G_L(r, \theta, t)e^{-pt}dt, \tag{7.198}$$

where

$$G_L(r, \theta, t) = \frac{1}{\pi}\mathscr{I}\left\{\left[\frac{(s_T^2-2\eta^2)^2}{R(\eta)}\right]_{\eta=\eta_{L+}}\frac{\partial\eta_{L+}}{\partial t}\right\}, \tag{7.199}$$

where we have taken into account the symmetry of the path of integration with respect to the real axis. For the integrand of $\bar{I}_T$ we need the path defined by (7.176). Now, depending on the value of θ the vertex $\eta = -s_T \cos\theta$ of the hyperbola defined by $\eta_{T\pm}$ may or may not be located in between the branch points $\eta = \pm s_L$. If θ lies in the range

$$\cos^{-1}(s_L/s_T) \leqq \theta \leqq \pi - \cos^{-1}(s_L/s_T) \tag{7.200}$$

the situation shown in figure 7.12a prevails, and the inverse transform can be written out as before. If $\theta < \cos^{-1}(s_L/s_T)$ or $\pi \geqq \theta > \pi - \cos^{-1}(s_L/s_T)$ we have to include the integral around the branch cut. For $0 < \theta < \cos^{-1}$

(s_L/s_T) the contour is shown in figure 7.12b.

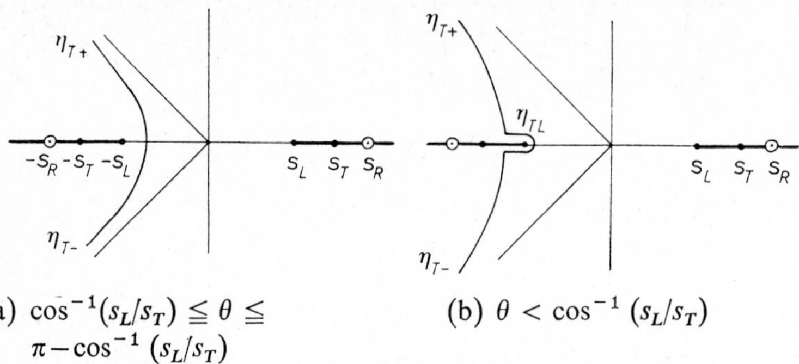

(a) $\cos^{-1}(s_L/s_T) \leqq \theta \leqq \pi - \cos^{-1}(s_L/s_T)$ (b) $\theta < \cos^{-1}(s_L/s_T)$

Fig. 7.12. Contours in the η-plane.

Since the deformation is symmetrical with respect to $x = 0$, we need to consider only the range $0 \leqq \theta \leqq \pi/2$. The additional path around the branch cut then consists of a circle of radius $\varepsilon(\varepsilon \to 0)$ centered at $\eta = -s_L$, and two segments represented by

$$\eta_{TL} = -\frac{t}{r}\cos\theta + \left(s_T^2 - \frac{t^2}{r^2}\right)^{\frac{1}{2}} \sin\theta \pm i\varepsilon. \tag{7.201}$$

The range of t is found by equating η_{LT} to $-s_L$ and $-s_T\cos\theta$, respectively, and we find

$$t_{TL} \leqq t \leqq s_T r, \tag{7.202}$$

where

$$t_{TL} = s_L r \cos\theta + r(s_T^2 - s_L^2)^{\frac{1}{2}} \sin\theta. \tag{7.203}$$

The integral $\bar{I}_T$ can then be converted into

$$\bar{I}_T = \int_{t_{TL}}^{s_T r} G_{TL}(r, \theta, t)e^{-pt}\mathrm{d}t + \int_{s_T r}^{\infty} G_T(r, \theta, t)e^{-pt}\mathrm{d}t, \tag{7.204}$$

where

$$G_{TL}(r, \theta, t) = \frac{1}{\pi}\mathscr{I}\left\{\left[\frac{4\eta^2\gamma_L(\eta)\gamma_T(\eta)}{R(\eta)}\right]_{\eta=\eta_{TL}} \frac{\partial\eta_{TL}}{\partial t}\right\} \tag{7.205}$$

$$G_T(r, \theta, t) = \frac{1}{\pi}\mathscr{I}\left\{\left[\frac{4\eta^2\gamma_L(\eta)\gamma_T(\eta)}{R(\eta)}\right]_{\eta=\eta_{T+}} \frac{\partial\eta_{T+}}{\partial t}\right\}. \tag{7.206}$$

By the use of the simple rule (7.163) the inverse Laplace transforms of (7.198) and (7.204) follow by inspection. Let us now return to eq. (7.195) and write out expressions for the stresses $\tau_y(x, y, t)$ due to the application of a line load of magnitude $Qf(t)$. In view of the symmetry with respect to $x = 0$ we need to consider only the range $0 \leqq \theta \leqq \pi/2$, but we have to distinguish between the ranges $0 \leqq \theta \leqq \cos^{-1}(s_L/s_T)$ and $\cos^{-1}(s_L/s_T) \leqq \theta \leqq \pi/2$. The results are

$$\tau_y(x, y, t) = -Q[(\tau_y)_L + (\tau_y)_{TL} + (\tau_y)_T]. \tag{7.207}$$

In the various time ranges the functions $(\tau_y)_L$, $(\tau_y)_{TL}$ and $(\tau_y)_T$ are shown in table 7.1.

The pattern of wavefronts is shown in figure 7.13 for $x \geqq 0$. At a time t after application of the line load the first term yields a disturbance in the region $r < c_L t$. In the range $0 \leqq \theta \leqq \cos^{-1}(s_L/s_T)$ the second term yields a disturbance in the shaded area. The wave motions in the shaded area are

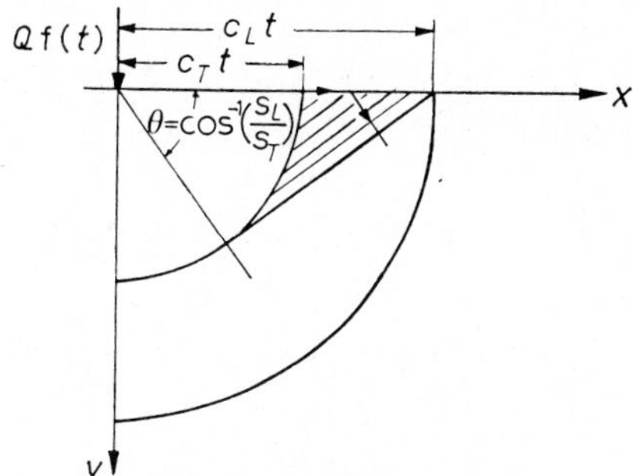

Fig. 7.13. Pattern of wavefronts.

called "head waves", and they are generated because the longitudinal wave cannot satisfy the boundary conditions of vanishing stresses at the free surface. At position (x, y) the arrival time of the wavefront of the head wave corresponds to a longitudinal wave traveling the distance $x - y[(s_T/s_L)^2 - 1]^{-\frac{1}{2}}$ with velocity c_L, and a transverse wave traveling the additional distance $y[1 - (s_L/s_T)^2]^{-\frac{1}{2}}$ with velocity c_T, as shown by arrows in figure 7.13.

As θ decreases to zero (or increases to π) the influence of the pole at $\eta = s_R$ becomes more pronounced. For $\theta = 0$ the hyperbolic paths fold around the branch cuts, and the pole then gives rise to a singularity which propagates along the free surface (and only along the surface) with the velocity of Rayleigh surface waves.

For an impulsive line load, i.e., a load applied as a δ-function in time,

TABLE 7.1

Components of eq. (7.207) for $0 \leqq \theta \leqq \pi/2$

t		$(\tau_y)_L$	$(\tau_y)_{TL}$	$(\tau_y)_T$
$t \leqq s_L r$		0	0	0
$s_L r \leqq t \leqq t_{TL}$		$\frac{df}{dt} * G_L\|_{s_L r}^{t}$	0	0
$t_{TL} \leqq t \leqq s_T r$	$0 \leqq \theta \leqq \cos^{-1}(s_L/s_T)$	$\frac{df}{dt} * G_L\|_{s_L r}^{t}$	$\frac{df}{dt} * G_{TL}\|_{t_{TL}}^{t}$	0
	$\cos^{-1}(s_L/s_T) \leqq \theta \leqq \pi/2$	$\frac{df}{dt} * G_L\|_{s_L r}^{t}$	0	0
$t \geqq s_T r$	$0 \leqq \theta \leqq \cos^{-1}(s_L/s_T)$	$\frac{df}{dt} * G_L\|_{s_L r}^{t}$	$\frac{df}{dt} * G_{TL}\|_{t_{TL}}^{s_T r}$	$\frac{df}{dt} * G_T\|_{s_T r}^{t}$
	$\cos^{-1}(s_L/s_T) \leqq \theta \leqq \pi/2$	$\frac{df}{dt} * G_L\|_{s_L r}^{t}$	0	$\frac{df}{dt} * G_T\|_{s_T r}^{t}$

Notation: $\frac{df}{dt} * G|_{t_1}^{t_2} = \int_{t_1}^{t_2} \frac{df(\tau)}{d\tau} G(x, y, t-\tau) d\tau$

numerical results were presented by Forrestal et al.[26] Some of the results obtained by these authors are shown in figures 7.14 and 7.15. Note that the horizontal displacement vanishes after the transverse wave has arrived, except for a δ-function propagating with the velocity of Rayleigh waves. The vertical displacement shows an infinite discontinuity propagating with the velocity of Rayleigh waves.

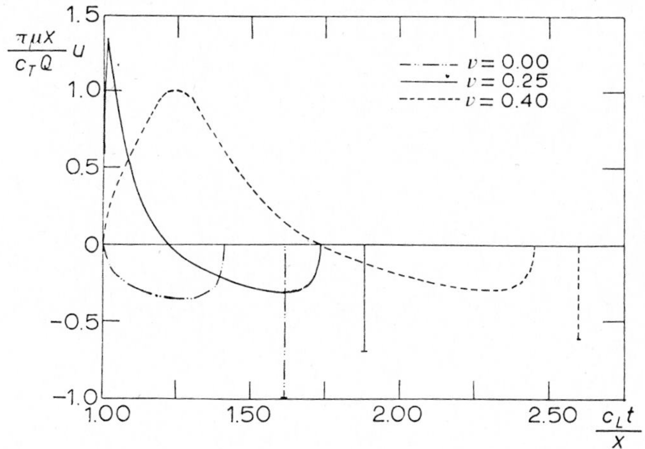

Fig. 7.14. Horizontal displacement at the free surface.

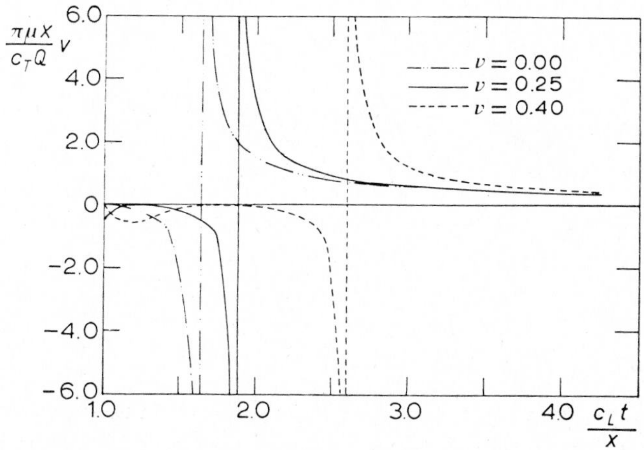

Fig. 7.15. Vertical displacement at the free surface.

[26] M. J. Forrestal, L. E. Fugelso, G. L. Neidhardt and R. A. Felder, *Proceedings Engineering Mechanics Division Specialty Conference, ASCE* (1966), p. 719.

7.12. Normal point load on a half-space

The first formulation and solution of this problem is due to Lamb[27], who synthesized the solution for the pulse from the one for a point load of harmonic time dependence. Here we prefer, however, to employ integral transform techniques. In this section we reproduce the expressions for the displacements at the free surface which were derived by Pekeris.[28] The details are worked out for the vertical displacement and for the special case of a load which varies in time as a Heaviside step function. By linear superposition over time the solution can of course be employed to determine the response for any other time dependence of the surface load.

7.12.1. *Method of solution*

We consider an homogeneous isotropic half-space $z \geqq 0$, whose free surface is subjected to a concentrated normal load of magnitude $QH(t)$. The wave motion generated by the point load is axially symmetric, and it is thus convenient to employ a system of cylindrical coordinates (see figure 7.16).

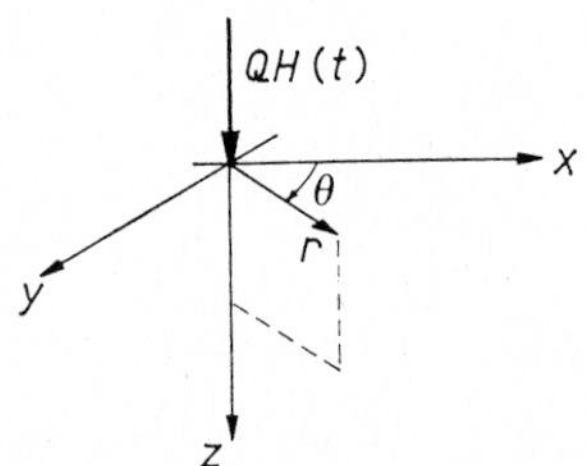

Fig. 7.16. Point load on half-space.

The boundary conditions at $z = 0$ may then be expressed as

$$\tau_{rz} = 0 \tag{7.208}$$

$$\tau_z = -QH(t)\frac{\delta(r)}{2\pi r}, \tag{7.209}$$

where we have used that in cylindrical coordinates the delta function $\delta(x)\delta(y)$ may be expressed as $\delta(r)/2\pi r$.

For an axially symmetric motion the vector potential ψ has a component ψ_θ only. For convenience of notation we write ψ rather than ψ_θ. The scalar potential φ and the single component of the vector potential depend on r,

[27] H. Lamb, fn. 16, p. 289.

[28] C. L. Pekeris, *Proceedings of the National Academy of Sciences* (*USA*) **41** (1955) 469.

z, and t only. By employing the formulas of section 2.13 the decomposition of the displacement vector may then be written as

$$u = \frac{\partial \varphi}{\partial r} - \frac{\partial \psi}{\partial z} \tag{7.210}$$

$$w = \frac{\partial \varphi}{\partial z} + \frac{1}{r}\frac{\partial (r\psi)}{\partial r}. \tag{7.211}$$

The two potentials satisfy the following wave equations:

$$\frac{\partial^2 \phi}{\partial r^2} + \frac{1}{r}\frac{\partial \phi}{\partial r} + \frac{\partial^2 \phi}{\partial z^2} = \frac{1}{c_L^2}\frac{\partial^2 \phi}{\partial t^2} \tag{7.212}$$

$$\frac{\partial^2 \psi}{\partial r^2} + \frac{1}{r}\frac{\partial \psi}{\partial r} + \frac{\partial^2 \psi}{\partial z^2} - \frac{\psi}{r^2} = \frac{1}{c_T^2}\frac{\partial^2 \psi}{\partial t^2}. \tag{7.213}$$

The relevant stress-displacement relations are

$$\tau_z = (\lambda+2\mu)\frac{\partial w}{\partial z} + \frac{\lambda}{r}\frac{\partial (ru)}{\partial r} \tag{7.214}$$

$$\tau_{zr} = \mu\left(\frac{\partial u}{\partial z} + \frac{\partial w}{\partial r}\right). \tag{7.215}$$

The formulation of the problem is completed by a statement of the initial conditions. If the half-space is at rest prior to $t = 0$, we have for $r^2+z^2 > 0$

$$\varphi(r, z, 0) = \dot{\varphi}(r, z, 0) = \psi(r, z, 0) = \dot{\psi}(r, z, 0) \equiv 0. \tag{7.216}$$

The appropriate integral transforms for the problem stated by (7.208)–(7.216) are the one-sided Laplace transform with respect to time, and the Hankel transform with respect to the radial variable r.

After application of the Laplace transform, (7.212) and (7.213) reduce to

$$\frac{\partial^2 \bar{\varphi}}{\partial r^2} + \frac{1}{r}\frac{\partial \bar{\varphi}}{\partial r} + \frac{\partial^2 \bar{\varphi}}{\partial z^2} = \frac{p^2}{c_L^2}\bar{\varphi} \tag{7.217}$$

$$\frac{\partial^2 \bar{\psi}}{\partial r^2} + \frac{1}{r}\frac{\partial \bar{\psi}}{\partial r} + \frac{\partial^2 \bar{\psi}}{\partial z^2} - \frac{\bar{\psi}}{r^2} = \frac{p^2}{c_T^2}\bar{\psi}. \tag{7.218}$$

As discussed in section 7.3, the Hankel transform of order n of a function $f(r)$ is defined as

$$f^{Hn}(\xi) = \int_0^\infty f(r)J_n(\xi r)r\,dr,$$

where $J_n(\xi r)$ is the ordinary Bessel function of order n. The inverse transform is

$$f(r) = \int_0^\infty f^{Hn}(\xi) J_n(\xi r)\xi \, d\xi. \tag{7.219}$$

In the present problem the Hankel transform of order zero must be applied to φ, and the Hankel transform of order unity to ψ. After integration by parts, (7.217) and (7.218) then reduce to the following ordinary differential equations:

$$\frac{d^2\bar{\varphi}^{H0}}{dz^2} - \alpha^2\bar{\varphi}^{H0} = 0 \tag{7.220}$$

$$\frac{d^2\bar{\psi}^{H1}}{dz^2} - \beta^2\bar{\psi}^{H1} = 0. \tag{7.221}$$

In eqs. (7.220) and (7.221),

$$\alpha = (\xi^2 + s_L^2 p^2)^{\frac{1}{2}}, \qquad \beta = (\xi^2 + s_T^2 p^2)^{\frac{1}{2}}, \tag{7.222a, b}$$

where as usual $s_L = 1/c_L$ and $s_T = 1/c_T$. Solutions of (7.220) and (7.221) which show the appropriate behavior for large values of z are

$$\bar{\varphi}^{H0} = \Phi(\xi, p)e^{-\alpha z}, \qquad \bar{\psi}^{H1} = \Psi(\xi, p)e^{-\beta z}. \tag{7.223a, b}$$

Application of the Laplace and Hankel transforms to the displacements and the stresses yields

$$\bar{w}^{H0} = \frac{d\bar{\varphi}^{H0}}{dz} + \xi\bar{\psi}^{H1} \tag{7.224}$$

$$\bar{u}^{H1} = -\xi\bar{\varphi}^{H0} - \frac{d\bar{\psi}^{H1}}{dz} \tag{7.225}$$

$$\bar{\tau}_z^{H0} = \mu\left\{(s_T^2 p^2 + 2\xi^2)\bar{\varphi}^{H0} + 2\xi\frac{d\bar{\psi}^{H1}}{dz}\right\} \tag{7.226}$$

$$\bar{\tau}_{zr}^{H1} = -\mu\left\{2\xi\frac{d\bar{\varphi}^{H0}}{dz} + (s_T^2 p^2 + 2\xi^2)\bar{\psi}^{H1}\right\}. \tag{7.227}$$

The boundary conditions at $z = 0$, eqs. (7.208) and (7.209), are transformed into

$$\tau_z^{H0} = -\frac{Q}{2\pi}\frac{1}{p} \tag{7.228}$$

$$\tau_{zr}^{H1} = 0. \tag{7.229}$$

If the expressions for $\bar{\varphi}^{H0}$ and $\bar{\psi}^{H1}$, (7.223a) and (7.223b), are now used in conjunction with (7.226) and (7.227), the boundary conditions (7.228) and (7.229) yield the following equations for $\Phi(\xi, p)$ and $\Psi(\xi, p)$:

$$(s_T^2 p^2 + 2\xi^2)\Phi - 2\beta\xi\Psi = -\frac{Q}{2\pi}\frac{1}{\mu}\frac{1}{p}$$

$$-2\alpha\xi\Phi + (s_T^2 p^2 + 2\xi^2)\Psi = 0.$$

The solutions of this system are

$$\Phi = -\frac{Q}{2\pi}\frac{1}{\mu}\frac{1}{p}\frac{s_T^2 p^2 + 2\xi^2}{D_H(\xi, p)} \tag{7.230}$$

$$\Psi = -\frac{Q}{2\pi}\frac{1}{\mu}\frac{1}{p}\frac{2\alpha\xi}{D_H(\xi, p)}, \tag{7.231}$$

where

$$D_H(\xi, p) = (s_T^2 p^2 + 2\xi^2)^2 - 4\xi^2\alpha\beta. \tag{7.232}$$

By means of eqs. (7.222)–(7.225), (7.230) and (7.231), the Laplace-Hankel transforms of the displacements may be written as

$$\bar{w}^{H0} = \frac{Q}{2\pi}\frac{1}{\mu}\left[(2\xi^2 + s_T^2 p^2)e^{-\alpha z} - 2\xi^2 e^{-\beta z}\right]\frac{\alpha}{D_H}\frac{1}{p} \tag{7.233}$$

$$\bar{u}^{H1} = \frac{Q}{2\pi}\frac{1}{\mu}\left[(2\xi^2 + s_T^2 p^2)e^{-\alpha z} - 2\alpha\beta e^{-\beta z}\right]\frac{\xi}{D_H}\frac{1}{p}. \tag{7.234}$$

7.12.2. *Normal displacement at $z = 0$*

Let us consider the displacements at the surface $z = 0$ in some detail. We begin by employing (7.219) to write out the inverse Hankel transforms, and then we proceed to introduce the substitution

$$\xi = p\eta.$$

The results are

$$\bar{w}(r, 0, p) = \frac{Q}{2\pi}\frac{s_T^2}{\mu}\int_0^\infty \frac{\eta(\eta^2 + s_L^2)^{\frac{1}{2}}}{D(\eta)} J_0(p\eta r)d\eta \tag{7.235}$$

$$\bar{u}(r, 0, p) = \frac{Q}{2\pi}\frac{1}{\mu}\int_0^\infty \frac{\eta^2[s_T^2 + 2\eta^2 - 2(\eta^2 + s_L^2)^{\frac{1}{2}}(\eta^2 + s_T^2)^{\frac{1}{2}}]}{D(\eta)} J_1(p\eta r)d\eta, \tag{7.236}$$

where $D(\eta)$ is defined as

$$D(\eta) = (2\eta^2+s_T^2)^2-4\eta^2(\eta^2+s_L^2)^{\frac{1}{2}}(\eta^2+s_T^2)^{\frac{1}{2}}.$$

By the substitution $i\eta$ for η this expression becomes just the same as the corresponding denominator for the line load, which is given by eq. (7.194). The roots of $D(\eta) = 0$ follow immediately from the roots of the Rayleigh equation which were discussed in section 5.11, i.e.,

$$\eta = \pm i s_R = \pm \frac{i}{c_R},$$

where c_R is the velocity of Rayleigh waves.

The next step consists in evaluating the inverse Laplace transform by means of the Cagniard-de Hoop scheme. Here the details will be worked out for the vertical displacement only. In the inversion procedure we employ the following representation for the Bessel function $J_0(x)$:

$$J_0(x) = \frac{2}{\pi}\mathscr{I}\int_1^\infty \frac{e^{ixs}}{(s^2-1)^{\frac{1}{2}}}\,ds.$$

This representation was earlier stated in chapter 4, prior to eq. (4.71). The Laplace transform of the vertical surface displacement can then be written

$$\bar{w}(r, 0, p) = \frac{Q}{\pi^2}\frac{s_T^2}{\mu}\mathscr{I}\int_0^\infty \frac{\eta(\eta^2+s_L^2)^{\frac{1}{2}}}{D(\eta)}\,d\eta\int_1^\infty \frac{e^{ip\eta rs}}{(s^2-1)^{\frac{1}{2}}}\,ds.$$

In the upper half of the complex η-plane we have a pole at $\eta = is_R$, and we have branch points at $\eta = is_L$ and $\eta = is_T$, respectively (see figure 7.17).

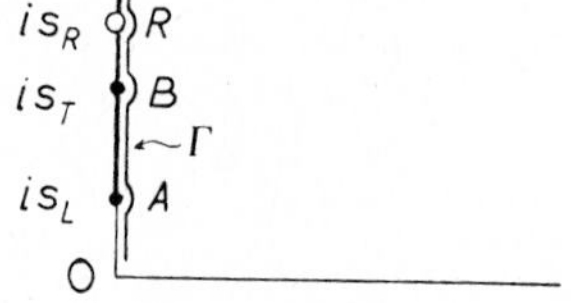

Fig. 7.17. η-plane.

Let us effect a change of contour from the real axis to the contour indicated by Γ in figure 7.17. The contour Γ consists of the imaginary axis plus three indentations around the points is_L, is_T and is_R, respectively. It is easily seen that the integrations around the branch points is_L and is_T vanish as we shrink the indentations. The integration around the pole $\eta = is_R$ also vanishes, because the contribution to the integral is real-valued. Thus we only need to consider integrations along parts of the imaginary axis, the only complication being that the principal part must be taken for integrals across $\eta = is_R$.

Since we are left only with an integration along the imaginary axis, we introduce the substitution $\eta = iv$, to obtain

$$\bar{w}(r, 0, p) = -\frac{Q}{\pi^2}\frac{s_T^2}{\mu}\mathscr{I}\int_0^\infty m(iv)v\,dv\int_1^\infty \frac{e^{-pvrs}}{(s^2-1)^{\frac{1}{2}}}\,ds,$$

where

$$m(\eta) = \frac{(\eta^2+s_L^2)^{\frac{1}{2}}}{D(\eta)}.$$

It is now recognized that the second integral is of the form of a Laplace transform. Indeed, it is easily concluded that

$$L^{-1}\int_1^\infty \frac{e^{-pvrs}}{(s^2-1)^{\frac{1}{2}}}\,ds = \frac{H(t-vr)}{(t^2-v^2r^2)^{\frac{1}{2}}}.$$

Since the Laplace transform parameter does not appear in the first integral we may write

$$w(r, 0, t) = -\frac{Q}{\pi^2}\frac{s_T^2}{\mu}\mathscr{I}\int_0^{t/r} \frac{vm(iv)}{(t^2-v^2r^2)^{\frac{1}{2}}}\,dv. \tag{7.237}$$

Let us now examine $m(\eta)$ at various intervals along the imaginary axis in the η-plane. We immediately find that $m(\eta)$ is real along OA. Thus, in view of (7.237) it is concluded that $w(r, 0, t) \equiv 0$ if $t/r < s_L$, or $t < s_L r$. Along AB (see figure 7.17) we have

$$m(iv) = \frac{i(v^2-s_L^2)^{\frac{1}{2}}}{(s_T^2-2v^2)^2+4iv^2(v^2-s_L^2)^{\frac{1}{2}}(s_T^2-v^2)^{\frac{1}{2}}}.$$

On the remaining part of the imaginary axis ($v > s_T$) we have

$$m(iv) = \frac{i(v^2-s_L^2)^{\frac{1}{2}}}{(s_T^2-2v^2)^2-4v^2(v^2-s_L^2)^{\frac{1}{2}}(v^2-s_T^2)^{\frac{1}{2}}}.$$

The foregoing results show that at a fixed position on the free surface, defined by a value of the radius r, the vertical displacement is expressed in a different form in three different time intervals. We have

$$t \leqq s_L r \qquad w(r, 0, t) = 0$$

$$s_L r \leqq t \leqq s_T r \qquad w(r, 0, t) = -\frac{Q}{\pi^2}\frac{s_T^2}{\mu}F_1\left(\frac{t}{r}\right)$$

$$s_T r \leqq t \qquad w(r, 0, t) = -\frac{Q}{\pi^2}\frac{s_T^2}{\mu}\left[F_1(s_T)+F_2\left(\frac{t}{r}\right)\right],$$

where

$$F_1\left(\frac{t}{r}\right) = \int_{s_L}^{t/r} \frac{v(v^2-s_L^2)^{\frac{1}{2}}(s_T^2-2v^2)^2(t^2-v^2r^2)^{-\frac{1}{2}}}{(s_T^2-2v^2)^4+16v^4(v^2-s_L^2)(s_T^2-v^2)}dv$$

$$F_2\left(\frac{t}{r}\right) = P\int_{s_T}^{t/r} \frac{v(v^2-s_L^2)^{\frac{1}{2}}(t^2-v^2r^2)^{-\frac{1}{2}}}{(s_T^2-2v^2)^2-4v^2(v^2-s_L^2)^{\frac{1}{2}}(v^2-s_T^2)^{\frac{1}{2}}}dv.$$

Here the symbol P indicates that the principal value of the integral must be taken.

7.12.3. *Special case $\lambda = \mu$*

For the special case $\lambda = \mu$ we have $c_L^2 = 3c_T^2$, or $s_T^2 = 3s_L^2$. The roots of $D(\eta) = 0$ can now be expressed explicitly as

$$\frac{\eta}{s_T} = \pm i\tfrac{1}{2}(3+\sqrt{3})^{\frac{1}{2}}.$$

In terms of the dimensionless time

$$\tau = \frac{t}{s_T r}$$

the displacement at the surface of the half-space can then be written as

$$\tau \leqq \frac{1}{\sqrt{3}}: \qquad w(\tau) = 0$$

$$\frac{1}{\sqrt{3}} \leqq \tau < 1: \qquad w(\tau) = -\frac{Q}{\pi^2}\frac{1}{\mu}\frac{3}{r}G_1(\tau)$$

$$1 \leqq \tau: \qquad w(\tau) = -\frac{Q}{\pi^2}\frac{1}{\mu}\frac{3}{r}[G_1(\tau)+G_2(\tau)],$$

where

$$G_1(\tau) = P\int_{1/\sqrt{3}}^{\tau} \frac{s(s^2-\frac{1}{3})^{\frac{1}{2}}(1-2s^2)^2 ds}{(\tau^2-s^2)^{\frac{1}{2}}(3-24s^2+56s^4-32s^6)}$$

$$G_2(\tau) = P\int_{1}^{\tau} \frac{s^3(s^2-1)^{\frac{1}{2}}(4s^2-\frac{4}{3})ds}{(\tau^2-s^2)^{\frac{1}{2}}(3-24s^2+56s^4-32s^6)}.$$

In this form the solution was obtained by Pekeris[29], who also further evaluated the integrals. According to Pekeris the vertical displacement can be expressed as

[29] C. L. Pekeris, fn. 28, p. 310.

$$\tau \leqq \frac{1}{\sqrt{3}}: \quad w(\tau) = 0 \tag{7.238}$$

$$\frac{1}{\sqrt{3}} \leqq \tau \leqq 1:$$

$$w(\tau) = \frac{Q}{32\pi}\frac{1}{\mu}\frac{1}{r}\left\{6 - \frac{\sqrt{3}}{(\tau^2-\frac{1}{4})^{\frac{1}{2}}} - \frac{(3\sqrt{3}+5)^{\frac{1}{2}}}{(\frac{3}{4}+\frac{1}{4}\sqrt{3}-\tau^2)^{\frac{1}{2}}} + \frac{(3\sqrt{3}-5)^{\frac{1}{2}}}{(\tau^2+\frac{1}{4}\sqrt{3}-\frac{3}{4})^{\frac{1}{2}}}\right\} \tag{7.239}$$

$$1 \leqq \tau < \tfrac{1}{2}(3+\sqrt{3})^{\frac{1}{2}} \quad w(\tau) = \frac{Q}{16\pi}\frac{1}{\mu}\frac{1}{r}\left\{6 - \frac{(3\sqrt{3}+5)^{\frac{1}{2}}}{(\frac{3}{4}+\frac{1}{4}\sqrt{3}-\tau^2)^{\frac{1}{2}}}\right\} \tag{7.240}$$

$$\tau \geqq \tfrac{1}{2}(3+\sqrt{3})^{\frac{1}{2}} \quad w(\tau) = \frac{3Q}{8\pi}\frac{1}{\mu}\frac{1}{r}. \tag{7.241}$$

In these expressions $\tau = \frac{1}{2}(3+\sqrt{3})^{\frac{1}{2}}$ is the time of arrival of the disturbance which travels with the velocity of Rayleigh waves. This disturbance is usually called the Rayleigh wave.

The vertical displacement $(\pi\mu r/Q)w(\tau)$ is plotted versus τ in figure 7.18.

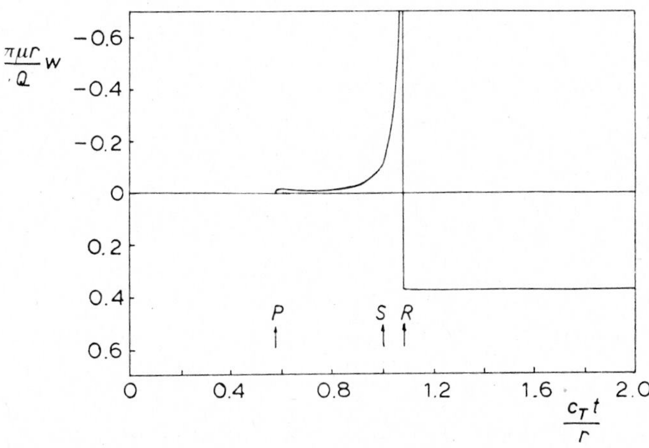

Fig. 7.18. Vertical displacement at the free surface according to eqs. (7.238)–(7.241). P denotes arrival time of the longitudinal wave, S of the transverse wave, and R of the Rayleigh wave. (After Pekeris.)

For a fixed value of r this figure shows the displacement as a function of time. We see that the first disturbance arrives at $t = r/c_L$. At $t = r/c_T$ the displacement shows a discontinuity in the slope of $w(\tau)$. The displacement becomes infinite at $t = r/c_R$, where c_R is the velocity of Rayleigh waves. For $t > r/c_R$ the displacement remains constant and equal to the static solution.

Although the evaluation of the integrals is somewhat more complicated, the radial displacement can be computed in a similar manner. The expressions can be found in the previously cited paper by Pekeris. The propagation of transient waves generated by a normal point load which is suddenly applied and then moves rectilinearly at a constant speed along the surface was investigated by Gakenheimer and Miklowitz.[30] The displacements are derived for the interior of the half-space and for all load speeds. For the limitcase of zero load velocity, which corresponds to the problem discussed in this section, curves showing the displacements in the interior of the half-space were presented by Gakenheimer.[31]

The wave motion generated in a half-space by a tangential point load was treated by Chao.[32]

7.13. Surface waves generated by a normal point load

The expressions for the normal displacement, eqs. (7.238)–(7.241), show that the largest disturbances at the free surface arrive with the velocity of Rayleigh waves. The expressions also show a geometrical attenuation with increasing r. In view of these results it is to be expected that at a large distance from the point of application of the load only the disturbances arriving with the velocity of surface waves will be of appreciable magnitude. This is well confirmed by experiments and seismological experience.

Analytically the Rayleigh wave effects correspond to the contributions from certain poles in the integrands of the inversion integrals. These contributions are generally not difficult to extract from the integral representation of the complete solution. Thus, it may very well be possible to find a simple expression for the surface wave for cases that closed-form expressions for the complete solution are not attainable.

For the dynamic response of a half-space to a point load we will examine the displacements and the stresses corresponding to the surface wave. These surface wave effects were investigated by Chao et al.[33]

By employing eqs. (7.233) and (7.234) and the formal inversion integrals, which are defined by eqs. (7.21) and (7.32), the displacements may be written as

[30] D. C. Gakenheimer and J. Miklowitz, *Journal of Applied Mechanics* **36** (1969) 505.

[31] D. C. Gakenheimer, *Journal of Applied Mechanics* **37** (1970) 522.

[32] C. C. Chao, *Journal of Applied Mechanics* **27** (1960) 559.

[33] C. C. Chao, H. H. Bleich and J. Sackman, *Journal of Applied Mechanics* **28** (1961) 300.

$$w = \frac{Q}{4\pi^2 \mu i}\int_0^\infty J_0(\xi r)\xi\,\mathrm{d}\xi \int_{\varepsilon-i\infty}^{\varepsilon+i\infty} \left[(2\xi^2+s_T^2 p^2)e^{-\alpha z} - 2\xi^2 e^{-\beta z}\right]\frac{\alpha}{D_H}\, e^{pt}\,\frac{\mathrm{d}p}{p} \tag{7.242}$$

$$u = \frac{Q}{4\pi^2 \mu i}\int_0^\infty J_1(\xi r)\xi^2\mathrm{d}\xi \int_{\varepsilon-i\infty}^{\varepsilon+i\infty} \left[(2\xi^2+s_T^2 p^2)e^{-\alpha z} - 2\alpha\beta e^{-\beta z}\right]\frac{1}{D_H}\, e^{pt}\,\frac{\mathrm{d}p}{p}. \tag{7.243}$$

In the complex p-plane poles are located at the zeros of the equation $D_H = 0$. This equation has two zeros corresponding to Rayleigh waves. For the special case $\lambda = \mu$, when $c_L^2 = 3c_T^2$, these roots can be written out explicitly:

$$p = \pm \frac{i\xi}{s_T \gamma}, \qquad \text{where} \quad \gamma = \tfrac{1}{2}(3+\sqrt{3})^{\frac{1}{2}}.$$

By computing the residues in the standard manner the contributions from the Rayleigh poles in the p-plane can now easily be determined. Let us just illustrate the computations by considering the term

$$\frac{1}{2\pi i}\int_0^\infty J_0(\xi r)\xi\,\mathrm{d}\xi \int_{\varepsilon-i\infty}^{\varepsilon+i\infty} \frac{\alpha(2\xi^2+s_T^2 p^2)}{D_H}\, e^{-\alpha z+pt}\,\frac{\mathrm{d}p}{p}.$$

The contributions from the Rayleigh poles are

$$I_1 = \int_0^\infty \xi J_0(\xi r)\left[\frac{\alpha(2\xi^2+s_T^2p^2)}{p\,\partial D_H/\partial p}\, e^{-\alpha z+pt}\right]_{p=+i\xi/s_T\gamma} \mathrm{d}\xi + \int_0^\infty \xi J_0(\xi r)\left[\frac{\alpha(2\xi^2+s_T^2p^2)}{p\,\partial D_H/\partial p}\, e^{-\alpha z+pt}\right]_{p=-i\xi/s_T\gamma} \mathrm{d}\xi.$$

Some simple computations show that

$$\left[p\,\frac{\partial D_H}{\partial p}\right]_{p=\pm i\xi/s_T\gamma} = -\frac{8\xi^4}{\gamma^2\sqrt{3}}.$$

We introduce

$$m = \frac{1}{\gamma}\left[\tau + i\left(\frac{z}{r}\right)(\gamma^2-\tfrac{1}{3})^{\frac{1}{2}}\right],$$

where $\tau = t/s_T r$. The integral then becomes

$$I_1 = -\frac{(3\gamma^2-1)^{\frac{1}{2}}(2\gamma^2-1)}{4\gamma}\,\mathscr{R}\int_0^\infty J_0(\xi r)e^{im\xi r}\mathrm{d}\xi.$$

The integral over the Bessel function can be evaluated[34] and we obtain

$$I_1 = \frac{(3\gamma^2-1)^{\frac{1}{2}}(1-2\gamma^2)}{4\gamma}\,\frac{1}{r}\,\mathscr{R}\left[\frac{1}{(1-m^2)^{\frac{1}{2}}}\right].$$

This result agrees with what was derived by Chao et al.

The second term in (7.242) as well as the two terms in (7.243) can be evaluated similarly. In this manner the contributions from the surface disturbance to $w(r, z, t)$ and $u(r, z, t)$ were obtained by Chao et al. as

$$w = \frac{Q}{\pi\mu r}\,\frac{(3\gamma^2-1)^{\frac{1}{2}}}{8\gamma}\,\mathscr{I}\left\{\frac{1-2\gamma^2}{(m^2-1)^{\frac{1}{2}}}+\frac{2\gamma^2}{(n^2-1)^{\frac{1}{2}}}\right\} \tag{7.244}$$

$$u = -\frac{Q}{\pi\mu r}\,\frac{\sqrt{3}}{8}\,\mathscr{R}\left\{(1-2\gamma^2)\left[1-\frac{m}{(m^2-1)^{\frac{1}{2}}}\right]\right.$$
$$\left.+2(\gamma^2-1)^{\frac{1}{2}}(\gamma^2-\tfrac{1}{3})^{\frac{1}{2}}\left[1-\frac{n}{(n^2-1)^{\frac{1}{2}}}\right]\right\}, \tag{7.245}$$

where $n = (1/\gamma)[\tau+i(z/r)(\gamma^2-1)^{\frac{1}{2}}]$.

Since only the Rayleigh phase is being considered, these expressions are applied only to shallow depths ($z/r \ll 1$) and to values of the time near the arrival time of the Rayleigh wave ($\tau \doteq \gamma$). Using these restrictions Chao et al. simplified (7.244) and (7.245) to the final form

$$w = K_1(\gamma^2-\tfrac{1}{3})^{\frac{1}{2}}\mathscr{I}[(1-2\gamma^2)Z_1^{-\frac{1}{2}}+2\gamma^2Z_2^{-\frac{1}{2}}] \tag{7.246}$$

$$u = K_1\gamma\mathscr{R}[(1-2\gamma^2)Z_1^{-\frac{1}{2}}+2(\gamma^2-1)^{\frac{1}{2}}(\gamma^2-\tfrac{1}{3})^{\frac{1}{2}}Z_2^{-\frac{1}{2}}], \tag{7.247}$$

where

$$K_1 = \frac{Q}{\pi\mu}\,\frac{\sqrt{3}}{8\sqrt{2}}\left(\frac{1}{\gamma}\right)^{\frac{1}{2}}\left(\frac{1}{rz}\right)^{\frac{1}{2}}$$

$$Z_1 = \frac{r(\tau-\gamma)}{z}+i(\gamma^2-\tfrac{1}{3})^{\frac{1}{2}}$$

$$Z_2 = \frac{r(\tau-\gamma)}{z}+i(\gamma^2-1)^{\frac{1}{2}}.$$

From these expressions for the displacements, expressions for the stress components may be derived by routine differentiations; where $z/r \ll 1$, and $\tau \doteq \gamma$ are used again. Here we list just $\tau_z(r, z, t)$; the other stress components can be found in the article by Chao et al.

[34] See N. W. McLachlan, *Bessel functions for engineers*. London, Oxford University Press (1934), p. 160, nr. 44.

$$\tau_z = K_2(2\gamma^2 - 1)^2 \mathscr{R}[Z_1^{-\frac{3}{2}} - Z_2^{-\frac{3}{2}}], \tag{7.248}$$

where

$$K_2 = \frac{Q}{\pi} \frac{\sqrt{3}}{16\sqrt{2}} \left(\frac{1}{\gamma}\right)^{\frac{1}{2}} \left(\frac{1}{r}\right)^{\frac{1}{2}} \left(\frac{1}{z}\right)^{\frac{3}{2}}.$$

It is noted that the expressions for the displacements and the stress are in terms of a single parameter, the nondimensional time τ. In figure 7.19 the stress τ_z is shown as a function of $r(\tau - \gamma)/z$.

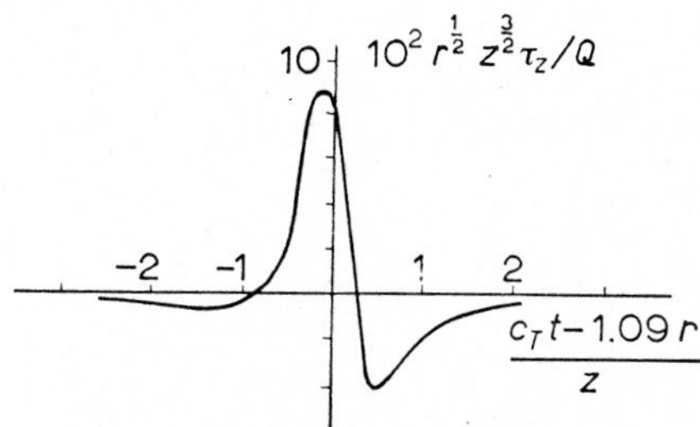

Fig. 7.19. Surface effect.

As can be seen from (7.246)–(7.248), the surface effects attenuate with distance from the point of application as $r^{-\frac{1}{2}}$, as opposed to the other components of the solution which decay as r^{-1}. The physical explanation is that the surface effects are essentially cylindrical waves, while the body waves are essentially spherical waves. The difference in geometrical attenuation is the reason why at a large distance from the point of application the surface effects form the major part of the propagating disturbance.

7.14. Problems

7.1. The wave motion generated in an initially undisturbed elastic half-space by the application of a spatially uniform surface pressure $p(t)$ was analyzed in section 1.3. Re-derive the solution by applying the one-sided Laplace transform with respect to time.

7.2. A spherical cavity in an unbounded elastic medium is rapidly pressurized. The governing equations are stated in section 4.3. Obtain the solution for the potential $\varphi(r, t)$ by applying the one-sided Laplace transform with respect to time.

7.3. An unbounded medium is subjected to a time-harmonic concentrated line load which may be represented by

$$\boldsymbol{F} = \boldsymbol{j}Q\delta(x)\delta(y)\sin \omega t.$$

Examine the steady-state displacement response by employing the exponential Fourier transform over the coordinate x.

(a) Evaluate the inversion integrals rigorously.

(b) Employ the method of steepest descent to obtain approximations for the integrals.

7.4. Suppose an unbounded elastic medium is subjected to an antiplane line load. The line load generates horizontally polarized shear motion of the medium. The motion is governed by

$$\frac{\partial^2 w}{\partial x^2} + \frac{\partial^2 w}{\partial y^2} + \delta(x)\delta(y)\mathrm{f}(t) = \frac{1}{c_T^2}\frac{\partial^2 w}{\partial t^2},$$

where $\mathrm{f}(t)$ determines the strength of the line source as a function of time; it is assumed that $\mathrm{f}(t) = 0$ for $t < 0$. It is further assumed that the medium is at rest prior to $t = 0$.

Determine $w(x, y, t)$ by means of the Cagniard-de Hoop method.

7.5. An elastic half-space is subjected to a uniform pressure distribution over half of its surface, i.e.,

$$y = 0 \qquad \tau_y = -QH(x)\mathrm{f}(t)$$

$$\tau_{yx} = \tau_{yz} \equiv 0.$$

Determine the displacement component v on the surface $y = 0$ for $x < 0$.

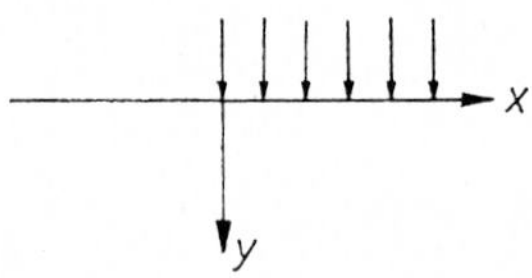

7.6. A quarter-space is subjected to a concentrated point load which is applied at a distance a from the edge. The boundary conditions on the surface $x = 0$ are

$$\tau_{xz} = \tau_{xy} \equiv 0 \quad \text{and} \quad u = 0.$$

The Poisson's ratio of the material is $\nu = 0.25$. Determine the vertical displacement at $x = 0$, $y = 0$.

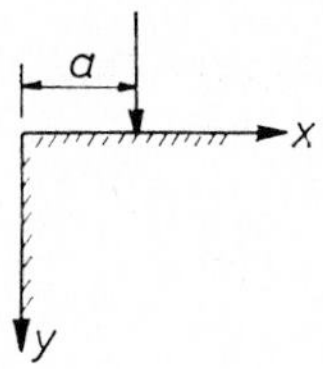

7.7. An initially undisturbed unbounded medium is subjected to a point load. State the problem in cylindrical coordinates and use the Hankel transform and the one-sided Laplace transform to determine the displacement in the radial direction. Invert the transforms by means of the Cagniard-de Hoop method. Check the result with the expression stated in problem 3.3.

7.8. A circular cylindrical hole of radius a in an unbounded medium is subjected to shear tractions in the plane $z = 0$. The shear tractions are uniformly distributed along the circumference. The boundary conditions at $r = a$ are

$$\tau_{r\theta} = \tau_0 \, \delta(z) H(t)$$
$$\tau_{rr} = \tau_{rz} \equiv 0.$$

Prior to time $t = 0$ the medium is at rest.

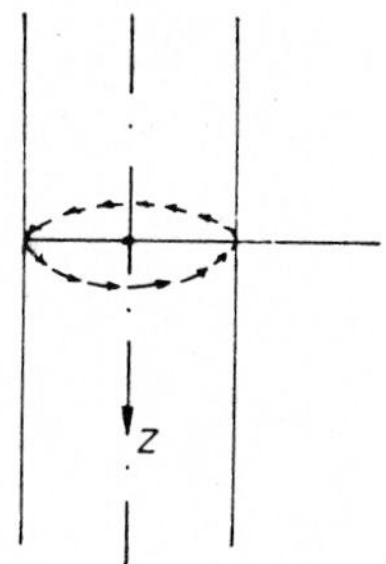

(a) Show that the wave motion involves the circumferential displacement $v(r, z, t)$ only.

(b) Write the equation governing $v(r, z, t)$ and solve the problem by application of integral transform techniques.

7.9. The surface of a circular cylindrical cavity of radius a in an unbounded medium is subjected to an antiplane shear load which is independent of the axial coordinate. The boundary conditions are $(r = a)$

$$\tau_{rz} = \tau_0\, \delta(\theta) H(t)$$
$$\tau_{rr} = \tau_{r\theta} \equiv 0.$$

The medium is initially at rest. Find expressions for the axial displacement $w(r, \theta, t)$.

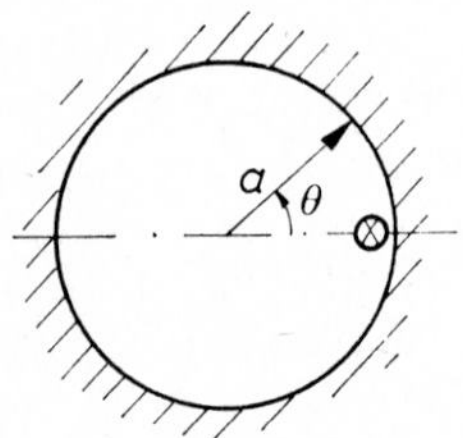

Hint: Note the symmetry with respect to the plane $\theta = 0$. Restate the problem for the domain $-\pi \leqq \theta \leqq \pi$ and relate the problem to a problem for the region $-\infty < \theta < \infty$.

7.10. Consider two quarter-spaces of distinct solids which are perfectly joined along a common boundary defined by $x = 0$. The surface $y = 0$ is subjected to surface tractions of the forms

$$x \leqq 0 \qquad \tau_{yz} = T_1\, H(t)$$
$$x \geqq 0 \qquad \tau_{yz} = T_2\, H(t).$$

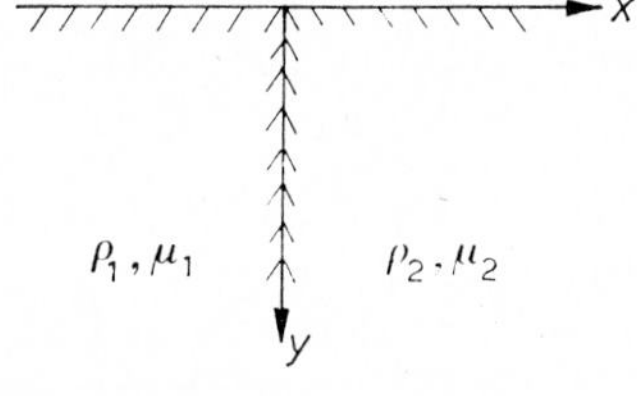

Define

$$\eta = \frac{y}{(c_T)_2\, t} \qquad m = \frac{(c_T)_1}{(c_T)_2} \qquad k = \frac{\mu_2}{\mu_1}$$

and show that the interface stress at $x = 0$ is given by

$0 < \eta \leqq m$:

$$\frac{\pi}{2}\frac{\tau_{xz}}{T_1} = -c \ln \frac{1+(1-\eta^2)^{\frac{1}{2}}}{\eta} + \frac{c(a^2-b^2)}{2b(m^2-b^2)^{\frac{1}{2}}}\left\{\arcsin \frac{b+m\eta}{m+b\eta} + \arcsin \frac{b-m\eta}{m-b\eta}\right\}$$

$$+ck \ln \frac{m+(m^2-\eta^2)^{\frac{1}{2}}}{\eta} - \frac{ck(a^2-b^2)}{2b(1-b^2)^{\frac{1}{2}}}\left\{\arcsin \frac{mb+\eta}{m+b\eta} + \arcsin \frac{mb-\eta}{m-b\eta}\right\},$$

$m \leqq \eta \leqq 1$:

$$\frac{\pi}{2}\frac{\tau_{xz}}{T_1} = -c \ln \frac{1+(1-\eta^2)^{\frac{1}{2}}}{\eta} - \frac{c(a^2-b^2)}{2b(m^2-b^2)^{\frac{1}{2}}}\left\{\arcsin \frac{b-m\eta}{b\eta-m} - \arcsin \frac{b+m\eta}{b\eta+m}\right\},$$

where

$$a^2 = \frac{km^2-n}{k-n}, \qquad b^2 = \frac{k^2m^2-1}{k^2-1}, \quad c = \frac{k-n}{k^2-1}, \qquad n = \frac{T_2}{T_1}.$$

Examine the singularity at $\eta = 0$.

CHAPTER 8

TRANSIENT WAVES IN LAYERS AND RODS

8.1. General considerations

The simplicity and elegance of the analysis of transient waves in unbounded media does not extend to bodies of finite dimensions. The complications are caused by the reflections of the wave motion at the bounding surfaces. For horizontally polarized shear waves reflections can still be analyzed in a simple manner on the basis of considerations of symmetry or antisymmetry with respect to the reflecting surface. It is, however, implicit in the discussion of section 3.11 that symmetry considerations are of no use for the reflection of more general types of wave motions, except if the physically less realistic mixed conditions are assumed to hold at the reflecting surface. Consequently it is rather complicated to determine the transient wave motion in simple waveguides such as layers and rods where a myriad of reflections take place. Simple closed form expressions for the field variables generally can be obtained only for short times, or for long times and at large distances of the external disturbances, or for any time but then only near the wavefront separating the disturbed from the undisturbed part of the layer. The short-time solutions are obtained by tracing a few of the reflections. For the long-time and far-field solutions it is assumed that a steady-state pattern has been established across the cross section of the waveguide. Within the context of transient motion in plane strain of a layer the near- and far-field solutions were recently reviewed by Miklowitz.[1]

In this section we will consider transient waves in a layer and in a rod. It is instructive to start with an analysis of two-dimensional antiplane shear motions in a layer. For shear waves the expressions for the field variables, which are simple in form, provide some interesting insights in the forced motions of waveguides. In subsequent sections we consider the more complicated cases of wave motions in plane strain in a layer and axially symmetric wave motions in a rod.

[1] J. Miklowitz, in: *Wave propagation in solids*, ed. by J. Miklowitz. American Society of Mechanical Engineers (1969), p. 44.

8.2. Forced shear motions of a layer

In an (x, y, z)-coordinate system a semi-infinite layer occupies the domain defined by $x \geqq 0$, $-h \leqq y \leqq h$, and $-\infty < z < \infty$. The geometry is

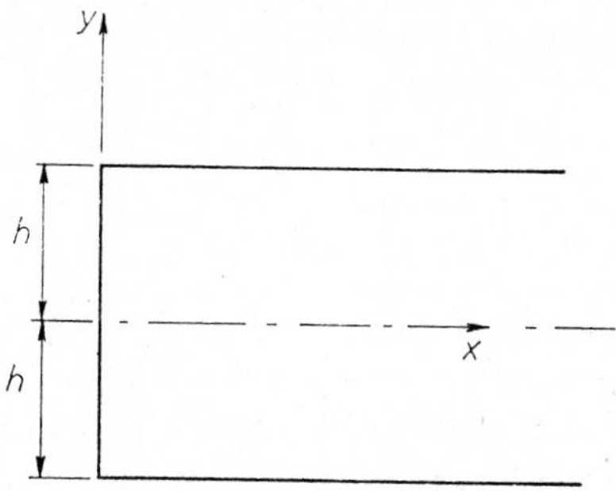

Fig. 8.1. Semi-infinite layer.

shown in figure 8.1. We consider a layer which is free of tractions on the faces of the layer, i.e.,

$$\text{at} \quad y = \pm h \qquad \tau_{yz} = \mu \frac{\partial w}{\partial y} \equiv 0. \tag{8.1}$$

Prior to the application of external disturbances the layer is at rest

$$w(x, y, 0) = \dot{w}(x, y, 0) \equiv 0 \qquad \text{for} \quad x > 0.$$

Suppose that the semi-infinite layer is subjected at $x = 0$ to time-varying shear tractions (at $x = 0$):

$$\tau_{xz} = \mu \frac{\partial w}{\partial x} = T(y)\mathrm{f}(t), \tag{8.2}$$

where $\mathrm{f}(t) \equiv 0$ for $t < 0$.

The dynamic response of the layer can be computed from the response of an unbounded medium by invoking symmetry considerations, as discussed in section 3.11 and illustrated in figure 3.3 for the case of a line load. In that manner the field variables are obtained as infinite series, each higher-order term representing a reflected wave. Only for very small times, or very close to the wavefront separating the disturbed from the undisturbed part of the layer, will the series consist of a small and therefore manageable number of terms.

Alternatively the field variables can be expressed as a superposition of terms which show a trigonometric dependence on the thickness coordinate y, and which are selected to satisfy the boundary conditions on the faces $y = \pm h$. These superpositions imply the assumption that a stationary pattern has been established across the thickness of the layer. The superpositions can be obtained by superposing modes of wave propagation for a time-harmonic excitation, or by application of integral transform techniques. As a preliminary it is instructive to examine the steady-state response of the layer.

8.2.1. *Steady-state harmonic motions*

The results of chapter 6 show that a rather special external disturbance must be applied to produce a specific single mode of wave propagation in a waveguide. Consider, for example, a symmetric mode of horizontally polarized shear motion in a layer,

$$w_n = A_n \cos\left(\frac{n\pi y}{2h}\right) \exp\left[i(k_n x - \omega t)\right], \tag{8.3}$$

where $n = 0, 2, 4, \ldots$, and

$$k_n = \left[\left(\frac{\omega}{c_T}\right)^2 - \left(\frac{n\pi}{2h}\right)^2\right]^{\frac{1}{2}}. \tag{8.4}$$

It is evident that (8.3) can represent the steady-state wave motion generated in a semi-infinite layer $x \geqq 0$ by a displacement applied at $x = 0$, of harmonic time-dependence with frequency ω, and distributed across the thickness as $\cos(n\pi y/2h)$. Whether this particular mode propagates unattenuated or is evanescent depends on whether k is real or imaginary, which in turn depends on the forcing frequency, the mode number n and the parameters c_T and h, as can be seen from eq. (8.4). If the externally applied displacement is time-harmonic, but distributed across the layer thickness in a manner which does not coincide with any of the trigonometric mode functions, an infinite number of modes is excited simultaneously. The relative amplitudes of each mode can be determined by a simple Fourier analysis.

Let us consider the case that $T(y)$ is symmetric with respect to $y = 0$, but otherwise arbitrary. It is then to be expected that the traveling wave can be expressed by a summation over the symmetric modes

$$w(x, y, t) = \sum_{n=0, 2, 4, \ldots} A_n \cos\left(\frac{n\pi y}{2h}\right) \exp\left[i(k_n x - \omega t)\right], \tag{8.5}$$

where A_n is the amplitude of each mode and k_n is defined by (8.4). Thus given ω, the wavenumber of each mode, which may be real or imaginary, can be computed from (8.4). The larger the forcing frequency, the more propagating modes are excited.

The solution (8.5) clearly satisfies the condition (8.1) that the surfaces $y = \pm h$ are free of tractions. The expression for τ_{xz} corresponding to (8.5) may be written as

$$\tau_{xz}(x, y, t) = \sum_{n=0,\, 2,\, 4, \ldots,} B_n \cos\left(\frac{n\pi y}{2h}\right) \exp\left[i(k_n x - \omega t)\right]. \tag{8.6}$$

Thus, we find that (8.2) is satisfied if

$$T(y) = \sum_{n=0,\, 2,\, 4, \ldots,} B_n \cos\left(\frac{n\pi y}{2h}\right). \tag{8.7}$$

To determine the coefficients B_n both sides of this equation are multiplied by $\cos(m\pi y/2\pi)$, where m is an integer, and both sides are subsequently integrated between $y = -h$ and $y = +h$. Employing the orthogonality of the trigonometric functions, we find

$$B_0 = \frac{1}{h}\int_0^h T(y)\mathrm{d}y \tag{8.8}$$

$$B_n = \frac{2}{h}\int_0^h T(y)\cos\left(\frac{n\pi y}{2h}\right)\mathrm{d}y, \qquad n = 2, 4, \ldots, \tag{8.9}$$

where the symmetry with respect to $y = 0$ has also been employed. For the special case $T(y) = T_0$ the integrals in (8.9) vanish. Only the lowest mode is then generated, and $B_0 = T_0$.

As an example we consider

$$T(y) = \left[\alpha - \frac{\beta}{h}|y|\right] T_0, \tag{8.10}$$

where $\beta \leqq \alpha$. The coefficients are obtained as

$$B_0 = (\alpha - \tfrac{1}{2}\beta)T_0$$

$$B_n = \left(\frac{4}{n\pi}\right)^2 \beta T_0 \qquad n = 2, 6, 10, \ldots,$$

$$B_n = 0 \qquad n = 4, 8, 12, \ldots,$$

The relative importance of the higher modes depends to some extent on the ratio α/β. If $\alpha/\beta > 1$, we have $B_0 > \frac{1}{2}\beta T_0$, and B_n/B_0 is smaller than $2(4/n\pi)^2$.

8.2.2. *Transient motions*

One way of determining the transient motion is based on the representation of the function $f(t)$ appearing in eq. (8.1) by a Fourier integral. Analogously to the examples treated in sections 1.9 and 4.3 the induced wave motion then takes the form of a summation over integrals, where the frequency ω is the integration variable. Although these integrals can be evaluated we will not pursue that approach. Here a somewhat more direct approach is employed, whereby at the outset the displacement is assumed in the form

$$w(x, y, t) = \sum_{n=0,2,4,\ldots} W_n(x, t) \cos\left(\frac{n\pi y}{2h}\right). \tag{8.11}$$

This expression satisfies the boundary conditions (8.1) at $y = \pm h$. Substitution of the series (8.11) into the governing equation for horizontally polarized shear motion shows that $W_n(x, t)$ must satisfy the partial differential equation

$$\frac{\partial^2 W_n}{\partial x^2} - \left(\frac{n\pi}{2h}\right)^2 W_n = \frac{1}{c_T^2}\frac{\partial^2 W_n}{\partial t^2}. \tag{8.12}$$

Let us now apply the one-sided Laplace transform over time t, to obtain

$$\frac{d^2 \overline{W}_n}{dx^2} - \left[\left(\frac{p}{c_T}\right)^2 + \left(\frac{n\pi}{2h}\right)^2\right] \overline{W}_n = 0. \tag{8.13}$$

Assuming that $T(y)$ can be represented by

$$T(y) = \sum_{n=0,2,4,\ldots} B_n \cos\left(\frac{n\pi y}{2h}\right), \tag{8.14}$$

the solution of eq. (8.13) which displays the appropriate behavior at $x = 0$ and $x \to \infty$ is

$$\overline{W}_n(x, p) = -\frac{c_T B_n \bar{f}(p)}{\mu(p^2 + a_n^2)^{\frac{1}{2}}} e^{-(p^2+a_n^2)^{\frac{1}{2}}(x/c_T)}, \tag{8.15}$$

where

$$a_n = \frac{n\pi c_T}{2h}. \tag{8.16}$$

In a table of Laplace transforms[2] the inverse of

$$\frac{e^{-(p^2+a_n)^{\frac{1}{2}}(x/c_T)}}{(p^2+a_n^2)^{\frac{1}{2}}}$$

can be found as

$$0 \quad \text{when} \quad 0 < t < \frac{x}{c_T}$$

$$J_0\left[\left(t^2-\frac{x^2}{c_T^2}\right)^{\frac{1}{2}} a_n\right] \quad \text{when} \quad t \geqq \frac{x}{c_T}.$$

In view of the convolution theorem for one-sided Laplace transforms $W_n(x, t)$ can thus be written as

$$W_n(x, t) = -\frac{c_T}{\mu} B_n \int_{x/c_T}^{t} \mathrm{f}(t-s) J_0\left[\left(s^2-\frac{x^2}{c_T^2}\right)^{\frac{1}{2}} a_n\right] \mathrm{d}s.$$

For the special case of an impulsively applied stress, when

$$\mathrm{f}(t) = \delta(t),$$

the displacement for $t > x/c_T$ may be expressed as

$$w(x, y, t) = -\frac{c_T}{\mu} \sum_{n=0,2,4} B_n J_0\left[\left(t^2-\frac{x^2}{c_T^2}\right)^{\frac{1}{2}} a_n\right] \cos\left(\frac{n\pi y}{2h}\right).$$

This solution shows that the displacement pulse is subjected to dispersion. At the wavefront, where $t = x/c_T$, the magnitude remains unchanged. Due to the unrealistic nature of the forcing function the displacement is discontinuous, with a jump of magnitude

$$[w] = -\frac{c_T}{\mu} T(y).$$

8.3. Transient in-plane motion of a layer

For the study of impact phenomena in layers it is of great interest to investigate the transient waves generated in a layer by the application of a normal line load of rapid time variation. The geometry is shown in figure 8.2. If the line load is independent of the z-coordinate, the induced wave

[2] *Tables of integral transforms*, ed. by A. Erdélyi et al., Vol. 1. New York, McGraw-Hill Book Co. (1953), p. 248, No. 24.

motion is in plane strain. The governing equations then are the same as those used in section 7.7 for the investigation of waves generated in a half-space by a line load.

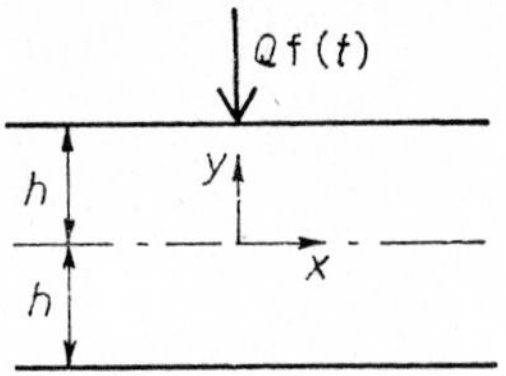

Fig. 8.2. Line load on a layer.

8.3.1. *Method of solution*

As the problem is depicted in figure 8.2 the boundary conditions are

$$\text{at } y = +h: \qquad \tau_y = -Qf(t)\delta(x), \qquad \tau_{yx} = 0 \tag{8.17}$$

$$\text{at } y = -h: \qquad \tau_y = 0, \qquad \tau_{yx} = 0. \tag{8.18}$$

Rather than work with the boundary conditions (8.17) and (8.18), we decompose the problem into two separate problems with inhomogeneous boundary conditions on both $y = +h$ and $y = -h$. The decomposition is indicated in figure 8.3. It is clear that the problem depicted in figure 8.3(A) is antisymmetric with respect to $y = 0$, while the problem of figure 8.3(B) is symmetric with respect to $y = 0$.

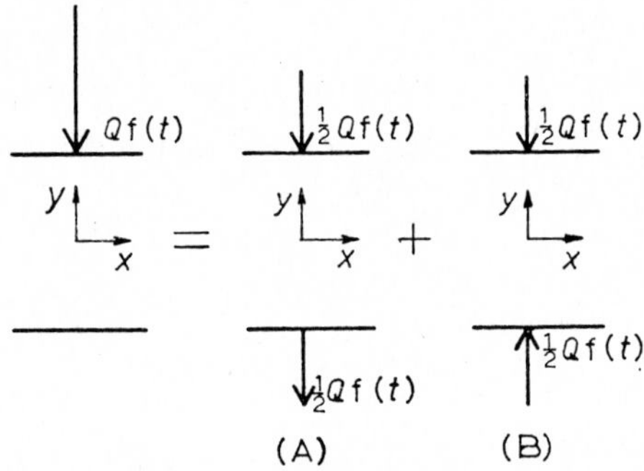

Fig. 8.3. Decomposition into antisymmetric and symmetric problems.

The respective boundary conditions are

Antisymmetric problem, figure 8.3(A):

$$y = +h: \qquad \tau_y = -\tfrac{1}{2}Qf(t)\delta(x), \qquad \tau_{yx} = 0 \tag{8.19}$$

$$y = -h: \qquad \tau_y = \tfrac{1}{2}Qf(t)\delta(x), \qquad \tau_{yx} = 0 \tag{8.20}$$

Symmetric problem, figure 8.3(B):

$$y = +h: \qquad \tau_y = -\tfrac{1}{2}Qf(t)\delta(x), \qquad \tau_{yx} = 0 \tag{8.21}$$

$$y = -h: \qquad \tau_y = -\tfrac{1}{2}Qf(t)\delta(x), \qquad \tau_{yx} = 0. \tag{8.22}$$

Since the layer is at rest prior to time $t = 0$, the initial conditions are

$$\varphi(x, y, 0) = \dot{\varphi}(x, y, 0) = \psi(x, y, 0) = \dot{\psi}(x, y, 0) \equiv 0. \tag{8.23}$$

The two problems are solved by applying integral transform techniques. We apply the one-sided Laplace transform with respect to time and the exponential Fourier transform with respect to the variable x. These transforms are defined by eqs. (7.20) and (7.7), respectively. Denoting the Laplace-Fourier transforms of $\varphi(x, y, t)$ and $\psi(x, y, t)$ by $\bar{\varphi}^*(\xi, y, p)$ and $\bar{\psi}^*(\xi, y, p)$, respectively, we find from (7.108) and (7.109)

$$\frac{d^2\bar{\varphi}^*}{dy^2} - (\xi^2 + s_L^2 p^2)\bar{\varphi}^* = 0$$

$$\frac{d^2\bar{\psi}^*}{dy^2} - (\xi^2 + s_T^2 p^2)\bar{\psi}^* = 0,$$

where p is the Laplace transform parameter and ξ is the parameter of the exponential Fourier transform. For a layer appropriate solutions of these equations are

$$\bar{\varphi}^*(\xi, y, p) = A_1 \sinh(\alpha y) + A_2 \cosh(\alpha y) \tag{8.24}$$

$$\bar{\psi}^*(\xi, y, p) = B_1 \sinh(\beta y) + B_2 \cosh(\beta y), \tag{8.25}$$

wherein

$$\alpha = (\xi^2 + s_L^2 p^2)^{\frac{1}{2}}, \qquad \beta = (\xi^2 + s_T^2 p^2)^{\frac{1}{2}}. \tag{8.26a, b}$$

The expressions for $\bar{\varphi}^*(\xi, y, p)$ and $\bar{\psi}^*(\xi, y, p)$ show some obvious similarities with the expressions for the potentials $\Phi(x_2)$ and $\Psi(x_2)$ which appear in the study of free time-harmonic wave motion of a layer, see eqs. (6.61) and (6.62).

For the solution of the antisymmetric problem we obviously need the terms of $\bar{\varphi}^*$ and $\bar{\psi}^*$ which correspond to antisymmetric motions. These are

$$\bar{\varphi}^* = A_1 \sinh(\alpha y), \qquad \bar{\psi}^* = B_2 \cosh(\beta y). \tag{8.27a, b}$$

Upon application of the Laplace and the exponential Fourier transforms to eqs. (7.112), (7.113), (8.19) and (8.20), we find that the boundary conditions (8.19) and (8.20) are satisfied if

$$[\lambda(-\xi^2+\alpha^2)+2\mu\alpha^2]A_1 \sinh(\alpha h)+2i\mu\xi\beta B_2 \sinh(\beta h) = -\tfrac{1}{2}Q\bar{f}(p)$$
$$-2i\xi\alpha A_1 \cosh(\alpha h)+(\xi^2+\beta^2)B_2 \cosh(\beta h) = 0.$$

The solutions for A_1 and B_2 are

$$A_1 = -\frac{1}{2}\frac{Q}{\mu}\frac{(\xi^2+\beta^2)\cosh(\beta h)}{D_a}\bar{f}(p)$$

$$B_2 = -\frac{1}{2}\frac{Q}{\mu}\frac{2i\xi\alpha\cosh(\alpha h)}{D_a}\bar{f}(p),$$

where

$$D_a = (\beta^2+\xi^2)^2 \sinh(\alpha h)\cosh(\beta h)-4\xi^2\alpha\beta\sinh(\beta h)\cosh(\alpha h). \quad (8.28)$$

For the symmetric problem defined by (8.21) and (8.22) the appropriate expressions for φ^* and ψ^* are

$$\bar{\varphi}^* = A_2\cosh(\alpha y), \qquad \bar{\psi}^* = B_1\sinh(\beta y).$$

The symmetric boundary conditions yield

$$[\lambda(-\xi^2+\alpha^2)+2\mu\alpha^2)]A_2 \cosh(\alpha h)+2i\mu\xi\beta B_1 \cosh(\beta h) = -\tfrac{1}{2}Q\bar{f}(p)$$
$$-2i\xi\alpha A_2 \sinh(\alpha h)+(\xi^2+\beta^2)B_1 \sinh(\beta h) = 0.$$

The solutions of these equations are

$$A_2 = -\frac{1}{2}\frac{Q}{\mu}\frac{(\xi^2+\beta^2)\sinh(\beta h)}{D_s}\bar{f}(p) \quad (8.29)$$

$$B_2 = -\frac{1}{2}\frac{Q}{\mu}\frac{2i\xi\alpha\sinh(\alpha h)}{D_s}\bar{f}(p), \quad (8.30)$$

where

$$D_s = (\xi^2+\beta^2)^2\sinh(\beta h)\cosh(\alpha h)-4\xi^2\alpha\beta\sinh(\alpha h)\cosh(\beta h). \quad (8.31)$$

We will work out some of the details of the inversion process for the longitudinal stress τ_x at the free surface $y = +h$. By employing (7.111) the Laplace-Fourier transform of τ_x is obtained as

$$\bar{\tau}_x^* = (\beta^2+\xi^2)\bar{\varphi}^*-2i\mu\xi\frac{\mathrm{d}\bar{\psi}^*}{\mathrm{d}y}.$$

At $y = +h$ we have

Antisymmetric problem:

$$\bar{\tau}_x^* = -\tfrac{1}{2}QN_a(\xi, p)\bar{f}(p)/D_a(\xi, p),$$

where

$$N_a(\xi, p) = (\xi^2+\beta^2)^2 \sinh(\alpha h)\cosh(\beta h)+4\xi^2\alpha\beta\cosh(\alpha h)\sinh(\beta h), \quad (8.32)$$

and $D_a(\xi, p)$ is defined by eq. (8.28).

Symmetric problem:

$$\bar{\tau}_x^* = -\tfrac{1}{2}QN_s(\xi, p)\bar{f}(p)/D_s(\xi, p),$$

where

$$N_s(\xi, p) = (\xi^2+\beta^2)^2 \cosh(\alpha h)\sinh(\beta h)+4\xi^2\alpha\beta\sinh(\alpha h)\cosh(\beta h).$$

8.3.2. *Inversion of the transforms*

By the use of the inversion integrals defined by eqs. (7.8) and (7.21) of chapter 7, formal representations for the stress component $\tau_x(x, y, t)$ can be written as:

Antisymmetric problem:

$$\tau_x = -\frac{Q}{8\pi^2 i}\int_{-\infty}^{\infty} e^{-i\xi x}\mathrm{d}\xi \int_{\gamma-i\infty}^{\gamma+i\infty} \frac{N_a(\xi, p)}{D_a(\xi, p)} e^{pt}\mathrm{d}p \quad (8.33)$$

Symmetric problem:

$$\tau_x = -\frac{Q}{8\pi^2 i}\int_{-\infty}^{\infty} e^{-i\xi x}\mathrm{d}\xi \int_{\gamma-i\infty}^{\gamma+i\infty} \frac{N_s(\xi, p)}{D_s(\xi, p)} e^{pt}\mathrm{d}p. \quad (8.34)$$

In both (8.33) and (8.34) we have used

$$\bar{f}(p) = 1,$$

which corresponds to

$$f(t) = \delta(t),$$

where $\delta(t)$ is the Dirac delta function. Thus from now on we consider the case that the loads are applied impulsively. Stresses generated by loads varying in a more general fashion can be determined by the use of the superposition principle.

It is relatively simple to evaluate the integrals in the p-plane. Inspection of the terms in (8.33) and (8.34) shows that the integrands are even in α and β. Consequently the integrands do not have branch points, even though the radicals α and β are themselves multivalued. The absence of branch points renders the evaluation of the integrals by contour integration in the p-plane a rather straightforward matter. By closing the contour in the left-half plane it follows that the integral along the path defined by $\mathscr{R}(p) = \gamma$

is equal to $2\pi i$ times the sum of the residues, since the integral along the closing contour in the left-half plane vanishes as $|p|$ increases beyond bounds. The positions of the poles in the complex p-plane are the zeros of $D_a(\xi, p)$ and $D_s(\xi, p)$ in eqs. (8.33) and (8.34), respectively.

By introducing the substitution

$$p = i\omega,$$

we find that the equations $D_a(\xi, p) = 0$ and $D_s(\xi, p) = 0$ become

$$(\xi^2 - q^2)^2 \sin(ph)\cos(qh) + 4\xi^2 pq \sin(qh)\cos(ph) = 0 \tag{8.35}$$

and

$$(\xi^2 - q^2)^2 \sin(qh)\cos(ph) + 4\xi^2 pq \sin(ph)\cos(qh) = 0, \tag{8.36}$$

respectively, where

$$p^2 = \frac{\omega^2}{c_L^2} - \xi^2, \qquad q^2 = \frac{\omega^2}{c_T^2} - \xi^2.$$

Eqs. (8.35) and (8.36) are recognized as the Rayleigh-Lamb frequency equations, which were earlier presented in chapter 6 as eqs. (6.69) and (6.68), respectively. These equations and the corresponding frequency spectrum were examined in some detail as part of the study of straight-crested time-harmonic waves in a layer. Thus we know that for every value of ξ, there is an infinite number of solutions for ω_n, corresponding to the frequency of each mode for the wavenumber that is considered. The implications of these results for the evaluation of the integrals in the p-plane are that we have an infinite number of poles located on the imaginary axis at

$$p_n = \pm i\omega_n,$$

where ω_n are the roots of the Rayleigh-Lamb frequency equation, with ξ as a variable. The poles are all simple poles and the residues can be determined by employing a well-established formula. For the antisymmetric problem the resulting expression for $\tau_x(x, h, t)$ is

$$\tau_x = -\frac{Q}{4\pi}\int_{-\infty}^{\infty} e^{-i\xi x}\sum_{n=1}^{\infty}\left\{\left[\frac{e^{pt}N_a}{\partial D_a/\partial p}\right]_{p=i\omega_n} + \left[\frac{e^{pt}N_a}{\partial D_a/\partial p}\right]_{p=-i\omega_n}\right\}\mathrm{d}\xi. \tag{8.37}$$

A completely analogous expression can be derived from (8.34) for the stress τ_x of the symmetric problem.

The further evaluation of (8.37) is still a major task which will be carried out by approximation.

8.3.3. *Application of the method of stationary phase*

An exact analytical evaluation of the integral (8.37) representing τ_x as a function of x and t does not appear to be possible. The two alternatives are numerical integration or approximate evaluation of the integral by one of the methods discussed in chapter 7. By the use of an electronic computer numerical evaluation is very well feasible although it requires a rather extensive programming effort and also considerable computer time. If one wishes to have detailed information on $\tau_x(x, h, t)$ over the whole range of time, numerical integration is, however, a necessity. If, on the other hand, it is sufficient to know the stress at relatively large values of time, a simple approximation can be obtained by employing the method of stationary phase.

Since $p^{-1}\partial D_a/\partial p$ and N_a are clearly even functions of p, eq. (8.37) can be written as

$$\tau_x = -\frac{Q}{2\pi}\int_{-\infty}^{\infty} e^{-i\xi x}\mathrm{d}\xi \sum_{n=1}^{\infty} \frac{\sin(\omega_n t)}{\omega_n}\left[\frac{pN_a}{\partial D_a/\partial p}\right]_{p=i\omega_n}. \tag{8.38}$$

At $p = i\omega_n$ the Rayleigh-Lamb frequency equation $D_a(i\omega, \xi) = 0$ holds, and it follows from eqs. (8.32) and (8.28) that N_a may then be written as

$$[N_a]_{p=\pm i\omega_n} = [2(\beta^2+\xi^2)^2 \sinh(\alpha h)\cosh(\beta h)]_{p=\pm i\omega_n}.$$

By carrying out the differentiation of D_a with respect to p it is subsequently found that

$$\frac{1}{pN_a}\frac{\partial D_a}{\partial p} = \frac{M(p, \xi)}{\beta^2+\xi^2}\frac{1}{c_L^2},$$

where $M(p, \xi)$ is a dimensionless function defined by

$$M(p, \xi) = \frac{1}{\alpha h}\frac{(\beta h)^2+(\xi h)^2}{\sinh(2\alpha h)} - \frac{\kappa^2}{\beta h}\frac{(\beta h)^2+(\xi h)^2}{\sinh(2\beta h)} + 2\kappa^2 + \frac{1}{2}\frac{(\beta h)^2+(\xi h)^2}{(\alpha h)^2(\beta h)^2}[(\beta h)^2+\kappa^2(\alpha h)^2].$$

In this expression κ is c_L/c_T. The dimensionless stress $(h^2/Qc_L)\tau_x$ may then be expressed in the form

$$\left(\frac{h^2}{Qc_L}\right)\tau_x = -\frac{1}{4\pi i}\sum_{n=1}^{\infty}(I_{1n}-I_{2n}), \tag{8.39}$$

where

$$I_{1n} = \int_{-\infty}^{\infty} \frac{2(\xi h)^2-(\omega_n h/c_T)^2}{M(i\omega_n, \xi)} \frac{c_L}{\omega_n h} e^{i(c_L t/h)w_1(\xi h)} \mathrm{d}(\xi h) \tag{8.40}$$

$$I_{2n} = \int_{-\infty}^{\infty} \frac{2(\xi h)^2-(\omega_n h/c_T)^2}{M(i\omega_n, \xi)} \frac{c_L}{\omega_n h} e^{-i(c_L t/h)w_2(\xi h)} \mathrm{d}(\xi h). \tag{8.41}$$

In these equations

$$w_1(\xi h) = \frac{\omega_n h}{c_L} - \frac{x}{c_L t} \xi h \tag{8.42}$$

$$w_2(\xi h) = \frac{\omega_n h}{c_L} + \frac{x}{c_L t} \xi h. \tag{8.43}$$

The integrals (8.40) and (8.41) are suitable for evaluation by the method of stationary phase, which was discussed in section 7.5.

Let us consider the integral in eq. (8.40). According to the stationary phase approximation, for large values of $c_L t/h$ the major contribution to the integral (8.40) occurs in the vicinity of $\xi = \xi_s$, where ξ_s satisfies the stationary phase condition

$$\frac{\partial w_1}{\partial \xi} = 0. \tag{8.44}$$

Assuming that there is one such value of ξ, the approximation to the integral I_{1n} follows from eq. (7.57) as

$$I_{1n} = \left(\frac{2\pi h}{c_L t}\right)^{\frac{1}{2}} \left[\frac{i}{w_1''(\xi_s h)}\right]^{\frac{1}{2}} \frac{2(\xi_s h)^2-(\omega_{ns} h/c_T)^2}{M(i\omega_{ns}, \xi_s)} \frac{c_L}{\omega_{ns} h} e^{iS_n(x,t)} \tag{8.45}$$

Here ω_{ns} is ω_n at $\xi = \xi_s$, and

$$w_1''(\xi_s h) = \left.\frac{\mathrm{d}^2 w_1(\xi h)}{\mathrm{d}(\xi h)^2}\right|_{\xi=\xi_s} \tag{8.46}$$

$$S_n(x, t) = (c_L t/h) w_1(\xi_s h). \tag{8.47}$$

In obtaining the approximation it is assumed that $w''(\xi_s h) > 0$, and also that $c_L t/h$ is large, so that the expansion of $w_1(\xi h)$ discussed in section 7.5 need not be taken beyond the term containing $w''(\xi_s h)$.

The stationary phase condition (8.44) takes the form

$$\frac{x}{t} = c_{gn}, \tag{8.48}$$

where $c_{gn} = d\omega_n/d\xi$ was earlier defined as the group velocity of the nth mode. At a given point we find from (8.48) at each time a value of ξ_s; thus $\xi_s = \xi_s(x, t)$. The quantity ξ_s is sometimes called the local wavenumber of the disturbance. This terminology can be motivated by differentiating

$$S_n(x, t) = t\omega_n[\xi_s(x, t)] - x\xi_s(x, t)$$

with respect to x and t, respectively. Using (8.48), we obtain

$$\frac{\partial S_n}{\partial x} = -\xi_s, \qquad \frac{\partial S_n}{\partial t} = \omega_{ns}.$$

For the various modes the group velocities may be computed from the frequency spectrum by the use of finite difference formulas. For the first four antisymmetric modes the curves showing the variation of the dimensionless group velocity with ξh are shown in figure 8.4. These curves, which were computed for a value of Poisson's ratio of approximately 0.29, are from a paper by Jones[3], in which the antisymmetric transient motion of a layer is investigated.

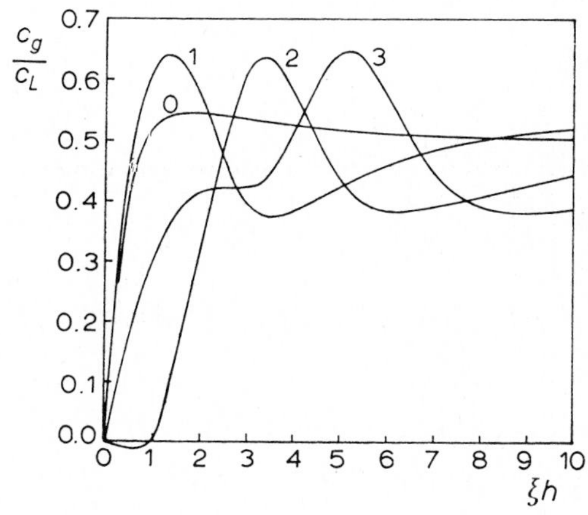

Fig. 8.4. Group velocities for the first four antisymmetric modes, $\nu = 0.29$.

For all modes the group velocity is zero at $\xi h = 0$. As ξh increases, the group velocity of the lowest mode, labeled $n = 0$ in the figure, approaches the velocity of Rayleigh surface waves. The group velocities of the higher modes approach the velocity c_T as ξh increases.

[3] R. P. N. Jones, *Quarterly Journal of Mechanics and Applied Mathematics XVII* (1964) 401.

In writing the approximation of I_{1n} in the form (8.45) it is tacitly taken into account that the function $M(i\omega_n, \xi)$ does not have zeros for real values of ξ. If there were zeros the integrand of (8.40) would have poles on the path of integration and additional contributions would come from the vicinities of the poles. The denominator of (8.40) would vanish if $\partial D_a(i\omega, \xi)/\partial\omega$ were zero at any point of the ω versus ξ curves which comprise the frequency spectrum of the layer. This does, however, not occur for $-\infty < \xi h < \infty$, as can be shown by the following argument, which is similar to that presented by Skalak[4] in the treatment of an analogous problem for a rod. Along any branch of the frequency spectrum $\partial D_a(i\omega, \xi)/\partial s_1 = 0$, if s_1 is the direction along the branch. If at any point on a curve defined by $D_a(i\omega, \xi) = 0$, $\partial D_a(i\omega, \xi)/\partial\omega$ were also zero, then $\partial D_a/\partial s$ would vanish for any direction s through this point, provided the curve $D_a(i\omega, \xi) = 0$ does not have a tangent parallel to the ω-axis at this point. The condition $\partial D_a/\partial s = 0$ means that the function $D_a(i\omega, \xi)$ is either a maximum, a minimum or a saddle point. If it is a maximum or a minimum, a curve $D_a = 0$ could not pass through it. If it were a saddle point, there would be two curves $D_a(i\omega, \xi) = 0$ passing through it. Inspection of the frequency spectrum shows, however, that the branches do not cross. Also, for real-valued ξ there is no point where a ω_n versus ξ curve is parallel to the ω-axis. Hence there are no points on the ω_n versus ξ curves at which $\partial D_a(i\omega, \xi)/\partial\omega$ vanishes. In the paper by Jones, where a solution is obtained by direct mode superposition another argument is used, based on the observation that the denominator is proportional to a kinetic energy function, which is necessarily positive.

Returning to the stationary phase condition (8.48) it is noted that for any value of $x/c_L t$ the value of $\xi_s h$ can immediately be determined from the curves of figure 8.4. Over a range of values of $x/c_L t$ there is more than one root of (8.48). In this range there will be a contribution of the form (8.45) for each root, and the complete solution is obtained by summation.

The assumptions made in the derivation of (8.45) are satisfied in the region $0 < x < 0.37c_L t$. Since the group velocity is odd in ξh, the stationary point of $w_2(\xi h)$ is located at $\xi = -\xi_s$, and the integral I_{2n} can be evaluated in just the same manner as I_{1n}. The stress is then obtained as

$$\left(\frac{h^2}{Qc_L}\right)\tau_x = \frac{1}{2\pi}\left(\frac{2\pi h}{c_L t}\right)^{\frac{1}{2}} \sum_{n=1}^{\infty} [\pm\, w_1''(\xi_s h)]^{-\frac{1}{2}} G_{ns} \sin\left(\xi_s x - \omega_{ns} t \mp \frac{\pi}{4}\right), \tag{8.49}$$

[4] R. Skalak, *Journal of Applied Mechanics* **24** (1957) 59.

where the alternative sign applies if $\mathrm{d}^2\omega_n/\mathrm{d}\xi^2$ is negative, and where

$$G_{ns} = \frac{2(\xi_s h)^2 - (\omega_{ns} h/c_T)^2}{M(i\omega_{ns}, \xi_s)} \frac{c_L}{\omega_{ns} h},$$

$$\omega_{ns} = \omega_n \quad \text{at} \quad \xi = \xi_s .$$

For each mode the local wavelength $2\pi/\xi_s$ of the disturbance decreases with increase of $x/c_L t$. When $x = 0.37\ c_L t$ the values of $\xi_s h$ in the lowest (fundamental) mode is 0.45, corresponding to a wavelength of approximately seven times the depth of the layer.

In the region $x > 0.65\ c_L t$ approximately, there is no value of ξh satisfying the stationary phase condition and the stress τ_x at $y = +h$ is therefore zero, to the order of accuracy implicit in the approximation. It is evident that the stresses cannot be identically zero throughout the whole of this region, since disturbances are propagated with the velocity c_L, the velocity of longitudinal waves. Consequently we should expect the disturbance in the layer to extend as far as $x = c_L t$. The stationary phase approximation indicates, however, that the stresses in the region $0.65\ c_L t < x < c_L t$ are of order smaller than $(c_L t/h)^{-\frac{1}{2}}$.

Consider finally the region $0.37\ c_L t < x < 0.65\ c_L t$. It is noted from figure 8.4 that in this region there are values of ξh for which the group velocity is stationary, so that $w_1(\xi_s h) = 0$, and the approximation (8.45) breaks down. From figure 8.4 we observe that for the higher modes the group velocity shows both a maximum and a minimum for finite values of ξh. The derivative $\mathrm{d}c_g/\mathrm{d}\xi$ vanishes for all modes as ξh increases beyond bounds. It can now be shown that for the points $\mathrm{d}c_g/\mathrm{d}\xi$ at finite values of ξh a stationary phase approximation in the form of Airy integrals can be obtained by the use of a modified expansion for $w_1(\xi_s h)$ in the vicinity of $\xi = \xi_s$. For transient antisymmetric motion of a layer the appropriate expressions were worked out in the previously cited paper by Jones. The most noteworthy feature of these contributions is that the stress varies as $(c_L t/h)^{-\frac{1}{3}}$, as opposed to the contributions represented by (8.49) which vary as $(c_L t/h)^{-\frac{1}{2}}$. Thus at large values of the time the "Airy phase" becomes predominant. Finally, the contributions at $x = c_R t$ and $x = c_T t$ are also discussed by Jones.

In the discussion presented here we have included the contributions of the higher modes. Actual computations show, however, that especially in the region $0 < x < 0.37\ c_L t$ the major part of the response arises from the fundamental mode.

The approximate plate theories of section 6.12 have been used extensively for transient problems of the type discussed in this section. There are of course definite ranges of x and t beyond which the approximate theories are not valid. For a discussion we refer to a paper by Nigul.[5]

8.4. The point load on a layer

The application of a normal point load generates axially symmetric wave motions in a layer. In the same manner as for a line load, the total response of the layer can conveniently be considered as the superposition of the responses to an antisymmetric and a symmetric pair of loads, respectively, as sketched in figure 8.3.

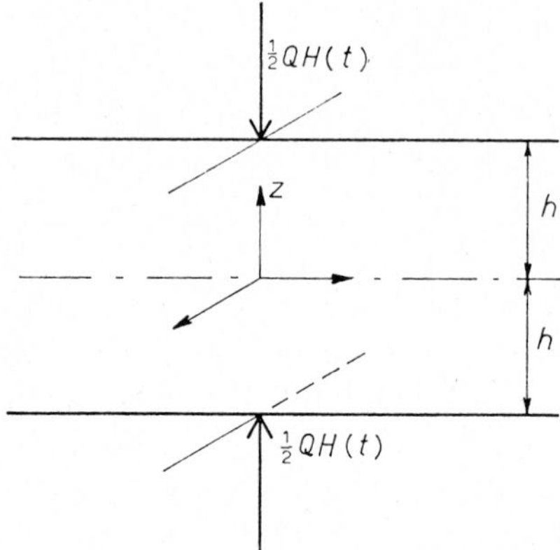

Fig. 8.5. Point load on a layer.

We choose a system of cylindrical coordinates with the z-axis coinciding with the direction of the load, i.e., normal to the faces of the layer, and with the origin in the midplane of the layer. The field variables are axially symmetric. The expressions relating the displacements and the potentials then follow from the equations of section 2.13 of chapter 2. In the present section the analysis of the symmetric problem is outlined. The geometry is shown in figure 8.5. Let us consider the case that the load is suddenly applied and thereafter maintained at a constant value, so that the boundary conditions may be expressed as

$$\tau_z = -\tfrac{1}{2}QH(t)\frac{\delta(r)}{2\pi r}$$

$$\tau_{zr} = 0,$$

[5] U. Nigul, *International Journal of Solids and Structures* **4** (1969) 607.

where we have used the representation of the Dirac delta function in polar coordinates. A further condition is that the displacements and the stresses, hence the potentials vanish at infinity ($r \to \infty$). Since the layer is at rest prior to $t = 0$ the initial conditions are

$$\varphi(r, z, 0) = \dot{\varphi}(r, z, 0) = \psi(r, z, 0) = \dot{\psi}(r, z, 0) \equiv 0.$$

The symmetric problem defined above was treated by Miklowitz.[6] The most efficient method of solution is again by the use of integral transform techniques. In addition to the one-sided Laplace transform with respect to time, Hankel transforms with respect to the radius r should be employed for this problem with axial symmetry. These transforms and their inverses are defined by eqs. (7.20), (7.21) and (7.31), (7.32), respectively. By application of these transforms the equations governing the Laplace-Hankel transforms of the displacement potentials reduce to simple ordinary differential equations of the second order. Their solutions are exactly of the same forms as (8.24) and (8.25). Using the terms of these expressions corresponding to symmetric deformations of the layer, we can subsequently employ the boundary conditions to solve for the unknown constants. In this manner expressions analogous to (8.29) and (8.30) are obtained.

By the use of the inversion integrals defined by eqs. (7.21) and (7.32) of chapter 7 a formal representation of, for example, the radial displacement $u(r, z, t)$ can be written as

$$u(r, z, t) = \frac{1}{2\pi i}\int_0^\infty \xi J_1(\xi r)\mathrm{d}\xi \int_{\gamma - i\infty}^{\gamma + i\infty} \frac{U_s(\xi, z, p)}{pD_s(\xi, p)} e^{pt}\mathrm{d}p. \tag{8.50}$$

In this expression $U_s(\xi, z, p)$ is a rather complicated function of the Hankel transform variable ξ, the Laplace transform variable p and the coordinate z. It is noteworthy that the denominator is just the same as for the line-load problem, i.e., $D_s(\xi, p)$ is given by eq. (8.31).

Just as for the line-load problem the evaluation of the integral in the complex p-plane is simple. Since there are no branch points in the p-plane the integral reduces to $2\pi i$ times the sum of the residues of the poles located in the half-plane $\mathscr{R}(p) < \gamma$. We find

[6] J. Miklowitz, *Journal of Applied Mechanics* **29** (1962) 53.

$$u(r, z, t) = \int_0^\infty \xi J_1(\xi r) \left\{ \lim_{|p| \to 0} \left[\frac{\partial U_s / \partial p}{\partial D_s / \partial p} \right] \right\} d\xi$$
$$+ \int_0^\infty \xi J_1(\xi r) \sum_{n=1}^\infty \left\{ \left[\frac{e^{pt} U_s}{p \partial D_s / \partial p} \right]_{p = i\omega_n} + \left[\frac{e^{pt} U_s}{p \partial D_s / \partial p} \right]_{p = -i\omega_n} \right\} d\xi, \tag{8.51}$$

where ω_n are the roots of the Rayleigh-Lamb frequency equation for symmetric modes, $D_s(\xi, i\omega) = 0$.

A further evaluation of the integrals in (8.51) can be achieved by approximate methods. Miklowitz first replaces the Bessel function by the leading term of its asymptotic expansion for large r/h and then proceeds to approximate the resulting integrals by the method of stationary phase. For details we refer to the previously cited paper by Miklowitz. As main result of the analysis it is found that the head of the propagating disturbance is composed predominantly of the low-frequency long waves from the lowest mode of symmetric wave transmission.

8.5. Impact of a rod

In engineering applications, for example in machinery, it is often required to transmit a signal from one position to another. Very often the transmission is accomplished via a cylindrical rod. In such cases it may be necessary to take into account wave propagation effects if the signal varies rapidly in time. The most elementary theory capable of describing the propagation of longitudinal pulses in a rod is governed by the equation

$$\frac{\partial^2 w}{\partial z^2} = \frac{\rho}{E} \frac{\partial^2 w}{\partial t^2}. \tag{8.52}$$

Here w is the axial displacement; ρ and E are the mass density and Young's modulus, respectively. Eq. (8.52), which was derived earlier, is based on the assumption of a one-dimensional state of stress in the rod.

Since (8.52) is a simple one-dimensional wave equation, it predicts that a pulse does not change shape as it propagates along the rod. Several years ago it was, however, already shown experimentally by Davies[7] that this prediction proves to be true only if the external disturbance producing the pulse is applied relatively slowly. If the disturbance is applied rapidly the resulting pulse shows dispersion. This observation, which was subsequently

[7] R. M. Davies, *Philosophical Transactions of the Royal Society* **A240** (1948) 375.

verified by other experimentalists, has motivated analytical investigations of transient axially symmetric wave propagation in a circular cylindrical rod by the use of the equations of the theory of elasticity.

Several investigators have examined the response of a semi-infinite circular rod of radius a to prescribed conditions at the end surface $z = 0$. General methods of solutions for arbitrary end conditions still are being investigated.[8] For the axially symmetric case a solution can, however, be obtained by integral transform methods if the conditions at $z = 0$ are specified by mixed conditions. Here we will discuss the case that a constant pressure is suddenly applied, while the points in the plane $z = 0$ are restrained from motion in the radial direction. The solution to this problem was presented by Folk et al.[9]

We choose a system of cylindrical coordinates with the z-axis coinciding with the axis of the cylinder. The expressions relating the radial displacement $u(r, z, t)$ and the axial displacement $w(r, z, t)$ to the displacement potentials follow from section 2.13 as

$$u = \frac{\partial \varphi}{\partial r} - \frac{\partial \psi}{\partial z} \tag{8.53}$$

$$w = \frac{\partial \varphi}{\partial z} + \frac{1}{r}\frac{\partial}{\partial r}(r\psi), \tag{8.54}$$

respectively, where the single component of the vector potential is denoted by ψ rather than by ψ_θ. The potentials must satisfy the following wave equations

$$\nabla^2 \varphi = \frac{1}{c_L^2}\frac{\partial^2 \varphi}{\partial t^2} \tag{8.55}$$

$$\nabla^2 \psi - \frac{1}{r^2}\psi = \frac{1}{c_T^2}\frac{\partial^2 \psi}{\partial t^2}, \tag{8.56}$$

where

$$\nabla^2 = \frac{\partial^2}{\partial r^2} + \frac{1}{r}\frac{\partial}{\partial r} + \frac{\partial^2}{\partial z^2}. \tag{8.57}$$

[8] See J. Miklowitz, *loc. cit.*, p.

[9] R. Folk, G. Fox, C. A. Shook and C. W. Curtis, *Journal of the Acoustical Society of America* **30** (1958) 552.

The pertinent components of the stress tensor are

$$\tau_r = \lambda\nabla^2\varphi + 2\mu \frac{\partial}{\partial r}\left[\frac{\partial\varphi}{\partial r} - \frac{\partial\psi}{\partial z}\right] \tag{8.58}$$

$$\tau_{rz} = \mu\left[\frac{\partial}{\partial z}\left(\frac{\partial\varphi}{\partial r} - \frac{\partial\psi}{\partial z}\right) + \frac{\partial}{\partial r}\left[\frac{\partial\varphi}{\partial z} + \frac{1}{r}\frac{\partial}{\partial r}(r\psi)\right]\right] \tag{8.59}$$

$$\tau_z = \lambda\nabla^2\varphi + 2\mu \frac{\partial}{\partial z}\left[\frac{\partial\varphi}{\partial z} + \frac{1}{r}\frac{\partial}{\partial r}(r\psi)\right]. \tag{8.60}$$

Solutions to these equations must satisfy boundary conditions at the end of the rod and at the lateral surface. We have

$$\text{at} \quad z = 0, \quad 0 \leqq r \leqq a \qquad \tau_z(r, 0, t) = -PH(t) \tag{8.61}$$

$$u(r, 0, t) = 0 \tag{8.62}$$

$$\text{at} \quad r = a, \quad z \geqq 0 \qquad \tau_r(a, z, t) = 0 \tag{8.63}$$

$$\tau_{rz}(a, z, t) = 0. \tag{8.64}$$

A further condition is that the displacements and the stresses, hence the potentials vanish at infinity ($z \to \infty$). Since the rod is at rest prior to time $t = 0$ the initial conditions are

$$\varphi(r, z, 0) = \dot{\varphi}(r, z, 0) = \psi(r, z, 0) = \dot{\psi}(r, z, 0) \equiv 0. \tag{8.65}$$

Boundary conditions on the potentials at the face $z = 0$ can easily be extracted from (8.61) and (8.62). From (8.62) and (8.53) it is concluded that at $z = 0$

$$\frac{\partial\varphi}{\partial r} = \frac{\partial\psi}{\partial z}. \tag{8.66}$$

Using this relation to eliminate ψ in the expression for τ_z, see eq. (8.60), we obtain from (8.61) at $z = 0$

$$(\lambda + 2\mu)\nabla^2\varphi = -PH(t),$$

where ∇^2 is defined by (8.57). Since φ satisfies the wave equation (8.55), we can also write (at $z = 0$)

$$\rho\ddot{\varphi} = -PH(t). \tag{8.67}$$

In view of this result (8.66) reduces to

$$\frac{\partial\psi}{\partial z} = 0. \tag{8.68}$$

Eqs. (8.67) and (8.68) provide boundary conditions at $z = 0$ on the potentials $\varphi(r, z, t)$ and $\psi(r, z, t)$.

8.5.1. *Exact formulation*

Integral transform techniques are again the appropriate method of solution for the problem stated by eqs. (8.53)–(8.60), (8.63)–(8.65) and (8.67)–(8.68). For the present semi-infinite domain ($z \geqq 0$) Fourier sine and cosine transforms are used with respect to z. These transforms were defined by eqs. (7.26) and (7.28) of chapter 7 as

$$\text{sine transform:} \qquad f^S(\xi) = \int_0^\infty f(z) \sin(\xi z) \mathrm{d}z$$

$$\text{cosine transform:} \qquad f^C(\xi) = \int_0^\infty f(z) \cos(\xi z) \mathrm{d}z.$$

The form of the boundary conditions (8.67) and (8.68) suggests the use of a cosine transform for ψ and a sine transform for φ. In addition the one-sided Laplace transform is used to eliminate the dependence on the time t.

Upon applying the integral transforms to the wave equations (8.55) and (8.56), whereby the initial conditions (8.65) and the boundary conditions (8.67) and (8.68) must be taken into account, these wave equations reduce to

$$\frac{\mathrm{d}^2\bar{\varphi}^S}{\mathrm{d}r^2} + \frac{1}{r}\frac{\mathrm{d}\bar{\varphi}^S}{\mathrm{d}r} + \alpha^2\bar{\varphi}^S = \frac{P}{\rho}\frac{\xi}{p^3} \tag{8.69}$$

$$\frac{\mathrm{d}^2\bar{\psi}^C}{\mathrm{d}r^2} + \frac{1}{r}\frac{\mathrm{d}\bar{\psi}^C}{\mathrm{d}r} + \left(\beta^2 - \frac{1}{r^2}\right)\bar{\psi}^C = 0, \tag{8.70}$$

where

$$\alpha^2 = -\frac{p^2}{c_L^2} - \xi^2, \qquad \beta^2 = -\frac{p^2}{c_T^2} - \xi^2. \tag{8.71a, b}$$

Solutions of (8.69) and (8.70) that are bounded at the center of the rod are

$$\bar{\varphi}^S = AJ_0(\alpha r) + \frac{P}{a^2\rho}\frac{\xi}{p^3} \tag{8.72}$$

$$\bar{\psi}^C = BJ_1(\beta r). \tag{8.73}$$

The conditions that remain to be satisfied are that the stress components τ_r and τ_{rz} vanish at the cylindrical surface defined by $r = a$. Applying the Laplace transform and the Fourier sine transform to eq. (8.58), and employing (8.72) and (8.73) we obtain at $r = a$

$$\left[-(\beta^2-\xi^2)J_0(\alpha a)+2\frac{\alpha}{a}J_1(\alpha a)\right]A+2\left[\beta\xi J_0(\beta a)-\frac{\xi}{a}J_1(\beta a)\right]B = -\frac{\lambda}{\lambda+2\mu}\frac{P}{\mu}\frac{\xi}{p\alpha^2}.$$

Similarly an application of the Laplace transform and the Fourier cosine transform to eq. (8.59) yields at $r = a$

$$2\alpha\xi J_1(\alpha a)A+(\beta^2-\xi^2)J_1(\beta a)B = 0.$$

The solutions of these equations are

$$A = -\frac{\lambda}{\lambda+2\mu}\frac{P}{\mu}\frac{\xi}{p}\frac{\beta^2-\xi^2}{\alpha^2}\frac{J_1(\beta a)}{D} \tag{8.74}$$

$$B = \frac{2\lambda}{\lambda+2\mu}\frac{P}{\mu}\frac{1}{p}\frac{\xi^2}{\alpha}\frac{J_1(\alpha a)}{D}, \tag{8.75}$$

where

$$D = \frac{2\alpha}{a}(\beta^2+\xi^2)J_1(\alpha a)J_1(\beta a)-(\beta^2-\xi^2)^2J_0(\alpha a)J_1(\beta a) -4\alpha\beta\xi^2J_1(\alpha a)J_0(\beta a). \tag{8.76}$$

The particle velocity in the axial direction,

$$V(r, z, t) = \frac{\partial w}{\partial t},$$

will be considered in some detail. In terms of the displacement potentials, V follows from (8.54) as

$$V(r, z, t) = \frac{\partial}{\partial z}\left(\frac{\partial \varphi}{\partial t}\right)+\frac{1}{r}\frac{\partial}{\partial r}\left(r\frac{\partial \psi}{\partial t}\right).$$

Application of the Fourier cosine transform and the Laplace transform results in

$$\overline{V}^C = \frac{P}{\rho}\frac{1}{p^2}+p\xi\overline{\varphi}^S+\frac{p}{r}\frac{\partial}{\partial r}(r\overline{\psi}^C),$$

which can be rewritten as

$$\overline{V}^C = \frac{P}{\rho}\frac{1}{p^2+\xi^2c_L^2}-\frac{\lambda}{\lambda+2\mu}\frac{P}{\mu}\frac{\xi^2}{\alpha^2}\frac{N}{D}, \tag{8.77}$$

where

$$N = (\beta^2 - \xi^2) J_0(\alpha r) J_1(\beta a) - 2\alpha\beta J_0(\beta r) J_1(\alpha a) \tag{8.78}$$

and D is defined by eq. (8.76).

8.5.2. *Inversion of the transforms*

The inversion of the transforms is again the major task of the analysis. Formally, inverting the Laplace transform yields the result

$$V^C = \frac{1}{2\pi i} \int_{\gamma - i\infty}^{\gamma + i\infty} \bar{V}^C e^{pt} \mathrm{d}p. \tag{8.79}$$

Just as for the response of a layer to line loads and point loads the integration in the complex p-plane can be carried out by applying Cauchy's residue theorem. The path of integration is closed by a semicircle of infinite radius about the origin in the left-half plane.

If the numerator and the denominator of eq. (8.77) are each multiplied by β and the various Bessel functions are replaced by their respective infinite series, it is found that only even powers of α and β appear. Hence there are no branch points in the complex p-plane in spite of the fact that α and β are radicals. On the infinite semicircle the various Bessel functions may be replaced by their asymptotic forms for large arguments. In the limit it is then found that the integral over the infinite semicircle vanishes. In view of these observations the original integral is equal to $2\pi i$ times the sum of the residues in the left-half plane.

Both terms of (8.77) have poles at $p = \pm i\xi c_L$. It can, however, be verified that the residues of these poles cancel each other. The remaining poles are at the points of the p-plane where the function $D(p, \xi)$ defined by (8.76) vanishes. By means of the substitution

$$p = i\omega,$$

the equation $D(p, \xi) = 0$ is cast in a form which is identical to the frequency equation of longitudinal motions of a rod, which was given by eq. (6.131). Consequently for any value of ξ the integrand of eq. (8.79) has an infinite number of poles along the imaginary axis in the p-plane, at positions defined by $p = i\omega_n$, where the functions $\omega_n(\xi)$ are the circular frequencies of longitudinal waves in a rod. The contributions from these poles yield the following expression for V^C:

$$V^C = -\frac{\lambda}{\lambda+2\mu}\frac{P}{\mu}\sum_{n=1}^{\infty} 2i\xi^2 M_n(\omega_n, \xi)\sin(\omega_n t), \tag{8.80}$$

where

$$M_n(\omega_n, \xi) = \left[\frac{1}{\alpha^2}\frac{N}{\partial D/\partial p}\right]_{p=i\omega_n}. \tag{8.81}$$

This function is even in ξ.

By applying the inversion integral for Fourier cosine transforms we find

$$V = \frac{\lambda}{\lambda+2\mu}\frac{P}{\mu}\sum_{n=1}^{\infty} I_n, \tag{8.82}$$

where

$$I_n = -\frac{4i}{\pi}\int_0^\infty \xi^2 M_n(\omega_n \xi)\sin(\omega_n t)\cos(\xi z)\mathrm{d}\xi. \tag{8.83}$$

As in the problems for the layer the integrals in (8.82) cannot be evaluated rigorously by analytical means. A very satisfactory approximation can, however, be obtained by the method of stationary phase.

8.5.3. Evaluation of the particle velocity for large time

The integrals appearing in (8.82) may be cast in the form

$$I_n = \frac{2i}{\pi}\int_0^\infty \xi^2 M_n(\omega_n, \xi)\mathscr{I}[e^{i(\xi z-\omega_n t)} - e^{i(\xi z+\omega_n t)}]\mathrm{d}\xi. \tag{8.84}$$

For large values of a dimensionless time such as $c_L t/a$ an approximate evaluation can be carried out by the method of stationary phase. As was shown in section 8.3, the resulting contributions of the various branches of the frequency spectrum will be at most of order $(c_L t/a)^{-\frac{1}{2}}$, except at positions $x/c_L t$ of stationary group velocity, where the contributions are of order $(c_L t/a)^{-\frac{1}{3}}$.

Additional contributions to I_n may come from poles on the real ξ-axis. It can, however, be shown by an argument of the type presented in section 8.3 that $\partial D/\partial p$ cannot have zeros for real-valued ξ. For the rod the argument of nonvanishing $\partial D/\partial p$ is given by Skalak.[10] A vanishing value of $\partial D/\partial p$ is, however, not the only potential cause for a pole on the ξ-axis. In the present problem it so happens that for the contribution of the lowest mode ω_1 is proportional to ξ in the vicinity of $\xi = 0$. The presence of the term α^2 in

[10] R. Skalak, *Journal of Applied Mechanics* **24** (1957) 59.

the denominator then causes a simple pole at $\xi = 0$. An additional complication is that for the lowest mode the point $\xi = 0$ is a point of stationary phase. Moreover, the group velocity is stationary at $\xi = 0$. These circumstances combine to yield a contribution which predominates the other contributions since it does not decay in time. This contribution will now be computed in a manner which follows by and large the previously cited work of Skalak.

According to eq. (6.135) of chapter 6, for small values of ξ the frequency of the lowest longitudinal mode is

$$\omega_1 = c_b \xi - \gamma \xi^3, \tag{8.85}$$

where c_b is the bar velocity

$$c_b = (E/\rho)^{\frac{1}{2}}, \tag{8.86}$$

and γ is defined as

$$\gamma = \tfrac{1}{4}\nu^2 c_b a^2. \tag{8.87}$$

Substituting $\omega_1 = c_b \xi$ in $M_1(\omega_1, \xi)$, the following limit can be derived:

$$\lim_{\xi \to 0} \xi^3 M_1(\omega_1, \xi) = \frac{i}{2} \frac{\mu}{\lambda} \frac{c_L^2}{c_b}. \tag{8.88}$$

The essential contribution to I_1 can now be obtained by limiting the range of integration of (8.84) to $\xi < \varepsilon$, where ε is a small number

$$I_1 = -\frac{1}{\pi} \frac{\mu}{\lambda} \frac{c_L^2}{c_b} \mathscr{I} \int_0^{+\varepsilon} \frac{1}{\xi} [e^{i(\xi z - \omega_1 t)} - e^{i(\xi z + \omega_1 t)}] d\xi. \tag{8.89}$$

The upper limits of these integrals are next again extended to ∞ because the integrals so added are of order $(c_L t/a)^{-\frac{1}{3}}$. If subsequently just the first term of (8.85) is substituted, i.e. $\omega_1 = c_b \xi$, the integrals can be evaluated, with as result the solution according to the elementary theory

$$V(z, t) = \frac{1}{c_b} \frac{P}{\rho} H(c_b t - z). \tag{8.90}$$

A better approximation is obtained by employing eq. (8.85) for ω_1. The result may be written as

$$I_1 = -\frac{\mu}{\lambda} \frac{c_L^2}{c_b} [J_1 + J_2],$$

where

$$J_1 = \frac{1}{\pi}\int_0^\infty \frac{\sin[\xi(z-c_b t)+\gamma\xi^3 t]}{\xi}\,d\xi$$

$$J_2 = \frac{1}{\pi}\int_0^\infty \frac{\sin[-\xi(z+c_b t)+\gamma\xi^3 t]}{\xi}\,d\xi.$$

By a change of the variable of integration these integrals become

$$J_1 = \frac{1}{\pi}\int_0^\infty \frac{\sin[q_1\eta+\eta^3/3]}{\eta}\,d\eta \tag{8.91}$$

$$J_2 = \frac{1}{\pi}\int_0^\infty \frac{\sin[q_2\eta+\eta^3/3]}{\eta}\,d\eta, \tag{8.92}$$

where

$$q_1 = \frac{z-c_b t}{(3\gamma t)^{\frac{1}{3}}}, \qquad q_2 = -\frac{z+c_b t}{(3\gamma t)^{\frac{1}{3}}}.$$

In the forms (8.91) and (8.92) the integrals are recognized as integrals of Airy's integral $Ai(q)$

$$Ai(q) = \frac{1}{\pi}\int_0^\infty \cos(q\eta+\eta^3/3)d\eta.$$

We obtain

$$J_1 = \int_0^{q_1} Ai(s)ds+\tfrac{1}{6}$$

$$J_2 = \int_0^{q_2} Ai(s)ds+\tfrac{1}{6},$$

where the terms $\frac{1}{6}$ enter as the values of J_1 and J_2 for $q_1 = 0$ and $q_2 = 0$, respectively. For large values of t the integrals approach $-\frac{2}{3}$, and the elementary solution $V = P/\rho c_b$ is obtained.

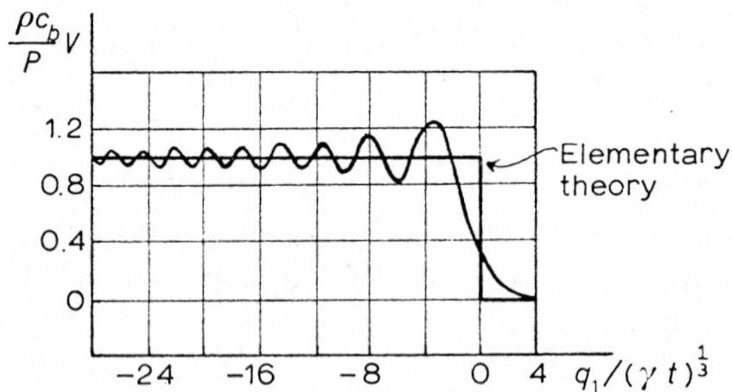

Fig. 8.6. Particle velocity near $z = c_b t$.

For small values of t the integrals vanish. Thus, the results show that some distance ahead and some distance behind the wavefront $z = c_b t$ the improved approximation agrees with the elementary theory. Numerical results are shown in figure 8.6.

In the natural coordinates z and t the wave form shown in figure 8.6. gradually spreads out around the wavefront $z = c_b t$.

8.6. Problems

8.1. Suppose that a thin rod of length l is fixed at the end $x = 0$ and that a force $P(t)$ acts at the other end. Assume that the wave motion of the rod is governed by the equation

$$\frac{\partial^2 u}{\partial x^2} = \frac{1}{c_b^2}\frac{\partial^2 u}{\partial t^2}, \qquad \text{where} \quad c_b^2 = \frac{E}{\rho}.$$

It is also assumed that the rod is undisturbed prior to time $t = 0$.

Show that the one-sided Laplace transform $\bar{u}(x, p)$ can be expressed in the form

$$\bar{u}(x, p) = \frac{\bar{P}(p)}{EA}\frac{c_b}{p}\frac{e^{px/c_b} - e^{-px/c_b}}{e^{pl/c_b} + e^{-pl/c_b}}.$$

Invert the transform by employing the expansion

$$\frac{1}{1 + e^{-2pl/c_b}} = 1 - e^{-2pl/c_b} + e^{-4pl/c_b} - \dots$$

What is the physical interpretation of the terms?

An alternative method is to evaluate the inverse Laplace transform of

$$\frac{1}{p}\frac{\sinh px/c_b}{\cosh pl/c_b}$$

by contour integration in the complex p-plane. What is the form of $u(x, t)$ which is obtained in that manner? What is the physical interpretation of the solution?

8.2. The system of equations governing the motions of a Timoshenko beam is given by eqs. (6.150) and (6.151) of chapter 6. Derive the expressions defining the characteristics and determine the differential equations along the characteristic curves.

8.3. A semi-infinite circular cylindrical rod ($0 \leq z < \infty$) of radius a is subjected to a prescribed displacement distribution over the cross section at $z = 0$. The prescribed displacement distribution is of the form

$$v(r, 0, t) = \frac{r^2}{a} \sin \omega t,$$

where v is the circumferential displacement. The lateral surface is free of tractions. Determine the steady-state displacement response of the rod.

8.4. The semi-infinite rod of problem 8.3 is now subjected to the following axially symmetric boundary conditions at $z = 0$:

$$\tau_z(r, 0, t) = \tau_0 \sin \omega t$$
$$u(r, 0, t) = 0.$$

Determine the steady-state displacement response.

8.5. An elastic layer of thickness $2h$ is subjected to a suddenly applied anti-plane line load. Choose the coordinate system as shown in the figure. The boundary conditions then are

$$y = 0 \qquad \tau_{yz} = T_0\, \delta(x) H(t)$$
$$y = -2h \qquad \tau_{yz} = 0.$$

The layer is at rest prior to time $t = 0$.

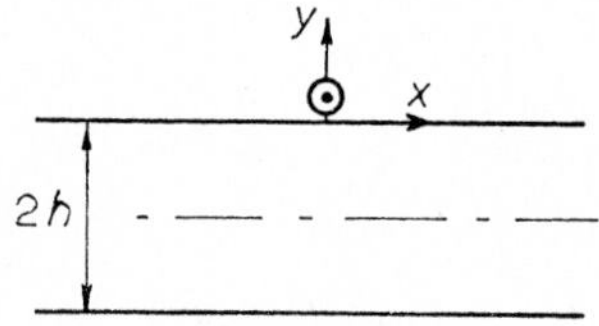

Apply the one-sided Laplace transform with respect to time and the two-sided Laplace transform with respect to x to obtain $\bar{w}^*$. Observe that the Cagniard-de Hoop method cannot be applied directly. For large enough values of p the exponentials appearing in the denominator of $\bar{w}^*$ can, however, be removed by using the same type of expansion as in problem 8.1. Invert the resulting sum of exponentials term by term by means of the Cagniard-de Hoop scheme. What is the physical significance of the terms?

8.6. The solution as suggested by problem 8.5 is not suitable for large values of time. Reexamine the problem in a coordinate system whose origin is in the midplane of the layer. Split the problem up in a symmetric and an anti-

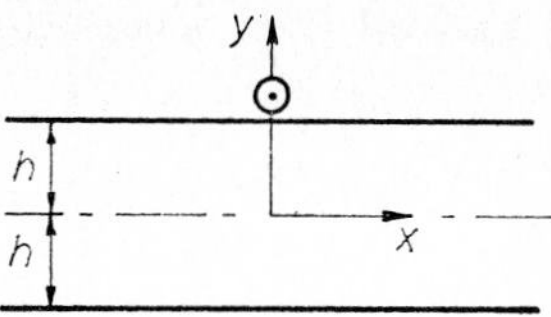

symmetric problem. Find an expression for the transformed displacement by applying the one-sided Laplace transform with respect to time, and the exponential Fourier transform with respect to x. Invert the Laplace transform by contour integration. Consider the contribution from the lowest mode and evaluate the integral which determines the inverse of the exponential Fourier transform.

8.7. A semi-infinite three-layered sandwich construction is subjected to a uniform antiplane shear traction at $x = 0$, i.e., at $x = 0$ $\tau_{xz} = \tau_0 H(t)$.

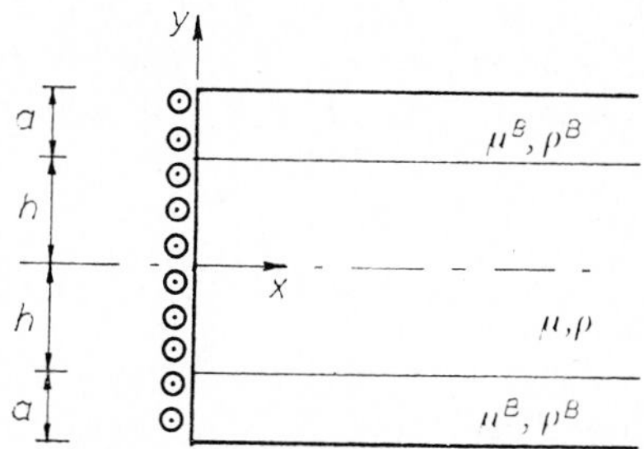

The surfaces at $y = \pm(a+h)$ are free of tractions. Use the results of problem 6.5 to investigate the wave motion in the layer. In particular find an expression for large values of x for the interface stress τ_{yz} at $y = \pm h$.

8.8. Consider a circular cylindrical rod of radius a and infinite length, whose cylindrical surface is loaded by a distribution of circumferential tractions which is independent of θ. As an idealization we assume that the tractions are applied as a line load. Also the load is suddenly applied and then maintained at a constant level. The boundary condition may be expressed as

$$\text{at} \quad r = a \qquad \tau_{r\theta} = T\delta(z)H(t) \qquad \tau_{rz} = \tau_{rr} \equiv 0.$$

The rod is at rest prior to time $t = 0$.

(a) Express the transient response of the rod in terms of a summation over integrals. The first term of the summation can easily be worked out, and this should be done. The remaining terms in the summation may be left in the form of integrals.

(b) Now suppose that the rod is of finite length, $|z| < l$. In what time interval is the solution of (a) valid for the finite rod? If the rod is *clamped* at $z = \pm l$, how can you use the result of (a) to determine the solution at any time?

8.9. A rod of radius a and infinite length is loaded by a ring load of normal surface tractions:

$$r = a: \quad |z| < b \qquad \tau_{rr} = \tau_0 H(t)$$

$$|z| > b \qquad \tau_{rr} = 0$$

$$-\infty < z < \infty \qquad \tau_{rz} = \tau_{r\theta} \equiv 0.$$

Determine an expression for the particle velocity $\dot{w}(r, z, t)$ for large values of z.

8.10. A semi-infinite elastic layer of thickness $2h$ is free of tractions on the lateral surfaces $y = \pm h$. The cross section at $x = 0$ is subjected to the following boundary conditions:

$$\tau_x(0, y, z, t) = T_0 \delta(z) H(t)$$

$$v(0, y, z, t) = w(0, y, z, t) \equiv 0.$$

The layer is at rest prior to time $t = 0$. Determine the particle velocity $\dot{u}(x, y, 0, t)$ for large values of x.

CHAPTER 9

DIFFRACTION OF WAVES BY A SLIT

9.1. Mixed boundary-value problems

In general terms a mixed boundary-value problem is a problem for which the boundary conditions are of different types on complementary parts of the boundary. For example, for a body B with boundary S an elastodynamic mixed boundary-value problem is defined by the conditions

$$u_i = U_i(\boldsymbol{x}, t) \quad \text{on} \quad S_1 \tag{9.1}$$

and

$$\tau_{ji} n_j = t_i(\boldsymbol{x}, t) \quad \text{on} \quad S - S_1. \tag{9.2}$$

Mixed boundary-value problems usually are exceedingly difficult to solve. The standard techniques of elastic wave analysis which were employed in the preceding chapters cannot be applied directly. It is, in fact, necessary to call upon another class of methods of applied mathematics, which may be classified as methods to solve integral equations. To exemplify, let us assume that the elastic fields inside a body due to both normal and tangential point loads on the surface S are known. The displacement solutions of the mixed boundary-value problem defined by eqs. (9.1) and (9.2) can then be expressed in the form of integrals over the known tractions $t_i(\boldsymbol{x}, t)$ on $S-S_1$, and over the unknown tractions acting on the part S_1, where the displacement field is prescribed. It is evident that the integral representations and the boundary condition (9.1) provide us with a set of integral equations for the unknown tractions acting on S_1. If it is possible to solve this set of integral equations the thus obtained surface tractions can be substituted in the integral representations of the solution.

With the exception of the simplest cases it is, unfortunately, rather difficult to solve the type of integral equations appearing in elastodynamic mixed boundary-value problems. When the body is a half-space and the complementary parts of the boundary, S_1 and $S-S_1$, are half-planes, the system of integral equations can in some cases be solved by means of integral transform techniques in conjunction with an application of the

Wiener-Hopf technique. For these cases it is, however, often more efficient to apply integral transforms directly to the boundary conditions and the governing equations.

Examples of physical situations which lead to the formulation of mixed boundary-value problems of elastodynamics are rapid indentation of a body and diffraction of elastic waves by cracks. In this chapter we will direct most of the attention toward the analysis of the diffraction of plane transient waves by semi-infinite cracks. These problems are amenable to treatment by the Wiener-Hopf technique. For a discussion of diffraction by cylindrical and spherical cavities we refer to the monograph by Mow and Pao.[1]

9.2. Antiplane shear motions

The simplest elastodynamic mixed boundary-value problems are concerned with two-dimensional horizontally polarized shear motions of a half-space. These motions are governed by the two-dimensional wave equation, which for the present purpose is written in the form

$$\frac{\partial^2 w}{\partial x^2} + \frac{\partial^2 w}{\partial y^2} = \frac{\partial^2 w}{\partial s^2}, \tag{9.3}$$

where

$$s = c_T t, \qquad c_T = \left(\frac{\mu}{\rho}\right)^{\frac{1}{2}}. \tag{9.4}$$

In eq. (9.3), $w(x, y, s)$ is the displacement normal to the (xy)-plane. The nonvanishing stresses are

$$\tau_{xz} = \mu \frac{\partial w}{\partial x}, \quad \text{and} \quad \tau_{yz} = \mu \frac{\partial w}{\partial y}. \tag{9.5a, b}$$

The following boundary conditions are considered:

$$y = 0, \quad x < 0: \quad \tau_{yz} = \tau(x, s) \tag{9.6}$$

$$y = 0, \quad x \geqq 0: \quad w = 0. \tag{9.7}$$

These boundary conditions apply if the surface of a half-space is prevented from motion over half of the surface and is subjected to tractions over the other half. It is assumed that the half-space is at rest prior to $t = 0$.

[1] C.-C. Mow and Y.-H. Pao, *The diffraction of elastic waves and dynamic stress concentrations*. Report R-482-PR, The Rand Corporation (1971).

In this section the problem defined by eqs. (9.3)–(9.7) is solved by employing the Green's function for the Neumann boundary conditions. We first compute the displacement at time t and position x, y for the case that the antiplane shear traction on the boundary $y = 0$ is independent of z and behaves as a delta function at position $x = x_0$ and time $s = s_0$. This displacement field is called the Green's function $G(x-x_0, y; s-s_0)$.

9.2.1. *Green's function*

The Green's function is the solution of eq. (9.3) with homogeneous initial conditions, and a boundary condition of the form

$$y = 0: \quad \tau_{yz} = \mu \frac{\partial w}{\partial y} = \delta(x-x_0)\delta(s-s_0). \tag{9.8}$$

The boundary condition (9.8) represents a concentrated impulsive antiplane shear load applied at time $s = s_0$ at position $x = x_0$. For a half-space which is initially at rest, the displacement wave generated by (9.8) can easily be worked out as

$$G(x-x_0, y; s-s_0) = -\frac{1}{\pi\mu R} H\{(s-s_0)-[(x-x_0)^2+y^2]^{\frac{1}{2}}\}, \tag{9.9}$$

where $H\{\ \}$ is the Heaviside step function and R is defined as

$$R = [(s-s_0)^2-(x-x_0)^2-y^2]^{\frac{1}{2}}. \tag{9.10}$$

Eq. (9.9) shows that a cylindrical wave emanates from $x = x_0$, $y = 0$. The wave front is shown in figure 9.1. A position $\bar{x}, \bar{y}$ is reached when $[(\bar{x}-x_0)^2+\bar{y}^2]^{\frac{1}{2}} = \bar{s}-s_0$. Conversely, at time $s = \bar{s}$ we can define a region of dependence for the position $\bar{x}, \bar{y}$ as a circle with radius $\bar{s}-s_0$ centered at $\bar{x}, \bar{y}$. At time $s = \bar{s}$, external disturbances applied at $s = s_0$ affect the field variables at $\bar{x}, \bar{y}$ only if they were applied inside the circle. It is now easily

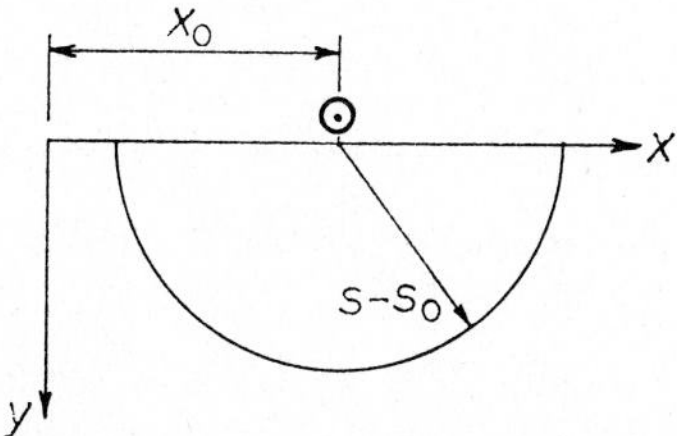

Fig. 9.1. Cylindrical shear wave.

seen that in more general terms the domain of dependence in the time-space domain for prescribed $\bar{x}$, $\bar{y}$ and $\bar{s}$ is a cone defined by

$$(\bar{s}-s)-[(\bar{x}-x)^2+(\bar{y}-y)^2]^{\frac{1}{2}} \geqq 0, \qquad \bar{s} \geqq s \geqq 0. \tag{9.11}$$

The cone is shown in figure 9.2.

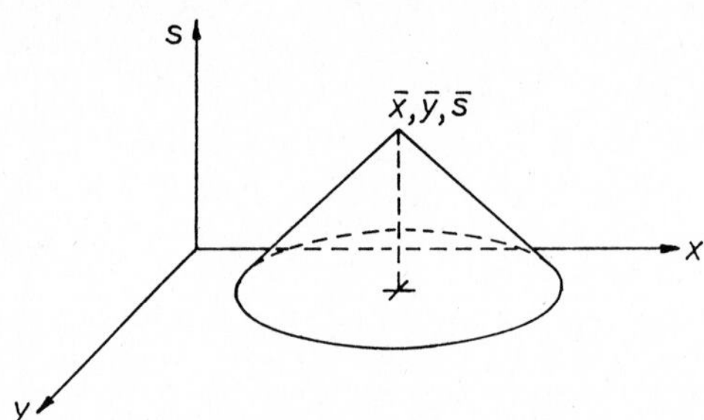

Fig. 9.2. Cone in time-space domain.

If the surface $y = 0$ is subjected to a distribution of surface tractions of the form

$$y = 0, \qquad \tau_{yz} = \tau(x, s), \tag{9.12}$$

linear superposition can be employed to write the displacement in the half-space $y \geqq 0$ in the form

$$w(x, y, s) = -\frac{1}{\pi\mu}\int_A\int \frac{\tau(x_0, s_0)}{R}\, \mathrm{d}x_0\, \mathrm{d}s_0, \tag{9.13}$$

where R is defined by eq. (9.10). It follows from the previous discussion that A is the area in the (xs)-plane which falls inside the cone (9.11). For $y > 0$, the intersection of the cone and the (xs)-plane is a hyperbola. In figure 9.3,

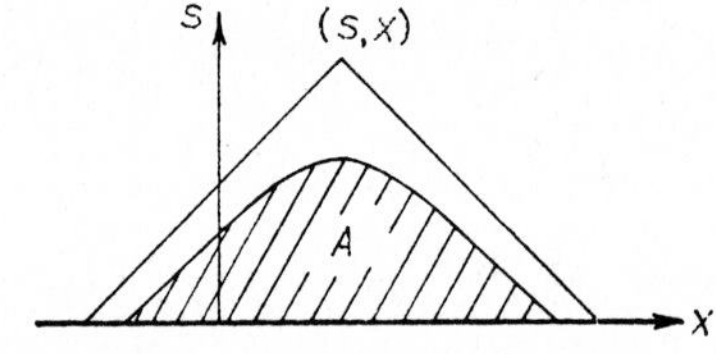

Fig. 9.3. Projection of cone on the xs-plane.

we show the projection of the cone on the (xs)-plane, as well as the intersection of the cone and the (xs)-plane. The integration in eq. (9.13) is over the shaded area. The integral may be written as

$$w(x, y, s) = -\frac{1}{\pi\mu}\int_0^{s-y} \mathrm{d}s_0 \int_{x-[(s-s_0)^2-y^2]^{\frac{1}{2}}}^{x+[(s-s_0)^2-y^2]^{\frac{1}{2}}} \frac{\tau(x_0, s_0)}{R}\,\mathrm{d}x_0. \tag{9.14}$$

Let us consider the special case that the surface traction is uniformly distributed over the shaded area, say $\tau(x, s) = \tau_0 f(s)$. The integral over x_0 then reduces to π, and we find

$$w(x, y, s) = -\frac{1}{\mu}\int_0^{s-y} f(s_0)\mathrm{d}s_0. \tag{9.15}$$

This is just a plane wave solution. Indeed, as far as position x, y at time s is concerned, the motion is nothing but a plane wave propagating into the medium.

A second example is concerned with a surface traction of the form

$$\tau_{yz} = H(-x)f(s).$$

Suppose the x-coordinate of the point at which we wish to compute the displacement is positive. The arrangement then is as shown in figure 9.3, and the displacement may be written as

$$w(x, y, s) = -\frac{1}{\pi\mu}\int_0^{s-r} f(s_0)\mathrm{d}s_0 \int_{x-[(s-s_0)^2-y^2]^{\frac{1}{2}}}^{0} \frac{\mathrm{d}x_0}{R},$$

where

$$r^2 = x^2+y^2. \tag{9.16}$$

The integration over x_0 can easily be worked out to yield

$$w(x, y, s) = \frac{1}{\pi\mu}\int_0^{s-r} \left\{\sin^{-1}\left[\frac{x}{[(s-s_0)^2-y^2]^{\frac{1}{2}}}\right] - \frac{\pi}{2}\right\} f(s_0)\mathrm{d}s_0. \tag{9.17}$$

If the surface disturbance varies with time as a Dirac delta function, $f(t) = \tau_0\delta(t)$, the displacement can, of course, be written out without further integration. Also introducing polar coordinates through

$$x = r\cos\theta, \quad \text{and} \quad y = r\sin\theta,$$

where r is defined by (9.16), and employing the relation $s = c_T t$, we find in that case

$$w(r, \theta, t) = \frac{c_T\tau_0}{\pi\mu}\left\{\sin^{-1}\left[\frac{r\cos\theta}{(c_T^2t^2-r^2\sin^2\theta)^{\frac{1}{2}}}\right] - \frac{\pi}{2}\right\} H(c_T t-r). \tag{9.18}$$

This expression is valid in the range $0 \leqq \theta \leqq \frac{1}{2}\pi$. Clearly the displacement wave is a cylindrical wave. It may be checked that in the domain $\frac{1}{2}\pi \leqq \theta \leqq \pi$

there is also a cylindrical wave as well as a plane wave. The pattern of wave fronts is shown in figure 9.4. The cylindrical wave is generated because the surface traction is discontinuous.

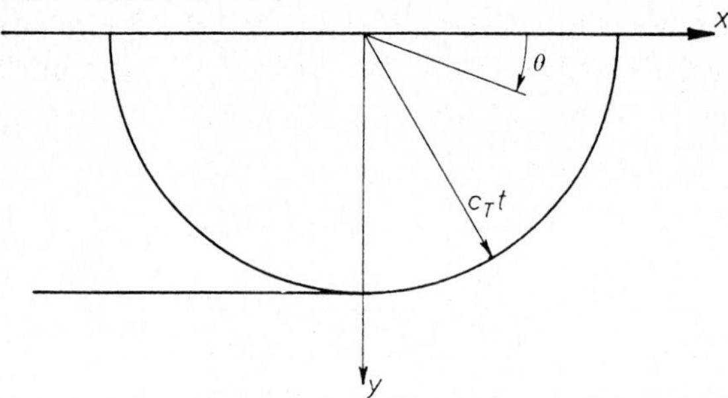

Fig. 9.4. Pattern of wave fronts for $\tau(x, 0, t) = \tau_0 H(-x) f(t)$.

The displacement (9.18) can be employed to compute the shear stress $\tau_{\theta z} = (\mu/r)\partial w/\partial\theta$. The result is

$$0 \leqq \theta \leqq \tfrac{1}{2}\pi: \frac{\tau_{\theta z}}{\tau_0} = -\frac{c_T}{\pi} \frac{(c_T^2 t^2 - r^2)^{\frac{1}{2}} \sin\theta}{c_T^2 t^2 - r^2 \sin^2\theta} H(c_T t - r). \qquad (9.19)$$

The stress shows a square-root singularity at the wavefront as θ approaches $\frac{1}{2}\pi$.

9.2.2. *The mixed boundary-value problem*

Now we return to the mixed conditions (9.6) and (9.7). Since the stress-distribution for $y = 0$, $x \geqq 0$ is unknown, the superposition integral (9.13)

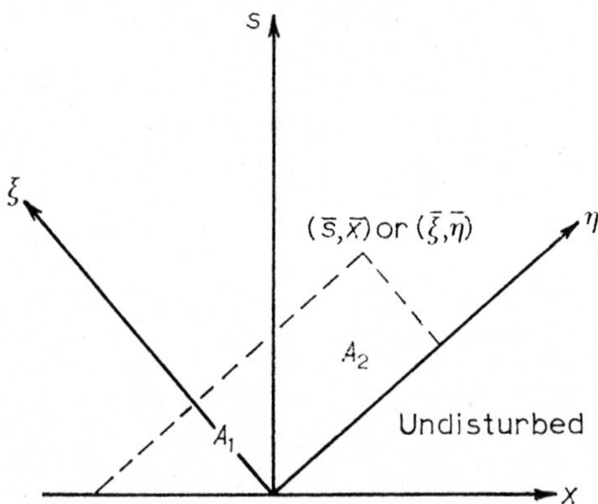

Fig. 9.5. Domain of influence in the (xs)-plane.

cannot be applied directly to express the displacement. The integral does, however, provide us with an integral equation to solve for the unknown stresses in the region $y = 0$, $x \geqq 0$.

For $y = 0$, the region of integration A in eq. (9.13) reduces to a triangular region because the vertex of the cone is in the (xs)-plane. The triangular region is indicated by dashed lines in figure 9.5. Suppose the unknown stresses for $x \geqq 0$ are denoted by

$$y = 0, \quad x \geqq 0\colon \tau_{yz} = \sigma(x, s). \tag{9.20}$$

The condition (9.7) that the displacement vanishes for $y = 0$, $\bar{x} \geqq 0$ and $\bar{s} \geqq 0$ then yields the following integral equation for $\sigma(x, s)$:

$$\bar{s} \geqq 0,\, \bar{x} \geqq 0\colon \int_{A_1}\!\!\int \frac{\tau(x, s)\mathrm{d}x\,\mathrm{d}s}{[(\bar{s}-s)^2-(\bar{x}-x)^2]^{\frac{1}{2}}} + \int_{A_2}\!\!\int \frac{\sigma(x, s)\mathrm{d}x\,\mathrm{d}s}{[(\bar{s}-s)^2-(\bar{x}-x)^2]^{\frac{1}{2}}} = 0. \tag{9.21}$$

The regions A_1 and A_2 are indicated in figure 9.5. It is *a priori* presumed that the line $s = x$, which represents the wave front of the cylindrical wave, separates the disturbed from the undisturbed part of the upper half of the (xs)-plane.

The integral equation (9.21) can be solved in a surprisingly simple manner by introducing the following characteristic coordinates in the (xs)-plane:

$$\xi = \frac{s-x}{\sqrt{2}}, \qquad \eta = \frac{s+x}{\sqrt{2}}. \tag{9.22a, b}$$

The denominators in (9.21) then reduce to

$$(\bar{s}-s)^2-(\bar{x}-x)^2 = 2(\bar{\xi}-\xi)(\bar{\eta}-\eta), \tag{9.23}$$

which has the advantage that the variables are separated. We consider a point $\bar{x}$, $\bar{s}$, or $\bar{\xi}$, $\bar{\eta}$, as indicated in figure 9.5. By introducing (9.22a, b) and (9.23) into (9.21), the integral equation (9.21) can be rewritten as

$$\int_0^{\bar{\xi}} \frac{\mathrm{d}\xi}{(\bar{\xi}-\xi)^{\frac{1}{2}}} \int_{-\xi}^{\xi} \frac{\tau(\xi, \eta)\mathrm{d}\eta}{(\bar{\eta}-\eta)^{\frac{1}{2}}} + \int_0^{\bar{\xi}} \frac{\mathrm{d}\xi}{(\bar{\xi}-\xi)^{\frac{1}{2}}} \int_{\xi}^{\bar{\eta}} \frac{\sigma(\xi, \eta)\mathrm{d}\eta}{(\bar{\eta}-\eta)^{\frac{1}{2}}} = 0, \tag{9.24}$$

where $\tau(\xi, \eta)$ follows from (9.6). Eq. (9.24) is evidently satisfied if

$$\int_{\xi}^{\bar{\eta}} \frac{\sigma(\xi, \eta)\mathrm{d}\eta}{(\bar{\eta}-\eta)^{\frac{1}{2}}} = -\int_{-\xi}^{\xi} \frac{\tau(\xi, \eta)\mathrm{d}\eta}{(\bar{\eta}-\eta)^{\frac{1}{2}}}. \tag{9.25}$$

The right-hand side of (9.25) is known. Eq. (9.25) is recognized as an integral equation for $\sigma(\xi, \eta)$ of the Abel type. This class of integral equations can be solved analytically. Upon multiplication of both sides by $(\zeta-\bar{\eta})^{-\frac{1}{2}}$ and integration over $\bar{\eta}$, we have

$$\int_{\xi}^{\zeta} \frac{\mathrm{d}\bar{\eta}}{(\zeta-\bar{\eta})^{\frac{1}{2}}} \int_{\xi}^{\bar{\eta}} \frac{\sigma(\xi, \eta)\mathrm{d}\eta}{(\bar{\eta}-\eta)^{\frac{1}{2}}} = -\int_{\xi}^{\zeta} \frac{\mathrm{d}\bar{\eta}}{(\zeta-\bar{\eta})^{\frac{1}{2}}} \int_{-\xi}^{\xi} \frac{\tau(\xi, \eta)\mathrm{d}\eta}{(\bar{\eta}-\eta)^{\frac{1}{2}}}.$$

By interchanging orders of integration and evaluating integrals we find

$$\sigma(\xi, \eta) = -\frac{1}{\pi} \frac{1}{(\eta-\xi)^{\frac{1}{2}}} \int_{-\xi}^{\xi} \frac{\tau(\xi, u)(\xi-u)^{\frac{1}{2}}}{\eta-u} \mathrm{d}u. \tag{9.26}$$

Eq. (9.26) expresses the stress in the region $\xi \leqq \eta$, which corresponds to $0 \leqq x \leqq c_T t$ in the physical variables.

Let us investigate in some detail the case that the applied traction is uniform in x and constant in time:

$$\tau(x, s) = \tau_0 H(s), \quad \text{or} \quad \tau(\xi, \eta) = \tau_0. \tag{9.27}$$

The integral in (9.26) can then be evaluated to yield

$$\sigma(\xi, \eta) = -\frac{2\tau_0}{\pi} \left\{ \left(\frac{2\xi}{\eta-\xi}\right)^{\frac{1}{2}} - \tan^{-1}\left(\frac{2\xi}{\eta-\xi}\right)^{\frac{1}{2}} \right\}.$$

In terms of the coordinates x and s, the stress is

$$\sigma(x, s) = -\frac{2\tau_0}{\pi} \left\{ \left(\frac{s-x}{x}\right)^{\frac{1}{2}} - \tan^{-1}\left(\frac{s-x}{x}\right)^{\frac{1}{2}} \right\}. \tag{9.28}$$

This expression applies of course for $x \leqq s$. It should be noted that the stress shows a square-root singularity at $x = 0$. For very small x we have

$$\sigma(x, t) \sim -\frac{2\tau_0}{\pi} \left(\frac{c_T t}{x}\right)^{\frac{1}{2}}. \tag{9.29}$$

If the complete displacement solution is desired, (9.28) together with (9.27) must be substituted into the integral representation (9.13), and the integrals must be evaluated.

The boundary conditions (9.27) and (9.7) are also pertinent to two problems of horizontally polarized wave motion in an unbounded medium due to the presence of a semi-infinite slit. The first of these is concerned with the transient waves generated by the sudden opening of a semi-infinite crack in an unbounded medium which is in a state of antiplane shear. The second problem concerns the diffraction by a semi-infinite slit of a plane wave whose wavefront is parallel to the slit.

In this section, a method was discussed to solve two-dimensional mixed boundary-value problems when the conditions are prescribed on semi-infinite parts of the boundary. It can be shown that this method can be extended to cases when boundary conditions are prescribed on finite parts of the boundary. There are, however, complications for two-dimensional problems with in-plane motions. We will, therefore, also discuss an alternative method to treat mixed boundary-value problems, which is known as the Wiener-Hopf technique.

9.3. The Wiener-Hopf technique

This ingenious scheme was devised to solve integral equations of the general form

$$f(x)+\int_0^\infty k(x-\eta)f(\eta)\mathrm{d}\eta = g(x), \qquad \text{for} \quad 0 \leqq x < \infty, \tag{9.30}$$

where the kernel $k(x)$ is defined in the interval $-\infty < x < \infty$. The function $g(x)$ is defined for $0 \leqq x < \infty$, but $g(x)$ is undefined for $-\infty < x < 0$. The function $f(x)$ is identically zero for $-\infty < x < 0$, and $f(x)$ is to be determined in the range $0 \leqq x < \infty$.

In the method of solution the two-sided Laplace transform, which was discussed in section 7.2, is used. Particularly instrumental in solving (9.30) is the convolution theorem which states that

$$\int_{-\infty}^\infty e^{-\zeta x}\mathrm{d}x \int_{-\infty}^\infty k(x-\eta)f(\eta)\mathrm{d}\eta = k^* f^*, \tag{9.31}$$

where $k^*(\zeta)$ and $f^*(\zeta)$ are the two-sided Laplace transforms of $k(x)$ and $f(x)$, respectively.

The following definitions are now introduced:

$$f_+(x) = \begin{cases} f(x) & x \geqq 0 \\ 0 & x < 0 \end{cases} \tag{9.32}$$

$$h_-(x) = \begin{cases} 0 & x \geqq 0 \\ h(x) & x < 0. \end{cases} \tag{9.33}$$

Eq. (9.30) would not be difficult to solve if the integral equation were defined over the whole range of x, i.e., $-\infty < x < \infty$. For this reason it appears opportune to extend the range of x and to rewrite (9.30) as

$$f_+(x)+\int_{-\infty}^\infty k(x-\eta)f_+(\eta)\mathrm{d}\eta = g_+(x)+h_-(x). \tag{9.34}$$

It is noted that for $x \geqq 0$, eq. (9.34) just reduces to (9.30). For $x < 0$, we have

$$\int_{-\infty}^{\infty} k(x-\eta)f_+(\eta) = h_-(x). \tag{9.35}$$

The price that was paid for extending the range of x comes in the form of the appearance of another unknown function, $h_-(x)$. Formally, eq. (9.34) can, however, be solved immediately by application of the two-sided Laplace transform, provided that $k(x)$ and the other functions satisfy some rather weak conditions.

Suppose that a real number γ can be found such that $k(x)\exp(-\gamma x)$ is absolutely integrable for $-\infty < x < \infty$. It is also assumed that $f_+(x)\exp(-\gamma x)$ and $g_+(x)\exp(-\gamma x)$ are absolutely integrable over $0 < x < \infty$ and, finally, that $h_-(x)\exp(-\gamma x)$ is absolutely integrable over $-\infty < x < 0$. We then define the transforms

$$k^*(\zeta) = \int_{-\infty}^{\infty} e^{-\zeta x}k(x)\mathrm{d}x \tag{9.36}$$

$$f_+^*(\zeta) = \int_{0}^{\infty} e^{-\zeta x}f_+(x)\mathrm{d}x \tag{9.37}$$

$$g_+^*(\zeta) = \int_{0}^{\infty} e^{-\zeta x}g_+(x)\mathrm{d}x \tag{9.38}$$

$$h_-^*(\zeta) = \int_{-\infty}^{0} e^{-\zeta x}h_-(x)\mathrm{d}x. \tag{9.39}$$

On the basis of the discussion of section 7.2, we conclude that $k^*(\zeta)$ is regular for $\mathscr{R}(\zeta) = \gamma$. Furthermore, $f_+^*(\zeta)$ and $g_+^*(\zeta)$ are regular for $\mathscr{R}(\zeta) > \gamma$, and h_-^* is regular for $\mathscr{R}(\zeta) < \gamma$. Furthermore, by virtue of the convolution theorem it follows from (9.34) that

$$f_+^* + k^* f_+^* = g_+^* + h_-^* \tag{9.40}$$

on the line $\mathscr{R}(\zeta) = \gamma$. It would seem that very little progress has been made, because h_-^* depends on f_+^*, as is clear from (9.35). By means of the Wiener-Hopf method it is, however, possible to solve for f_+^* without explicit knowledge of h_-^*.

We assume that $1+k^*(\zeta)$ does not have roots on $\mathscr{R}(\zeta) = \gamma$.[2] Under certain weak conditions, which will be touched upon in the next section,

[2] If there are roots, they can be factored out on both sides of eq. (9.40).

$1+k^*(\zeta)$ can be *factored*, i.e., $1+k^*(\zeta)$ can be written as

$$1+k^* = m_+^* m_-^*, \tag{9.41}$$

where m_+^* and $(m_+^*)^{-1}$ are bounded on $\mathscr{R}(\zeta) = \gamma$, and regular for $\mathscr{R}(\zeta) > \gamma$, while m_-^* and $(m_-^*)^{-1}$ are also bounded on $\mathscr{R}(\zeta) = \gamma$, but regular on $\mathscr{R}(\zeta) < \gamma$. Eq. (9.40) can then be rewritten as

$$m_+^* f_+^* = \frac{g_+^*}{m_-^*} + \frac{h_-^*}{m_-^*}. \tag{9.42}$$

Subsequently the term g_+^*/m_-^* can generally be *split up* in the following manner:

$$\frac{g_+^*}{m_-^*} = n_+^* + n_-^*, \tag{9.43}$$

where n_+^* is regular in $\mathscr{R}(\zeta) > \gamma$ and n_-^* is regular in $\mathscr{R}(\zeta) < \gamma$. Substitution of (9.43) into (9.42) yields

$$m_+^* \mathrm{f}_+^* - n_+^* = \frac{h_-^*}{m_-^*} + n_-^* \quad \text{on} \quad \mathscr{R}(\zeta) = \gamma. \tag{9.44}$$

Now, we define the functions

$$e_+^*(\zeta) = m_+^* \mathrm{f}_+^* - n_+^* \tag{9.45}$$

and

$$e_-^*(\zeta) = \frac{h_-^*}{m_-^*} + n_-^*. \tag{9.46}$$

Then, $e_+^*(\zeta)$ is regular in $\mathscr{R}(\zeta) > \gamma$ and $e_-^*(\zeta)$ is regular in $\mathscr{R}(\zeta) < \gamma$, while on $\mathscr{R}(\zeta) = \gamma$ the functions $e_+^*(\zeta)$ and $e_-^*(\zeta)$ are continuous and equal to each other. On the basis of a theorem of functional analysis[3], $e_+^*(\zeta)$ may then be considered as the analytic continuation of $e_-^*(\zeta)$, and vice versa. The functions $e_+^*(\zeta)$ and $e_-^*(\zeta)$ thus represent one and the same *entire* function $e^*(\zeta)$ (a function which is regular everywhere in the ζ-plane). In the next step, Liouville's theorem is employed, which states: If $e^*(\zeta)$ is entire and $|e^*(\zeta)|$ is bounded for all values of ζ in the complex plane, then $e^*(\zeta)$ *must be a constant*. To determine the actual value of the constant it is necessary to determine the value of either $e_+^*(\zeta)$ or $e_-^*(\zeta)$ at one particular value of ζ. Often it can be shown that $e^*(\zeta) = o(1)$ for $|\zeta| \to \infty$; then it follows from

[3] E. C. Titchmarsh, *The theory of functions*, 2nd ed. Oxford, Oxford University Press (1950), p. 157.

Liouville's theorem that $e^*(\zeta) \equiv 0$, and we find from (9.45)

$$f_+^*(\zeta) = \frac{n_+^*}{m_+^*}\,. \tag{9.47}$$

The function $f(x)$ can subsequently be obtained by the use of the inversion integral of the two-sided Laplace transform.

Let us summarize the various steps leading to the solution of an integral equation of the type (9.30):

(1) Extend the range of definition of the integral equation to $-\infty < x < \infty$.

(2) Apply the two-sided Laplace transform.

(3) Determine the line of juncture (often there is a strip of overlap).

(4) Carry out the factorization

$$1+k^* = m_+^* m_-^*,$$

and rewrite

$$m_+^* f_+^* = \frac{g_+^*}{m_-^*} + \frac{h_-^*}{m_-^*}$$

(5) Split up the term

$$\frac{g_+^*}{m_-^*} = n_+^* + n_-^*,$$

and observe that

$$\begin{aligned} m_+^* f_+^* + n_+^* &= \frac{h_-^*}{m_-^*} + n_-^* \\ &= \text{entire function.} \end{aligned}$$

(6) Apply Liouville's theorem to conclude

$$\text{entire function} = C$$

(7) Determine C from the behavior of $f(x)$ at small x. This behavior can usually be gleaned from physical considerations.

(8) Evaluate the inverse transform.

The factorization generally is the most difficult part of the procedure.

It may be worthwhile to emphasize that in the present exposition of the Wiener-Hopf technique it is not necessary that the two half-planes of regularity overlap. All that is required is that the two half-planes border along the line $\mathscr{R}(\zeta) = \gamma$. It should also be noted that the entire function need not be bounded, but may be $O(\zeta^{\frac{1}{2}})$ as $|\zeta| \to \infty$. An extension of Liouville's theorem may then be invoked to conclude that the function is a constant.

9.4. The decomposition of a function

The difficult task in most applications of the Wiener-Hopf technique is the factoring and splitting of functions into components which are regular in adjoining half-planes. Generally only the very simplest problems allow an identification by inspection of the factors and the terms in eqs. (9.41) and (9.42), respectively. It is, therefore, necessary to devise a general method for carrying out the required decomposition of functions.

9.4.1. General procedure

Under rather general conditions the splitting

$$L(\zeta) = L_-(\zeta)+L_+(\zeta) \tag{9.48}$$

can be carried out by means of an integration in the complex ζ-plane. We do not require that the equality hold in a strip, but rather just along a straight line $\mathscr{R}(\zeta) = \gamma$, which forms the intersection of the two half-planes.[4]

Let us consider the function $L(\zeta)$ satisfying

$$L(\zeta) = O(|\zeta|^{-p}) \quad \text{for} \quad |\zeta| \to \infty, \tag{9.49}$$

where $p > 0$, and let us define the functions

$$L_+(\zeta) = -\frac{1}{2\pi i}\int_{\gamma-i\infty}^{\gamma+i\infty} \frac{L(z)}{z-\zeta}\,\mathrm{d}z \qquad \text{for} \quad \mathscr{R}(\zeta) > \gamma \tag{9.50}$$

and

$$L_-(\zeta) = \frac{1}{2\pi i}\int_{\gamma-i\infty}^{\gamma+i\infty} \frac{L(z)}{z-\zeta}\,\mathrm{d}z \qquad \text{for} \quad \mathscr{R}(\zeta) < \gamma. \tag{9.51}$$

It is assumed that $L(z)$ is such that these integrals are absolutely convergent provided $\mathscr{R}(\zeta) \neq \gamma$. By considering closed contours in the half-planes $\mathscr{R}(z) > \gamma$ and $\mathscr{R}(z) < \gamma$, respectively, and employing eq. (9.49) and Jordan's lemma, it becomes apparent that $L_+(\zeta)$ and $L_-(\zeta)$ are regular for $\mathscr{R}(\zeta) > \gamma$ and $\mathscr{R}(\zeta) < \gamma$, respectively.

For a point $\zeta_0 = \gamma+i\omega_0$ on $\mathscr{R}(\zeta) = \gamma$, we have by definition

$$L_+(\zeta_0) = \lim_{\zeta\to\zeta_0} L_+(\zeta), \tag{9.52}$$

[4] In this we follow by and large a report by A. T. de Hoop, *Wiener-Hopf Techniek* (in Dutch). Technological University of Delft (1963).

where ζ approaches ζ_0 in the right half-plane, and

$$L_-(\zeta_0) = \lim_{\zeta\to\zeta_0} L_-(\zeta), \tag{9.53}$$

where ζ approaches ζ_0 in the left half-plane. We will now prove that

$$L(\zeta_0) = L_+(\zeta_0)+L_-(\zeta_0). \tag{9.54}$$

The proof is provided by an actual evaluation of the limits in (9.52) and (9.53). If $\Omega_1 < \omega_0 < \Omega_2$, then

$$\begin{aligned} L_+(\zeta_0) = &-\frac{1}{2\pi i}\int_{\gamma-i\infty}^{\gamma+i\Omega_1} \frac{L(z)}{z-\zeta_0}\,\mathrm{d}z - \frac{1}{2\pi i}\lim_{\zeta\to\zeta_0}\int_{\gamma+i\Omega_1}^{\gamma+i\Omega_2}\frac{L(z)}{z-\zeta}\,\mathrm{d}z \\ &-\frac{1}{2\pi i}\int_{\gamma+i\Omega_2}^{\gamma+i\infty}\frac{L(z)}{z-\zeta_0}\,\mathrm{d}z. \end{aligned} \tag{9.55}$$

Furthermore,

$$\begin{aligned} -\frac{1}{2\pi i}\lim_{\zeta\to\zeta_0}\int_{\gamma+i\Omega_1}^{\gamma+i\Omega_2}\frac{L(z)}{z-\zeta}\,\mathrm{d}z = &-\frac{1}{2\pi i}\int_{\gamma+i\Omega_1}^{\gamma+i\Omega_2}\frac{L(z)-L(\zeta_0)}{z-\zeta_0}\,\mathrm{d}z \\ &-\frac{L(\zeta_0)}{2\pi i}\lim_{\zeta\to\zeta_0}\int_{\gamma+i\Omega_1}^{\gamma+i\Omega_2}\frac{\mathrm{d}z}{z-\zeta}. \end{aligned} \tag{9.56}$$

The second integral on the right-hand side can, however, be written as

$$-\frac{L(\zeta_0)}{2\pi i}\lim_{\zeta\to\zeta_0}\int_{\gamma+i\Omega_1}^{\gamma+i\Omega_2}\frac{\mathrm{d}z}{z-\zeta} = \tfrac{1}{2}L(\zeta_0)-\frac{1}{2\pi i}\,P\int_{\gamma+i\Omega_1}^{\gamma+i\Omega_2}\frac{L(z)}{z-\zeta_0}\,\mathrm{d}z, \tag{9.57}$$

if ζ approaches ζ_0 in the right half-plane. The symbol P indicates that we must take the Cauchy principal value of the integral. By substituting (9.56) and (9.57) into (9.55) and letting Ω_1 and Ω_2 approach ω_0, we obtain

$$L_+(\zeta_0) = \tfrac{1}{2}L(\zeta_0)-\frac{1}{2\pi i}\,P\int_{\gamma-i\infty}^{\gamma+i\infty}\frac{L(z)}{z-\zeta_0}\,\mathrm{d}z. \tag{9.58}$$

In an analogous manner we can show that

$$L_-(\zeta_0) = \tfrac{1}{2}L(\zeta_0)+\frac{1}{2\pi i}\,P\int_{\gamma-i\infty}^{\gamma+i\infty}\frac{L(z)}{z-\zeta_0}\,\mathrm{d}z. \tag{9.59}$$

These two equations are special cases of the formulas of Plemelj. An addition of the two equations yields (9.54).

On $\mathscr{R}(\zeta) = \gamma$ we next consider the decomposition

$$K(\zeta) = K_+(\zeta)K_-(\zeta), \tag{9.60}$$

where $K(\zeta)$ does not have zeros on $\mathscr{R}(\zeta) = \gamma$. We write

$$\ln K(\zeta) = \ln K_+(\zeta) + \ln K_-(\zeta), \tag{9.61}$$

and we observe that, provided that K_+ and K_- have no zeros in their half-planes of regularity, each of $\ln K_+$ and $\ln K_-$ is itself regular in the corresponding domain. Thus, assuming that $\ln K$ satisfies (9.49), the decomposition of $K(\zeta)$ as a product is equivalent to the decomposition of $\ln K$ as a sum.

An approximate manner of determining the factors, which does not require integrations in the complex plane, was developed by Koiter.[5] In this method the function $K(\zeta)$ is replaced by a much simpler function $K'(\zeta)$, which shows the same general characteristics as $K(\zeta)$ but which can be factored with little effort.

9.4.2. *Example: the Rayleigh function*

In applications of the Wiener-Hopf technique to elastodynamic problems the Rayleigh function

$$R(\zeta) = (s_T^2 - 2\zeta^2)^2 + 4\zeta^2(s_L^2 - \zeta^2)^{\frac{1}{2}}(s_T^2 - \zeta^2)^{\frac{1}{2}} \tag{9.62}$$

has to be factored on a line or in a strip in the complex ζ-plane. The function $\ln R(\zeta)$ does, however, not satisfy the condition (9.49). For that reason we consider instead of $R(\zeta)$

$$K(\zeta) = \frac{(s_T^2 - 2\zeta^2)^2 + 4\zeta^2(s_L^2 - \zeta^2)^{\frac{1}{2}}(s_T^2 - \zeta^2)^{\frac{1}{2}}}{2(s_T^2 - s_L^2)(s_R^2 - \zeta^2)}. \tag{9.63}$$

The function $K(\zeta)$ is nowhere zero, while

$$K(\zeta) = 1 + O(\zeta^{-2}) \quad \text{as} \quad |\zeta| \to \infty.$$

Now,

$$\ln K_+ = -\frac{1}{2\pi i}\int_{\gamma - i\infty}^{\gamma + i\infty} \frac{\ln K(z)}{z - \zeta}\, dz \qquad \text{for} \quad \mathscr{R}(\zeta) > \gamma \tag{9.64}$$

and

$$\ln K_- = \frac{1}{2\pi i}\int_{\gamma - i\infty}^{\gamma + i\infty} \frac{\ln K(z)}{z - \zeta}\, dz \qquad \text{for} \quad \mathscr{R}(\zeta) < \gamma. \tag{9.65}$$

[5] W. T. Koiter, *Proceedings Koninklijke Nederlandse Akademie van Wetenschappen*, Series B57 (1954) p. 558.

The logarithms are made single-valued by introducing branch cuts along $\mathscr{I}(z) = 0, s_L < |\mathscr{R}(z)| < s_T$, as shown in figure 9.6, and taking the

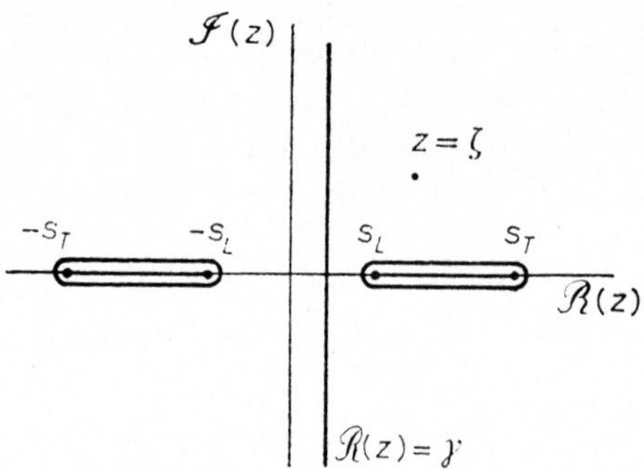

Fig. 9.6. Branch cuts for ln $K(z)$.

principal value of the integral. If ζ is not a real number the contours in (9.64) and (9.65) may be deformed into the loops Γ_+ and Γ_- around the branch cuts. We find

$$\ln K_+(\zeta) = -\frac{1}{\pi}\int_{s_L}^{s_T} \tan^{-1}\left[\frac{4z^2(z^2-s_L^2)^{\frac{1}{2}}(s_T^2-z^2)^{\frac{1}{2}}}{(s_T^2-2z^2)^2}\right]\frac{dz}{z+\zeta}, \tag{9.66}$$

while $K_-(\zeta)$ follows from the relation $K_-(\zeta) = K_+(-\zeta)$.

When the point ζ approaches the branch cuts in the z-plane from above or below, the path Γ_+ or Γ_- must be indented below or above ζ. The integrals defining K_+ or K_- then become singular and we may resort to principal values. The integrals defining K_+ and K_- are, however, obviously never singular simultaneously. Thus we may calculate these exceptional cases without resorting to principal values by employing

$$K_\pm(\zeta) = \frac{K(\zeta)}{K_\mp(\zeta)},$$

where we choose in the denominator the K-function which is not defined by a singular integral.

9.5. Diffraction of a horizontally polarized shear wave

The wave motion generated when a plane wave is diffracted by an obstacle of vanishing thickness and semi-infinite extent can be analyzed in an elegant fashion by employing integral transforms together with the Wiener-

Hopf technique and the Cagniard-de Hoop method. In this section we will investigate the diffraction of a plane horizontally polarized shear wave by a semi-infinite slit. The incident displacement wave is of the form

$$w^i(x, y, t) = G(t - s_T x \cos\alpha - s_T y \sin\alpha), \tag{9.67}$$

where

$$G(t) = H(t)\int_0^t g(s)\mathrm{d}s. \tag{9.68}$$

In eq. (9.68), $H(\ )$ denotes the Heaviside step function, α is the angle of the normal to the wavefront and the x-axis; and $s_T = 1/c_T$ is the slowness of transverse waves. The position of the wavefront prior to time $t = 0$ is shown in figure 9.7. Here the angle α is restricted to the range

$$0 \leqq \alpha \leqq \frac{\pi}{2}.$$

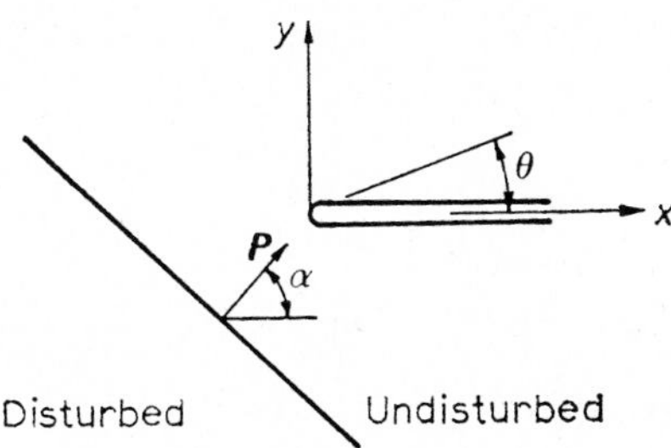

Fig. 9.7. Incident SH-wave.

The wavefront strikes the tip of the slit at time $t = 0$. The pattern of diffracted waves is, of course, two-dimensional in nature, since both the incident wave and the diffracting slit are independent of the z-coordinate.

If there were no slit at $y = 0$, $x > 0$, the incident wave would give rise to the following shear stress in the plane $y = 0$:

$$\tau_{yz}^i = -\mu s_T \sin\alpha\; g(t - s_T x \cos\alpha) H(t - s_T x \cos\alpha). \tag{9.69}$$

The solution to the diffraction problem is now obtained by superimposing on the incident wave the wave motion which is generated in an initially undisturbed medium by shear tractions that are equal and opposite to (9.69) and that are applied on both sides of the slit $y = \pm 0$, $x > 0$. Through the superposition the surface of the slit is rendered free of tractions. Since the shear tractions are equal but opposite on the two sides of the slit, the induced wave motion is antisymmetric with respect to the plane $y = 0$, and the displacement in the z-direction thus vanishes for $y = 0$, $x \leqq 0$. Considering the half-plane $y \geqq 0$, the superimposed wave motion must then satisfy the following conditions at $y = 0$:

$$x > 0 \qquad \tau_{yz} = \mu s_T \sin\alpha \, g(t - s_T x \cos\alpha) H(t - s_T x \cos\alpha) \tag{9.70}$$

$$x \leqq 0 \qquad w = 0. \tag{9.71}$$

The governing partial differential equation is

$$\frac{\partial^2 w}{\partial x^2} + \frac{\partial^2 w}{\partial y^2} = s_T^2 \frac{\partial^2 w}{\partial t^2}, \tag{9.72}$$

while the initial conditions are

$$w(x, y, 0) = \dot{w}(x, t, 0) \equiv 0.$$

Eqs. (9.70)-(9.72) define a mixed boundary-value problem.

The boundary conditions (9.70) and (9.71) may be recast in the form ($y = 0$, $-\infty < x < \infty$)

$$\tau_{yz} = \tau_- + \tau_+ \tag{9.73}$$

$$w = w_+, \tag{9.74}$$

where we employ the notation defined by (9.32) and (9.33), and where

$$\tau_+ = \mu s_T \sin\alpha \, g(t - s_T x \cos\alpha) H(t - s_T x \cos\alpha) H(x). \tag{9.75}$$

Upon application of the one-sided Laplace transform with respect to time, see eq. (7.20), and the two-sided Laplace transform with respect to x, see eq. (7.15), eq. (9.72) reduces to an ordinary differential equation. For $y \geqq 0$ the pertinent solution of that equation is

$$\bar{w}^* = A(p, \zeta) e^{-(s_T^2 p^2 - \zeta^2)^{\frac{1}{2}} y}, \tag{9.76}$$

where $A(p, \zeta)$ is to be determined. The boundary conditions (9.73) and (9.74) transform into

$$\bar{\tau}^*_{yz} = \bar{\tau}^*_- + \bar{\tau}^*_+ \tag{9.77}$$

$$\bar{w}^* = \bar{w}^*_+, \tag{9.78}$$

where we find from (9.75) that

$$\bar{\tau}^*_+ = \frac{\mu s_T \bar{g}(p) \sin\alpha}{\zeta + s_T p \cos\alpha}. \tag{9.79}$$

From (9.76) and (9.78) it follows that $A(p, \zeta) = \bar{w}_+^*$, whereupon substitution into (9.77) yields

$$-\mu(s_T^2 p^2 - \zeta^2)^{\frac{1}{2}}\bar{w}_+^* = \bar{\tau}_-^* + \bar{\tau}_+^*. \tag{9.80}$$

It is apparent that (9.80) is analogous to (9.40).

The usual Wiener-Hopf manipulations readily lead to an expression for $A(p, \zeta)$. First we make an observation on the analytic behavior of $\bar{\tau}_-^*$ in the ζ-plane. For the boundary conditions (9.70) and (9.71) the plane $y = 0$, $x < 0$ is undisturbed until sufficient time has passed for a cylindrical wave propagating with slowness s_T to arrive. Considering then $\tau_{yz}(x, 0, t)$ for some $x_1 < 0$, its Laplace transform over time can be written as

$$\begin{aligned}\bar{\tau}_{yz}(x_1, 0, p) &= \int_{t_1}^{\infty} e^{-pt}\tau_{yz}(x_1, 0, t)\mathrm{d}t \\ &= e^{-pt_1}\int_0^{\infty} e^{-ps}\tau_{yz}(x_1, 0, s+t_1)\mathrm{d}s,\end{aligned} \tag{9.81}$$

where $t_1 = -s_T x_1$ is the time at which the first disturbance reaches the position $y = 0$, $x = x_1 < 0$. From (9.81) it follows that $\bar{\tau}_{yz}$ has the asymptotic behavior

$$\bar{\tau}_{yz}(x, 0, p) \sim e^{ps_T x} \qquad \text{as} \quad x \to -\infty. \tag{9.82}$$

In view of (9.73) this also represents the asymptotic behavior of $\bar{\tau}_-$. It follows that $\bar{\tau}_-^*$ is regular in the half-plane $\mathscr{R}(\zeta) < s_T p$. With regard to the points on $y = 0$, $x > 0$, it is apparent that a particle at position $y = 0$, $x = x_2 > 0$ is undisturbed until time $t_2 = s_T x_2 \cos\alpha$. Thus, $\bar{w}_+$ is $O(\exp -ps_T x \cos\alpha)$, which implies that $\bar{w}_+^*$ is regular for $\mathscr{R}(\zeta) > -s_T p \cos\alpha$. Returning to (9.80) it is next noted that the factoring of $(s_T^2 p^2 - \zeta^2)^{\frac{1}{2}}$ is trivial. The factors are $(s_T p - \zeta)^{\frac{1}{2}}$ and $(s_T p + \zeta)^{\frac{1}{2}}$, which are regular in the half-planes $\zeta < s_T p$ and $\zeta > -s_T p$, respectively. Eq. (9.80) may then be rewritten as

$$-\mu(s_T p + \zeta)^{\frac{1}{2}}\bar{w}_+^* = \frac{\bar{\tau}_-^*}{(s_T p - \zeta)^{\frac{1}{2}}} + \frac{\bar{\tau}_+^*}{(s_T p - \zeta)^{\frac{1}{2}}}. \tag{9.83}$$

From (9.79) it follows that $\bar{\tau}_+^*$ is regular for $\mathscr{R}(\zeta) > -s_T p \cos\alpha$, and it is thus necessary to split the second term on the right-hand side of (9.83), since the radical in the denominator is regular for $\mathscr{R}(\zeta) < s_T p$. The splitting can be performed as

$$\frac{s_T p(1+\cos\alpha)}{(\zeta + s_T p \cos\alpha)(s_T p - \zeta)^{\frac{1}{2}}} = \frac{(s_T p - \zeta)^{\frac{1}{2}}}{\zeta + s_T p \cos\alpha} + \frac{1}{(s_T p - \zeta)^{\frac{1}{2}}}.$$

The first term on the right-hand side is regular for $\zeta < s_T p$, except for the pole at $\zeta = -s_T p \cos\alpha$. This pole can, however, be removed by writing

$$\frac{(s_T p-\zeta)^{\frac{1}{2}}}{\zeta+s_T p\cos\alpha} = \frac{(s_T p-\zeta)^{\frac{1}{2}}-(s_T p)^{\frac{1}{2}}(1+\cos\alpha)^{\frac{1}{2}}}{\zeta+s_T p\cos\alpha} + \frac{(s_T p)^{\frac{1}{2}}(1+\cos\alpha)^{\frac{1}{2}}}{\zeta+s_T p\cos\alpha}.$$

Eq. (9.83) can now be rearranged in the following manner:

$$-\mu(s_T p+\zeta)^{\frac{1}{2}}\bar{w}_+^* - \frac{\mu s_T \bar{g}(p)\sin\alpha}{(s_T p)^{\frac{1}{2}}(1+\cos\alpha)^{\frac{1}{2}}}\frac{1}{\zeta+s_T p\cos\alpha}$$
$$= \frac{\mu s_T \bar{g}(p)\sin\alpha}{s_T p(1+\cos\alpha)}\left[\frac{(s_T p-\zeta)^{\frac{1}{2}}-(s_T p)^{\frac{1}{2}}(1+\cos\alpha)^{\frac{1}{2}}}{\zeta+s_T p\cos\alpha} + \frac{1}{(s_T p-\zeta)^{\frac{1}{2}}}\right]$$
$$+\frac{\bar{\tau}_-^*}{(s_T p-\zeta)^{\frac{1}{2}}}. \tag{9.84}$$

The left-hand side of (9.84) is regular for $\mathscr{R}(\zeta) > -s_T p\cos\alpha$, while the right-hand side is regular for $\mathscr{R}(\zeta) < s_T p$. Because of the equality in the strip of overlap both sides of eq. (9.65) represent one and the same bounded entire function, say $e(\zeta)$. In view of Liouville's theorem, $e(\zeta)$ is a constant. The magnitude of the constant can be obtained from order conditions on $e(\zeta)$ as $|\zeta| \to \infty$, which in turn are obtained from order conditions on the dependent field variables in the vicinity of $x = 0$. It is well known that the stress may show a singularity of the square root variety near $x = 0$, see for example eq. (9.29). Referring to the boundary condition (9.73) we find that this result implies that $\bar{\tau}_-(x) = O(|x|^{-\frac{1}{2}})$ as $x \to 0^-$. By virtue of an Abelian theorem for one-sided Laplace transforms, see eqs. (7.24) and (7.25), it is then concluded that $\bar{\tau}_-^*$ is $O(|\zeta|^{-\frac{1}{2}})$ as $|\zeta| \to \infty$. As a consequence the entire function $e(\zeta)$ vanishes identically and we can solve for $\bar{w}_+^*$ from the left-hand side of eq. (9.84). Since $A(\zeta, p) = \bar{w}_+^*$, we find

$$A(\zeta, p) = -\frac{s_T^{\frac{1}{2}}\bar{g}(p)\sin\alpha}{p^{\frac{1}{2}}(1+\cos\alpha)^{\frac{1}{2}}}\frac{1}{(s_T p+\zeta)^{\frac{1}{2}}(\zeta+s_T p\cos\alpha)}. \tag{9.85}$$

By the use of the inversion integral for the two-sided Laplace transform, eq. (7.16), $\bar{w}(x, y, p)$ may be expressed as

$$\bar{w} = -\frac{\bar{g}(p)}{p}\overline{W}(x, y, p), \tag{9.86}$$

where

$$\overline{W} = \frac{1}{2\pi i}\frac{s_T^{\frac{1}{2}}p^{\frac{1}{2}}\sin\alpha}{(1+\cos\alpha)^{\frac{1}{2}}}\int_{\zeta_1-i\infty}^{\zeta_1+i\infty}\frac{e^{\zeta x-(s_T{}^2p^2-\zeta^2)^{\frac{1}{2}}y}}{(s_T p+\zeta)^{\frac{1}{2}}(\zeta+s_T p\cos\alpha)}\,d\zeta. \tag{9.87}$$

Eq. (9.87) is valid for $y \geqq 0$. The path of integration $\mathscr{R}(\zeta) = \zeta_1$ is restricted to the strip $-s_T p \cos\alpha < \zeta_1 < s_T p$. It follows from the convolution theorem for the one-sided Laplace transform that $w(x, y, t)$ can be expressed as a convolution integral of $W(x, y, t)$ and the inverse Laplace transform of $\bar{g}(p)/p$.

To invert the one-sided Laplace transform (9.87) the Cagniard-de Hoop method is used. Thus, following the steps discussed in section 7.9 we first introduce the substitution $\zeta = p\eta$, whereupon (9.87) becomes

$$\overline{W} = \frac{1}{2\pi i}\frac{s_T^{\frac{1}{2}}\sin\alpha}{(1+\cos\alpha)^{\frac{1}{2}}}\int_{\eta_1-i\infty}^{\eta_1+i\infty}\frac{e^{-p[(s_T^2-\eta^2)^{\frac{1}{2}}y-\eta x]}}{(s_T+\eta)^{\frac{1}{2}}(\eta+s_T\cos\alpha)}\,d\eta. \tag{9.88}$$

The singularities in the η-plane are a simple pole at $\eta = -s_T\cos\alpha$ and branch points at $\eta = \pm s_T$. Next we deform the path of integration from $\mathscr{R}(\eta) = \eta_1$ to a path along which the integral can be recognized as a one-sided Laplace transform. The appropriate path is given by eq. (7.176) as

$$\eta_{T\pm} = -\frac{t}{r}\cos\theta \pm i\left(\frac{t^2}{r^2} - s_T^2\right)^{\frac{1}{2}}\sin\theta, \tag{9.89}$$

where r and θ are defined by eqs. (7.168) and (7.169), respectively. When $s_T r < t < \infty$, eq. (9.89) represents a hyperbola whose point of intersection with the real axis is always located in between the branch points $\eta = -s_T$ and $\eta = +s_T$. Therefore, no difficulties arise in connection with the branch cuts. On the other hand the contribution of the pole $\eta = -s_T\cos\alpha$ has to be taken into account for values of θ in the region $0 \leqq \theta < \alpha$. The contribution from the pole is

$$\overline{W} = e^{-ps_T(x\cos\alpha + y\sin\alpha)}. \tag{9.90}$$

From (9.86) and (9.90) the inverse Laplace transform of the corresponding displacement wave is

$$w(x, y, t) = -H(t - s_T x\cos\alpha - s_T y\sin\alpha)\int_0^{t-s_T x\cos\alpha - s_T y\sin\alpha} g(v)\,dv. \tag{9.91}$$

Eq. (9.91) is a plane wave of the same form as the incident wave. It cancels the incident wave in the shadow zone $0 \leqq \theta < \alpha$.

The integral along the path defined by eq. (9.89) represents a cylindrical wave. Since the contour defined by eq. (9.89) is symmetric with respect to the real axis, and since p and t are both real, the cylindrical wave can be rewritten as

$$\overline{W}^c = \frac{1}{\pi}\frac{s_T^{\frac{1}{2}}\sin\alpha}{(1+\cos\alpha)^{\frac{1}{2}}}\int_{s_T r}^{\infty} e^{-pt}\mathscr{I}\left\{\frac{\partial\eta_{T+}/\partial t}{(s_T+\eta_{T+})^{\frac{1}{2}}(\eta_{T+}+s_T\cos\alpha)}\right\}\mathrm{d}t \tag{9.92}$$

By using the relation

$$\frac{1}{(s_T^2-\eta_{T+}^2)^{\frac{1}{2}}}\frac{\partial\eta_{T+}}{\partial t} = \frac{i}{(t^2-s_T^2r^2)^{\frac{1}{2}}}, \tag{9.93}$$

Eq. (9.92) can be further simplified to

$$\overline{W}^c = \frac{1}{\pi}\frac{s_T^{\frac{1}{2}}\sin\alpha}{(1+\cos\alpha)^{\frac{1}{2}}}\int_{s_T r}^{\infty}\frac{e^{-pt}}{(t^2-s_T^2r^2)^{\frac{1}{2}}}\mathscr{R}\left\{\frac{(s_T-\eta_{T+})^{\frac{1}{2}}}{\eta_{T+}+s_T\cos\alpha}\right\}\mathrm{d}t,$$

and thus

$$W^c(x, y, t) = \frac{1}{\pi}\frac{s_T^{\frac{1}{2}}(1-\cos\alpha)^{\frac{1}{2}}}{(t^2-s_T^2r^2)^{\frac{1}{2}}}\mathscr{R}\left\{\frac{(s_T-\eta_{T+})^{\frac{1}{2}}}{\eta_{T+}+s_T\cos\alpha}\right\}H(t-s_Tr).$$

Upon substitution of (9.89) this expression reduces to

$$W^c(x, y, t) = -\frac{1}{\pi}\left(\frac{s_Tr}{2}\right)^{\frac{1}{2}}\left\{\frac{\sin\frac{1}{2}(\alpha-\theta)}{t-s_Tr\cos(\alpha-\theta)}+\frac{\sin\frac{1}{2}(\alpha+\theta)}{t-s_Tr\cos(\alpha+\theta)}\right\}\times\frac{H(t-s_Tr)}{(t-s_Tr)^{\frac{1}{2}}}, \tag{9.94}$$

where $0 \leqq \theta \leqq \pi$. In view of (9.86) and (9.68) the corresponding displacement is

$$w^d(x, y, t) = -\int_{s_Tr}^{t} G(t-s)W^c(x, y, s)\mathrm{d}s, \tag{9.95}$$

where $G(t)$ is defined by eq. (9.68). Eq. (9.95) is called the diffracted wave. For $0 \leqq \theta \leqq \pi$ the complete solution of the diffraction problem can now be expressed as

$$0 \leqq \theta < \alpha \qquad w(x, y, t) = w^d(x, y, t)$$

$$\alpha < \theta \leqq \pi \qquad w(x, y, t) = w^i(x, y, t)+w^d(x, y, t),$$

where w^i and w^d are defined by eqs. (9.67) and (9.95), respectively. The solutions in the domain $\pi \leqq \theta \leqq 2\pi$ can be obtained in the same manner. For a time $t \geqq 0$ the pattern of wavefronts is shown in figure 9.8.

It is of interest to examine the stress singularities in the vicinity of the tip of the crack. The singularities come from the diffracted wave, which is

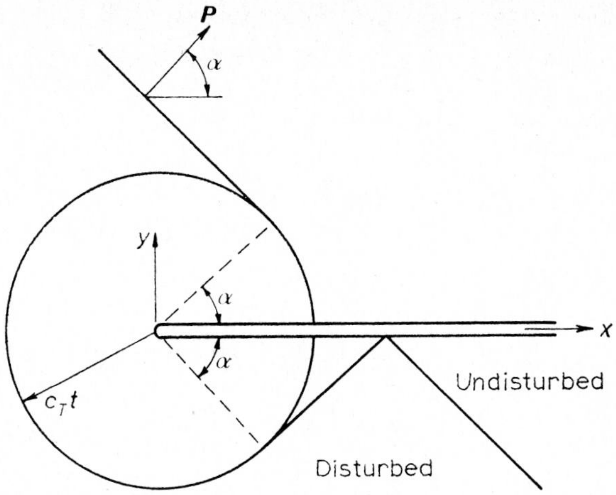

Fig. 9.8. Pattern of wavefronts for diffraction of an SH-wave by a slit.

given by eq. (9.95). By considering the limit as $r \to 0$ the shear stress $\tau_{\theta z}$ becomes[6]

$$\tau_{\theta z} = \frac{\mu}{r}\frac{\partial w}{\partial \theta} = \frac{\mu}{\pi}\left(\frac{s_T}{2r}\right)^{\frac{1}{2}}\left[\cos \tfrac{1}{2}(\alpha+\theta) - \cos \tfrac{1}{2}(\alpha-\theta)\right]F(t), \qquad (9.96)$$

where

$$F(t) = \tfrac{1}{2}\int_0^t \frac{G(t-\tau)-G(t)}{\tau^{\frac{3}{2}}}\,\mathrm{d}\tau - \frac{G(t)}{t^{\frac{1}{2}}}. \qquad (9.97)$$

It is noted that the stresses are of order $r^{-\frac{1}{2}}$ for small values of r.

The appearance of a stress singularity has implications from the point of view of fracture mechanics. It is generally assumed that the magnitude of the stress intensity factor is a criterion for extension of a crack. In the expression for the stress in the vicinity of the tip of a crack of length $2a$, the stress-intensity factor K is defined as

$$K = \lim_{r \to 0}\left(\frac{\pi r}{a}\right)^{\frac{1}{2}} |\tau_{\theta z}|.$$

[6] Details of the computation can be found in G. H. Handelman and L. A. Rubenfeld, *Journal of Applied Mechanics* **36** (1969) 873.

Now let us consider the case when the function $g(v)$ in eq. (9.68) is of the form

$$g(v) = \frac{\tau_0}{\mu} c_T,$$

which corresponds to an incident step-stress wave. The function $F(t)$ then is of the form

$$F(t) = -2\frac{\tau_0}{\mu} t^{\frac{1}{2}} c_T,$$

and the stress intensity factor for eq. (9.96) becomes

$$K = 2\tau_0 \left(\frac{c_T t}{2\pi a}\right)^{\frac{1}{2}} [\cos \tfrac{1}{2}(\alpha+\theta) - \cos \tfrac{1}{2}(\alpha-\theta)].$$

This stress intensity factor increases with time. If the crack is of finite length $2a$, the stress intensity factor increases until a wave diffracted from the opposite crack tip arrives. It is of interest that for the case of a step-stress wave the dynamic stress intensity factor shows an overshoot of $4/\pi-1$ as compared to the stress intensity factor of the corresponding static problem.[7] It is, therefore, conceivable that there are cases for which extension of a crack does not occur under a gradually applied system of loads, but where the material does indeed fracture when the same system of loads is rapidly applied and gives rise to waves.

9.6. Diffraction of a longitudinal wave

9.6.1. Formulation

The diffraction of a plane longitudinal wave by a semi-infinite slit is another problem that can conveniently be investigated by using integral transforms together with the Wiener-Hopf technique and the Cagniard-de Hoop method. Let the incident wave be represented by

$$u^i(x, y, t) = G(t - s_L x \cos\alpha - s_L y \sin\alpha) \cos\alpha \tag{9.98}$$

$$v^i(x, y, t) = G(t - s_L x \cos\alpha - s_L y \sin\alpha) \sin\alpha, \tag{9.99}$$

where

$$G(t) = H(t)\int_0^t g(s)\mathrm{d}s. \tag{9.100}$$

[7] For details we refer to J. D. Achenbach, *International Journal of Engineering Science* **8** (1970) 947.

In eq. (9.100), $H(\ )$ denotes the Heaviside step function, α is the angle of the normal to the wavefront and the x-axis, and $s_L = 1/c_L$ is the slowness of longitudinal waves. We restrict the angle α to the range

$$0 \leqq \alpha \leqq \frac{\pi}{2},$$

The position of the wavefront prior to time $t = 0$ is as shown in figure 9.7.

It can easily be verified that the wave motion generated by the diffraction of the longitudinal wave consists of the superposition of the incident wave and the solutions to two boundary-initial value problems which are symmetric and antisymmetric, respectively, with respect to the plane $y = 0$. The two problems are:

Symmetric problem:

$$y = 0: x > 0 \quad \tau_y = (\lambda + 2\mu \sin^2 \alpha) s_L H(t - s_L x \cos \alpha) g(t - s_L x \cos \alpha) \tag{9.101}$$

$$x \leqq 0 \qquad v = 0 \tag{9.102}$$

$$-\infty < x < \infty \quad \tau_{yx} = 0 \tag{9.103}$$

Antisymmetric problem:

$$y = 0: x > 0 \quad \tau_{yx} = \mu s_L \sin 2\alpha \quad H(t - s_L x \cos \alpha) g(t - s_L x \cos \alpha) \tag{9.104}$$

$$x \leqq 0 \qquad u = 0 \tag{9.105}$$

$$-\infty < x < \infty \quad \tau_y = 0. \tag{9.106}$$

Indeed, the superposition of (9.101), (9.103), (9.104) and (9.106) on the corresponding stresses due to the incident wave renders the surface $y = 0$, $x > 0$ free of tractions. The initial conditions for the problems defined by (9.101)–(9.106) are

$$u(x, y, 0) = \dot{u}(x, y, 0) = v(x, y, 0) = \dot{v}(x, y, 0) \equiv 0. \tag{9.107}$$

The symmetric and antisymmetric boundary-initial value problems are two-dimensional in nature. The problems are formulated and solved by using the displacement potentials φ and ψ. The pertinent equations are stated by eqs. (7.106)–(7.113) of section 7.7. The solutions to the transient problems are again obtained by applying the one-sided Laplace transform over time and the two-sided Laplace transform over the spatial variable x. Since the transforms will be inverted by means of the Cagniard-de Hoop method, we employ the pair of two-sided Laplace transforms stated by eqs. (7.182) and (7.183) of section 7.11. In the half-space $y \geqq 0$ the appropriate expressions for $\bar{\varphi}^*$ and $\bar{\psi}^*$ then are

$$\bar{\varphi}^* = \Phi(p, \eta)e^{-p\gamma_L y} \tag{9.108}$$

$$\bar{\psi}^* = \Psi(p, \eta)e^{-p\gamma_T y}, \tag{9.109}$$

where

$$\gamma_L = (s_L^2 - \eta^2)^{\frac{1}{2}}, \qquad \gamma_T^2 = (s_T^2 - \eta^2)^{\frac{1}{2}}. \tag{9.110a, b}$$

The corresponding transforms of the displacements and the stresses are given by eqs. (7.187)–(7.191) of section 7.11.

9.6.2. *Application of the Wiener-Hopf technique*

As a first step toward the solution of the symmetric and antisymmetric problems we define boundary conditions at $y = 0$ over the whole range of x by introducing pairs of unknown functions τ_{y-}, v_+ and τ_{xy-}, u_+, respectively. Upon applying the one-sided Laplace transform with respect to time and the two-sided Laplace transform with respect to x, the boundary conditions (9.101)–(9.103) then lead to the equations

$$y = 0, \; -\infty < x < \infty: \qquad \bar{\tau}_y^* = \bar{\tau}_{y-}^* + \bar{\tau}_{y+}^* \tag{9.111}$$

$$\bar{v}^* = \bar{v}_+^* \tag{9.112}$$

$$\bar{\tau}_{yx}^* = 0, \tag{9.113}$$

where $\bar{\tau}_{y+}^*$ follows from (9.101) as

$$\bar{\tau}_{y+}^* = \frac{(\lambda + 2\mu \sin^2 \alpha)s_L}{\eta + s_L \cos \alpha} \frac{\bar{g}(p)}{p}. \tag{9.114}$$

Similarly the boundary conditions (9.104)–(9.106) result in the equations

$$y = 0, \; -\infty < x < \infty: \qquad \bar{\tau}_{yx}^* = \bar{\tau}_{yx-}^* + \bar{\tau}_{yx+}^* \tag{9.115}$$

$$\bar{u}^* = \bar{u}_+^* \tag{9.116}$$

$$\bar{\tau}_y^* = 0, \tag{9.117}$$

where

$$\bar{\tau}_{yx+}^* = \frac{\mu s_L \sin 2\alpha}{\eta + s_L \cos \alpha} \frac{\bar{g}(p)}{p}. \tag{9.118}$$

By arguments that are completely analogous to those presented in the previous section we conclude that $\bar{\tau}_{y-}^*$ and $\bar{\tau}_{yx-}^*$ are regular in the half-plane $\mathscr{R}(\eta) < s_L$, while $\bar{v}_+^*$ and $\bar{u}_+^*$ are regular in the half-plane $\mathscr{R}(\eta) > -s_L \cos \alpha$.

Employing eqs. (9.108) and (9.109) in conjunction with the boundary conditions (9.111)–(9.113) we obtain three equations for the four unknowns $\Phi_s(p, \eta)$, $\Psi_s(p, \eta)$, $\bar{\tau}^*_{y-}$ and $\bar{v}^*_+$, where the subscript s is used for the symmetric problem. Eqs. (9.112) and (9.113) yield $\Phi_s(p, \eta)$ and $\Psi_s(p, \eta)$ in the forms

$$\Phi_s(p, \eta) = -\frac{s_T^2 - 2\eta^2}{p s_T^2 \gamma_L} \bar{v}^*_+ \tag{9.119}$$

$$\Psi_s(p, \eta) = -\frac{2\eta}{p s_T^2} \bar{v}^*_+, \tag{9.120}$$

whereupon the remaining equation reduces to the following Wiener-Hopf type equation relating $\bar{\tau}^*_{y-}$ and $\bar{v}^*_+$:

$$2\mu p \frac{s_T^2 - s_L^2}{s_T^2} \frac{\eta^2 - s_R^2}{\gamma_L} K(\eta) \bar{v}^*_+ = \bar{\tau}^*_{y-} + \bar{\tau}^*_{y+}, \tag{9.121}$$

where

$$K(\eta) = -\frac{(s_T^2 - 2\eta^2)^2 + 4\eta^2 \gamma_L(\eta)\gamma_T(\eta)}{2(s_T^2 - s_L^2)(\eta^2 - s_R^2)}. \tag{9.122}$$

The numerator of this expression is recognized as the function $R(\eta)$ for the slowness of Rayleigh surface waves, see eq. (5.96) of chapter 5. The roots of the equation $R(\eta) = 0$ are $s_R = \pm 1/c_R$, where c_R is the velocity of Rayleigh waves.

In the same manner eqs. (9.115)–(9.117) lead to three equations for the four unknowns $\Phi_a(p, \eta)$, $\Psi_a(p, \eta)$, $\bar{\tau}^*_{yx-}$ and $\bar{u}^*_+$, where the subscripts a are employed to label the antisymmetric problem. The functions $\Phi_a(p, \eta)$ and $\Psi_a(p, \eta)$ are subsequently expressed in terms of $\bar{u}^*_+$ as

$$\Phi_a(p, \eta) = \frac{2\eta}{p s_T^2} \bar{u}^*_+ \tag{9.123}$$

$$\Psi_a(p, \eta) = -\frac{s_T^2 - 2\eta^2}{p s_T^2 \gamma_T} \bar{u}^*_+, \tag{9.124}$$

whereupon the remaining eq. (9.115) relates $\bar{u}^*_+$ and $\bar{\tau}^*_{yx-}$ by

$$2\mu p \frac{s_T^2 - s_L^2}{s_T^2} \frac{\eta^2 - s_R^2}{\gamma_T} K(\eta) \bar{u}^*_+ = \bar{\tau}^*_{yx-} + \bar{\tau}^*_{yx+}, \tag{9.125}$$

where $K(\eta)$ is defined by eq. (9.122).

The Wiener-Hopf equations (9.121) and (9.125) hold in the strip defined by $-s_L \cos\alpha < \mathscr{R}(\eta) < s_L$. As a first step in determining the solutions to these equations $K(\eta)$ is factored as

$$K(\eta) = K_+(\eta) \cdot K_-(\eta), \tag{9.126}$$

where $K_+(\eta)$ and its reciprocal are regular for $\mathscr{R}(\eta) > -s_L$, and $K_-(\eta)$ and its reciprocal are regular for $\mathscr{R}(\eta) < s_L$. Explicit expressions for $K_+(\eta)$ and $K_-(\eta)$ are derived in section 9.4. Also factoring $\gamma_L(\eta)$ as

$$\gamma_L(\zeta) = \gamma_{L+} \cdot \gamma_{L-}, \tag{9.127}$$

where

$$\gamma_{L+} = (s_L+\eta)^{\frac{1}{2}}, \qquad \gamma_{L-} = (s_L-\eta)^{\frac{1}{2}}, \tag{9.128a, b}$$

Eq. (9.121) may be rewritten as

$$2\mu p\,\frac{s_T^2-s_L^2}{s_T^2}\,\frac{\eta+s_R}{\gamma_{L+}}\,K_+\,\bar{v}_+^* = \frac{\gamma_{L-}\,\bar{\tau}_{y-}^*}{(\eta-s_R)K_-} + \frac{\gamma_{L-}\,\bar{\tau}_{y+}^*}{(\eta-s_R)K_-}. \tag{9.129}$$

It remains to split the term $\gamma_{L-}\,\bar{\tau}_{y+}^*/K_-$. By employing (9.114) we write

$$\frac{\gamma_{L-}\,\bar{\tau}_{y+}^*}{(\eta-s_R)K_-} = \frac{(\lambda+2\mu\sin^2\alpha)s_L}{\eta+s_L\cos\alpha}\,\frac{\bar{g}(p)}{p}\left[\frac{\gamma_{L-}}{(\eta-s_R)K_-}-L\right] + \frac{(\lambda+2\mu\sin^2\alpha)s_L}{\eta+s_L\cos\alpha}\,\frac{\bar{g}(p)}{p}\,L, \tag{9.130}$$

where

$$L = \left.\frac{\gamma_{L-}(\eta)}{(\eta-s_R)K_-(\eta)}\right|_{\eta=-s_L\cos\alpha}. \tag{9.131}$$

Thus, L is independent of η. Eq. (9.129) may now be expressed in the form

$$2\mu p\,\frac{s_T^2-s_L^2}{s_T^2}\,\frac{\eta+s_R}{\gamma_{L+}}\,K_+\,\bar{v}_+^* - \frac{(\lambda+2\mu\sin^2\alpha)s_L}{\eta+s_L\cos\alpha}\,\frac{\bar{g}(p)}{p}\,L = \frac{\gamma_{L-}\,\bar{\tau}_{y-}^*}{(\eta-s_R)K_-} + \frac{(\lambda+2\mu\sin^2\alpha)s_L}{\eta+s_L\cos\alpha}\,\frac{\bar{g}(p)}{p}\left[\frac{\gamma_{L-}}{(\eta-s_R)K_-}-L\right]. \tag{9.132}$$

The left-hand side of this equation is regular for $\mathscr{R}(\eta) > -s_L\cos\alpha$, while the right-hand side is regular for $\mathscr{R}(\eta) < s_L$.

Similarly, eq. (9.125) may be expressed in the form

$$2\mu p\,\frac{s_T^2-s_L^2}{s_T^2}\,\frac{\eta+s_R}{\gamma_{T+}}\,K_+\,\bar{u}_+^* - \frac{\mu s_L\sin 2\alpha}{\eta+s_L\cos\alpha}\,\frac{\bar{g}(p)}{p}\,M = \frac{\gamma_{T-}\,\bar{\tau}_{yx-}^*}{(\eta-s_R)K_-} + \frac{\mu s_L\sin 2\alpha}{\eta+s_L\cos\alpha}\,\frac{\bar{g}(p)}{p}\left[\frac{\gamma_{T-}}{(\eta-s_R)K_-}-M\right], \tag{9.133}$$

where

$$M = \frac{\gamma_{T-}(\eta)}{(\eta - s_R)K_-(\eta)}\bigg|_{\eta=-s_L\cos\alpha}. \tag{9.134}$$

The usual reasoning now leads to the solutions

$$\bar{v}_+^* = \frac{\lambda+2\mu\sin^2\alpha}{2\mu}\,\frac{s_L s_T^2}{s_T^2-s_L^2}\,\frac{\bar{g}(p)}{p^2}\,\frac{\gamma_{L+}}{\eta+s_R}\,\frac{L}{\eta+s_L\cos\alpha}\,\frac{1}{K_+} \tag{9.135}$$

$$\bar{u}_+^* = \frac{s_L s_T^2\sin 2\alpha}{2(s_T^2-s_L^2)}\,\frac{\bar{g}(p)}{p^2}\,\frac{\gamma_{T+}}{\eta+s_R}\,\frac{M}{\eta+s_L\cos\alpha}\,\frac{1}{K_+}. \tag{9.136}$$

The four functions Φ_s, Ψ_s, Φ_a and Ψ_a subsequently follow from eqs. (9.119), (9.120), (9.123) and (9.124), respectively.

9.6.3. *Inversion of transforms*

The transforms can be inverted by means of the Cagniard-de Hoop method. We will carry out some of the manipulations to obtain the stress component $\tau_y(x, y, t)$.

The transform of the stress is

$$\bar{\tau}_y^* = (\bar{\tau}_y^*)_s + (\bar{\tau}_y^*)_a, \tag{9.137}$$

where subscripts s and a refer to the symmetric and the antisymmetric problems, respectively. By employing eqs. (7.190), (9.108), (9.109), (9.119) and (9.120), we can write

$$(\bar{\tau}_y^*)_s = \frac{\mu\bar{g}(p)}{p}\left[\Lambda_{Ls}e^{-p\gamma_L y} + \Lambda_{Ts}e^{-p\gamma_T y}\right], \tag{9.138}$$

where

$$\Lambda_{Ls} = -\frac{(s_T^2-2\eta^2)^2}{s_T^2\gamma_L}\,\frac{p^2}{\bar{g}(p)}\,\bar{v}_+^* \tag{9.139}$$

$$\Lambda_{Ts} = -\frac{4\eta^2\gamma_T}{s_T^2}\,\frac{p^2}{\bar{g}(p)}\,\bar{v}_+^*. \tag{9.140}$$

The corresponding expression for the antisymmetric problem is

$$(\bar{\tau}_y^*)_a = \frac{\mu\bar{g}(p)}{p}\left[\Lambda_{La}e^{-p\gamma_L y} + \Lambda_{Ta}e^{-p\gamma_T y}\right], \tag{9.141}$$

where

$$\Lambda_{La} = \frac{2\eta(s_T^2-2\eta^2)}{s_T^2}\,\frac{p^2}{\bar{g}(p)}\,\bar{u}_+^* \tag{9.142}$$

$$\Lambda_{Ta} = -\Lambda_{La}. \tag{9.143}$$

A formal application of the inversion integral for the two-sided Laplace transform, eq. (7.183), yields $\bar{\tau}_y(x, y, p)$ in the form

$$\bar{\tau}_y = \mu\bar{g}(p)(\bar{I}_L + \bar{I}_T), \tag{9.144}$$

where

$$\bar{I}_L = \frac{1}{2\pi i}\int_{\eta_1 - i\infty}^{\eta_1 + i\infty} (\Lambda_{Ls} + \Lambda_{La}) e^{-p(\gamma_L y - \eta x)} d\eta \tag{9.145}$$

$$\bar{I}_T = \frac{1}{2\pi i}\int_{\eta_1 - i\infty}^{\eta_1 + i\infty} (\Lambda_{Ts} + \Lambda_{Ta}) e^{-p(\gamma_T y - \eta x)} d\eta. \tag{9.146}$$

Following the usual steps of the Cagniard-de Hoop scheme we seek contours in the complex η-plane along which the exponentials in the integrand take the form exp $(-pt)$. As shown in section 7.11, such contours are given by

$$\eta = \eta_{L\pm}(r, \theta, t) = -\frac{t}{r}\cos\theta \pm i\left(\frac{t^2}{r^2} - s_L^2\right)^{\frac{1}{2}} \sin\theta \tag{9.147}$$

for the integral in eq. (9.145), where $s_L r \leqq t < \infty$, and

$$\eta = \eta_{T\pm}(r, \theta, t) = -\frac{t}{r}\cos\theta \pm i\left(\frac{t^2}{r^2} - s_T^2\right)^{\frac{1}{2}} \sin\theta \tag{9.148}$$

for the integral in eq. (9.146), where $s_T r \leqq t < \infty$. Eqs. (9.147) and (9.148) each define one branch of two distinct hyperbolae. The contours defined by $\eta_{L\pm}$ and $\eta_{T\pm}$ are discussed in some detail in section 7.11.

An examination of eqs. (9.139), (9.140), (9.142) and (9.143) reveals that the functions Λ_{Ls}, Λ_{Ts}, Λ_{La} and Λ_{Ta} asymptotically behave as $\zeta^{-\frac{1}{2}}$ as $|\zeta| \to \infty$. Thus by the residue theorem and Jordan's lemma we can replace the inversion integrals in (9.145) and (9.146) by the integrals along the hyperbolic paths $\eta_{L\pm}$ and $\eta_{T\pm}$, respectively, plus the residues due to any poles crossed in the changes of the paths of integration.

Let us first consider the integral in eq. (9.145). Since the point of intersection of the hyperbola (9.147) with the real axis is always located between $\eta = -s_L$ and $\eta = s_L$, no difficulties arise in connection with the branch cuts. In changing the path of integration the contribution of the pole at $\eta = -s_L \cos\alpha$ must be taken into account when $\cos\theta > \cos\alpha$, i.e., when $\theta < \alpha$ if $0 \leqq \theta \leqq \pi$. We have

$$\bar{I}_L = \int_{s_L r}^{\infty} D_L(r, \theta, t) e^{-pt} dt + R_L, \tag{9.149}$$

where

$$D_L = \frac{1}{\pi} \mathscr{I} \left\{ [\Lambda_{Ls} + \Lambda_{La}]_{\eta = \eta_{L+}} \frac{\partial \eta_{L+}}{\partial t} \right\}, \tag{9.150}$$

and the contribution from the pole, R_L, is

$$\theta < \alpha:\ R_L = [(\eta + s_L \cos\alpha)(\Lambda_{Ls} + \Lambda_{La})]_{\eta = -s_L \cos\alpha}\, e^{-ps_L(y \sin\alpha + x\cos\alpha)} \tag{9.151}$$

$$\theta > \alpha:\ R_L = 0. \tag{9.152}$$

Now we turn to the integral in eq. (9.146), and we encounter the same difficulty as in section 7.11, in that the vertex $\eta = -s_T \cos\theta$ of the hyperbola defined by $\eta_{T\pm}$ may or may not be located in between the branch points $\eta = \pm s_L$. Since the functions defined by eqs. (9.138)–(9.143) are regular in the right-half planes we need be concerned only with the branch cuts in the left half-plane. This implies that for $\cos^{-1}(s_L/s_T) \leqq \theta \leqq 0$ the branch cut emanating from $\eta = s_L$ must be encircled. We find ($0 \leqq \theta \leqq \cos^{-1}(s_L/s_T)$):

$$\bar{I}_T = \int_{t_{TL}}^{s_T r} D_{TL}(r, \theta, t) e^{-pt} dt + \int_{s_T r}^{\infty} D_T(r, \theta, t) e^{-pt} dt + R_T, \tag{9.153}$$

where t_{TL} is defined by eq. (7.203), and

$$D_{TL} = \frac{1}{\pi} \mathscr{I} \left\{ [\Lambda_{Ts} + \Lambda_{Ta}]_{\eta = \eta_{TL}} \frac{\partial \eta_{TL}}{\partial t} \right\} \tag{9.154}$$

$$D_T = \frac{1}{T} \mathscr{I} \left\{ [\Lambda_{Ts} + \Lambda_{Ta}]_{\eta = \eta_{T+}} \frac{\partial \eta_{T+}}{\partial t} \right\}, \tag{9.155}$$

and the contribution from the pole is

$$\theta < \alpha:\ R_T = [(\eta + s_L \cos\alpha)(\Lambda_{Ts} + \Lambda_{Ta})]_{\eta = -s_L \cos\alpha}\, e^{-ps_L(y \sin\alpha + x\cos\alpha)}, \tag{9.156}$$

$$\theta > \alpha:\ R_T = 0.$$

A little algebra leads to the result that the sum of the contributions (9.151) and (9.156) is

$$R = R_T + R_L = \frac{(\lambda + 2\mu \sin^2\alpha) s_L}{\mu} e^{-ps_L(y \sin\alpha + x\cos\alpha)}. \tag{9.157}$$

For $\theta \geqq \cos^{-1}(s_L/s_T)$ we have

$$\bar{I}_T = \int_{s_T r}^{\infty} D_T(r, \theta, t) e^{-pt} dt, \tag{9.158}$$

where D_T is defined by eq. (9.155).

The inverse one-sided Laplace transforms of eqs. (9.150), (9.154), (9.155), (9.157) and (9.158) can now be obtained by inspection. In view of eq. (9.144) the stresses τ_y are subsequently obtained as convolution integrals over these results and the function $g(t)$. Eq. (9.150) leads to a diffracted longitudinal wave, while (9.155) and (9.158) lead to diffracted transverse waves. Just as in section 7.11, eq. (9.155) is a head wave. Finally, eq. (9.157) cancels out the incident wave in the shadow of the slit. The corresponding results for $y \leqq 0$ can be obtained by minor modifications. The pattern of wavefronts is shown in figure 9.9.

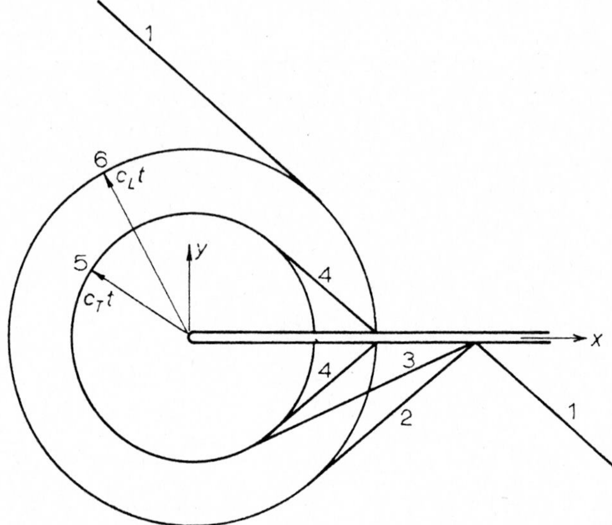

Fig. 9.9. Diffraction of a longitudinal wave by a slit. 1 = incident longitudinal wave. 2 = reflected longitudinal wave. 3 = reflected transverse wave. 4 = head waves. 5 = diffracted transverse wave. 6 = diffracted longitudinal wave.

For short times the analysis can be extended to diffraction by a crack of finite length. The computations were carried out by Thau and Lu.[8] The stresses at the crack tip are again singular, and for a step-stress wave the stress-intensity factor shows a dynamic overshoot of about 30 % over the corresponding static factor.

9.7. Problems

9.1. An unbounded elastic medium contains a semi-infinite slit in the plane

[8] S. A. Thau and T.-H. Lu, *International Journal of Solids and Structures* 7 (1971) 731.

defined by $x > 0$, $y = \pm 0$, $-\infty < z < \infty$. The faces of the slit are subjected to uniform antiplane surface tractions over the strip $0 < x \leqq a$, i.e.,

$$
\begin{aligned}
0 < x \leqq a,\ y = \pm 0: &\quad \tau_{yz} = \tau_0 H(t) \\
a < x < \infty,\ y = \pm 0: &\quad \tau_{yz} = 0 \\
0 < x < \infty,\ y = \pm 0: &\quad \tau_y = \tau_{yx} \equiv 0.
\end{aligned}
$$

The medium is at rest prior to time $t = 0$.

Use the method of Green's functions to derive an integral equation for the shear stress τ_{yz} in the plane $y = 0$. Solve the integral equation. Determine the stress singularity for small values of $|x|$.

9.2. An elastic half-space is clamped over $y = 0$, $x \leqq 0$, and free of surface tractions over $y = 0$, $x > 0$. A pair of incident and reflected plane transient

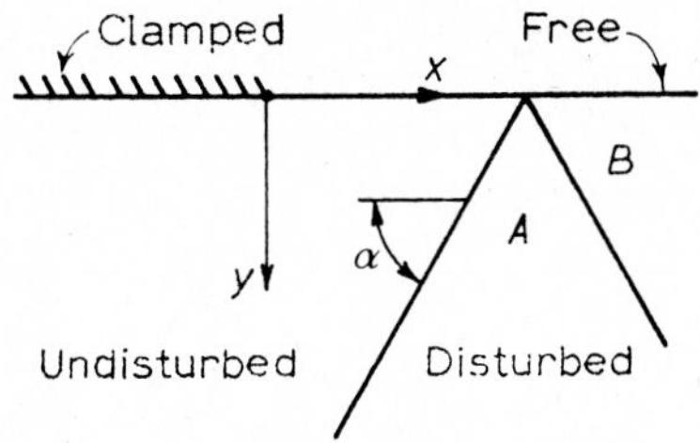

waves propagates along the free surface, as shown in the figure. The incident displacement wave is horizontally polarized and is of the general form

$$w_i(x, y, t) = H(c_T t + x \sin\alpha + y \cos\alpha) \int_0^{c_T t + x \sin\alpha + y \cos\alpha} g(v) \mathrm{d}v.$$

This expression represents the displacement in region A.

(a) Determine the displacement in region B.

(b) The system of waves reaches the point $x = 0$, $y = 0$ at time $t = 0$. For $t > 0$ compute the stress τ_{yz} in the plane $y = 0$ by the method of Green's functions.

(c) Determine the stress singularity in the plane $y = 0$.

9.3. In the domain $0 \leqq x < \infty$ a function $f(x)$ satisfies the following integral equation:

$$f(x) + 4 \int_0^\infty e^{-|x-\eta|} f(\eta) \mathrm{d}\eta = e^{-x}, \qquad 0 \leqq x < \infty.$$

Solve for $f(x)$ by applying the Wiener-Hopf technique.

9.4. An unbounded elastic medium contains a rigid semi-infinite screen which prevents displacement in any direction over the domain $x \geqq 0$, $y = 0$, $-\infty < z < \infty$. A horizontally polarized shear wave of the form eq. (9.67) strikes the screen at time $t = 0$. Determine $w(x, y, t)$ by applying integral transform techniques.

9.5. Examine the diffraction of a plane SV-wave by a semi-infinite slit. The wavefront of the incident wave is parallel to the slit, i.e., the incident wave is defined by

$$u^i(x, y, t) = G(t - s_T y), \qquad v^i = w^i \equiv 0,$$

where

$$G(t) = H(t)\int_0^t g(s)ds.$$

Determine the shear stress τ_{yx} in the plane of the slit.

CHAPTER 10

THERMAL AND VISCOELASTIC EFFECTS, AND EFFECTS OF ANISOTROPY AND NONLINEARITY

10.1. Thermal effects

Most materials undergo appreciable changes of volume when subjected to variations of the temperature. If the thermal expansions or contractions are not freely admitted, temperature variations give rise to thermal stresses. Conversely a change of volume is attended by a change of the temperature. When a given element is compressed or dilated, these volume changes are accompanied by heating and cooling, respectively. The study of the influence of the temperature of an elastic solid upon the distribution of stress and strain, and of the inverse effect of the deformation upon the temperature distribution is the subject of the theory of thermoelasticity.

The first effect, that of the temperature on the stresses, can be accounted for by modifying Hooke's law. The classical argument for the modification proceeds as follows[1]: Consider an isotropic elastic solid in an arbitrary state of stress, and let a small element be detached from its surroundings and subjected to a temperature change $T-T_0$, where T_0 is the reference temperature. The additional straining of the element is given by

$$(\varepsilon_{ij})_T = \alpha(T-T_0)\delta_{ij}, \tag{10.1}$$

where α is the coefficient of linear thermal expansion. It follows that if the distribution of strain in the heated solid is ε_{ij} then the strain produced by the mechanical forces is $\varepsilon_{ij}-(\varepsilon_{ij})_T$, and this tensor should replace ε_{ij} in Hooke's law. This leads to the following modification of eq. (2.28):

$$\tau_{ij} = \lambda\varepsilon_{kk}\delta_{ij}+2\mu\varepsilon_{ij}-\alpha(3\lambda+2\mu)(T-T_0)\delta_{ij}. \tag{10.2}$$

On the basis of intuitive arguments it is less simple to account for the effect of the deformation on the temperature in the equation governing the temperature distribution. Although it is plausible that a term proportional

[1] P. Chadwick, "Thermoelasticity, the dynamic theory", in: *Progress in solid mechanics*, Vol. 1, ed. by I. N. Sneddon and R. Hill. Amsterdam, North-Holland Publishing Co. (1960).

to the time rate of change of the dilatation should be included in the heat conduction equation, the form of the proportionality factor is not obvious. In any case it is more satisfactory to derive the equations governing the mechanical and thermal disturbances by a more fundamental approach based on thermodynamic considerations. Such a derivation is presented in the book by Boley and Weiner.[2] By employing an expansion of the free energy in terms of the temperature change and the principal invariants of the strain tensor, it is found that the stress is related to the strain and the variation in temperature by eq. (10.2), while the temperature is governed by

$$KT_{,mm} = \rho c_V \dot{T} + (3\lambda+2\mu)\alpha T_0 \dot{\varepsilon}_{kk}. \tag{10.3}$$

Here K is the thermal conductivity of the solid and c_V is the specific heat at constant deformation. In deriving (10.3) it is assumed that Fourier's law of heat conduction holds and that $T - T_0 \ll T_0$. Since eqs. (10.2) and (10.3) are based on an expansion from the reference state defined by the temperature distribution T_0, the material constants appearing in these equations are essentially isothermal constants. Eqs. (10.2) and (10.3) are supplemented by the equation of motion

$$\tau_{ij,j} + \rho f_i = \rho \ddot{u}_i, \tag{10.4}$$

and appropriate boundary and initial conditions. The complete system of equations defines the coupled thermoelastic theory. Substituting the stress-strain-temperature relation into the equation of motion we obtain in vector notation

$$(\lambda+\mu)\nabla\nabla\cdot\boldsymbol{u} + \mu\nabla^2\boldsymbol{u} - \alpha(3\lambda+2\mu)\nabla(T-T_0) + \rho\boldsymbol{f} = \rho\ddot{\boldsymbol{u}}. \tag{10.5}$$

10.2. Coupled thermoelastic theory

The coupling of the deformation and the temperature through the coupling term in eq. (10.3) does have some interesting implications for the propagation of waves. Let us start with an examination of harmonic waves.

10.2.1. Time-harmonic plane waves

A plane displacement wave of harmonic time dependence propagating with phase velocity c in a direction defined by the propagation vector $\boldsymbol{p}$ is represented by

$$\boldsymbol{u} = A\boldsymbol{d} \exp\left[ik(\boldsymbol{x}\cdot\boldsymbol{p} - ct)\right], \tag{10.6}$$

[2] B. A. Boley and J. H. Weiner, *Theory of thermal stresses*. New York, John Wiley and Sons, Inc. (1960), p. 27.

where k is the wavenumber and $\boldsymbol{d}$ is a unit vector defining the direction of motion. If the deformation affects the thermal state of the medium, a displacement wave is accompanied by a temperature wave, which is a scalar quantity, and which may be assumed of the form

$$T-T_0 = B\exp\left[ik(\boldsymbol{x}\cdot\boldsymbol{p}-ct)\right]. \tag{10.7}$$

Substituting (10.6) and (10.7) into the displacement equation of motion (10.5) and into the equation governing the temperature field (10.3), we obtain

$$(\mu-\rho c^2)k^2A\boldsymbol{d}+(\lambda+\mu)(\boldsymbol{p}\cdot\boldsymbol{d})k^2A\boldsymbol{p}+\alpha(3\lambda+2\mu)ikB\boldsymbol{p} = 0$$

$$\alpha(3\lambda+2\mu)(\boldsymbol{p}\cdot\boldsymbol{d})T_0\,k^2cA+(Kk^2-\rho c_V\,ikc) = 0.$$

By eliminating B we obtain

$$(\mu-\rho c^2)\boldsymbol{d}+(\lambda+\mu)(\boldsymbol{p}\cdot\boldsymbol{d})\boldsymbol{p}+\frac{(\lambda+2\mu)(\boldsymbol{p}\cdot\boldsymbol{d})c\varepsilon}{c+ik\kappa_V}\,\boldsymbol{p} = 0. \tag{10.8}$$

In eq. (10.8), we have introduced the dimensionless thermoelastic coupling constant ε, which is defined as

$$\varepsilon = \frac{(3\lambda+2\mu)^2\alpha^2T_0}{(\lambda+2\mu)\rho c_V}. \tag{10.9}$$

Also, the constant κ_V, which is defined as

$$\kappa_V = \frac{K}{\rho c_V}, \tag{10.10}$$

is the thermal diffusivity at constant deformation. For a number of materials the magnitude of ε is shown in table 10.1. It is noted that ε is much smaller than unity. Nevertheless the coupling effect cannot always be ignored.

TABLE 10.1

Coupling constant at 20 °C (after Chadwick)

	Coupling Constant ε
Aluminum	3.56×10^{-2}
Copper	1.68×10^{-2}
Iron	2.97×10^{-2}
Lead	7.33×10^{-2}

Eq. (10.8) is equivalent to eq. (4.5) of chapter 4. Since $\boldsymbol{p}$ and $\boldsymbol{d}$ are two different unit vectors, eq. (10.8) can be satisfied in two ways only:

$$\text{either} \quad \boldsymbol{d} = \pm \boldsymbol{p}, \quad \text{or} \quad \boldsymbol{p} \cdot \boldsymbol{d} = 0.$$

10.2.2. Transverse waves

If $\boldsymbol{d} \neq \pm \boldsymbol{p}$, both terms in (10.8) have to vanish independently. This implies

$$\boldsymbol{p} \cdot \boldsymbol{d} = 0 \quad \text{and} \quad c = \left(\frac{\mu}{\rho}\right)^{\frac{1}{2}}. \tag{10.11}$$

Eqs. (10.11) define transverse waves. The noteworthy observation is that transverse waves do not interact with the temperature field.

10.2.3. Longitudinal waves

If $\boldsymbol{d} = \pm \boldsymbol{p}$, we have $\boldsymbol{d} \cdot \boldsymbol{p} = \pm 1$, and eq. (10.8) becomes

$$\lambda + 2\mu - \rho c^2 + \frac{(\lambda+2\mu)c}{c + ik\kappa_V}\varepsilon = 0. \tag{10.12}$$

This rather complicated equation shows that the phase velocity depends on k, which means that thermoelastic waves are dispersive. Since the solution for c of (10.12) generally is complex-valued, coupled thermoelastic waves also suffer attenuation. If we consider the wavenumber k as the variable, the limit cases for $k \to 0$ and $k \to \infty$ are

$$\lim_{k \to 0} c = (1+\varepsilon)^{\frac{1}{2}} c_L \tag{10.13}$$

$$\lim_{k \to \infty} c = c_L. \tag{10.14}$$

Since the elastic constants λ, μ and ρ are isothermal constants, it can be concluded that strictly speaking longitudinal waves propagate with the phase c_L only if the wavelength approaches zero. If the wavelength increases beyond bounds the phase velocity is given by eq. (10.13). For very long waves the propagation of harmonic waves is essentially an adiabatic process, and the phase velocity depends on adiabatic material constants. On physical grounds the limitcases were explained by Deresiewicz.[3] The gist of the explanation is that heat is produced and consumed over a wavelength

[3] H. Deresiewicz, *Journal of the Acoustical Society of America* **29** (1957) 204.

and that thermal equilibrium is established rapidly when the wavelength is short and slowly when the wavelength is large.

A rather detailed examination of eq. (10.12) was carried out in the previously cited article by Chadwick. Following Chadwick we introduce the following dimensionless quantities

$$\xi = \frac{c_L k}{\omega^*} \tag{10.15}$$

$$\chi = \frac{\omega}{\omega^*}, \quad \text{where} \quad \omega^* = \frac{c_L^2}{\kappa_V}. \tag{10.16}$$

Eq. (10.12) then becomes

$$(\xi^2-\chi^2)(\chi+i\xi^2)+\varepsilon\xi^2\chi = 0. \tag{10.17}$$

For the case of most physical interest the frequency is presumed known and the wavenumber is to be computed. When χ is regarded as a real constant, eq. (10.17) is a quartic in ξ with roots $\pm\xi_1$ and $\pm\xi_2$. Exact expressions as well as expansions for small values of ε and/or small values of χ are given by Chadwick. Let us write

$$\xi_j = c_L\left(\frac{\chi}{v_j}+i\frac{q_j}{\omega^*}\right), \qquad j = 1, 2,$$

where v_1 and v_2 are phase velocities and q_1 and q_2 are attenuation constants. Expansions for these quantities are

$$v_1 = c_L\left[1+\frac{\varepsilon}{2(1+\chi^2)}+O(\varepsilon^2)\right]$$

$$q_1 = \frac{\omega^*}{c_L}\left[\frac{\varepsilon\chi^2}{2(1+\chi^2)}-O(\varepsilon^2)\right]$$

$$v_2 = c_L(2\chi)^{\frac{1}{2}}\left[1-\frac{\varepsilon(1+\chi)}{2(1+\chi^2)}+O(\varepsilon^2)\right]$$

$$q_2 = \frac{\omega^*}{c_L}(\tfrac{1}{2}\chi)^{\frac{1}{2}}\left[1+\frac{\varepsilon(1-\chi)}{2(1+\chi^2)}+O(\varepsilon^2)\right].$$

These results show that the roots $\pm\xi_1$ represent modified elastic waves while $\pm\xi_2$ represent modified thermal waves.

It is noted that q_1 is an increasing function of χ, but it can be shown that q_1 approaches a finite value as χ increases beyond bounds. It is, however, evident that the wave motion may be significantly attenuated at high frequencies. In the lower scale of frequencies the coupling between a longitudinal wave and the accompanying thermal disturbance is, however, weak.

In summary, for plane time-harmonic waves the coupling between the thermal and mechanical fields affects the essentially mechanical waves in that these waves become dispersed and attenuated.

10.2.4. Transient waves

The coupling between mechanical and thermal disturbances is also of some importance in problems of transient wave propagation. To display the effect of the coupling we will examine a one-dimensional initial boundary value problem. The influence of thermoelastic coupling should be most pronounced when the external excitation is of a mechanical nature, because in that case the deformation is purely mechanical if the coupling effect is neglected.

In a one-dimensional geometry the coupled thermoelastic equations reduce to

$$\frac{\partial^2 u}{\partial x^2} = \frac{(3\lambda+2\mu)\alpha}{\lambda+2\mu}\frac{\partial}{\partial x}(T-T_0)+\frac{1}{c_L^2}\frac{\partial^2 u}{\partial t^2} \tag{10.18}$$

$$\frac{\partial^2 T}{\partial x^2} = \frac{\rho c_V}{K}\frac{\partial T}{\partial t}+\frac{(3\lambda+2\mu)\alpha T_0}{K}\frac{\partial^2 u}{\partial x\,\partial t}. \tag{10.19}$$

We will consider a half-space $x \geqq 0$ which is initially at rest and at a uniform temperature

$$u(x, 0) = \dot{u}(x, 0) \equiv 0; \qquad T(x, 0) = T_0. \tag{10.20}$$

At time $t = 0$ the surface $x = 0$ is subjected to a uniform stress, while the temperature is maintained at the reference temperature

$$\tau_x(0, t) = -\tau_0 H(t) \tag{10.21}$$

$$T(0, t) = T_0. \tag{10.22}$$

Solutions of problems of transient wave propagation in coupled thermoelasticity are usually sought by means of integral transform techniques. Unfortunately it is generally rather cumbersome to obtain inverse transforms. In many cases it is necessary to resort to approximate methods. A number of problems for the half-space were solved by Boley and Tollins.[4]

[4] B. A. Boley and I. S. Tollins, *Journal of Applied Mechanics* **29** (1962) 637.

In figure 10.1 we have sketched the stress $\tau_x(x, t)$ for the problem defined by eqs. (10.18)–(10.22) as a function of $x/c_L t$ and for a specific time t. The solid line indicates the isothermal solution which is obtained from the uncoupled theory. The most notable feature of the solution which does include the effect of coupling is that the wavefront propagating with velocity

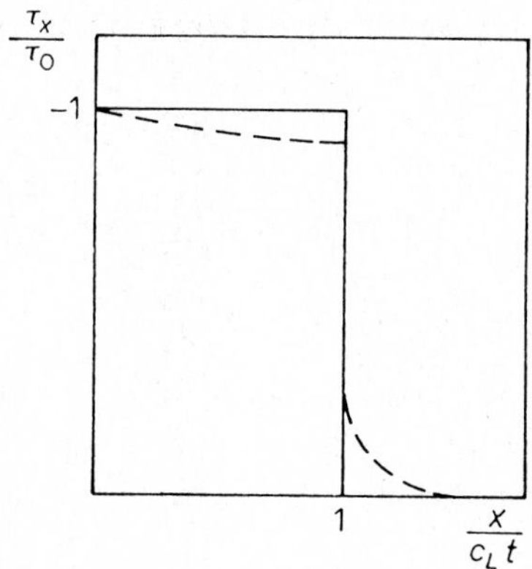

Fig. 10.1. Longitudinal stress. ——— uncoupled theory. - - - - coupled theory.

c_L is preceded by a disturbance which (in theory) extends to infinity. This disturbance is generated by the thermal field which is due to conduction of heat produced by the deformation behind the wavefront $x/c_L t = 1$. Since heat conduction according to Fourier's law is a diffusion process the thermal field instantaneously extends to infinity and thus causes a (very small) deformation which precedes the wavefront. If the surface pressure is suddenly applied a discontinuity of decreasing magnitude propagates with the velocity of longitudinal waves. In figure 10.1 we have also shown the corresponding uncoupled field. The thermal coupling thus causes a precursor effect and a damping effect.

By employing the analytical techniques of chapter 4 to determine the magnitude of a propagating discontinuity it can easily be shown that the stress jump at $x = c_L t$ is of the form

$$[\tau_x] = -\tau_0 \, e^{-\varepsilon(c_L/2\kappa_v)x}, \tag{10.23}$$

where ε is the coupling constant and κ_v is the diffusivity at constant deformation. For a derivation of (10.23) we refer to the book by Parkus.[5]

[5] H. Parkus, *Thermoelasticity*, Blaisdell Publishing Co. (1968), p. 101.

For most materials the exponent is a relatively large number and the discontinuity is damped out over a very short distance.

10.2.5. *Second sound*

The classical theory of the conduction of heat rests upon the hypothesis that the flux of heat is proportional to the gradient of the temperature distribution. As a consequence of this hypothesis, which is known as Fourier's law, the temperature distribution in a body is governed by a parabolic partial differential equation, which predicts that the application of a thermal disturbance in a finite region of a body instantaneously affects all points of the body. This behavior, which implies an "infinite speed of propagation" of thermal disturbances, has been reason to doubt the validity of Fourier's law for initial value problems and short times, and it has motivated proposals to modify Fourier's law to a relation of the type

$$q_i + \tau \dot{q}_i = -K \frac{\partial T}{\partial x_i}, \tag{10.24}$$

where q_i is the heat flux. Through this equation the temperature distribution is governed by a partial differential equation of the hyperbolic type, and heat conduction is described as a wave propagation phenomenon. Often the term "second sound" is used in referring to these thermal waves.

The implications of propagation of thermal effects on the initial-boundary value problem defined by (10.20)–(10.22) were examined by Achenbach.[6] If the transport of heat is governed by the modified Fourier's law (10.24) the application of a mechanical or a thermal disturbance in a one-dimensional geometry gives rise to two wavefronts. The speed of the fast wavefront is greater than the larger of c_L and c_t, where c_L and c_t are the speeds of mechanical and thermal wavefronts if thermomechanical coupling is ignored. The speed of the slow wavefront is less than the smaller of c_L and c_t. For $c_t/c_L < (1+\varepsilon)^{\frac{1}{2}}$, where ε is the coupling constant, the slow wave is essentially thermal and the fast wave is essentially mechanical. For $c_t/c_L > (1+\varepsilon)^{\frac{1}{2}}$ the opposite is the case and the fast and slow waves are essentially thermal and mechanical, respectively. If the externally applied disturbance is discontinuous in time, the temperature is discontinuous at the wavefronts unless Fourier's classical law of heat conduction is satisfied. The discontinuities in the mechanical and the thermal fields decay exponentially,

[6] J. D. Achenbach, *Journal of the Mechanics and Physics of Solids* **16** (1968) 273.

where the attenuation is faster at the wavefronts of the essentially thermal waves.

10.3. Uncoupled thermoelastic theory

If the thermal and mechanical fields are independent of time, the coupling in the heat equation between the temperature and the deformation vanishes. Even for fields that do vary with time, the coupling term is often neglected, so that eq. (10.3) reduces to

$$KT_{,mm} = \rho c_V \dot{T}. \tag{10.25}$$

By neglecting the influence of the deformation on the temperature field the temperature distribution can be solved first from eq. (10.25), whereupon the then known temperature field enters eq. (10.5) effectively as a body force. If the temperature variation is sufficiently rapid it may induce wave propagation effects. These thermally induced waves can be analyzed by the techniques discussed in chapter 3 for motions generated by distributions of body forces.

According to the uncoupled theory the thermal state is not affected in the event that the external disturbances are purely mechanical in nature. The temperature terms in eqs. (10.2) and (10.5) vanish and the analysis is purely isothermal. The theory and methods discussed in this book are pertinent to this case. It follows from the preceding section that a strictly isothermal analysis is, however, an approximation, because the heat generated by deformation is neglected.

10.4. The linearly viscoelastic solid

In this section we investigate the propagation of waves in a class of materials for which loads and deformations are linearly related, but for which the deformation depends not only on the present magnitude of the loads but also on the history of the loading process. These materials are called linearly viscoelastic.[7] For a more specific description of viscoelastic behavior it is convenient to examine the simple case of one-dimensional longitudinal stress.

10.4.1. Viscoelastic behavior

Let us consider an infinitesimal element of a material, and let us suppose that we instantaneously place the element in a state of homogeneous longitudinal

[7] The theory is discussed by R. M. Christensen, *Theory of viscoelasticity*. New York, Academic Press (1971).

stress defined by $\tau_x \neq 0$, $\tau_y = \tau_z \equiv 0$. In a perfectly elastic element the stress τ_x instantaneously gives rise to strains, in particular to a homogeneous extensional strain of magnitude $\varepsilon_x = \tau_x/E$, where E is the extensional, or Young's, modulus. In an element of a viscoelastic material the instantaneous response will, however, be followed by an additional strain which increases with time. This phenomenon, which is called creep, is characteristic of viscoelastic materials. The extensional strain response to a homogeneous longitudinal stress of unit magnitude is called the creep function $J_E(t)$. The creep function is sketched in figure 10.2. Conversely, if the element is instantaneously placed in a state of extensional strain ε_x, combined with

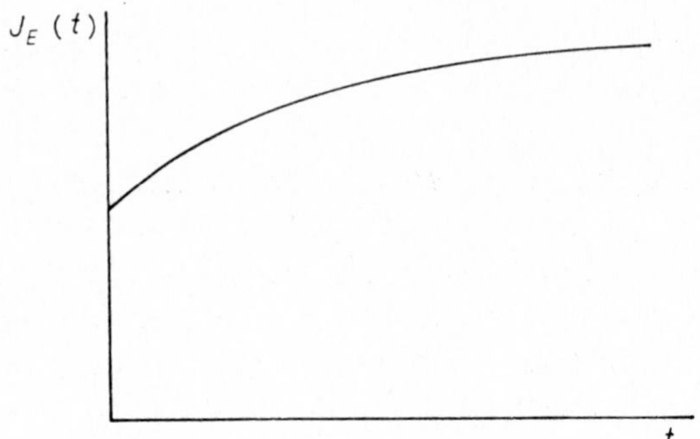

Fig. 10.2. Creep function.

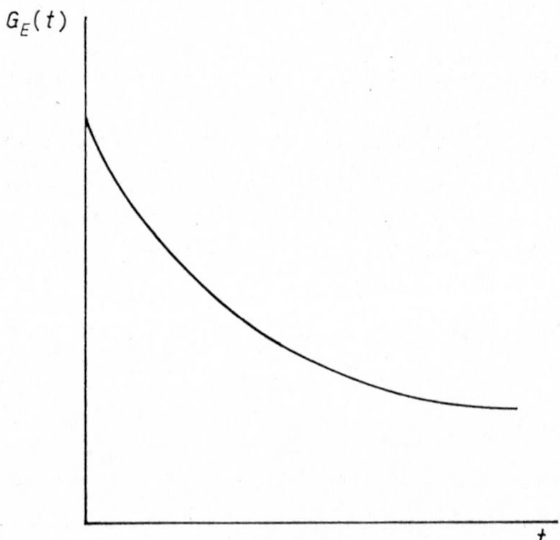

Fig. 10.3. Relaxation function.

$\tau_y = \tau_z \equiv 0$, the instantaneous stress response is followed by a decrease of the stress level. This phenomenon is called relaxation. The longitudinal stress response to a strain ε_x of unit magnitude is termed the relaxation function $G_E(t)$. The relaxation function is sketched in figure 10.3. By virtue of the linearity of the process the longitudinal stress due to an extensional strain of arbitrary time dependence may then be expressed as a superposition integral over $G_E(t)$ and $\varepsilon_x(t)$,

$$\tau_x(t) = \varepsilon_x(0)G_E(t-t_0) + \int_{t_0^+}^{t} G_E(t-s)\frac{d\varepsilon_x}{ds}\,ds, \tag{10.26}$$

where it is assumed that the process starts at time $t = t_0$. Eq. (10.26) can also be written in the form

$$\tau_x(t) = G_E(0)\varepsilon_x(t) + \int_{t_0^+}^{t} G_E'(t-s)\varepsilon_x(s)ds, \tag{10.27}$$

where a prime denotes a derivative with respect to the argument. Eqs. (10.26) and (10.27) show one way of representing linear viscoelastic constitutive behavior for the case of one-dimensional stress. A more compact way of writing (10.26) is

$$\tau_x(t) = \int_{t_0}^{t} G_E(t-s)d\varepsilon_x. \tag{10.28}$$

10.4.2. Constitutive equations in three dimensions

In an isotropic elastic solid the mechanical behavior can be completely described by two elastic constants. As shown in section 2.4 a convenient choice consists of the shear modulus μ and the bulk modulus B. The advantage of using these constants is that they have definite physical interpretations, and that they can be measured. In terms of μ and B the elastic stress-strain relations are

$$s_{ij} = 2\mu e_{ij}, \qquad \tfrac{1}{3}\tau_{kk} = B\varepsilon_{kk}, \tag{10.29a, b}$$

where the stress deviatior s_{ij} and the strain deviatior e_{ij} are defined by eqs. (2.32) and (2.33), respectively. In analogy to eq. (10.28) the viscoelastic relations corresponding to eqs. (10.29a, b) may be written in the forms

$$s_{ij} = 2\int_0^t G_S(t-s)de_{ij} \tag{10.30}$$

$$\tau_{kk} = 3\int_0^t G_B(t-s)d\varepsilon_{kk}, \tag{10.31}$$

where $G_S(t)$ and $G_B(t)$ are the relaxation functions in shear and in bulk, respectively. The corresponding relation between τ_{ij} and ε_{ij} is

$$\tau_{ij} = \delta_{ij}\int_0^t [G_B(t-s)-\tfrac{2}{3}G_S(t-s)]\mathrm{d}\varepsilon_{kk}+2\int_0^t G_S(t-s)\mathrm{d}\varepsilon_{ij}. \quad (10.32)$$

10.4.3. Complex modulus

Suppose that the strain history is specified as a harmonic function of time, so that the strain deviator may be written as

$$e_{ij} = e_{ij}^{*}e^{i\omega t}.$$

Assuming that a steady state has been reached the stress deviator is also time-harmonic with frequency ω. The stress deviator may then be written in the form

$$s_{ij} = 2G_S^{*}(\omega)e_{ij}^{*}e^{i\omega t}, \quad (10.33)$$

where the complex modulus $G_S^{*}(\omega)$ is a complex function of the frequency. We write

$$G_S^{*} = G_S'(\omega)+iG_S''(\omega), \quad (10.34)$$

or

$$G_S^{*} = |G_S^{*}|e^{i\varphi_S}, \quad (10.35)$$

where

$$|G_S^{*}| = \{[G_S'(\omega)]^2+[G_S''(\omega)]^2\}^{\frac{1}{2}} \quad (10.36)$$

and

$$\varphi_S(\omega) = \tan^{-1}\frac{G_S''(\omega)}{G_S'(\omega)}. \quad (10.37)$$

It follows that the stress and strain are out of phase, which implies that energy is dissipated during the steady-state oscillations. It can be shown that the real and imaginary parts of the complex modulus can be expressed in terms of the relaxation function.[8]

For steady-state time-harmonic oscillations eq. (10.32) becomes

$$\tau_{ij}^{*} = \delta_{ij}[G_B^{*}(\omega)-\tfrac{2}{3}G_S^{*}(\omega)]\varepsilon_{kk}^{*}+2G_S^{*}(\omega)\varepsilon_{ij}^{*}, \quad (10.38)$$

where

$$\tau_{ij} = \tau_{ij}^{*}e^{i\omega t}, \qquad \varepsilon_{ij} = \varepsilon_{ij}^{*}e^{i\omega t}. \quad (10.30a, b)$$

[8] The expressions are given in the previously cited book by Christensen.

10.5. Waves in viscoelastic solids

Viscoelasticity affects the propagation of waves in a rather significant manner. It is shown in this section that time-harmonic waves in an unbounded medium are subjected to dispersion and attenuation due to the viscoelastic constitutive behavior. A pronounced effect on transient waves is damping in the vicinity of the wavefront. We will restrict the attention to isothermal wave propagation.

10.5.1. Time-harmonic waves

It has been stated several times that a time-harmonic displacement wave is of the form

$$\boldsymbol{u} = A\boldsymbol{d} \exp\,[i(k\boldsymbol{x}\cdot\boldsymbol{p}-\omega t)], \tag{10.40}$$

where ω is the frequency, k is the wavenumber and $\boldsymbol{p}$ and $\boldsymbol{d}$ are unit vectors in the directions of propagation and motion, respectively. Substituting (10.40) into eq. (10.38), and subsequently substituting (10.38) and (10.40) into the equation of motion $\tau_{ij,j} = \rho\ddot{u}_i$, we obtain in vector notation

$$[G_S^*(\omega)k^2-\rho\omega^2]\boldsymbol{d}+[G_B^*(\omega)+\tfrac{1}{3}G_S^*(\omega)]k^2(\boldsymbol{p}\cdot\boldsymbol{d})\boldsymbol{p} = 0. \tag{10.41}$$

This equation, which is the equivalent for a viscoelastic solid of eq. (4.5), again can be satisfied in two ways only,

$$\text{either}\quad \boldsymbol{d} = \pm\boldsymbol{p}, \quad\text{or}\quad \boldsymbol{p}\cdot\boldsymbol{d} = 0.$$

In eq. (10.41) either k or ω can be considered as the real-valued independent variable. Considering the wavenumber k as the independent variable, the phase velocity $c = \omega/k$ can be computed from eq. (10.41). Evidently c must come out as a complex number,

$$c = c_1(k)+ic_2(k). \tag{10.42}$$

The dependence of c_1 on k indicates that viscoelastic waves are dispersive, while the presence of the imaginary term $c_2(k)$ shows that the amplitude decreases as k increases. In the remainder of this section we will, however, choose to consider the physically more realistic case when a forcing frequency is taken at the real-valued independent variable and k is to be computed.

10.5.2. Longitudinal waves

If $\boldsymbol{d} = \pm\boldsymbol{p}$ we obtain from eq. (10.41)

$$k = \left[\frac{\rho}{G_B^*(\omega)+\frac{4}{3}G_S^*(\omega)}\right]^{\frac{1}{2}}\omega. \tag{10.43}$$

Evidently k is complex, which implies that the amplitude decreases with increasing x. The attenuation rate clearly depends on the frequency.

10.5.3. Transverse waves

If $\boldsymbol{p} \cdot \boldsymbol{d} = 0$, we have

$$k = \left[\frac{\rho}{G_S^*(\omega)}\right]^{\frac{1}{2}} \omega. \tag{10.44}$$

Thus, transverse waves are also attenuated.

10.5.4. Transient waves

The proper statement of the dynamic problem for a viscoelastic body is provided by the equations of section 2.5, except that Hooke's law, eq. (2.40), must be replaced by the viscoelastic stress-strain relation, which is given by eq. (10.32). For the type of boundary conditions stated in section 3.2 the initial-boundary value problem can in a formal manner be solved without difficulty, by employing the one-sided Laplace transform over time. It is rather simple to obtain the Laplace transform of the solution of any viscoelastic problem provided that the Laplace transform of the solution of the corresponding elastic problem is available. A simple relation between the Laplace transforms exists by virtue of the form of viscoelastic constitutive laws, which contain convolution integrals, and by virtue of the property that the one-sided Laplace transform of a Riemann convolution of two functions is given by the product of the Laplace transforms of the two functions.

By applying the one-sided Laplace transform to eq. (10.32) it follows that $\bar{\tau}_{ij}$ and $\bar{\varepsilon}_{ij}$ are related by

$$\bar{\tau}_{ij} = \delta_{ij}[p\bar{G}_B(p) - \tfrac{2}{3}p\bar{G}_S(p)]\bar{\varepsilon}_{kk} + 2p\bar{G}_S(p)\bar{\varepsilon}_{ij}, \tag{10.45}$$

where p is the Laplace transform parameter. The Laplace transform of the corresponding elastic stress-strain relation is

$$\bar{\tau}_{ij} = \delta_{ij}[B - \tfrac{2}{3}\mu]\bar{\varepsilon}_{kk} + 2\mu\bar{\varepsilon}_{ij}. \tag{10.46}$$

Since the other governing equations are exactly the same for the elastic and the viscoelastic problems, it now follows that the Laplace transform of the viscoelastic solution can be obtained from the Laplace transform of the corresponding elastic solution by

replacing B by $p\bar{G}_B(p)$
replacing μ by $p\bar{G}_S(p)$.

By the corresponding elastic solution we mean the solution to the problem for an elastic body of the same dimensions which is subjected to the same initial conditions, body forces and boundary conditions as the viscoelastic body. The relation between the Laplace transforms of the corresponding elastic and viscoelastic solutions is called the elastic-viscoelastic correspondence principle.

Although the Laplace transforms of the viscoelastic solutions can be obtained in a simple manner, this provides us with little more than rather formal expressions for the field variables in the form of complex inversion integrals. Usually it is difficult to evaluate these integrals. It appears that it is possible to invert the transforms in a relatively simple manner only for the rather special case that the relaxation functions in bulk and in shear show the same time dependence. For details we refer to an article by Chao and Achenbach.[9]

As an example of the analysis of a transient viscoelastic problem we examine the propagation of longitudinal waves in a thin rod defined by $x \geqq 0$. The rod is originally at rest, i.e.,

$$u(x, 0) = \dot{u}(x, 0) \equiv 0, \tag{10.47}$$

where $u(x, t)$ is the displacement in the x-direction. At time $t = 0$ a stress τ_0 is suddenly applied at $x = 0$,

$$\tau_x(0, t) = -\tau_0 H(t). \tag{10.48}$$

According to the elementary rod theory the equation of motion for the corresponding elastic problem is

$$\frac{\partial^2 u}{\partial x^2} = \frac{1}{c_b^2}\frac{\partial^2 u}{\partial t^2}, \quad \text{where} \quad c_b = \left(\frac{E}{\rho}\right)^{\frac{1}{2}}.$$

It can be verified that the Laplace transform of the elastic stress is

$$\bar{\tau}_x(x, p) = -\frac{\tau_0}{p} e^{-px/c_b}. \tag{10.49}$$

By virtue of the correspondence principle the Laplace transform of the viscoelastic displacement is of the form

$$\bar{\tau}_x(x, p) = -\frac{\tau_0}{p} \exp\left\{-\frac{p^{\frac{1}{2}}\rho^{\frac{1}{2}}x}{[\bar{G}_E(p)]^{\frac{1}{2}}}\right\}. \tag{10.50}$$

[9] C. C. Chao and J. D. Achenbach, in: *Stress waves in inelastic solids*, ed. by H. Kolsky and W. Prager. Berlin, Springer-Verlag (1964), p. 222.

For the simplest viscoelastic model solid the relaxation function is of the form

$$G_E(t) = E_0 e^{-\alpha t}. \tag{10.51}$$

Thus, E_0 defines the instantaneous value of $G_E(t)$ at $t = 0$. The Laplace transform of (10.51) is

$$\bar{G}_E(p) = \frac{E_0}{p+\alpha}. \tag{10.52}$$

Substituting (10.52) into (10.50) yields

$$\bar{\tau}_x(x, p) = -\frac{\tau_0}{p} \exp\left[-(p^2+\alpha p)^{\frac{1}{2}}(x/c_b^0)\right], \tag{10.53}$$

where

$$c_b^0 = \left(\frac{E_0}{\rho}\right)^{\frac{1}{2}}. \tag{10.54}$$

The inversion of (10.53) by way of contour integration is not a trivial task. Fortunately $\tau_x(x, t)$ can be obtained with the aid of a table of Laplace transforms. For $t > x/c_b^0$ the result is

$$\tau_x(x, t) = -\frac{1}{2}\frac{x}{c_b^0}\alpha\tau_0 \int_{x/c_b{}^0}^{t} \frac{I_1\{\frac{1}{2}\alpha[s^2-(x/c_b^0)^2]^{\frac{1}{2}}\}}{[s^2-(x/c_b^0)^2]^{\frac{1}{2}}} e^{-\frac{1}{2}\alpha s} ds$$
$$-\tau_0 e^{-\frac{1}{2}\alpha(x/c_b{}^0)}, \tag{10.55}$$

while $\tau_x(x, t) \equiv 0$ for $t < x/c_b^0$. In eq. (10.55) $I_1(\;)$ is the modified Bessel function of the first kind of order one. It is of interest to note that the speed of the wavefront depends only on the magnitude of the relaxation function at time $t = 0$. The stress is discontinuous at the wavefront, but the jump

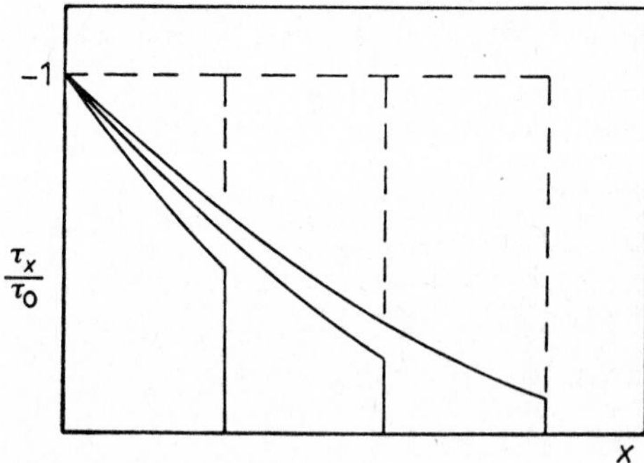

Fig. 10.4. Stress waves in a viscoelastic rod. ——— viscoelastic stress. - - - - elastic stress.

decays exponentially. For various values of time eq. (10.55) is sketched in figure 10.4, where the elastic solution for a solid with Young's modulus E_0 is also shown.

10.5.5. Propagation of discontinuities

There are not many viscoelastic materials whose relaxation function in one-dimensional longitudinal stress can with some degree of accuracy be represented by the single exponential given by eq. (10.51). Experimental measurements generally require the sum of several exponentials with different exponents for an acceptable curvefitting. If the relaxation function is represented by a summation of exponentials it becomes, however, difficult to carry out the inversion of the Laplace transform given by eq. (10.50), at least by analytical methods. It is therefore worthwhile to explore alternative methods of analysis for problems of transient wave propagation in viscoelastic solids. One such method is based on the computation of magnitudes of propagating discontinuities; it can be used for any analytical form that the relaxation function may have. This method allows us to compute the field variables in the vicinity of the wavefront.

Let us return to the problem defined by eqs. (10.47), (10.48), (10.27), and the equation of motion

$$\frac{\partial \tau_x}{\partial x} = \rho \frac{\partial^2 u}{\partial t^2}, \tag{10.56}$$

and let us assume that a wavefront propagates with a velocity c, whose magnitude still is to be determined. According to eq. (4.83) the possible jumps in τ_x and $\partial u/\partial t$ are related by

$$[\tau_x] = -\rho c \left[\frac{\partial u}{\partial t}\right]. \tag{10.57}$$

It follows from (10.27) that $[\tau_x]$ and $[\partial u/\partial x]$ are related by

$$[\tau_x] = G_E(0) \left[\frac{\partial u}{\partial x}\right]. \tag{10.58}$$

Since the displacement is continuous the kinematical condition of compatibility, eq. (4.87) yields

$$\left[\frac{\partial u}{\partial t}\right] = -c \left[\frac{\partial u}{\partial x}\right]. \tag{10.59}$$

Combining (10.57), (10.58) and (10.59) we find

$$c = c_b^0 = \left[\frac{G_E(0)}{\rho}\right]^{\frac{1}{2}}. \tag{10.60}$$

This result states in general terms what was already observed for a special case, namely, the velocity of the wavefront in the rod is governed by the initial value of the relaxation function.

In a similar manner it can be shown[10] that the wavefronts of longitudinal and transverse waves propagate with velocities $\{G_B(0)+\frac{4}{3}G_S(0)\}^{\frac{1}{2}}/\rho^{\frac{1}{2}}$ and $\{G_S(0)/\rho\}^{\frac{1}{2}}$, respectively.

Now let us examine the magnitude of the propagating jump in the particle velocity. According to the kinematical condition of compatibility we have

$$\frac{\mathrm{d}_D}{\mathrm{d}t}[\tau_x] = \left[\frac{\partial \tau_x}{\partial t}\right] + c\left[\frac{\partial \tau_x}{\partial x}\right]. \tag{10.61}$$

By differentiating (10.27) once with respect to t we obtain

$$\frac{\partial \tau_x}{\partial t} = G_E(0)\frac{\partial^2 u}{\partial t\,\partial x} + G_E'(0)\frac{\partial u}{\partial x} + \int_{t_0^+}^{t} G_E''(t-s)\frac{\partial u}{\partial x}\,\mathrm{d}s. \tag{10.62}$$

The corresponding relation between discontinuities is

$$\left[\frac{\partial \tau_x}{\partial t}\right] = G_E(0)\left[\frac{\partial^2 u}{\partial t\,\partial x}\right] + G_E'(0)\left[\frac{\partial u}{\partial x}\right]. \tag{10.63}$$

According to the equation of motion (10.56) we have

$$\left[\frac{\partial \tau_x}{\partial x}\right] = \rho\left[\frac{\partial^2 u}{\partial t^2}\right]. \tag{10.64}$$

Writing the kinematical condition of compatibility for $\partial u/\partial t$ yields

$$\frac{\mathrm{d}_D}{\mathrm{d}t}\left[\frac{\partial u}{\partial t}\right] = \left[\frac{\partial^2 u}{\partial t^2}\right] + c\left[\frac{\partial^2 u}{\partial x\,\partial t}\right], \tag{10.65}$$

or, in view of eqs. (10.57) and (10.64)

$$\frac{\mathrm{d}_D}{\mathrm{d}t}[\tau_x] = -c\left[\frac{\partial \tau_x}{\partial x}\right] - \rho c^2\left[\frac{\partial^2 u}{\partial x\,\partial t}\right]. \tag{10.66}$$

The discontinuities $[\partial \tau_x/\partial x]$, $[\partial^2 u/\partial x \partial t]$, and $[\partial u/\partial x]$ can subsequently be

[10] E.g. K. C. Valanis, *Journal of Mathematical Physics* **44**, (1965) 227.

eliminated from eqs. (10.58), (10.61), (10.63) and (10.66), whereupon the following ordinary differential equation for $[\tau_x]$ is obtained

$$\frac{\mathrm{d}_D}{\mathrm{d}t}[\tau_x] - \frac{1}{2}\frac{G'(0)}{G(0)}[\tau] = 0. \tag{10.67}$$

In view of the initial condition given by eq. (10.48) the appropriate solution of this equation is

$$[\tau_x] = -\tau_0 \exp\left[\frac{1}{2}\frac{G'(0)}{G(0)}t\right]. \tag{10.68}$$

This is the value of the stress *at the wavefront* as the wavefront travels through the rod. Since the position of the wavefront is defined by $x = c_b^0 t$, eq. (10.68) can also be written as

$$[\tau] = -\tau_0 \exp\left[\frac{1}{2}\frac{G'(0)}{G(0)}\frac{x}{c_b^0}\right]. \tag{10.69}$$

In this form the result agrees with what follows for a special case from eq. (10.55).

It is possible to extend the preceding analysis to the computation of the magnitudes of propagating discontinuities of the temporal derivatives of the stress. Such an extension makes it feasible to construct a Taylor expansion for the stress at a fixed position for short times after the wavefront has passed. For a detailed discussion we refer to a note by Achenbach and Reddy.[11]

An important constitutive theory which can describe the inelastic effects that are typical for crystalline materials is the theory of plasticity. The propagation of plastic waves is, however, a field of research by itself and it will not be discussed here. For a treatment of the main problems considered in the theory of dynamic deformation of plastic materials we refer to the book by Critescu.[12]

10.6. Waves in anisotropic materials

It was stated in section 2.4 that for a homogeneous elastic medium the general linear relations between the components of the stress tensor and the components of the strain tensor are

$$\tau_{ij} = C_{ijkl}\varepsilon_{kl}, \tag{10.70}$$

[11] J. D. Achenbach and D. P. Reddy, *Zeitschrift für angewandte Mathematik und Physik* **18** (1967) 141.

[12] N. Critescu, *Dynamic plasticity*. Amsterdam, North-Holland Publishing Co. (1968).

where the constants C_{ijkl} satisfy the relations

$$C_{ijkl} = C_{jikl} = C_{klij} = C_{ijlk},$$

so that only 21 of the 81 components of the tensor C_{ijkl} are independent. The tensor C_{ijkl} is positive definite in the sense that

$$C_{ijkl}\mathrm{f}_{ij}\mathrm{f}_{kl} \geqq 0$$

for all symmetric f_{ij} where the equality is satisfied only when $\mathrm{f}_{ij} \equiv 0$. Substitution of (10.70) into the stress-equation of motion yields

$$C_{ijkl}u_{k,lj} = \rho\ddot{u}_i. \tag{10.71}$$

In the preceding chapters we have investigated waves in isotropic solids for which there are only two independent elastic constants. In the present section we will comment briefly on wave propagation in anisotropic materials. This subject has a lengthy history, with some of the first contributions dating back to the middle of the 19th century. In recent years interest has been revived because of interest in the areas of seismology, ultrasonics and the interaction between deformation and electromagnetic fields. For a review of waves in anisotropic media we refer to an article by Musgrave.[13]

Of particular interest are the results of Synge[14] for time-harmonic waves propagating in a general anisotropic medium. Let us consider the components of a plane harmonic displacement wave in the form (4.18), i.e.,

$$u_m = Ad_m \exp\,[i\omega(x_p q_p - t)], \tag{10.72}$$

where ω is the (real-valued) angular frequency and q_i are the components of the slowness vector. Upon substitution of (10.72) into the displacement equation of motion (10.71) we obtain

$$(C_{ijkl}q_j q_l - \rho\delta_{ik})d_k = 0, \tag{10.73}$$

where δ_{ik} is the Kronecker delta. Eq. (10.73) is an equation for the components d_k. For a nontrivial solution the determinant of the coefficients of d_k must vanish, which gives

$$\det |C_{ijkl}q_j q_l - \rho\delta_{ik}| = 0. \tag{10.74}$$

[13] M. J. P. Musgrave, in *Progress in solid mechanics*, Vol. II, ed. by I. N. Sneddon and R. Hill. Amsterdam, North-Holland Publishing Company (1961), p. 61.

[14] J. L. Synge, *Journal of Mathematical Physics* **35** (1957) 323.

In section 4.2 the components of the slowness vector were defined as

$$q_j = p_j/c, \tag{10.75}$$

where p_i are the components of the vector which defines the direction of propagation and c is the phase velocity. If we let

$$\Gamma_{ik} = C_{ijkl}p_j p_l, \tag{10.76}$$

Eq. (10.74) can be rewritten as

$$\det |\Gamma_{ik} - \rho c^2 \delta_{ik}| = 0. \tag{10.77}$$

The constants Γ_{ik} are known as the Christoffel stiffnesses. If the components p_j are given, eq. (10.77) describes three velocity sheets in the space spanned up by p_j.

From the properties of the C_{ijkl} it follows that Γ_{ik} is a symmetric and positive definite matrix. That is,

$$\Gamma_{ik} = \Gamma_{ki}, \qquad \Gamma_{ik} d_i d_k \geqq 0 \qquad \text{for all } d_i.$$

It follows that all of the eigenvalues of Γ_{ik} are real and positive and their corresponding eigenvectors are orthogonal. The physical interpretation of these observations is that for a given direction of wave propagation defined by p_i there will be three phase velocities, c_{I}, c_{II} and c_{III}, and the three corresponding displacement vectors will be orthogonal. Contrary to the isotropic case the displacements are, however, neither truly longitudinal nor truly transverse in character.

For further details on wave propagation in anisotropic media we refer to the literature on the subject. Some aspects of a formal elastodynamic theory for anisotropic solids were discussed by Wheeler.[15] A more detailed discussion of slowness surfaces can be found in the previously cited papers by Synge and Musgrave. Both of these authors have also discussed the reflection of waves at plane surfaces and surface waves at the surface of an anisotropic half-space. Transient problems including the Green's function for an infinite medium and Lamb's problem for a half-space were discussed in a review article by Kraut.[16] Transient anisotropic waves in bounded elastic media were studied by Scott.[17]

[15] L. T. Wheeler, *Quarterly of Applied Mathematics* **XXVIII** (1970) 91.

[16] E. A. Kraut, *Reviews of Geophysics* **1** (1963) 401.

[17] R. A. Scott, in: *Wave propagation in solids*, ed. by J. Miklowitz. The American Society of Mechanical Engineers (1969), p. 71.

10.7. A problem of transient nonlinear wave propagation

The analysis of the propagation of waves of finite strain in elastic materials requires the surmounting of some difficult mathematical obstacles. Perhaps for that reason it has not been until quite recently that a substantial body of results on nonlinear dynamic elasticity has become available. The work on the propagation of nonlinear elastic waves can roughly be divided into four categories, namely: (1) studies of propagating singular surfaces, (2) simple wave solutions of boundary-initial value problems, (3) studies of propagating steady-state shocks, and (4) the analysis of periodic waves by means of asymptotic methods. A treatment of selected problems of nonlinear dynamic elasticity can be found in the book by Bland.[18]

Some peculiarly nonlinear effects such as the steepening of the head of a pulse and the subsequent formation of shocks can be exhibited by an analysis of the simplest one-dimensional nonlinear example. In section 1.2 we derived the equations governing the one-dimensional motion of an elastic continuum in the material description. Let us define $F(X, t)$ as the displacement gradient

$$F = \frac{\partial U}{\partial X}. \tag{10.78}$$

Following eq. (1.22) the relation between the stress $T(X, t)$ and deformation gradient $F(X, t)$ is taken in the form

$$T(X, t) = \mathscr{S}(F). \tag{10.79}$$

Representative curves for $\mathscr{S}(F)$ are sketched in figure 10.5. It will be shown in the sequel that the wave propagation response is very different for convex

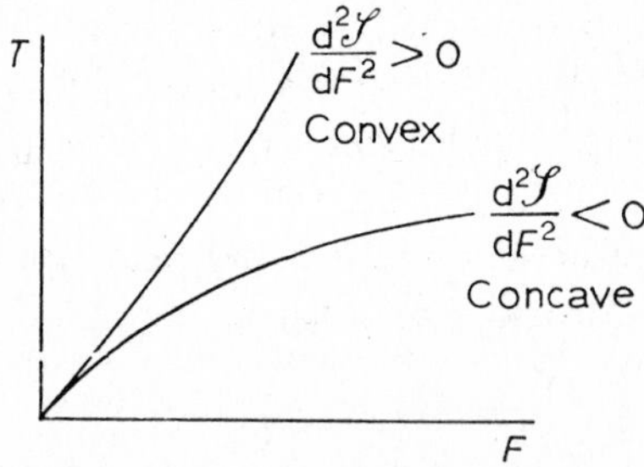

Fig. 10.5. Nonlinear relation between the stress and the deformation gradient.

[18] D. R. Bland, *Nonlinear dynamic elasticity*, Blaisdell Publishing Company (1969).

and concave stress-deformation-gradient curves. For rubbery-type of materials, for which nonlinear elasticity applies, the relation between the stress and the deformation gradient is generally concave.

According to eqs. (1.23) and (1.24) the equation of motion may be written as

$$C^2 \frac{\partial^2 U}{\partial X^2} = \frac{\partial^2 U}{\partial t^2}, \tag{10.80}$$

where

$$C(F) = \left(\frac{1}{\rho_0} \frac{\mathrm{d}\mathscr{S}}{\mathrm{d}F}\right)^{\frac{1}{2}}. \tag{10.81}$$

Let us consider a half-space $X \geqq 0$ which, prior to time $t = 0$, is in a state of homogeneous deformation defined by a displacement gradient F_0, i.e.,

$$t < 0: \qquad F = F_0. \tag{10.82}$$

At time $t = 0$ the surface of the half-space is subjected to a spatially uniform particle velocity:

$$t \geqq 0, X = 0: \qquad V(0, t) = f(t), \tag{10.83}$$

where $f(0) \equiv 0$. The linearized counterpart of this problem can immediately be solved by the method of section 1.3.

The problem as formulated by eqs. (10.80)–(10.83) can be treated by the method of characteristics which was already mentioned in section 4.7. Eq. (10.80) can be rewritten as

$$\frac{\partial V}{\partial t} - C^2 \frac{\partial F}{\partial X} = 0, \tag{10.84}$$

while it follows from eq. (10.78) that

$$\frac{\partial V}{\partial X} - \frac{\partial F}{\partial t} = 0. \tag{10.85}$$

In these equations $V(X, t)$ is the particle velocity in the material description. Following the development of section 4.7 the characteristic curves in the (Xt)-plane are easily obtained as

$$\frac{\mathrm{d}X}{\mathrm{d}t} = \pm C(F). \tag{10.86}$$

The lines in the (Xt)-plane defined by (10.86) are referred to as the C^+ and the C^- characteristics, respectively. Following the usual procedure we introduce new coordinates ξ and η such that

$$\xi = \text{constant along } C^+ \text{ curves} \tag{10.87}$$

$$\eta = \text{constant along } C^- \text{ curves.} \tag{10.88}$$

These definitions of ξ and η imply that

$$\frac{\partial X}{\partial \eta} = C(F)\frac{\partial t}{\partial \eta} \quad \text{along} \quad C^+ \tag{10.89}$$

and

$$\frac{\partial X}{\mathrm{d}\xi} = -C(F)\frac{\partial t}{\partial \xi} \quad \text{along} \quad C^-. \tag{10.90}$$

In the next step we multiply eq. (10.85) by $+C$ and $-C$, respectively, and we add the respective products to eq. (10.84) to obtain

$$\left(\frac{\partial V}{\partial t} + C\frac{\partial V}{\partial X}\right) - C\left(\frac{\partial F}{\partial t} + C\frac{\partial F}{\partial X}\right) = 0 \tag{10.91}$$

and

$$\left(\frac{\partial V}{\partial t} - C\frac{\partial V}{\partial X}\right) + C\left(\frac{\partial F}{\partial t} - C\frac{\partial F}{\partial X}\right) = 0, \tag{10.92}$$

respectively. Combining (10.91) with (10.89), and (10.92) with (10.90) yields

$$\frac{\partial V}{\partial \eta} - C(F)\frac{\partial F}{\partial \eta} = 0 \quad \text{along} \quad C^+(\xi = \text{const}) \tag{10.93}$$

and

$$\frac{\partial V}{\partial \xi} + C(F)\frac{\partial F}{\partial \xi} = 0 \quad \text{along} \quad C^-(\eta = \text{const}), \tag{10.94}$$

respectively. It is now convenient to introduce the quantity

$$Q(F) = \int_{F_0}^{F} C(F)\mathrm{d}F, \tag{10.95}$$

whereupon eqs. (10.93) and (10.94) further simplify to

$$\frac{\partial V}{\partial \eta} - \frac{\partial Q}{\partial \eta} = 0 \quad \text{along} \quad C^+ \tag{10.96}$$

and

$$\frac{\partial V}{\partial \xi} + \frac{\partial Q}{\partial \xi} = 0 \quad \text{along} \quad C^-. \tag{10.97}$$

This pair of equations is readily integrated to yield

$$V(\xi, \eta)-Q(\xi, \eta) = g_1(\xi) \quad \text{along} \quad C^+ \tag{10.98}$$

$$V(\xi, \eta)+Q(\xi, \eta) = g_2(\eta) \quad \text{along} \quad C^-. \tag{10.99}$$

The functions $g_1(\xi)$ and $g_2(\eta)$ are called the Riemann invariants.

Let us label the characteristic variables ξ and η such that $\xi = \eta = t$ at $X = 0$. The initial conditions (10.82) and the boundary conditions (10.83) then yield the following conditions in the $(\xi\eta)$-plane:

$$\xi < 0 \qquad V = 0, Q = 0 \tag{10.100}$$

$$\xi = \eta = t \qquad V = f(t) = f(\xi). \tag{10.101}$$

It now immediately follows from (10.99) and (10.100) that $g_2(\eta) \equiv 0$, while (10.98) and (10.101) subsequently yield $g_1(\xi) = 2f(\xi)$. The solutions for $V(\xi, \eta)$ and $Q(\xi, \eta)$ thus are

$$V = f(\xi) \tag{10.102}$$

$$Q = \int_{F_0}^{F} C(F)\mathrm{d}F = -f(\xi). \tag{10.103}$$

Hence both V and Q are functions of ξ only, which in turn leads to the following conclusions:

$$F = F(\xi) \quad \text{and} \quad C = C(\xi). \tag{10.104a, b}$$

In view of eq. (10.81) the latter of these implies that lines defined by $\xi =$ const are straight lines. Integration of $\mathrm{d}X/\mathrm{d}t = +C$ subject to the condition $\xi = t$ at $X = 0$ then gives

$$X = C(\xi)(t-\xi). \tag{10.105}$$

Since $f(\xi) \equiv 0$ for $\xi < 0$ the characteristic line defined by $\xi = 0$ corresponds to the wavefront whose velocity in the material description is given by

$$C_0 = C(F_0) = \left(\frac{1}{\rho_0}\frac{d\mathscr{S}}{\mathrm{d}F}\right)^{\frac{1}{2}}\Bigg|_{F=F_0}. \tag{10.106}$$

The speed of subsequent disturbances can be computed from the equation that results by rewriting eq. (10.103) in the form

$$\int_{C_0}^{C(\xi)} C\frac{\mathrm{d}F}{\mathrm{d}C}\mathrm{d}C = -f(\xi). \tag{10.107}$$

The solution provided by eqs. (10.102) and (10.103) is called a simple wave solution. If $C(\xi)$ can be solved from eq. (10.107), the particle velocity can be determined at any point in the (Xt)-plane, since $V(\xi)$ remains constant along lines ξ = constant. A geometrical construction is shown in figure 10.6. To obtain an explicit expression for the particle velocity in terms of X and t, the result for $C(\xi)$ must be used to express ξ in terms of X and t by using eq. (10.105).

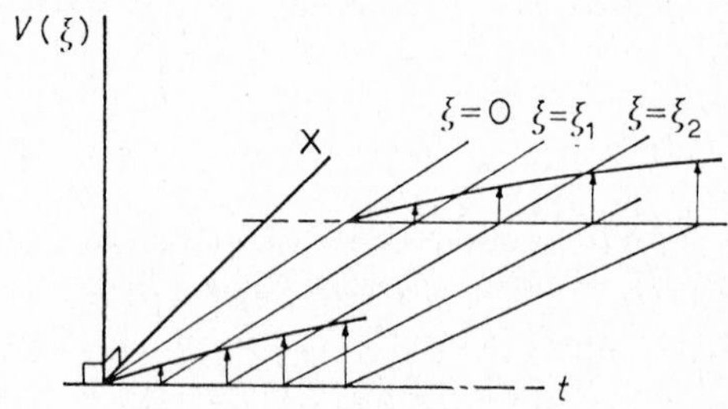

Fig. 10.6. The C^+ characteristics with the corresponding particle velocity.

The simple wave solutions (10.102) and (10.103) are valid as long as the C^+ characteristics do not intersect. Let us consider the case that $f(\xi)$ increases monotonically. Since $C(F) > 0$ it follows from eq. (10.103) that F decreases in magnitude. If $\mathrm{d}^2\mathscr{S}/\mathrm{d}F^2 > 0$, $C(\xi)$ decreases monotonically as F decreases, as follows from figure 10.6; the C^+ characteristics then never intersect and the solution as presented here is valid. If on the other hand $\mathrm{d}^2\mathscr{S}/\mathrm{d}F^2 < 0$, $C(\xi)$ increases as F decreases and intersection occurs. In physical terms the latter case means that earlier disturbance are overtaken by subsequent disturbances. At a point of intersection of two C^+ characteristics the solution breaks down because the intersection implies two solutions of the field variables at one point in space-time. At such a point the solution becomes discontinuous, i.e., it is a point of shock formation.

At points of intersection of C^+ characteristics the derivatives $\partial\xi/\partial t$ and $\partial\xi/\partial X$ become unbounded, or equivalently

$$\frac{\partial t}{\partial \xi} = \frac{\partial X}{\partial \xi} \equiv 0. \tag{10.108}$$

If these conditions hold the spatial and temporal derivatives of V and F increase beyond bounds, as can be seen by considering for example $\partial V/\partial X = (\mathrm{d}f/\mathrm{d}\xi)(\partial\xi/\partial X)$. The time at which a shock initiates is obtained from eq.

(10.105) and the conditions $\partial t/\partial \xi = 0$ and $\partial X/\partial \xi = 0$ as

$$t_s = \min_{\xi} t_d(\xi), \tag{10.109}$$

where

$$t_d(\xi) = \xi + \frac{C(\xi)}{\mathrm{d}C/\mathrm{d}\xi}. \tag{10.110}$$

If the minimum of $t_d(\xi)$ occurs at $\xi = \xi_s$, the value of ξ_s can be computed from

$$C(\xi)\frac{\mathrm{d}^2C}{\mathrm{d}\xi^2} = 2\left(\frac{\mathrm{d}C}{\mathrm{d}\xi}\right)^2. \tag{10.111}$$

An apparent additional condition is that t_s must be larger than ξ_s. A shock forms at the wavefront only if $\xi_s = 0$. Since we must have $t_s > \xi_s \geqq 0$, and since $C(\xi) > 0$ it follows from eqs. (10.109) and (10.110) that a necessary condition for shock formation at $t = t_s$ is $\mathrm{d}C/\mathrm{d}\xi > 0$ for $\xi \leqq \xi_s$, or equivalently by employing eqs. (10.81) and (10.103)

$$\left(\frac{\mathrm{d}^2\mathscr{S}}{\mathrm{d}F^2}\right)\left(\frac{\mathrm{d}f}{\mathrm{d}\xi}\right) < 0 \quad \text{for} \quad \xi \leqq \xi_s. \tag{10.112}$$

Thus, as was already observed earlier, if $f(0) = 0$ and $\mathrm{d}f/\mathrm{d}\xi > 0$ (or $\mathrm{d}f/\mathrm{d}t > 0$), shock formation can occur only if $\mathrm{d}^2\mathscr{S}/\mathrm{d}F^2 < 0$, i.e., if the stress-strain curve is concave.

Shock formation is a typically nonlinear effect which has been observed experimentally in stretched natural rubber.[19]

10.8. Problems

10.1. In the uncoupled thermoelastic theory the temperature distribution is computed from a separate (uncoupled) equation. Suppose the temperature distribution $T - T_0$ is known as a function of $\boldsymbol{x}$ and t. Introduce displacement potentials in eq. (10.5) to determine a general expression for $\boldsymbol{u}(\boldsymbol{x}, t)$.

10.2. The surface of a half-space ($x \geqq 0$) is suddenly heated to the temperature T_1 and then held at that temperature, i.e., we have

$$T(0, t) = T_1 H(t).$$

The half-space is initially at a uniform temperature T_0. Show that the

[19] H. Kolsky, *Nature* **224** (1969) 1301, no. 5226.

solution of the heat conduction equation is

$$T(x, t) = (T_1 - T_0) erfc \left[\frac{x}{(4\kappa_V t)^{\frac{1}{2}}} \right] + T_0,$$

where *erfc* is the error function complement. The surface of the half-space remains free of tractions. Assuming that the half-space is at rest prior to time $t = 0$, determine τ_x as a function of x and t.

10.3. A viscoelastic half-space is subjected to a time-harmonic normal traction $\tau_0 \sin \omega t$. Express the steady-state displacement response in the form

$$u(x, t) = Ue^{-\alpha x} \sin(\omega t - \beta x)$$

and determine U, α and β in terms of the frequency and the components of the complex moduli. Now suppose that the normal traction at $x = 0$ is an arbitrary function $\tau(t)$ of time. Write $\tau(t)$ as a superposition of harmonics, and express $u(x, t)$ as a real-valued integral.

10.4. In a transversely isotropic material the material properties are the same in all directions in planes that are under the same angle with an axis of symmetry. A transversely isotropic elastic material has five independent

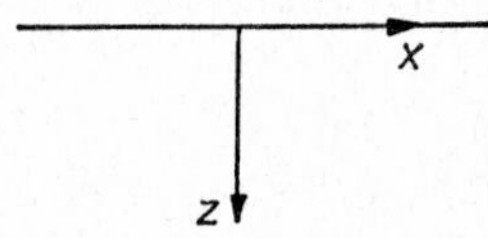

elastic constants. We consider a half-space $z \geqq 0$ and we assume that the axis of symmetry coincides with the z-axis. The stress-strain relations in plane strain are

$$\tau_x = A \frac{\partial u}{\partial x} + C \frac{\partial w}{\partial z}$$

$$\tau_y = B \frac{\partial u}{\partial x} + C \frac{\partial w}{\partial z}$$

$$\tau_z = C \frac{\partial u}{\partial x} + D \frac{\partial w}{\partial z}$$

$$\tau_{xz} = E \left(\frac{\partial u}{\partial z} + \frac{\partial w}{\partial x} \right).$$

Examine the propagation of straight-crested surface waves.

10.5. In the problem discussed in section 10.7 assume that $f(t)$ can be expanded in a Maclaurin series

$$V(0, t) = f(t) = \sum_{n=1}^{\infty} \frac{1}{n!} f_n t^n,$$

where f_n are constants. It is also assumed that $f_1 > 0$. For $t \geqq X/C_0$ we seek the motion at an arbitrary position X as a Taylor expansion about the time of arrival of the wavefront. Because the medium is at rest prior to arrival of the wavefront we may write

$$x = P(X, t) = F_0 X + \sum_{n=2}^{\infty} \frac{1}{n!} \left(t - \frac{X}{C_0}\right)^n \left[\frac{\partial^n P}{\partial t^n}\right],$$

where the brackets denote discontinuities across the wavefront. By employing the relations for propagating discontinuities show that $[\partial^2 P/\partial t^2]$ satisfies the nonlinear ordinary differential equation

$$2\rho_0 C_0 \frac{\mathrm{d}}{\mathrm{d}t} \left[\frac{\partial^2 P}{\partial t^2}\right] + \frac{2S_2}{C_0^2} \left\{\left[\frac{\partial^2 P}{\partial t^2}\right]\right\}^2 = 0,$$

where

$$S_n = \frac{1}{n!} \left.\frac{\mathrm{d}^n \mathscr{S}}{\mathrm{d}F^n}\right|_{F=F_0}.$$

What is the solution of the equation? What happens at $t = -\rho_0 C_0^3/f_1 S_2$? What is the physical meaning of $2S_2$?

AUTHOR INDEX

SUBJECT INDEX